CILIA and FLAGELLA

CILIA and FLAGELLA

Edited by

M. A. SLEIGH

Department of Zoology, University of Bristol, England

1974

ACADEMIC PRESS · LONDON · NEW YORK

A Subsidiary of Harcourt Brace Jovanovich, Publishers

ACADEMIC PRESS INC. (LONDON) LTD
24/28 Oval Road
London NW1

United States Edition published by
ACADEMIC PRESS INC.
111 Fifth Avenue
New York, New York 10003

Library of Congress Catalog Card Number: 73-9476
ISBN: 0-12-648150-4

PRINTED IN GREAT BRITAIN BY
T. AND A. CONSTABLE LTD, EDINBURGH

List of Contributors

EDWARD AIELLO Department of Biological Sciences, Fordham University, Bronx, New York 10458, U.S.A.

V. C. BARBER Department of Biology, Memorial University, St John's, Newfoundland, Canada.

C. J. BROKAW Division of Biology, California Institute of Technology, Pasadena, California 91109, U.S.A.

ROGER ECKERT Department of Biology, University of California, Los Angeles, California 90024, U.S.A.

STUART F. GOLDSTEIN Department of Zoology, University of Minnesota, Minneapolis, Minnesota 55455, U.S.A.

YUKIO HIRAMOTO Biological Laboratory, Tokyo Institute of Technology, O-Okayama, Meguro-Ku, Tokyo, Japan.

M. E. J. HOLWILL Department of Physics, Queen Elizabeth College, London, England.

*HANS MACHEMER Department of Biology, University of California, Los Angeles, California 90024, U.S.A.

YUTAKA NAITOH Department of Biology, University of California, Los Angeles, California 90024, U.S.A.

DAVID M. PHILLIPS The Population Council, The Rockefeller University, New York, N.Y., U.S.A.

DOROTHY R. PITELKA Department of Zoology and its Cancer Research Laboratory, University of California, Berkeley, California 94720, U.S.A.

PETER SATIR Department of Physiology–Anatomy, University of California, Berkeley, California 94720, U.S.A.

MICHAEL A. SLEIGH Department of Zoology, University of Bristol, England.

R. E. STEPHENS Marine Biological Laboratory, Woods Hole, Massachusetts, and Department of Biology, Brandeis University, Waltham, Massachusetts, U.S.A.

FRED D. WARNER Department of Biology, Biological Research Laboratories, Syracuse University, Syracuse, New York, U.S.A

*On leave of absence from: Zoologisches Institut, Fachbereich Biologie, Universität Tübingen, D-74 Tübingen, Federal Republic of Germany.

Preface

The widespread occurrence of cilia in the bodies of animals has been recognized for more than a century. From the viewpoint of the human physiologist, and indeed of zoologists studying almost any of the larger animals, the functions performed by cilia in the body are overshadowed by muscular activities, yet these cellular organelles demand the serious attention of physiologists. The human body employs cilia extensively for moving fluids over epithelial surfaces in respiratory and reproductive systems, a substantial proportion of human sense cells contain a ciliary component, and the human sperm is propelled by a flagellum. In many invertebrate animals cilia perform a much wider variety of functions concerned, for example, with the maintenance of water currents for feeding and respiration, the transport of food, gametes and excretory products, the cleansing of surfaces and, in small organisms, locomotion. In spite of this diversity of function cilia still receive little or no consideration in physiological textbooks. Neglect in textbooks largely reflects past neglect in research, resulting in this instance principally from the lack of suitable techniques. The contributions in this book show that research on a wide range of structural, physiological and hydrodynamic aspects of these organelles is now being performed successfully using a variety of new techniques. It is hoped that the new insights into structure and function described here will lead to a wider general understanding of these organelles, help biologists to better interpretation of observations on cilia and encourage further research.

This book came to be written because the late John Cruise detected a growing interest in the functioning of cilia; if it accelerates this growth, his vision and the efforts of the authors will be justly rewarded.

M.A.S.

October, 1973

Contents

Basal structures attached to cilia

Chapter 1

Introduction

MICHAEL A. SLEIGH
Department of Zoology, University of Bristol, England

It is almost 300 years since cilia were first seen by such early microscopists as van Leeuwenhoek and de Heide. By 1835 microscopes had been considerably improved, and two reviews published in that year, by Purkinje and Valentin and by Sharpey, contained accounts of the occurrence and functioning of cilia in a wide range of animals. Observations described in the later years of the nineteenth century provided an extensive knowledge of the structure and functions of ciliary systems, and physiological studies, principally by Engelmann and Verworn, directed attention to the major problems of ciliary functioning that we believe we are close to solving in research described in this book. An excellent digest of this early work can be found in Gray's book "Ciliary Movement", published in 1928. This, together with an account of ciliated epithelium by Lucas (1932) remained essential reading for anyone interested in cilia up to the early 1960s when the information provided by electron microscopy and new experimental studies diminished their completeness; however, Gray's book is still a valuable source of information on many aspects of ciliary activity.

Two new books concerned with cilia appeared in 1962, a bibliographic review on "Cilia, Ciliated Epithelium and Ciliary Activity" by Rivera, directed principally at those concerned with work on ciliated epithelia of vertebrates, and a more fundamental review for a wider audience of biologists on "The Biology of Cilia and Flagella" by Sleigh. The latter book provided an account of the fine structure of cilia, their movement and metachronism, and of physical and chemical factors affecting these; it also presented evidence and discussed theories about the mechanisms of contractility and co-ordination of cilia. In the decade since these books were published the intensity of research activity concerned with cilia has increased greatly, with an associated increase in published literature. This decade has seen, for example, successful studies of the molecular structure of cilia, pioneered by Gibbons (1965), and the formulation of the sliding fibril hypothesis of ciliary bending by Satir (1965), to cite but two of the many important advances that have made this book necessary. In the chapters which follow, a number of workers active in research on cilia and flagella have contributed accounts of recent research in their own specialisms, to provide an up-to-date survey of most aspects of knowledge about the structure and functioning of cilia and flagella.

Cilia and flagella are living motile projections from the surfaces of cells; they are characterized by the possession of a bundle of protein fibrils, called the axoneme, which consists of a precise array of 9 double fibrils surrounding a pair of single fibrils, enclosed in a membrane that is continuous with the cell membrane. While cilia and flagella are identical in basic structure, as such workers as Prenant (1913) believed long before electron microscopy provided proof, the two different names were coined for organelles showing distinctive patterns of movement. Flagella are beyond doubt the original $9+2$ organelles; they typically show symmetrical planar or helical undulations that pass along the organelle, and in so doing exert on the water a propulsive force whose resultant effect is to move water along the flagellar axis. Cilia show a more eccentric beat, whose typical form includes an effective stroke in which the extended organelle moves rapidly towards one side, followed by a recovery stroke in which it moves more slowly back in the other direction as a bent organelle, either remaining in the same plane or moving back at one side of the plane followed in the effective stroke; this cycle of activity produces a resultant movement of water in the direction taken by the effective stroke, and not along the axis of the organelle. Such typical patterns of movement are compared in Fig. 1, and are described in more detail in Chapter 4 (p. 79). There is considerable variety in the movement of cilia and flagella, and intermediates between the typical patterns occur which are difficult to assign to one category or the other. The identical structure of cilia and flagella, and the problem of deciding whether, for example, the eccentric beat of an organelle borne on a flagellate protozoan required that the organelle be called a cilium, have led some people to seek a single name to describe both categories; thus the name undulipodium has been suggested (Shmagina, 1948), but has not found very wide acceptance. Even if the term undulipodium becomes widely used, it will still be necessary to retain both existing words, for the use of the word cilium saves the need for explanation that one is considering the specialized type of flagellum that has an eccentric rather than a symmetrical beat.

The flagellar appendages of bacteria bear only a superficial similarity to these more complex organelles of eukaryote cells; bacterial flagella are only 12–20 nm thick, no larger than a single fibril of the eukaryote flagellum; but bundles of these flagella form helical aggregates that are visible in the light microscope, and are reminiscent of the appearance of some eukaryote flagella at rather lower magnifications. These bacterial flagella are often composed entirely of globular protein molecules, without any surrounding membrane, and are anchored to a basal granule within the cell membrane of the bacterium. It is unfortunate that these flagella carry the same name as those of eukaryotes; they will not be considered further in this book, but some recent information on their structure and movement may be obtained from a paper by Lowy and Spencer (1968).

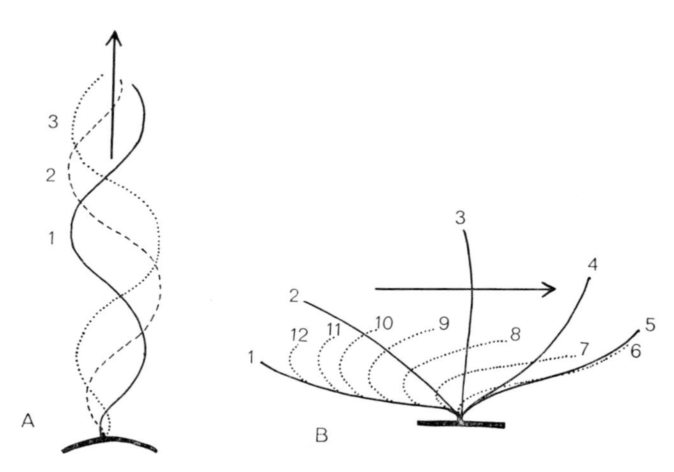

Fig. 1. Comparison of typical patterns of beating of flagella and cilia and of the direction of propulsion of water caused by activity of the organelles. A. Three stages in the undulatory movement of a flagellum along which waves are propagated towards the tip. B. Stages in the cycle of beating of a cilium, with stages of the effective stroke shown as solid lines and stages of the recovery stroke as dotted lines. The arrows indicate the resultant water flow.

The patient application of improved techniques of electron microscopy to cilia and flagella has led to the discovery of further structural details of axoneme components. There is now reasonable agreement about at least some features of the molecular organization of the longitudinal fibrils, and the arrangement of arms, radial links and circumferential links in association with peripheral fibrils and of lateral projections on the central fibrils is well documented. Warner summarizes knowledge on the structure of the axoneme in Chapter 2 (p. 11), and reviews the evidence upon which present views are based. The fractionation of ciliary components and biochemical analysis of the isolated fractions has been an area of fruitful research closely associated with ultrastructural studies, and biochemical characterization of several of the major components has been possible; the techniques used and the results obtained are discussed by

Stephens in Chapter 3 (p. 39). This improved knowledge of structure and biochemical organization has made possible some more meaningful comparisons between cilia and other cellular mechanisms, such as muscle.

Cilia and flagella of different organisms move in distinctive ways. While some of these patterns of beating have been known in general terms for some years, the application of high-speed ciné photography and of improved microscopic techniques has made possible more precise studies of the form of movement and permitted reliable measurements of parameters of beating. It has been possible to show, as described in Chapter 4 (p. 79), that both cilia and flagella may perform either a planar or a three-dimensional beat, and that in some flagella the waves of bending travel in an unconventional direction, from the flagellar tip towards the base. A more detailed analysis by Brokaw of the movement of the flagella of some invertebrate sperm tails under a variety of physical and chemical conditions is presented in Chapter 5 (p. 93). Consideration of such information has led to computer simulation studies of flagellar movement (Brokaw, 1972), as an aid in the evaluation of assumptions about the processes going on within the axoneme of the active organelle.

Among the most informative experiments on flagellar functioning have been those concerned with the reactivation of "model" flagella extracted with glycerol or treated with detergent; these experiments are described by Goldstein in Chapter 6 (p. 111), where the effects of such treatments as laser irradiation of flagella are also considered. In the years since Satir (1965) produced the first evidence in support of a hypothesis of ciliary bending by the sliding of fibrils of constant length, a number of important structural and experimental observations bearing upon the sliding fibril hypothesis have been published, and the re-examination of the evidence by Satir in Chapter 7 (p. 131) shows how well the hypothesis has stood the test of time.

The functioning of cilia and flagella depends upon the transmission of energy from the organelles to the surrounding fluid, so that when the organelle moves the fluid is propelled. The development of a hydrodynamic approach to problems of this type was necessary before any confident estimates could be made of forces exerted or energy consumed in ciliary beating. The principles and results of hydrodynamic analysis of ciliary and flagellar movements are discussed in Chapter 8 (p. 143) by Holwill. These results are important in considerations of the mechanics of ciliary beating, as outlined by Hiramoto in Chapter 9 (p. 177), for estimates of the bending forces required and of the stiffness of the ciliary axis both depend upon estimates of the forces exerted externally against the fluid, or, in some experiments, against other forms of resistance. It had been planned to include a chapter concerned with the mechanics of flagellar motion, with particular regard to mammalian spermatozoa, but unfortunately this chapter could not be completed in the time available; important recent articles in this area that supplement the contributions in this volume by Brokaw (p. 93), Holwill (p. 143) and Hiramoto (p. 177) are those by Rikmenspoel (1965, 1971), Nelson (1967) and Lubliner and Blum (1972).

Ciliary functioning also involves the co-ordination of cilia, both with one another and in relation to the overall physiological integration of the organism. The early microscopists were intrigued by the wave-like appearance of tracts of active cilia. Improvements in our understanding of the patterns of beating of the cilia that make up these waves, together with the results of experimental observations, lend support to the view that the co-ordination within metachronal waves is mechanical, depending upon hydrodynamic interaction, and does not involve any internal conduction process. The first comprehensive review of the evidence for this in Protozoa is presented by Machemer in Chapter 10 (p. 199) and in Chapter 11 (p. 287) Sleigh discusses those aspects of ciliary metachronism in metazoan systems that are not exhibited by Protozoa. Exciting developments in the study of the control of ciliary activity have been taking place in research on both Protozoa and Metazoa. In the former, the rate and direction of beating of cilia show relationships with the ionic distribution and membrane potential of ciliate cells as described by Naitoh and Eckert in Chapter 12 (p. 305), in which other aspects of protozoan motile behaviour are also considered. The more extensive tracts of cilia in Metazoa, used for water circulation in feeding or respiration, for example, must often be controlled, usually by activation or inhibition by nerves and/or hormones; the evidence for the mechanisms of control in these cases is reviewed by Aiello in Chapter 13 (p. 353).

The specialized pattern of movement of cilia, which is only efficient in locomotion when many organelles move together in co-ordinated fashion, apparently does not permit any structural specialization of the organelles, although cilia are often formed into compound structures of tens, hundreds or even thousands of ciliary units, making up, for example, conical cirri, fan-shaped membranelles or palisade-like undulating membranes in ciliated Protozoa and paddle-shaped comb plates in ctenophores. The usual advantage conferred by compounding cilia together is to provide the increased stiffness necessary for the functioning of a longer cilium. Solitary flagella, by contrast, are modified in a variety of structural ways, normally by the addition of hairs or scales outside the flagella or of further longitudinal elements within the flagella. External appendages are common on flagella of various groups of unicellular organisms, and some of these also have internal modifications: the sperm tails of animals, both invertebrate and vertebrate, commonly have additional internal structures, but they seldom show external specializations other than increased diameter or sometimes one or more lateral fins. Structural modifications of invertebrate sperm flagella are described by Phillips in Chapter 14 (p. 379); the structure of flagella of plant cells was discussed by Manton (1965) and references to more recent work may be found in a book by Sleigh (1973). Recent information on vertebrate sperm will be found in papers by Fawcett (1970) and Phillips (1972) and on sperm of all types in a book edited by Baccetti (1970). The functional significance of these structural specializations of flagella has only been established in one or two cases, for example the reversal of water flow caused by the presence of lateral rows of stiff hairs on certain flagella (Holwill and Sleigh, 1967), but examination of further specialized forms could be valuable in a general understanding of flagellar movement. A group of ciliary specializations, not necessarily associated with movement, is characteristic of the cilia that occur in the sense organs of animals; it must be admitted that in many cases we do not understand the role of the ciliary component in the functioning of the sensory cell, but the structural variants are numerous and the associated sensory functions are very diverse, as described by Barber in Chapter 15 (p. 403).

The basal attachments of cilia have been rather neglected in research. It has long been known that a basal body of a fairly standardized pattern is always present, but the diversity of associated fibres or "root" structures is very wide, and a variety of possible functions has been ascribed to them. Clearly, anchorage of the cilium is a vital function, but the role of ciliary roots in a variety of conducting functions is much less certain. The state of present knowledge concerning the structure and replication of basal bodies, and the forms of roots attached to these is discussed by Pitelka in Chapter 16 (p. 437). One role of the basal body that is undoubted is the function served in the development of cilia, when the outer fibrils of the axoneme appear as extensions of the fibrils in the basal body; in recent years much new information has been obtained about the processes of development of cilia and flagella, and an account of flagellar development in *Chlamydomonas* by Rosenbaum *et al.* (1969) refers to many earlier studies.

References

Baccetti, B. (1970). "Comparative Spermatology." Accademia Nationale dei Lincei, Rome.

Brokaw, C. J. (1972). *Biophys. J.* **12**, 564–586.

Fawcett, D. W. (1970). *Biol. Reprod.* **2** (Suppl. 2), 90–127.

GIBBONS, I. R. (1965). *Archs Biol., Liège* **76**, 317–352.
GRAY, J. (1928). "Ciliary Movement." Cambridge University Press, London.
HOLWILL, M. E. J. and SLEIGH, M. A. (1967). *J. exp. Biol.* **47**, 267–276.
LOWY, J. and SPENCER, M. (1968). *Symp. Soc. exp. Biol.* **22**, 215–236.
LUBLINER, J. and BLUM, J. J. (1972). *J. theor. Biol.* **34**, 515–534.
LUCAS, A. M. (1932). *In* "Special Cytology" (E. V. Cowdry, ed.), 2nd edition, pp. 409–473. Hoeber, New York.
MANTON, I. (1965). *Adv. bot. Res.* **2**, 1–34.
NELSON, L. (1967). *In* "Fertilization" (C. B. Metz and A. Monroy, eds), Vol. 1, pp. 27–97. Academic Press, New York.
PHILLIPS, D. M. (1972). *J. Cell Biol.* **53**, 561–573.
PRENANT, A. (1913). *J. Anat. Physiol., Paris* **49**, 88–108, 344–382, 506–553 and 565–617.
PURKINJE, J. E. and VALENTIN, G. (1835). "Commentatio Physiologica de Phenomeno Motus Vibratorii etc.", Wratislav. (Quoted from Sharpey, 1835.)
RIKMENSPOEL, R. (1965). *Biophys. J.* **5**, 365–392.
RIKMENSPOEL, R. (1971). *Biophys. J.* **11**, 446–463.
RIVERA, J. A. (1962). "Cilia, Ciliated Epithelium and Ciliary Activity." Pergamon Press, Oxford.
ROSENBAUM, J. L., MOULDER, J. E. and RINGO, D. L. (1969). *J. Cell Biol.* **41**, 600–619.
SATIR, P. (1965). *J. Cell Biol.* **26**, 805–834.
SHARPEY, W. (1835). *In* "Cyclopaedia of Anatomy and Physiology" (R. B. Todd, ed.), pp. 606–638. Longman, Brown, Green, Longmans and Roberts, London.
SHMAGINA, A. P. (1948). "The Ciliary Movement." Medgiz, Moscow.
SLEIGH, M. A. (1962). "The Biology of Cilia and Flagella." Pergamon Press, Oxford.
SLEIGH, M. A. (1973). "The Biology of Protozoa." Edward Arnold, London.

Components of the axoneme

Chapter 2

The fine structure of the ciliary and flagellar axoneme

FRED D. WARNER

Department of Biology, Biological Research Laboratories, Syracuse University, Syracuse, New York, U.S.A.

I. Introduction

Some 15 years have elapsed since the first extensive electron microscopic observations of cilia and flagella were published. The ensuing period has seen increasingly refined methods applied to analyses of the organelles' fine structure and biochemistry: indeed, the delicate lattice comprising the motile organelle has only recently been resolved. What emerges is the remarkable constancy with which a cellular process, motility, can be expressed in a single kind of structural unit.

Motile cilia and flagella, and herein the names are used interchangeably, either propel cells through an aqueous environment or, if necessary, propel that environment over a cell's surface. To accomplish these functions, a variety of ciliary movement or bending patterns have developed among the different kinds of cells. Bending is nevertheless accomplished by a single kind of structure that we term the ciliary or flagellar axoneme and which is delineated by the well known microtubule array in 9-fold radial symmetry.

Because of their structural position within motile organelles, microtubules have evoked considerable interest regarding their functional and biochemical properties. The ciliary axoneme comprises 3 kinds of tubules, each consisting of a specific number and arrangement of linear subunits or protofilaments (Warner and Satir, 1973). Microtubules are generally acknowledged to consist of a minimum of 2 proteins or tubulins each (Olmsted *et al.*, 1971) and evidence suggests that individual protofilaments may comprise 2 or more tubulins (Meza *et al.*, 1972; Witman *et al.*, 1972b). Satir (1968) has shown that ciliary microtubules slide with respect to one another as the organelle bends, and recently, Summers and Gibbons (1971) demonstrated in model axonemes that the force for sliding is generated between adjacent doublet tubules. Microtubules are joined in the axoneme lattice by an array of delicate and periodic linkages (Warner, 1970a, b). Evidence suggests that not only do these linkages participate in maintaining the geometrical integrity of the axoneme, but that they also participate in bend formation.

This chapter presents in detail the known fine structural properties of the ciliary and flagellar axoneme in order to provide, at least in part, a structural basis for the numerous aspects of ciliary biology which are explored in this volume.

II. Substructure of the axoneme

A. *Axoneme geometry*

The 9+2 ciliary or flagellar axoneme comprises an orderly geometrical array of 9 interconnected doublet microtubules surrounding and joined by crossbridges to 2 centrally positioned microtubules. Variations in axoneme morphology have been discussed in detail elsewhere (Warner, 1972) and they typically occur in the axial region occupied by the 2 central microtubules. The 2 tubules can be replaced by a single opaque core (9+0 axoneme) or by 1, 3, or 7 microtubules ($9+1_n$ axoneme). In either case, all other structural aspects remain, so far as is known, similar to what will be described for the 9+2 axoneme (Fig. 1). Peripheral to the outer doublet microtubules, sperm flagella often contain structures such as accessory microtubules and fibers, and mitochondrial specializations (Fig. 2). Accessory structures, while they may contribute to motility, are not regarded as part of the basic motile mechanism.

The length of a cilium or flagellum is generally greater than 5 μm and less than 100 μm. Individual flagella of some insect spermatozoa may,

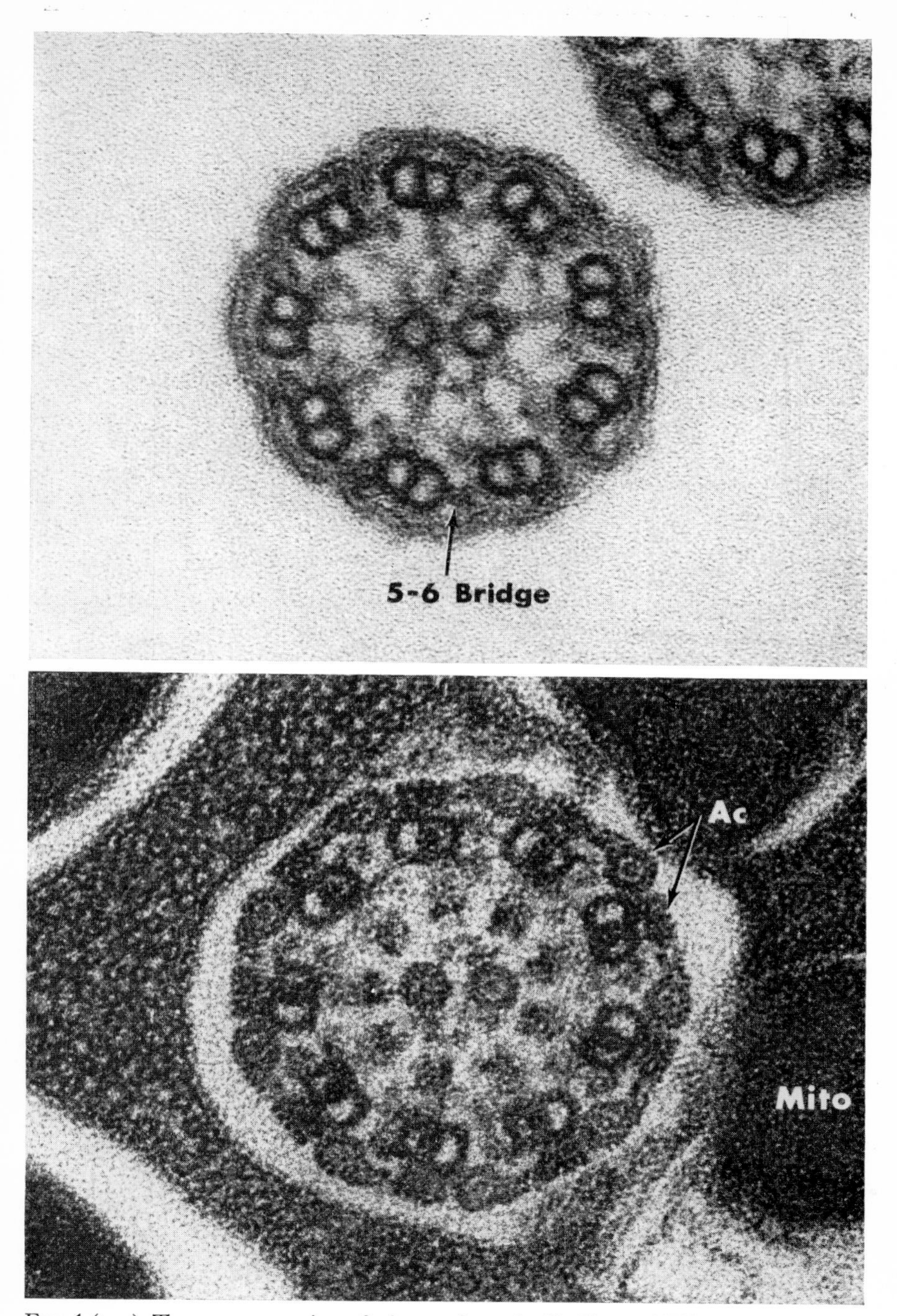

FIG. 1 (top). Transverse section of a latero-frontal gill cilium of *Elliptio complanatus*. The view is from axoneme base-to-tip and doublet number 1 lies in the 12 o'clock position; doublets 5 and 6 lie directly opposite. $\times$ 270,000. (From Warner and Satir, 1973.)

FIG. 2 (bottom). Transverse section of a spermatid flagellum from the blow fly *Sarcophaga bullata*. Accessory microtubules and fibers (Ac) and mitochondrial derivatives (Mito) surround the 9 + 2 axoneme. The view is from axoneme base-to-tip. $\times$ 270,000.

however, reach lengths of 1–2 mm. Within the organelle the 9 doublet microtubules are equidistantly positioned around the 2 central tubules, resulting in the 9-fold or 40-degree radial symmetry of the axoneme (Figs 1–3) and a diameter of about 0·2 μm at the centrifugal surface of subfiber B. Each of the 9 doublets is skewed and lies in a plane about 10 degrees

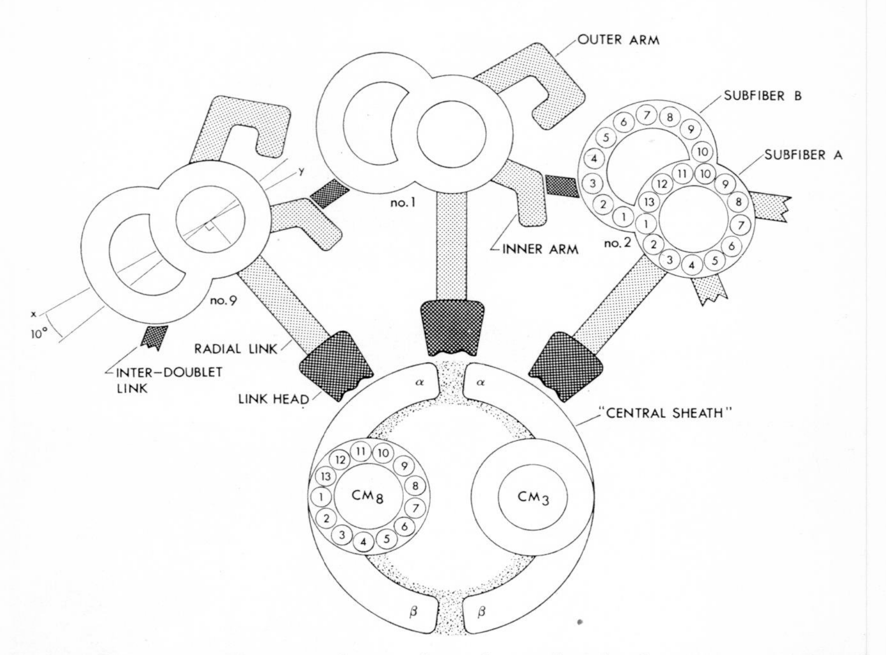

FIG. 3. Interpretive diagram of a portion of a typical 9+2 axoneme as viewed from base-to-tip. The outer doublet tubules (nos 1, 2, 9) are joined to the central sheath or paired projections (α and β) attached to each of the 2 central microtubules (CM_3 and CM_8). Drawn approximately to scale.

to the tangent of the axoneme radius as extended through the center of subfiber A (Fig. 3). This gives the axoneme an inside diameter of about 1460 Å at the centripetal surface of subfiber A.

Each doublet consists of an A- and a B-subfiber or microtubule. Subfiber A is a complete tubule while subfiber B comprises a lesser number of protofilaments and shares as a common wall, part of the A-subfiber (see section III.A). The doublet subfibers are continuous with the A- and B-subfibers of the basal body triplet microtubules. In some insect sperm flagella the 9 accessory microtubules (Fig. 2) are continuous with the 9 C-subfibers of the basal body triplets (Warner, 1971).

Approximately 175–200 Å separates the A- and B-subfibers of adjacent doublets and in this space occur the paired ATPase or dynein arms attached to subfiber A (see section II.B). The dynein arms, coupled with the 10-degree doublet skewing, result in the property of axoneme enantiomorphism. If the axoneme is viewed in transverse section and from the base of the organelle to its tip, the enantiomorphic form reveals the doublet skew and arms directed in a clockwise manner (Figs 1, 2): a tip-to-base view of the axoneme reverses this direction to counterclockwise (Gibbons and Grimstone, 1960; Satir, 1965).

In addition to dynein arms, subfiber A of each doublet bears radial links or spokes (Figs 1, 2) that join the doublets to the so-called sheath around the 2 central microtubules (see section II.B). Two laterally adjacent radial links delineate 1 of the axoneme's 9 40-degree sectors. The central pair microtubules are separated by about 90 Å and lie in a plane related to the bend direction of the motile organelle. A single bend in an axoneme can be regarded as planar: the bend plane passes through the axoneme axis and is roughly perpendicular to the plane of the central pair tubules (Gibbons, 1961; Satir, 1968; Tamm and Horridge, 1970). With the aid of reference structures, each doublet can be assigned a specific and, because of the enantiomorphic profile, unchanging number (Azfelius, 1959). A prominent bridge (see section II.B) between the 2 doublets lying in the plane and direction of the effective stroke provides the necessary marker for numbering the doublets of some cilia (Afzelius, 1959; Gibbons, 1961; Satir, 1965). The 2 bridged doublets are numbered 5 and 6 (Fig. 1) and doublet number 1 lies directly opposite, perpendicular to the central tubule plane. Numbering then proceeds clockwise if the axoneme is viewed from base-to-tip. Because of the numbering method, the central microtubules can also be specifically identified. That tubule nearest doublet number 3 is designated CM_3 and that nearest doublet 8, CM_8 (Satir, 1973). Similarly, all structures lying within a given 40-degree axoneme sector can, if necessary, be assigned the number of the doublet to which they are attached. Only cilia from a few kinds of organisms have been described as having the 5–6 bridge, so number assignments must be used with caution and preferably with the aid of other reference structures.

B. *Inter-microtubule linkage*

1. *Radial linkage*

The axoneme microtubules are maintained in their geometrical lattice by a system of delicate linkages or crossbridges that both join adjacent doublets peripherally and join the doublets to the axial region. The structurally more prominent crossbridges are termed radial links and they join the A-subfiber of each doublet with the central sheath or paired rows of projections attached to each of the 2 central tubules (see below). The links

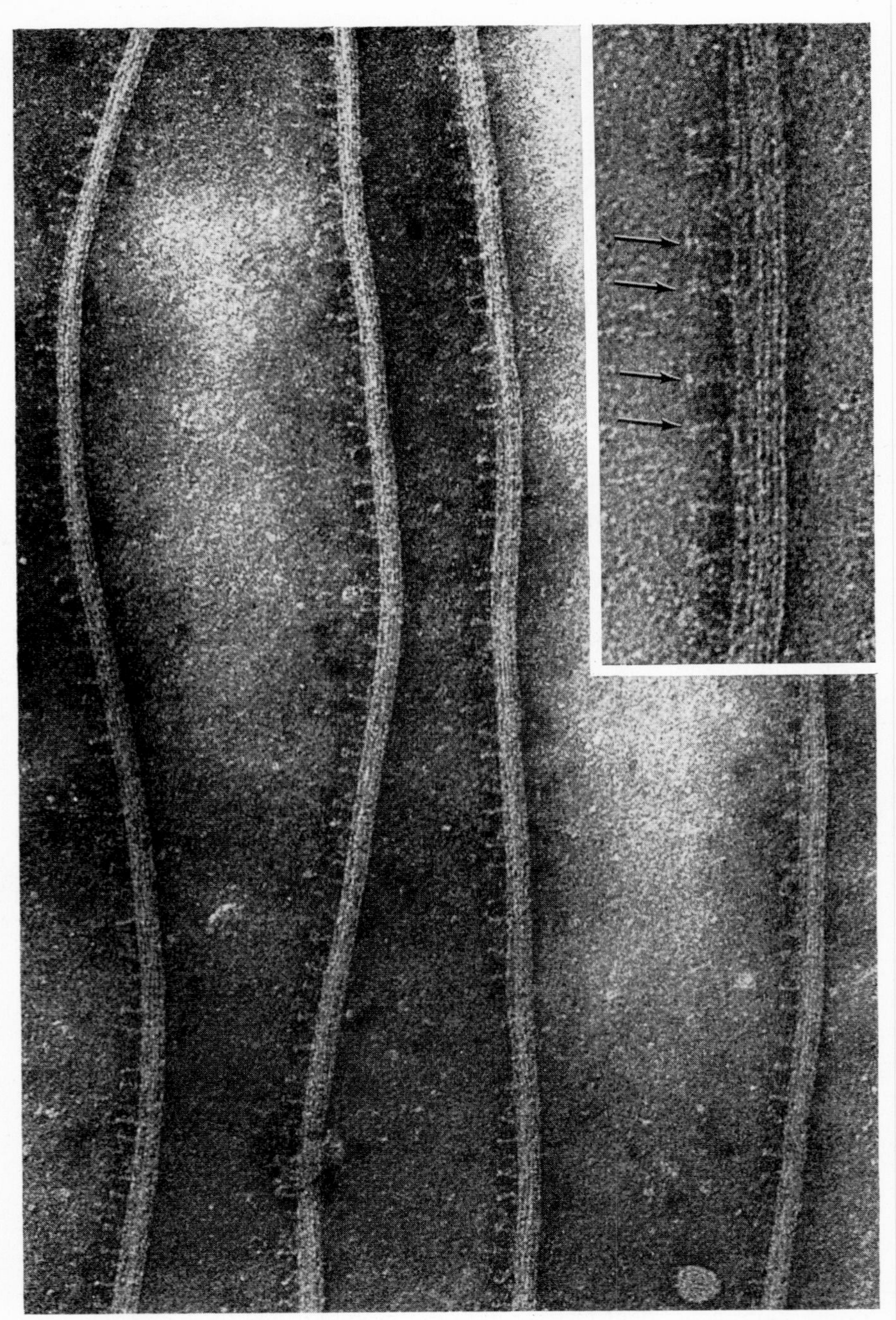

Fig. 4. Phosphotungstic acid negatively stained doublet microtubules from a spermatid flagellum of *Sarcophaga bullata*. The doublets are lying so that only the A-subfibers are visible: paired radial links (arrows) lie along the subfiber with a spacing of about 320/560 Å. × 90,000. Inset, × 200,000. (From Warner, 1970a.)

are attached to the centripetal surface of subfiber A and project radially towards the axoneme axis. Each radial link is about 360 Å long and terminates in an opaque knob or link head near the surface of the central sheath. The link head is about 200 Å across in insect spermatozoa (Fig. 2) or only 125 Å across in mussel gill cilia (Fig. 1). The substructure of the head is uncertain, but in *Chlamydomonas* flagella heads have been observed detached from the link (Hopkins, 1970). Often the head is solubilized in negatively stained preparations while the link remains intact (Fig. 4). The radial links observed in a variety of cilia and flagella show 2 patterns of grouping along subfiber A, although their three-dimensional organization is similar in all organelles (see below). The radial links occur in groups of either 2 or 3 links: the grouping appears to be a species-specific property. Figure 4 shows negatively stained doublets from spermatids of the blow fly *Sarcophaga bullata* (Warner, 1970a) and the paired radial links lie with an alternate spacing of 320/560–640 Å along subfiber A. The pairs thus have a basic repeat of 880–960 Å. The pairing of the enlarged link heads is apparent in longitudinally sectioned flagella (Fig. 6): the actual links are obscured because of amorphous material surrounding them. Paired radial links also have been extensively described in *Chlamydomonas* flagella (Hopkins, 1970).

When the radial links occur in triplets rather than in pairs, their spacing changes to about 220/280/360 Å although the same 880–960 Å group repeat prevails. The triplet links are clearly visible in very thin sections of *Elliptio* gill cilia (Fig. 5). From similar sections taken near the base of the axoneme we have determined that the wider or 360 Å space between links of one group is always nearest the base of the organelle (Warner and Satir, unpublished results). This provides a valuable marker for determining the orientation of cilia sectioned some distance from the base. Triplet radial links have also been described in negatively stained *Tetrahymena* cilia (Chasey, 1972). The characteristic spacing of both the triplet and paired links eliminates the possibility of their being confused, in negatively stained axonemes, with the other tubule-attached appendages.

In both *Sarcophaga* and *Chlamydomonas* (Warner, 1970a) and in *Elliptio* the radial links lie helically organized, but not joined, in the axoneme matrix (Figs 6, 7). The links inscribe a double, probably left-handed helix with a period of 880–960 Å and a repeat of 1760–1920 Å. The angular pitch is about 50 degrees at the link head and 35 degrees at the link base (for a 960 Å repeat). Although the helix generally must be reconstructed from thin sections (Warner, 1970a), it is sometimes visible in a thicker section ($\sim$1000 Å) that includes most of the axial region. Figure 7 shows such a section: it is important to emphasize that this is not a helical central sheath (see below) but rather is the image of the helically organized link heads attached to the sheath.

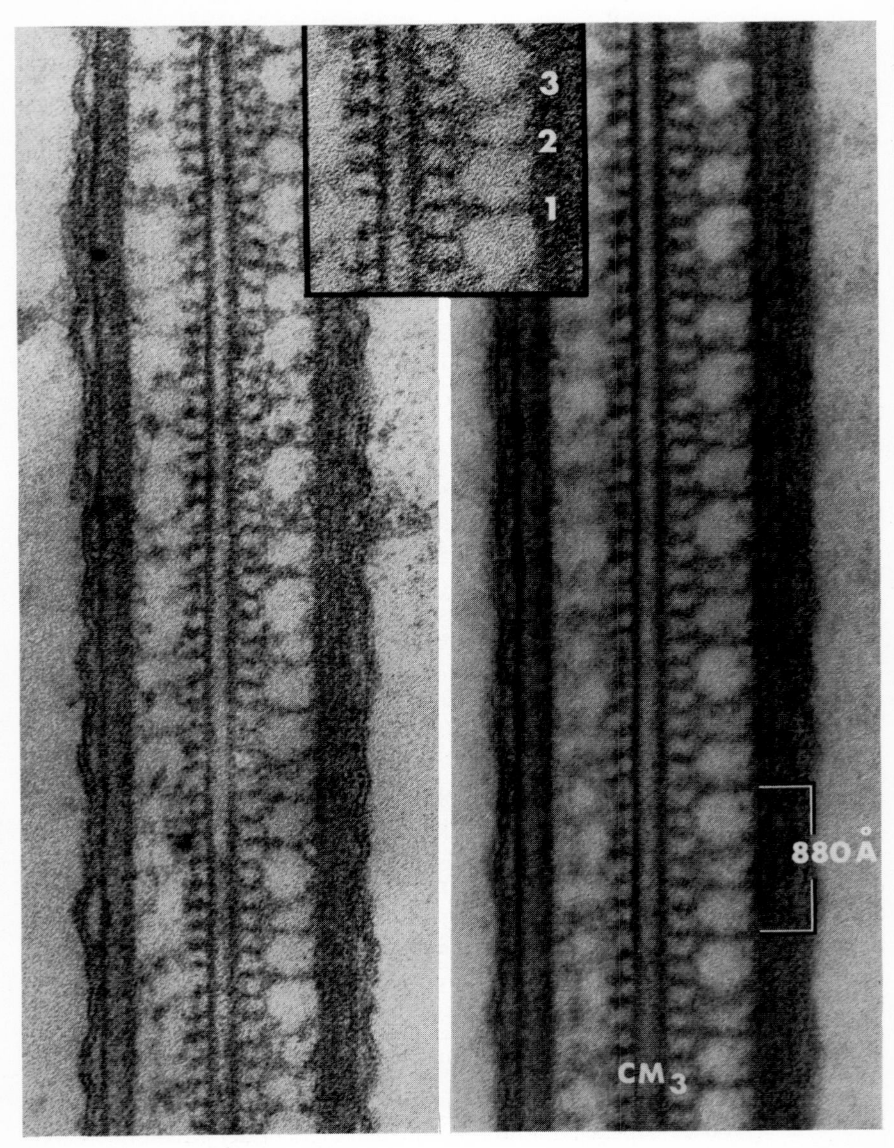

FIG. 5. Longitudinal thin section of a gill cilium of *Elliptio complanatus* (left). The triplet radial links (1, 2, 3) have a basic repeat of 880 Å along subfiber A and join the subfiber with the 2 rows of projections along the central microtubules (CM_3). The right-hand figure is a linear translation of the left, which reinforces both the radial links and the central tubule projections. The translation period used was the 880 Å link repeat. × 200,000 (left); × 230,000 (right); × 300,000 (inset).

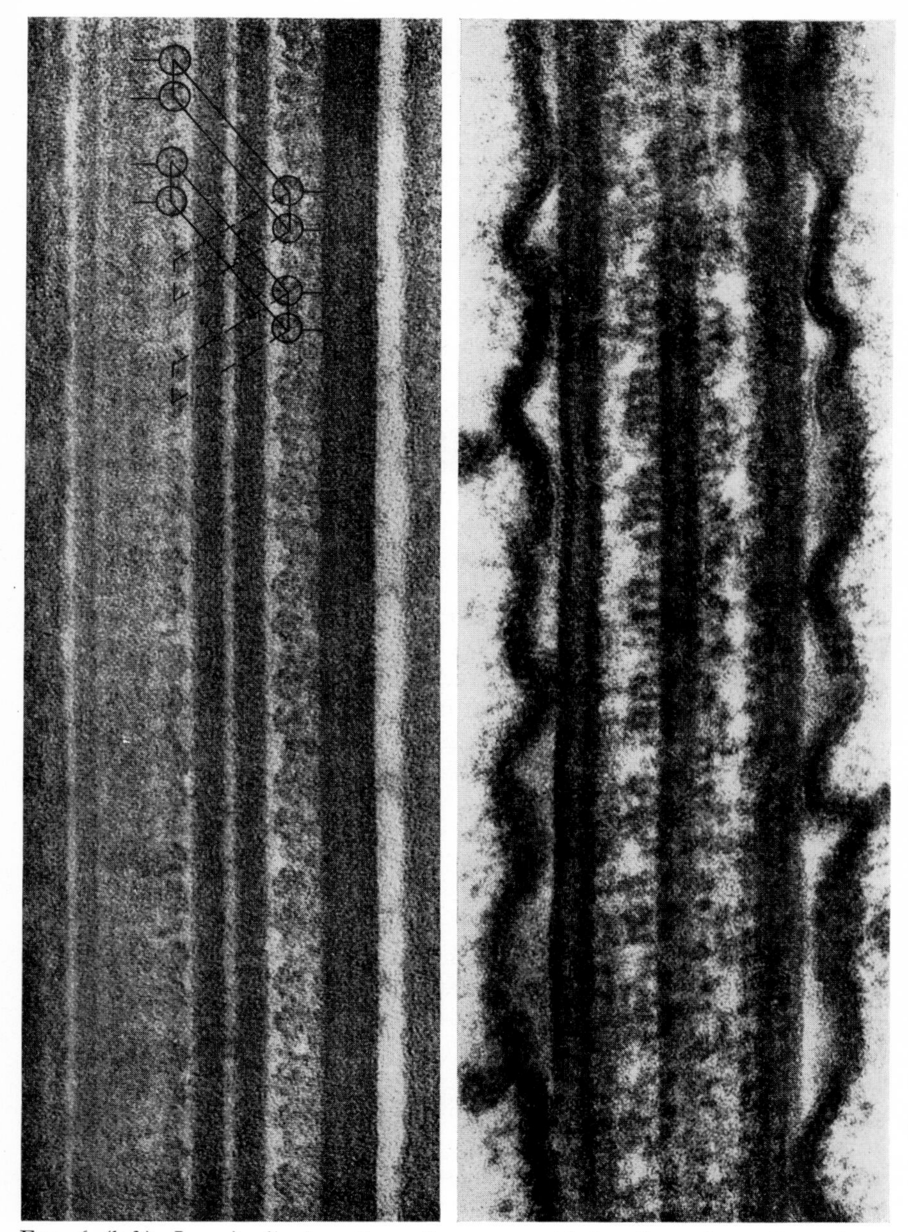

FIG. 6 (left). Longitudinal section of a spermatid flagellum from *Sarcophaga bullata*. The enlarged radial link heads (circles) are helically organized in the axoneme matrix. ×150,000. (From Warner, 1970a.)

FIG. 7 (right). Thick longitudinal section of a cilium from *Spirostomum ambiguum*. The radial link heads attached to the central sheath form a helix in the axoneme matrix. ×150,000. (Micrograph by courtesy of N. B. Gilula.)

2. *The central sheath enigma*

Fine structure studies of cilia and flagella have often indicated that a helical sheath or fiber(s) surrounds the 2 central microtubules (Gibbons and Grimstone, 1960; Pedersen, 1970; Warner, 1970a). In view of the helical axial core of platyhelminth 9+0 sperm flagella (Silveira, 1969), this interpretation has seemed reasonable. Recently, however, Chasey (1969) and Hopkins (1970) described negatively stained rows of projections attached to the central tubules: it appears that the projections are either the remnants of a sheath or that a true helical sheath does not occur. Chasey describes 2 rows of projections with a 160 Å repeat along only 1 of the central tubules in *Tetrahymena* cilia. Similarly, Hopkins describes 2 rows of projections along 1 tubule and only 1 row along the second tubule in *Chlamydomonas* flagella. In both organisms the projections are angled at about 10 degrees off perpendicular and extend about 180 Å from the tubule surface. As discussed previously (Warner, 1972) the organization of the projections might, in sectioned axonemes, result in the structure that we have commonly interpreted as a helical central sheath.

Recently, we have determined from careful sectioning of *Elliptio* gill cilia that the sheath in gill cilia does consist of 2 rows of projections along *each* of the central microtubules (Warner and Satir, unpublished results). The projections repeat at 140–160 Å intervals and lie in 2 parallel rows on each tubule: 1 row toward the number 1 doublet side (α side) and a second row toward the doublets 5–6 side (β side) of the axoneme (see interpretive diagram, Fig. 3). The projections are inclined about 10 degrees off perpendicular and extend about 180 Å from the tubule surface (Fig. 5). The inclination is polarized on each of the 2 tubules: on CM_8 the inclination of the 2 rows is directed distally while on CM_3 it is directed to the base of the axoneme. This results in the angular cross-striations often interpreted as a helical sheath. In the studies by Chasey (1969) and Hopkins (1970), it is possible that 1 or 2 rows of the projections were solubilized during negative staining.

Importantly, the central tubule projections are the structures to which the radial link heads attach (Fig. 5). By contrast, they join with the surface of the helical axial core in 9+0 flagella (Silveira, 1969): their relationship to the axial region of $9+1_n$ flagella has not been determined. In *Elliptio* each radial link appears to be attached to 2 of the central tubule projections (Fig. 5). Further, there appears to be a precise vernier relationship between the link spacings and the 140–160 Å repeat of the projections. When linear translational image reinforcement is applied at the radial link repeat, not only do the triplet links reinforce but the projections reinforce as well (Fig. 5). Conversely, if translation is applied to the 140 Å repeat, both the links and the projections reinforce, although the links can be given a false spacing depending on the number of 140 Å periods used in each translation. Accordingly, the photographic translation technique must be used with some caution.

3. *Inter-doublet linkage: dynein arm complex*

All motile cilia and flagella have paired projections or arms attached to subfiber A of each doublet. The arms comprise at least part of the axoneme ATPase dynein (Gibbons, 1965). Each of the 2 arms has a characteristic morphology when the axoneme is seen in transverse section (Figs 1–3): the most obvious difference between them is the terminal hook-like region on the outer or most centrifugally positioned arm (Allen, 1968; Warner, 1970a). In longitudinal sections the arms are spaced at regular intervals that vary between 160–220 Å depending on the organism being viewed. The arms of both rows are apparently positioned in transverse register along the microtubule wall, since when a thin section with only 1 row of arms is obtained (Fig. 8), their periodicity remains unchanged. If the arms were staggered with respect to one another, their period in such a section would necessarily be doubled. Although the arms typically do not preserve well in negatively stained axonemes, Chasey's (1972) description of negatively stained arms of *Tetrahymena* cilia agrees well with studies on sectioned organelles. Shadowcast preparations of monomeric or 14 S dynein extracted from isolated flagella show single 90–140 Å spherical particles (Gibbons and Rowe, 1965), while the polymer appears as a linear aggregate of the spherical particles that can exceed lengths of 1500 Å. The overall size of a single intact arm is about 90×300 Å.

Because of the arms' enzymatic composition and their position between sliding doublets, the presence of crossbridge-forming activity in association with the arms has often been sought. Studies of thin-sectioned cilia and flagella have shown that there is a fine linkage, termed an interdoublet link (Warner, 1970b) that appears to connect the terminal portion of the inner arm to the adjacent B-subfiber (Allen, 1968; Kiefer, 1970; Williams and Luft, 1968). The link is often visible in transverse sections of the axoneme (Figs 1–3) and is particularly clear in axonemes from which some of the arms have been solubilized (Fig. 9). The link is seen only occasionally in longitudinal sections, where it seems to repeat at about 160–200 Å along subfiber A (Fig. 8). However, it has not been determined whether the link is joined to the arm or is positioned between adjacent arms. As mentioned previously, some cilia have a structural crossbridge between doublets numbers 5 and 6. The bridge may be part of the arm structure between these doublets; that is, it appears to join both the inner and outer rows of arms with the adjacent B-subfiber (Fig. 1) and accentuates the regularity of the arm repeat in longitudinal sections of the 5 and 6 doublets.

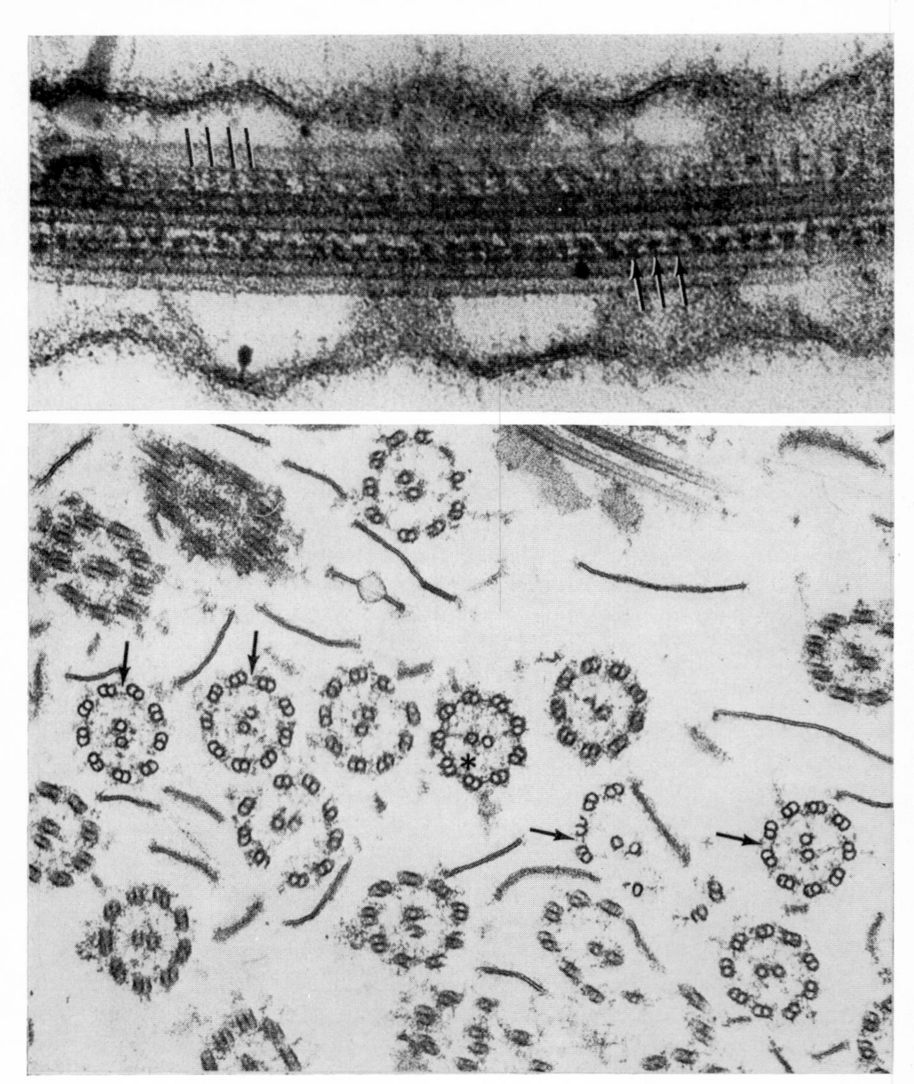

FIG. 8 (top). Longitudinally sectioned flagellum from *Chlamydomonas reinhardi*. Regularly spaced arms (outer row) are visible along the lower doublet (arrows) while the upper doublet has similarly spaced links connecting it to the adjacent B-subfiber (lines). *Chlamydomonas* is not known to have the doublet 5–6 bridge. × 160,000. (Micrograph by courtesy of G. K. Ojakian.)

FIG. 9 (bottom). Transverse section of isolated and de-membranated flagellar axonemes of *Chlamydomonas reinhardi*. Thin linkages connect adjacent doublets near the inner arm position in several of the axonemes (arrows). The axoneme marked with the asterisk shows a kind of linkage that is restricted to the basal region of the axoneme. × 62,000. (Micrograph by courtesy of J. L. Rosenbaum.)

Isolated flagellar axonemes from which the outer dynein arms have been selectively extracted (Gibbons and Fronk, 1972) still retain the interdoublet links described above. However, Stephens (1970) refers to these structures as nexin links and has apparently located their contributing band on urea-acrylamide gels. Stephens reports, from observation of negatively stained A-subfibers, that the nexin links repeat at 960 Å along the subfiber and join A-subfibers to A-subfibers, rather than joining A to B. If the interdoublet or nexin links (assuming they are the same structure) do repeat with a period similar to the dynein arms, it is possible that the links are part of the hypothetical dynein arm crossbridge mechanism between adjacent doublets. If this is not their function, then the links probably represent a structure that maintains the geometrical integrity of the axoneme, regardless of their individual periodicity. In either case, any linkage or crossbridge that joins 2 sliding surfaces in the axoneme will have to be broken or displaced in order to accommodate microtubule sliding during motility (Warner, 1972).

C. *The ciliary membrane*

Although generally it is either solubilized or ignored, cilia and flagella do have a surrounding membrane. The membrane is an extension of the cell membrane and may bear structures such as the mastigonemes of algal flagella. The membrane is regarded as lying closely apposed to the doublet microtubules (Fig. 1) and in well preserved organelles, it has a smooth profile. Fixative tonicities, however, usually cause varying degrees of membrane swelling or shrinkage although the enclosed axoneme appears to be unaffected. In flagella containing accessory components, the membrane can normally lie at some distance from the axoneme (Fig. 2).

Linkages between the doublet tubules and adjacent membrane have been postulated although none has ever been demonstrated. If the membrane is disrupted, either mechanically or by detergents, membrane fragments sometimes can be seen adhering to the doublets. If membrane–doublet linkages do occur, they must be of a tenuous nature, since the rapid, three-dimensional movements of the axoneme would seem to require considerable membrane flexibility or fluidity. Observations of well preserved, bent cilia do not show obvious differences in membrane conformation on either side of the bend.

Recent freeze-fracture studies of the ciliary membrane (Gilula and Satir, 1972) have shown particles on the membrane surface that can lie in longitudinal rows corresponding in position to the underlying doublets (Fig. 10). Such particles might be a membrane manifestation of junctional sites with the doublets. More interesting, however, Gilula and Satir have

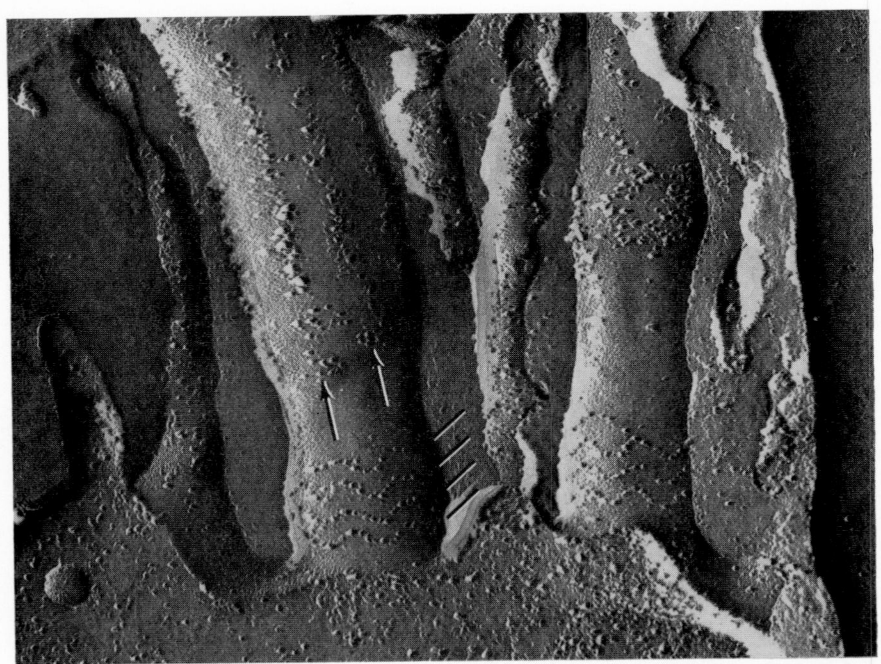

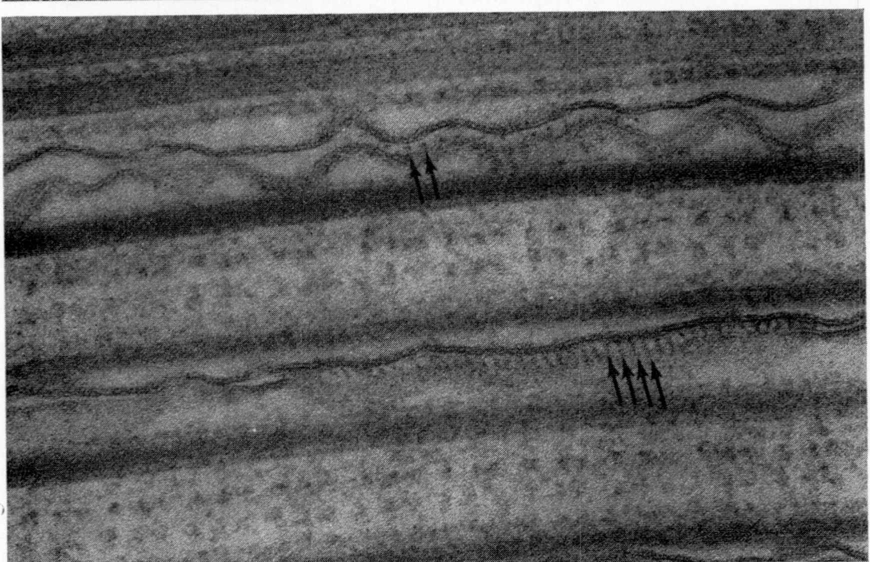

FIG. 10 (top). Freeze-fractured gill cilia of *Mytilus edulis*. Longitudinal rows of membrane particles (arrows) correspond in position to the underlying doublets. Four circumferential rows of particles near the ciliary base (lines) form the ciliary necklace. ×90,000. (Micrograph by courtesy of N. B. Gilula and P. Satir, 1972.)

FIG. 11 (bottom). Longitudinally sectioned latero-frontal gill cilia (cirrus) of *Elliptio complanatus*. Periodic bridges join adjacent ciliary membranes (arrows). ×120,000.

described a new structure, the ciliary necklace, which is related to membrane linkage with the doublet tubules. The necklace occurs just below the basal plate (basal terminus of the central tubules) in the swollen region of the axoneme that occurs distal to the basal body triplet microtubules. In transverse sections, a thin link can be seen joining the A-subfibers with the nearby membrane. Freeze-fracture preparations reveal 2–6 rows of membrane particles, the ciliary necklace, circumscribing the cilium (Fig. 10) and lying in a position that corresponds to the underlying links. The function of the links and membrane particles, if of more than structural significance, has so far not been demonstrated.

In compound cilia (cirri) where the cilia are closely apposed and function as a unit, fine connections have been observed between adjacent ciliary membranes (Warner, unpublished observations). They have been seen only in the latero-frontal cirri of the *Elliptio* gill where they lie with a repeat of about 150 Å and span a similar distance between adjacent membranes (Fig. 11). These linkages might also be related to the longitudinally oriented membrane particles revealed by the freeze-fracture technique.

III. Macromolecular architecture of microtubules

A. *Substructure of microtubules*

The microtubules of cilia and flagella, as well as those from cytoplasmic organelles, consist of a number of linear protofilaments extending, unbranched, the length of the tubule (Fig. 12). While the question of the number of protofilaments comprising microtubules has caused much debate, we have recently been able to count accurately the number of filaments in each of the 3 kinds of ciliary microtubules (Warner and Satir, 1973). Negatively stained *Elliptio* gill cilia reveal that doublet subfiber A comprises 13 filaments; subfiber B, 10 filaments (Fig. 13); and the central tubules, 13 filaments each (Fig. 12). Since most cytoplasmic and flagellar microtubules have been reported to have 13 protofilaments, 13 is probably an accurate number for most single tubules. However, exceptions do occur: Kaye (1970) has observed 16 subunits in the accessory tubules of cricket sperm flagella.

Most microtubules have a circular outline when seen in cross-section: doublet subfiber B has an elliptical profile, probably because it is an

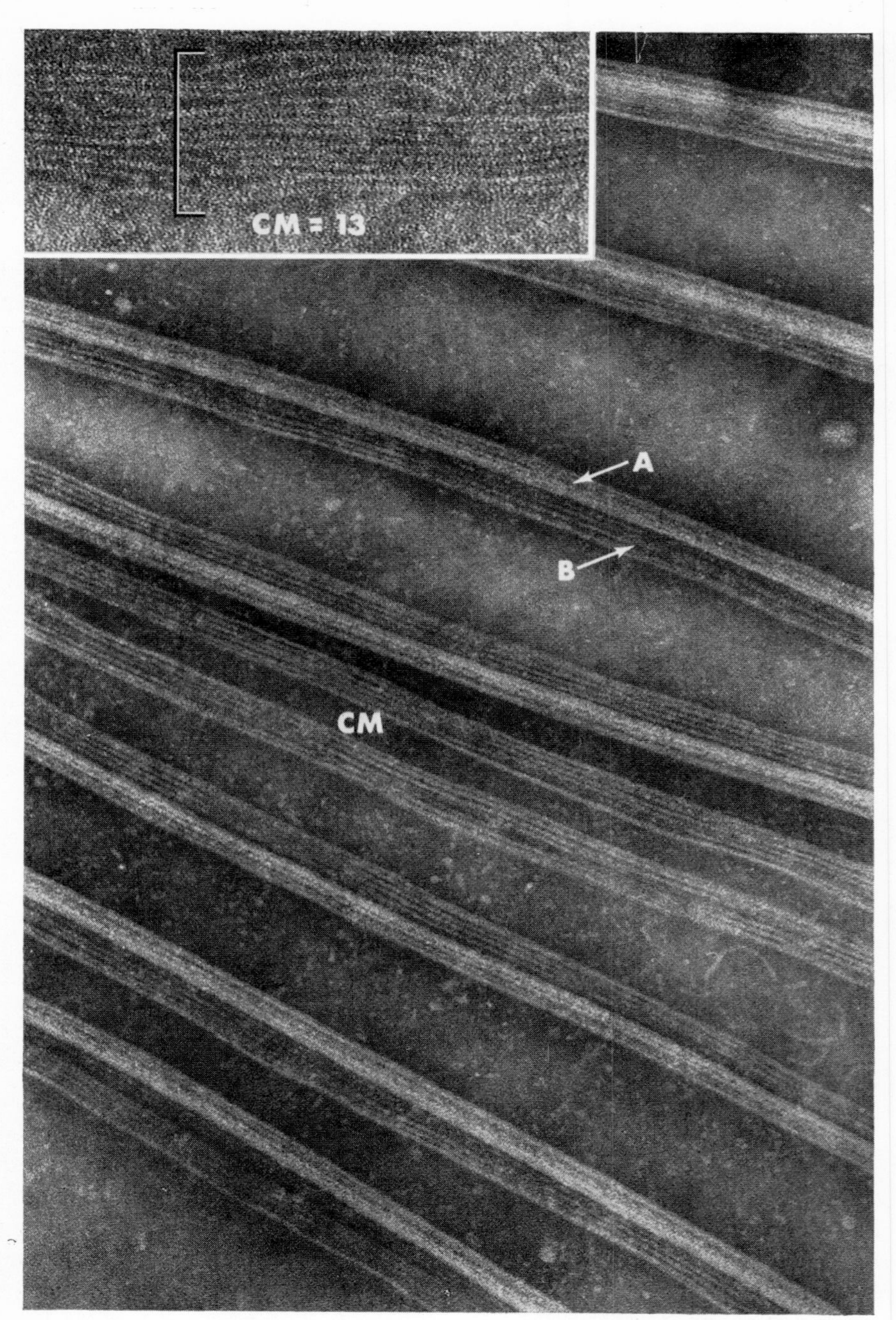

FIG. 12. Phosphotungstic acid negatively stained gill cilium from *Elliptio complanatus*. The A subfibers of each doublet are distinguishable from the B-subfibers by their tendency to remain intact. The central microtubules (CM) each comprise 13 protofilaments (inset). ×180,000. Inset, ×270,000. (From Warner and Satir, 1973.)

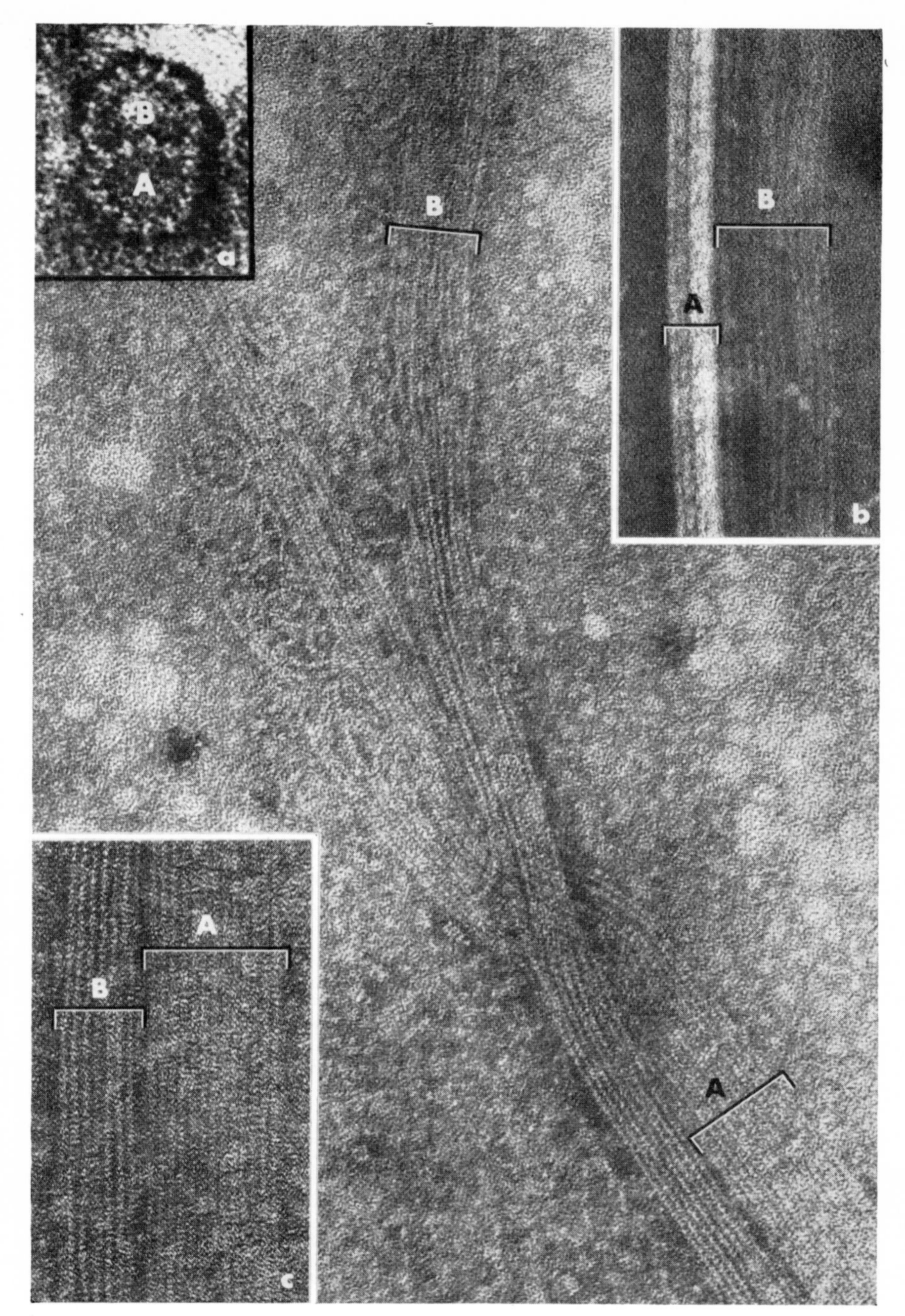

FIG. 13. Phosphotungstic acid negatively stained doublet microtubules from the gill cilia of *Elliptio complanatus*. The doublets are spread so that all of their protofilaments are visible. Thirteen protofilaments occur in subfiber A (inset c) and 10 filaments occur in subfiber B (inset b). Inset a shows a sectioned, thiourea-treated doublet where the tubule wall protofilaments stand out in negative relief. × 200,000. Inset a, × 620,000; inset b, × 260,000; inset c, × 260,000. (From Warner and Satir, 1973.)

incomplete microtubule that shares part of the A-subfiber as a common middle wall (Phillips, 1966). Four or 5 protofilaments occur in the shared region of subfiber A (Ringo, 1967; Warner and Satir, 1973), although 3 filaments lie completely free from the A–B-subfiber junction (Figs 3, 13). These 3 filaments are resistant to the detergent action of Sarkosyl and remain as intact groups of 3 (Fig. 14) after the other 20 filaments of the doublet have been solubilized (Meza *et al.*, 1972; Witman *et al.*, 1972a).

Single microtubules can vary considerably in diameter: the reported range is about 180–300 Å. The typical diameter is about 240–260 Å, although it is uncertain if the extremes of variation are due to differences in protofilament number or due to the staining properties of the tubules. Within the range of 200–280 Å the latter is probably true. In stained thin sections, microtubules occasionally have their protofilaments revealed in negative relief (Kaye, 1970; Phillips, 1966; Warner and Satir, 1973). The heavy metal stain accumulates either on the filament surface or stains material around the filaments (Fig. 13a) rather than penetrating the filaments and positively staining the wall (Fig. 1). When this kind of negative staining occurs, the tubule diameter measured from the protofilament axis is uniformly about 240 Å. The tubule wall thickness varies, however, between 55 and 85 Å depending on the amount of stain that has accumulated. The thickness of individual protofilaments in traditional negatively stained cilia (Fig. 13) is generally 35–40 Å.

While the lumen of a microtubule usually appears translucent, it can contain electron-opaque material. The central and accessory tubules of insect sperm flagella often have a single filament lying in their lumen (Fig. 2). In cricket sperm flagella the entire lumen of all the axoneme tubules is filled with filament-like material (Kaye, 1970). In *Chlamydomonas* flagella, small projections are attached to the inner surface of the walls of both doublet subfibers (Witman *et al.*, 1972a). The microtubules of *Elliptio* gill cilia (Warner, unpublished observations) always have amorphous material filling their lumen, but only in the more proximal regions of the axoneme. In all cases, the biochemical and functional properties of lumen material have not been determined.

B. *Microtubule helicity: the wall lattice*

Witman (in press) shows that assembly of regenerating flagellar microtubules apparently proceeds by addition of protein subunits near the distal ends of the tubules. However, assembly or disassembly of tubules probably does not occur in a helical manner as might be predicted for a hypothetical self-assembly system. Morphogenetic stages of flagellar doublet formation (Warner, 1971) reveal lengthy regions of incomplete or C-shaped micro-

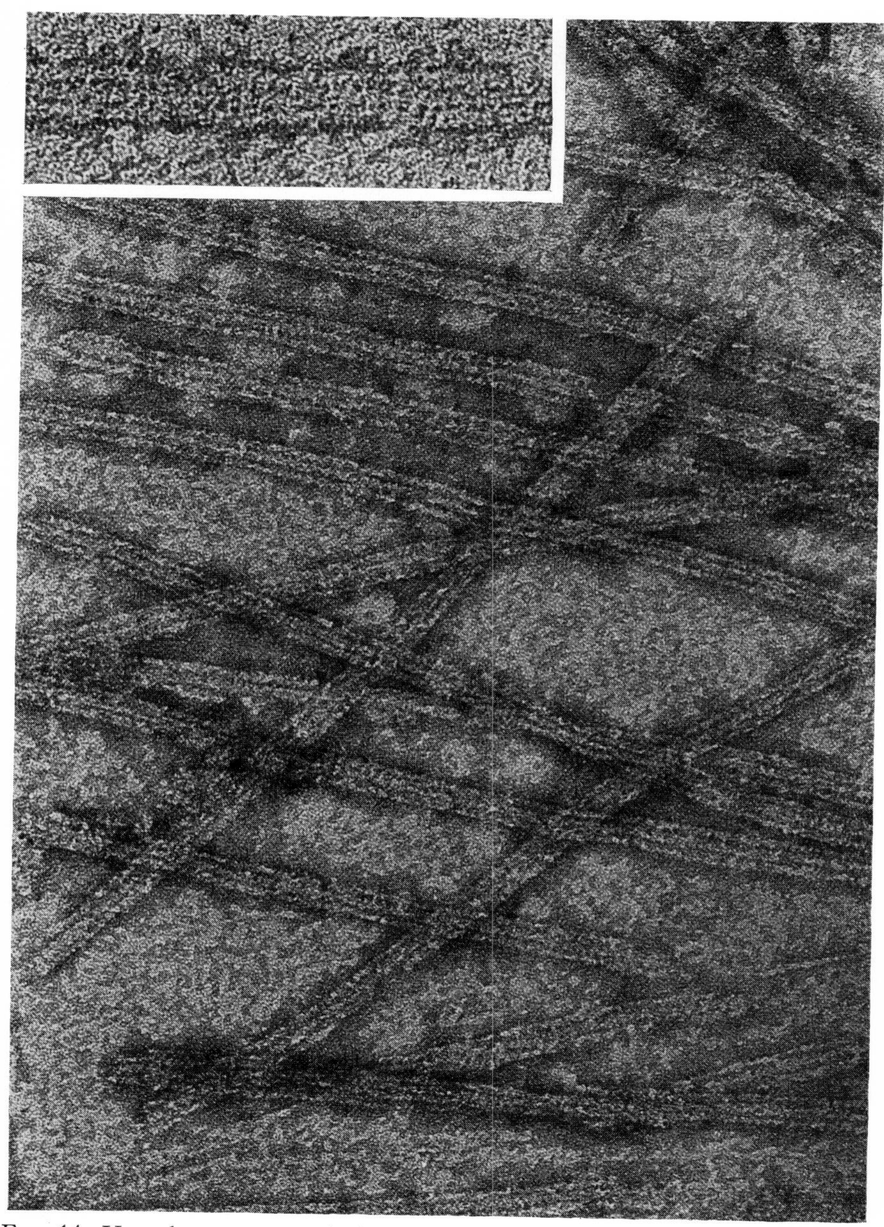

FIG. 14. Uranyl acetate negatively stained protofilaments obtained by Sarkosyl solubilization of doublet tubules from sperm flagella of the sea urchin *Strongylocentrotus purpuratus*. The 3 intact protofilaments probably comprise part of the region of doublet subfiber A that is shared as a common middle wall by the B-subfiber. ×315,000. Inset ×560,000. (From Warner and Meza, 1972.)

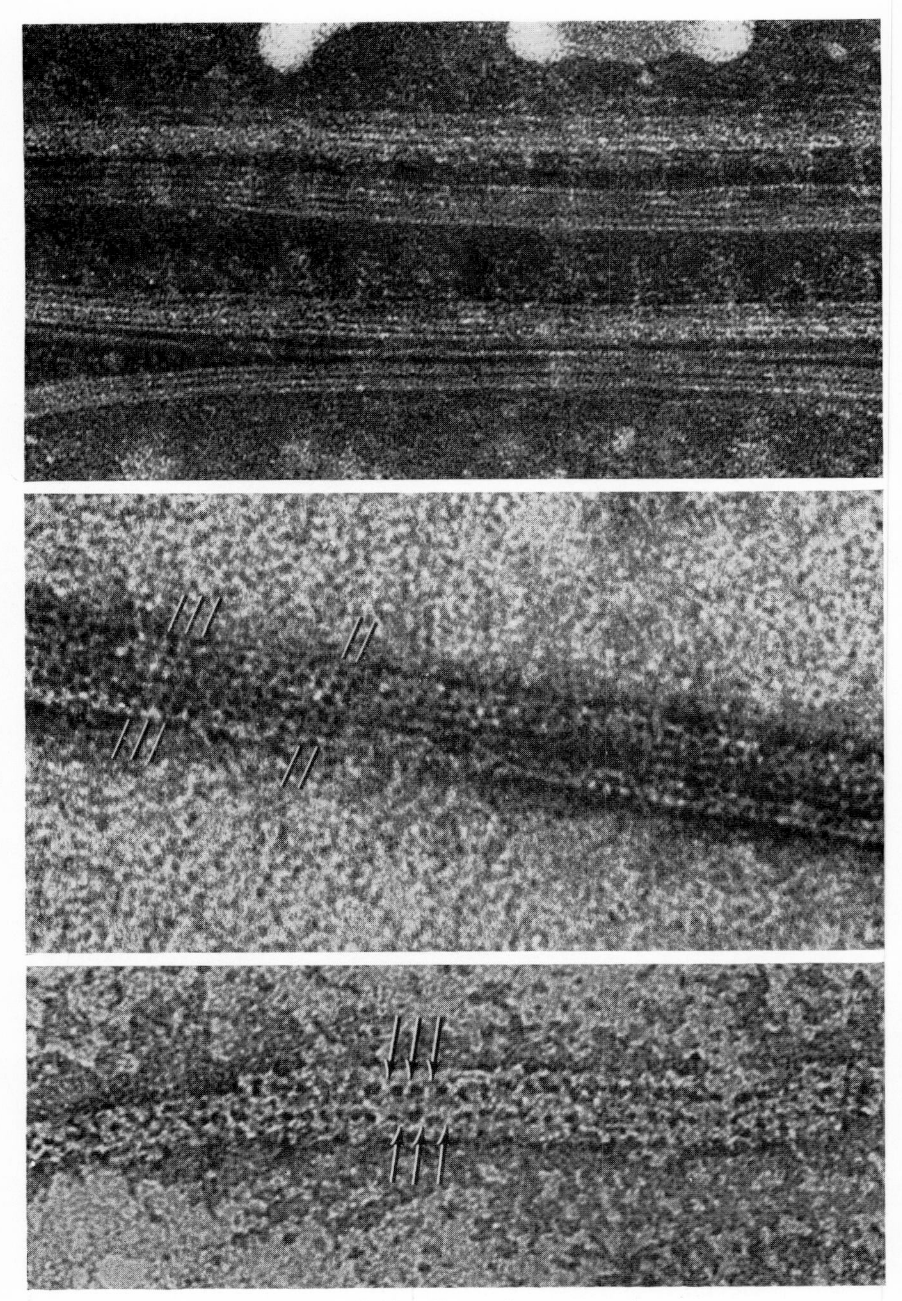

FIGS 15, 16 and 17.—Legends on opposite page.

tubules, suggesting that assembly follows a circular path, protofilament by protofilament, around the forming doublet. How far the incomplete regions extend along the tubule is not known, but they appear to exceed 2000 Å. If assembly pursued a helical course around the tubule, observed regions of incomplete tubules would be few, if any. Moreover, the solubilization sequence of outer doublets (Witman *et al.*, 1972b) closely parallels the morphogenetic sequence. Doublets are solubilized by sequential dissociation of laterally adjacent protofilaments, although this might indicate that the bonding between adjacent filaments is weaker than the linear association of subunits.

The doublet morphogenetic and solubilization sequences suggest that any helical properties of microtubules result from lateral associations of adjacent protofilaments. Negatively stained filaments have always suggested, by the very fact that filaments are observed, that the stronger intermolecular bonding between tubule subunits occurs along the filament axis and a weaker co-lateral bonding holds adjacent filaments into the wall lattice. Further, curling of broken protofilaments suggests that axial bonding produces considerable tension along the filament (Fig. 13). High resolution electron microscopy of isolated protofilaments (Warner and Meza, 1972) indicates that the filament is the structurally cohesive unit and that a visible junction occurs between adjacent filaments. Figure 17 shows 3 isolated protofilaments from sperm flagellar doublets of the sea urchin and in several regions, periodic junctions are visible repeating at about 40 Å between filaments.

An X-ray diffraction study of hydrated microtubules (Cohen *et al.*, 1971) has shown that the basic 40 Å axial repeat or subunit (Fig. 15) of a protofilament (see below) is displaced laterally by about 20 Å between adjacent filaments. This results in an angular packing between laterally adjacent subunits, although other surface lattice conformations are possible in both a lateral and radial direction. Angular packing of tubule wall subunits is indicated by direct observation of negatively stained microtubules (Grimstone and Klug, 1966) where a 15–25-degree lateral deflection of the 40 Å subunits is often visible (Fig. 16).

FIG. 15 (top). Phosphotungstic acid negatively stained doublet microtubules from a spermatid flagellum of *Sarcophaga bullata*. The protofilaments appear to consist of linear aggregates of 40 Å spherical subunits. × 200,000.

FIG. 16 (middle). Uranyl acetate negatively stained doublet tubule from a sperm flagellum of *Strongylocentrotus purpuratus*. The 40 Å subunits of the protofilaments are in angular alignment across the tubule wall (lines). × 375,000.

FIG. 17 (bottom). Uranyl acetate negatively stained protofilaments isolated from doublet tubules of *Strongylocentrotus purpuratus* sperm flagella. Visible bonding spaced at about 40 Å occurs between the 3 adjacent filaments (arrows). × 800,000. (From Warner and Meza, 1972.)

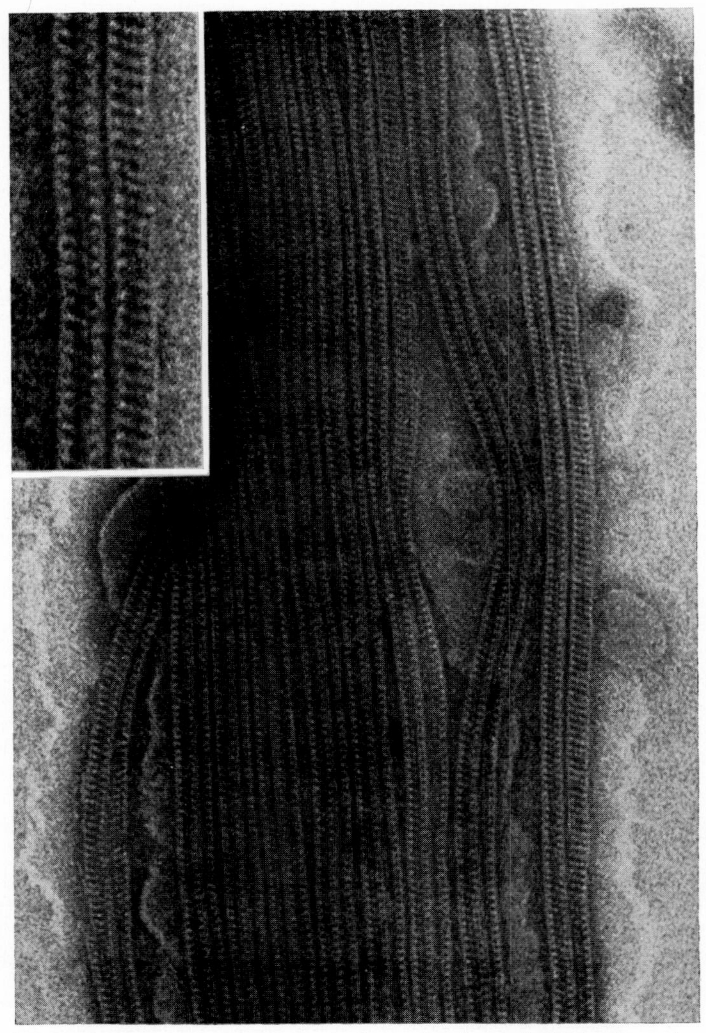

FIG. 18. Phosphotungstic acid negatively stained cortical singlet microtubules from the sperm flagella of *Dugesia tigrina*. A prominent helical wall substructure repeats at about 80 Å along the tubule, although the wall protofilaments are visible in regions. ×120,000. Inset, ×250,000. (Micrographs by courtesy of P. R. Burton and M. Silveira, 1971.)

Certain kinds of microtubules have a major helical repeat of the wall lattice. The cortical singlet tubules from platyhelminth sperm flagella (Burton and Silveira, 1971; Thomas and Henley, 1971) have a prominent 80 Å repeat inclined 15–25 degrees across the tubule wall (Fig. 18). In fact, the repeat is so pronounced that, when negatively stained, the edge of the tubule appears grooved and the wall protofilaments are mostly obscured. The filaments are visible in regions (Fig. 18) and in macerated tubules Thomas and Henley (1971) have counted as many as 12. A definite transition from the helical wall configuration to the protofilament configuration results from tubule maceration during negative staining. Interestingly, Thomas and Henley report that dissociation of the 80 Å helical wall lattice occurs in the following sequence:

(a) a lateral dissociation of wall subunits which reveals individual protofilaments with an 80 Å axial repeat,
(b) a change occurs in the 80 Å axial repeat that results in a 40 Å axial repeat along the protofilament.

The nature of the helical association of wall subunits and their transition to the protofilament configuration is not understood but could be explained by several models of both lateral and radial subunit associations (Thomas and Henley, 1971). Nevertheless, the cortical singlet microtubules show the first clear evidence of an 80 Å structural repeat along a protofilament: the 80 Å repeat might be related to the hypothetical structure of the microtubule protein dimer (see below).

C. *Protofilament substructure*

Since protofilaments are the largest single structural unit of a microtubule and are, necessarily, the structures to which the various kinds of tubule appendages attach, determination of their individual substructure and biochemistry is very important.

It is generally acknowledged that individual axoneme microtubules, and probably cytoplasmic tubules as well, consist of a minimum of 2 proteins (tubulins) in a 1 : 1 ratio (Bryan and Wilson, 1971; Olmsted *et al.*, 1971; Witman *et al.*, 1972a, b). Further, there is evidence that the 2 tubulins (α and β or 1 and 2) may both occur in individual protofilaments (Meza *et al.*, 1972). When the 3 midwall filaments of sea urchin sperm flagellar doublets are isolated by Sarkosyl solubilization (Meza *et al.*, 1972), they still show a 1 : 1 ratio of α and β tubulins as determined by dye binding on urea-acrylamide gel electrophoresis. However, midwall preparations from *Chlamydomonas* flagella (Witman *et al.*, 1972a, b) reveal only a single tubulin band on urea-acrylamide gels. Nevertheless, it is pertinent to ask how the microtubule protein(s) are incorporated into protofilaments.

When negatively stained, protofilaments from a variety of microtubules often appear to consist of a linear chain of 40 Å beads or spheres (Grimstone and Klug, 1966). The 40 Å subunits (Fig. 15) are, so far, the smallest subunits of a protofilament to be observed and they are thought to represent the globular and monomeric form of microtubule protein of molecular weight ~55,000. The biochemically distinct tubulin dimer was thus thought to be a ~40 × 80 Å unit, or 2 of the 40 Å subunits in axial association along the filament (Shelanski and Taylor, 1968). Since dimeric tubulin appears to consist of biochemically non-identical subunits (Bryan, 1972), it is possible that the dimer comprises 1 monomer of each of the 2 tubulins. However, other models based on a homopolymeric protofilament have been proposed (Witman *et al.*, 1972b).

Optical diffraction studies of negatively stained flagellar tubules support the idea that the tubulin dimer is a 40 × 80 Å subunit. Grimstone and Klug (1966) demonstrated a major 80 Å layer line in the diffraction pattern although 40, 160, and 480 Å repeats were also detected. Physico-chemical studies of dimeric tubulin also support a 40 × 80 Å subunit (Shelanski and Taylor, 1968). X-ray diffraction of intact microtubules (Cohen *et al.*, 1971), however, reveals only 1 linear diffracting unit with a repeat of about 40 Å.

High resolution electron microscopy of the tubulin dimer obtained after thermal fractionation of doublet microtubules (Warner and Meza, 1972; Warner and Meza, manuscript in press) indicates that the dimer is an approximate 35 × 80 Å, figure-8 shaped subunit; that is, the negative stain has penetrated into the center of each spherical protofilament subunit. Examination of isolated protofilaments reveals a dual-strandedness to the filament: each protofilament appears to consist of 2 side-by-side, ~17 Å thick subfilaments (Fig. 19) that have periodic lateral associations resulting in the 40 Å repeat along the protofilament. It appears that the tubulin monomer may have either an elongated or bilobed configuration that results in the protofilament strandedness and the figure-8 shaped dimer. Such a configuration can account, in part, for the longstanding inability to isolate the spherical 40 Å protofilament subunit as both the structural and biochemically active subunit of microtubules.

IV. Concluding remarks

Although we have seen with relative clarity the complex geometrical arrangement of structures comprising the motile axoneme of cilia and flagella, it is clear that microtubule-based motility or sliding occurs as a result of interaction between neighboring tubules. It will doubtless be shown that microtubules are simply the skeletal framework upon which shear forces act to produce sliding of the tubules. However, several questions pertaining to axoneme substructure and biochemistry, and hence motility, remain unresolved:

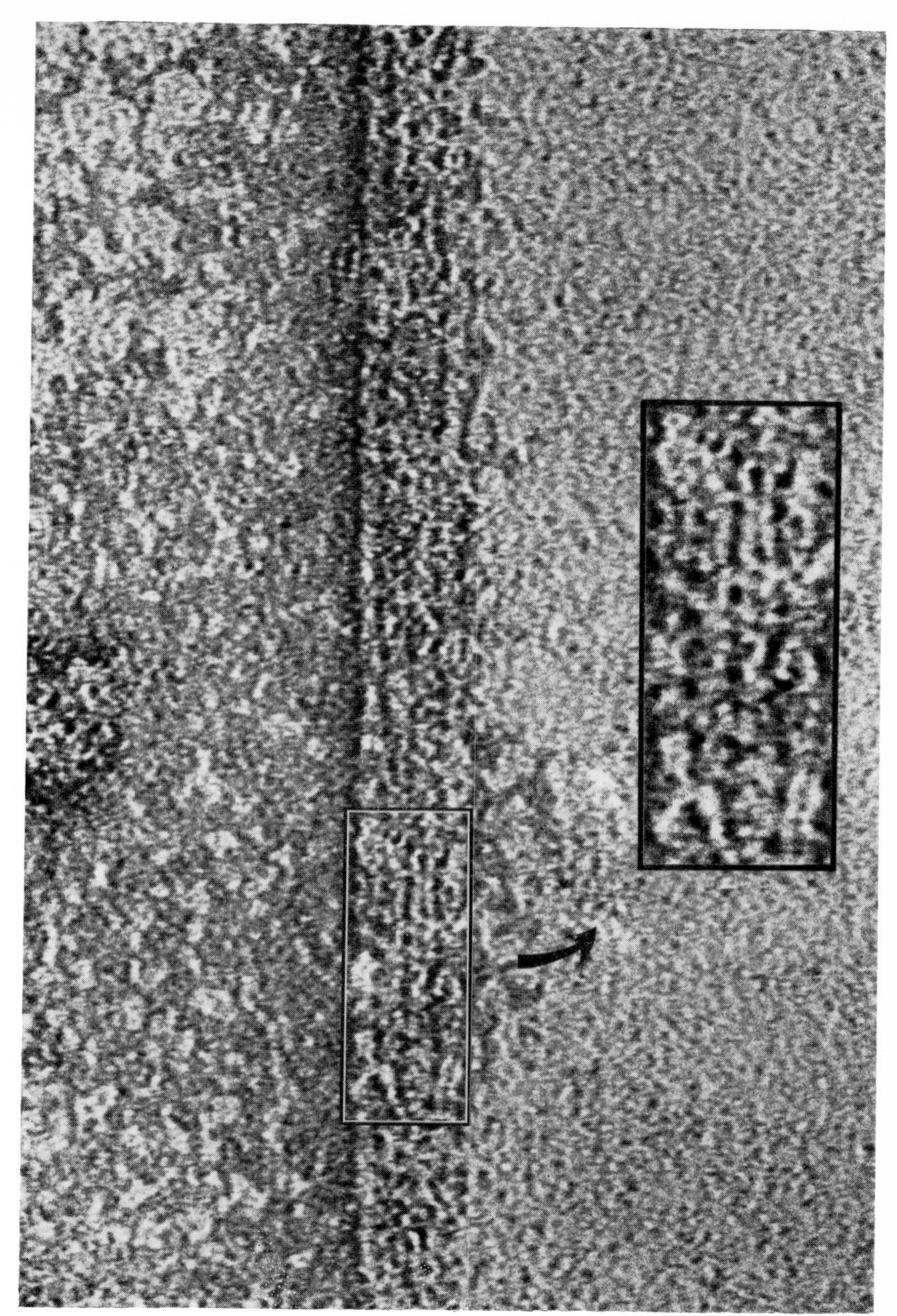

FIG. 19. High-resolution electron micrograph of 4 uranyl acetate negatively stained protofilaments from doublet tubules of *Strongylocentrotus purpuratus* sperm flagella. Each of the 40 Å thick protofilaments can be seen, in regions, to comprise 2 subfilaments of about 17 Å thickness. ×1,000,000. Inset, ×1,500,000. (From Warner and Meza, 1972.)

1. Structural interaction of dynein or the ATPase arms with the neighboring B-subfiber has not been demonstrated but can be assumed on the basis of Summers' and Gibbons' (1971) observations. Clearly, elucidation of the hypothetical crossbridge mechanism between adjacent doublets will be of major importance in defining the nature of microtubule-based motility.

2. Doublet sliding is obviously translated into bending forces in the axoneme. How is bending mediated? The position of the radial links connecting doublet tubules with the central tubules suggests that the links may be involved in the bending process. The links are not passively attached to the central tubules. In bend regions they undergo systematic angular changes with respect to the tubules, while distal to a bend they remain perpendicular to the tubules (Warner and Satir, unpublished results).

3. The numerous tubule-attached appendages seemingly require specific attachment sites along the microtubule wall. Are these sites specified by the different tubulins occurring in the tubule or might other components such as carbohydrates or bound nucleotides also be involved?

Resolution of these questions and others will involve understanding of both the number and conformation of the tubulin polypeptides and their integration into protofilaments. It should then be possible to specifically explore the biochemical and structural interaction of all axoneme components.

Acknowledgements

This paper was written while the author was a Post-doctoral Fellow at the University of California, Berkeley, and was supported by Post-doctoral Fellowship 41008 from the National Science Foundation and research grant HL 13849 from the United States Public Health Service.

Figures 4, 6, and 10 are reproduced by courtesy of the Rockefeller University Press and the *Journal of Cell Biology*; Fig. 18 by courtesy of Academic Press and the *Journal of Ultrastructure Research*; and Figs 1, 12, and 13 by courtesy of the Company of Biologists and the *Journal of Cell Science*.

References

Afzelius, B. A. (1959). *J. biophys. biochem. Cytol.* **5**, 269–278.
Allen R. D. (1968). *J. Cell Biol.* **37**, 825–831.
Bryan, J. (1972). *J. molec. Biol.* **66**, 157–168.
Bryan, J. and Wilson, L. (1971). *Proc. natn. Acad. Sci. U.S.A.* **68**, 1762–1766.

Burton, P. R. and Silveira, M. (1971). *J. Ultrastruct. Res.* **36**, 757–767.
Chasey, D. (1969). *J. Cell Sci.* **5**, 453–458.
Chasey, D. (1972). *Expl Cell Res.* **74**, 140–146.
Cohen, C., Harrison, S. C. and Stephens, R. E. (1971). *J. molec. Biol.* **59**, 375–380.
Gibbons, I. R. (1961). *J. biophys. biochem. Cytol.* **11**, 179–205.
Gibbons, I. R. (1965). *Archs Biol., Liège* **76**, 317–352.
Gibbons, I. R. and Fronk, E. (1972). *J. Cell Biol.* **54**, 365–381.
Gibbons, I. R. and Grimstone, A. V. (1960). *J. biophys. biochem. Cytol.* **4**, 697–716.
Gibbons, I. R. and Rowe, A. J. (1965). *Science, Wash.* **149**, 424–426.
Gilula, N. B. and Satir, P. (1972). *J. Cell Biol.* **53**, 494–509.
Grimstone, A. V. and Klug, A. (1966). *J. Cell Sci.* **1**, 351–362.
Hopkins, J. M. (1970). *J. Cell Sci.* **7**, 823–839.
Kaye, J. S. (1970). *J. Cell Biol.* **45**, 416–430.
Kiefer, B. I. (1970). *J. Cell Sci.* **6**, 177–194.
Meza, I., Huang, B. and Bryan, J. (1972). *Expl. Cell Res.* **74**, 535–540.
Olmsted, J. B., Witman, G. B., Carlson, C. and Rosenbaum, J. L. (1971). *Proc. natn. Acad. Sci. U.S.A.* **68**, 2273–2277.
Pedersen, H. (1970). *J. Ultrastruct. Res.* **33**, 451–462.
Phillips, D. M. (1966). *J. Cell Biol.* **31**, 635–638.
Ringo, D. L. (1967). *J. Ultrastruct. Res.* **17**, 266–277.
Satir, P. (1965). *J. Cell Biol.* **26**, 805–834.
Satir, P. (1968). *J. Cell Biol.* **39**, 77–94.
Satir, P. (1973). *In* "Behaviour of Microorganisms" (A. Perez-Mirvete, ed.), pp. 214–228. Plenum Press, London.
Shelanski, M. L. and Taylor, E. W. (1968). *J. Cell Biol.* **38**, 304–315.
Silveira, M. (1969). *J. Ultrastruct. Res.* **26**, 274–288.
Stephens, R. E. (1970). *Biol. Bull.* **139**, 438.
Summers, K. E. and Gibbons, I. R. (1971). *Proc. natn. Acad. Sci. U.S.A.* **68**, 3092–3096.
Tamm, S. L. and Horridge, G. A. (1970). *Proc. R. Soc.* B **175**, 219–233.
Thomas, M. B. and Henley, C. (1971). *Biol. Bull.* **141**, 592–601.
Warner, F. D. (1970a). *J. Cell Biol.* **47**, 159–182.
Warner, F. D. (1970b). *J. Cell Biol.* **47**, 220a.
Warner, F. D. (1971). *J. Ultrastruct. Res.* **35**, 210–232.
Warner, F. D. (1972). *In* "Advances in Cell and Molecular Biology" (E. J. DuPraw, ed.), Vol. 2, pp. 193–235. Academic Press, New York.
Warner, F. D. and Meza, I. (1972). *J. Cell Biol.* **55**, 273a.
Warner, F. D. and Satir, P. (1973). *J. Cell. Sci.* **12**, 313–326.
Williams, N. E. and Luft, J. H. (1968). *J. Ultrastruct. Res.* **25**, 271–292.
Witman, G. B., Carlson, K., Berliner, J. and Rosenbaum, J. L. (1972a). *J. Cell Biol.* **54**, 507–539.
Witman, G. B., Carlson, K. and Rosenbaum, J. L. (1972b). *J. Cell Biol.* **54**, 540–555.

Chapter 3

Enzymatic and structural proteins of the axoneme

R. E. STEPHENS

The Marine Biological Laboratory, Woods Hole, Massachusetts, U.S.A.
and
The Department of Biology, Brandeis University, Waltham, Massachusetts, U.S.A.

I. Introduction

The "9+2" structure of the axoneme holds within its unique array an assemblage of contractile, control, and structural proteins which together hold the key to the biochemical mechanism of ciliary and flagellar beat. Since the axoneme contains an ATPase and an "actin-like" protein, numerous studies have attempted to relate, at both the chemical and the morphological level, the motility of cilia and flagella to muscle contraction. However, ambiguities in isolation and extraction conditions, equivocal protein chemistry, and non-axonemal sources have essentially ruled out all such claims. Much recent work, based primarily on Gibbons' (1963, 1965a) fractionation of isolated *Tetrahymena* cilia into unique ATPase and microtubule components, has led to the structural and chemical identification of several characteristic axonemal proteins whose nature demonstrates quite clearly that cilia and flagella constitute a distinct motile system bearing only superficial similarity to muscle. Both are dual systems involving interaction of a characteristic high molecular weight ATPase with a hydrophobic, nucleotide-containing, globular protein in polymeric form: myosin with actin in muscle and dynein with tubulin in the axoneme. The structural and biochemical uniqueness of these two systems, contrasted with their many basic similarities, point to the co-evolution of two distinct contractile systems, one serving primarily for intracellular or cellular movement and the other for cellular or, in a sense, extracellular propulsion.

It is the goal of this chapter to outline the various isolation and fractionation procedures for the biochemical characterization of ciliary and flagellar proteins, to discuss in detail the properties of these components, to relate the isolated proteins to distinct axonemal structures, to examine possible enzymatic interactions in the functioning motile system, to compare these axonemal proteins with those of other motile systems, and finally, to explore some of the many questions yet to be answered in these most widely distributed of contractile systems.

II. Sources of cilia and flagella for biochemical work

A. Cilia

Historically, the first mass isolation of cilia for purposes of characterization was carried out by Child (1959), using the ciliate *Tetrahymena pyriformis*. The cells were treated with 40% ethanol at $-10°C$ and then transferred to 0·1 M KCl; stirring caused detachment of the cilia which were then separated from intact cell bodies by differential centrifugation. He also demonstrated that treatment with 20% glycerol at $-8°C$, followed by stirring at 0°C for 5–10 min, caused a similar detachment. Child (1959) attributed the isolation to a combination of osmotic shock and mechanical shear, causing breakage at the kinetosome but leaving the latter in the pellicle.

Later, Watson and Hopkins (1962) devised a procedure whereby *Tetrahymena* cilia could be removed under somewhat milder conditions and without mechanical disruption. Treatment with 10% ethanol containing EDTA, followed by addition of $CaCl_2$ to a concentration of 25 mM, caused spontaneous detachment of the organelle; low speed centrifugation removed cell bodies while centrifugation at high speed harvested nearly pure cilia. This procedure was then modified by Gibbons (1965a); the final calcium concentration was halved and the isolation was conducted about 1·5 pH units above that used by Watson and Hopkins (*ca* pH 7·5 *vs.* 6). The Gibbons modification, now a standard for *Tetrahymena* cilia isolation, will yield about 100 mg of total ciliary protein from 10 l of a stationary culture.

Auclair and Siegel (1966) have used this same basic methodology to isolate cilia from the blastula or gastrula stage of sea urchin embryos, but in addition they have devised an alternative method involving treatment of the embryos with hypertonic media, in this case sea water having twice the normal NaCl concentration. This latter method is sufficiently gentle to allow up to five successive regenerations from the same cells (Iwaikawa, 1967). Depending upon egg diameter and cell number, 10 ml of packed sea urchin embryos will yield up to 1 mg of cilia with either the ethanol–calcium or hypertonic methods.

In terms of quantity and simplicity, the hypertonic salt procedure is most advantageously applicable to the gills of lamellibranch molluscs (Stephens and Linck, 1969; Linck, 1970, 1971, 1973a, b). Thorough sea water washing, followed by a 10 min treatment with hypertonic NaCl (sea water, with 30 g l^{-1} excess NaCl) at room temperature causes release of essentially pure cilia; filtration or low speed centrifugation removes the gill fragments while high-speed centrifugation harvests the cilia. This procedure has been successfully applied to *Aequipecten irradians*, *Placto-pecten magellanicus*, *Mytilus edulis*, *Mactra solidissima*, and *Pinna fragilis*; these represent all of the species studied thus far. Starting with a wet weight of 100 g of gill material, 50–100 mg of cilia can be obtained by the hypertonic treatment; a similar yield can be obtained from direct application of the ethanol–calcium procedure of Watson and Hopkins (Linck, 1971, 1973a).

Since cilia produced directly by the ethanol–calcium method do not become motile when exogenous ATP is added, Gibbons (1965b) devised a procedure whereby cilia capable of reactivation could be obtained from *Tetrahymena* by treatment with 60% glycerol at -20°C. Brief vortex-mixing detached the cilia, the cell bodies were removed by low-speed centrifugation, and the cilia were harvested by ultracentrifugation at -10°C. Interestingly, ethanol–calcium-isolated cilia will show reasonably good reactivation if treated with glycerol after isolation (Winicur, 1967); this is evidently due to increased membrane permeability. The direct glycerination method is also applicable to gill cilia, but the yield is substantially lessened apparently due to concomitant disruption of epithelial cells (Linck, 1971, 1973a).

B. *Flagella*

Flagella from protists or from sperm often represent an abundant source of protein, with protists being amenable to the ethanol–calcium methodology while sperm flagella require only mechanical shear to detach them from the sperm head.

Chlamydomonas reinhardii serves as an excellent source of algal flagella; two procedures are available for isolation in large quantity (Witman *et al.*, 1972a). One of these is based on a pH shock, wherein cells were suspended in 5% sucrose and the pH was rapidly reduced to 4·5; the flagella were detached immediately, the pH was restored to neutrality, and the flagella were then isolated by differential centrifugation on a sucrose step-gradient. The other is a modification of the Gibbons (1965a) ethanol–calcium procedure for *Tetrahymena* cilia; here 9% ethanol treatment in the presence of EDTA or EGTA, followed by $CaCl_2$ addition to a concentration of 15 mM resulted in deflagellation. No specific data on yields were given.

By far the most simple and convenient source of axonemal proteins from flagella is that of the sperm tail, most readily obtainable by decapitation of sea urchin or other echinoderm sperm through mechanical shear (Stephens *et al.*, 1967; Mohri, 1968a). Low-speed centrifugation at 1000 *g* for 5–10 min sediments the head fraction, while successive high-speed spins at 10,000 *g* for 5–10 min separates the tails from the mid-piece and mitochondrial fraction (see illustration in Stephens, 1970b). The basic procedure has been applied to molluscan sperm with only minor modification (Linck, 1971, 1973b). Depending on the ratio of head to tail volume, 1 ml of packed sperm will yield about 10 mg of isolated flagella.

C. *Morphological basis for isolation*

It was recently pointed out by Blum (1971) that all of the isolation procedures for protozoan and metazoan cilia and for algal flagella seem to have one feature in common. They all appear to operate by detachment of the organelle at the *transition zone*, that is, at the point where the triplet microtubules of the basal body become the doublets of the axoneme, where the central pair and dynein arms begin, and where often the maximum diametrical constriction of the membrane sheath occurs. Blum postulated that this region was particularly sensitive to the effects of osmotic shock, alcohol or calcium-induced stiffening, and mechanical shear, presumably because of some inherent mechanical weakness or discontinuity.

This certainly appears to be the case with cilia isolated from the gill epithelium (Linck, 1971) and from the sea urchin blastula (Stephens, unpublished). The isolated organelles characteristically show a remarkably clean break precisely at the basal plate, an amorphous disc which cuts across the axoneme at the transition point between triplets and doublets. Such cilia, when disrupted chemically, splay out into a fountain-like arrangement of outer fibers rigidly attached to the basal plate and bound tightly together for the first 2000 Å of axonemal length. This region is particularly difficult to dissolve, even in denaturing solvents, implying that it may be chemically cross-linked (Linck, 1971).

Blum (1971) makes the further argument that in such isolations the basal body (or its equivalent) remains intact and viable, citing the fact that complete and rapid regeneration will occur in ciliates, flagellates, and sea urchin embryos shorn of their cilia or flagella. Supporting this idea is the fact that, after hypertonic deciliation, Triton-treated molluscan epithelial cells will yield whole, intact basal apparatuses, consisting of morphologically quite normal basal bodies in orderly array, having firmly attached bifurcating striated rootlets (Stephens, 1974).

The situation is less clear in the case of sperm tails, but the fact that mild shear will initially yield free flagella that quite closely approximate the intact flagellum in length argues in favor of Blum's contention that breakage first takes place at the sensitive basal region. The fact that some sperm have basal bodies buried deeply within the sperm head while others do not leaves us with the ambiguous situation where some sperm must, of necessity, break at a point other than at the transition zone while still others may yield flagella with intact integral basal bodies. The possible necessity of either a basal body or a basal plate for normal motility may explain why some isolated flagella can be reactivated while others cannot (cf. Gibbons and Fronk, 1972).

III. Fractionation of the axoneme

For convenience, many of the various fractionation schemes referred to throughout this section are illustrated in Fig. 1.

A. *Dynein: arms and energy*

1. *Isolation and localization*

The impermeable and impervious nature of the ciliary or flagellar membrane posed nearly insurmountable problems to early investigators of axonemal proteins. The key discovery in this regard was made by Gibbons

(1963) when he found that digitonin treatment removed the membrane from *Tetrahymena* cilia, thus making available to differential extraction all of the "9+2" enzymatic and structural components. Gibbons (1963, 1965a) found that either low ionic strength dialysis in the presence of EDTA or high salt extraction released into solution nearly all of the ATPase bound to the axoneme. Ultracentrifugal analysis of this solubilized ATPase

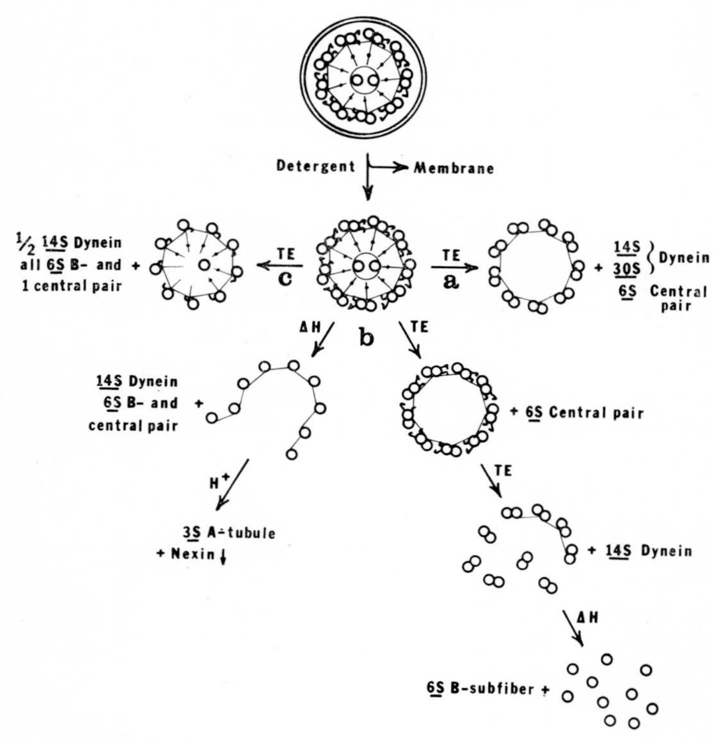

FIG. 1. Flow diagram for representative fractionation schemes. (a) *Tetrahymena* cilia; (b) sea urchin and other flagella; (c) *Aequipecten* cilia; TE = dialysis against 1 mM Tris-HCl, pH 8·0, and 0·1 mM EDTA.

fraction revealed three components with corrected sedimentation coefficients of 4–6 S, 14 S, and 30 S. Subsequent analysis of this material on a sucrose density gradient indicated that the latter two components contained all of the ATPase activity. On the other hand, electron microscopic examination of the insoluble fraction from these extraction procedures revealed that the axoneme consisted of outer fiber doublet microtubules in cylindrical array, often missing one or both members of the central pair, and having no detectable *arms* (Afzelius, 1959; Gibbons and Grimstone, 1960) extending from the A-subfiber of the doublet.

Since EDTA appeared to ease the release of ATPase into solution and since distinct structures were missing from the axoneme after extraction, Gibbons (1963, 1965a) attempted to reconstitute the cilium by addition of either the purified 14 S or 30 S ATPase to the outer fiber array in the presence of magnesium. When added in stoichiometric amount (i.e. approximating the initial ratios in intact *Tetrahymena* cilia), little addition of the 14 S ATPase took place, but in the case of the 30 S component nearly two-thirds of the ATPase activity was rebound to the outer fibers. When such reconstituted preparations were examined in the electron microscope, it was found that many of the arms previously removed by extraction had been restored to their proper position and, furthermore, that the number of arms which were reconstituted correlated with the percentage of ATPase which had reassociated with the outer fibers. Short-term or interrupted dialysis, wherein only part of the ATPase went into solution, likewise showed excellent correspondence between the number of observed arms and the amount of ATPase remaining. Gibbons (1963, 1965a) thus concluded that the arms of the "9+2" structure were in fact composed of at least the 30 S ATPase, but that the reconstitution experiments did not prove that all of the ATPase resided in the arms since the 14 S component showed little recombination and no obvious localization.

The relationship between the 14 S and 30 S ATPase was considerably clarified by the studies of Gibbons and Rowe (1965). These workers first determined molecular weights for the two components, using the Archibald method with the 30 S ATPase to obtain a value of 5,400,000 ± 1,000,000 daltons and using the short-column equilibrium method with the 14 S protein to obtain a value of 600,000 ± 100,000 daltons. Electron microscopic observation of the 14 S particle revealed an ellipsoid with axes of 85, 90, and 140 Å, giving an estimated molecular weight of 540,000. The 30 S form was found to consist of a linear array of globular particles with a diameter of 70–90 Å and a repeat period of about 140 Å, varying in length from 400 to 5000 Å. Such length heterogeneity was not inconsistent with the single 30 S peak observed in the ultracentrifuge since the boundary was hypersharp and could thus accommodate this degree of variation through a strong dependence of sedimentation coefficient upon concentration. Observation of a linear 30 S polymer made up of units the same size as those of the 14 S ATPase, coupled with Gibbons' (1963) demonstration that the 30 S particle could be converted to a 10 S subunit by alkaline treatment and the fact that the two ATPases had very similar properties, drew Gibbons and Rowe to the conclusion that the 30 S polymer was made up of the 14 S ATPase subunits. Since these proteins are evidently related as monomer to polymer and are obviously involved somehow in the production of the motile force in cilia, Gibbons and Rowe

(1965) named them collectively *dynein*. Preliminary experiments by Gibbons (1967) indicated that 5 M guanidine hydrochloride and 1% mercaptoethanol could, in turn, dissociate dynein into subunits of about 220,000 daltons, accompanied by some lower molecular weight protein. Denaturation of 14 S dynein in sodium dodecyl sulfate and reducing agents, on the other hand, resulted in an electrophoretically determined molecular weight of about 500,000 daltons, with no evidence for smaller subunits (Linck, 1972; see also section III. C.3).

Applying the same dialysis procedures to sea urchin sperm flagellar axonemes, Gibbons (1965a) found that essentially all of the ATPase could be obtained in the 14 S form. Mohri (1964) had demonstrated earlier that when the Gibbons dialysis procedure was applied to sea urchin flagellar axonemes, nearly all of the ATPase was solubilized, but when applied to fish sperm flagella, one-half remained bound to the axoneme; the form of the dynein—14 S or 30 S—was not reported. The soluble sea urchin dynein was later purified by column chromatography and characterized as a 14 S form which apparently underwent considerable aggregation with no basic change in enzymatic properties (Ogawa and Mohri, 1972). When previously glycerinated sea urchin sperm flagella were extracted with high salt, both 14 S and 30 S dynein resulted but showed considerable variation in relative proportions (Brokaw and Benedict, 1971).

The equivalence of the dynein arms and their mode of attachment to the outer fibers is presently unclear in several respects. Firstly, the apparent need for EDTA or high salt to release dynein during dialysis or the requirement of magnesium for re-addition of dynein would point to a magnesium linkage or related ionic bond, between the A-tubule and dynein. On the other hand, Raff and Blum (1969) found that 20 mM ATP extraction released both 14 S and 30 S dynein from *Tetrahymena* axonemes in the same manner as did low ionic strength dialysis or high salt extraction (Gibbons, 1965a) and paralleling, in a sense, the dissociation of myosin from actin in actomyosin. However, all of these dynein extractions are consistent with some type of ionic protein–protein bonding. Raff and Blum (1969), in fact, postulate that the action of high concentrations of ATP is one of plasticizing the axoneme, perhaps making it amenable to the chelating action of ATP. Secondly, using radial reinforcement of electron micrographs of *Tetrahymena* axonemes, Allen (1968) was able to demonstrate that the two arms on each outer fibre were not morphologically equivalent. The outer arm appeared to be doubly lobed and somewhat larger than the inner hooked arm, opening the possibility that the monomer–polymer concept of dynein might be an over-simplification (i.e. there may be two different dyneins) or that the two sites on the A-tubule are not the same (i.e. differentially binding or having accessory or connecting elements). Both of these alternatives have received some circumstantial support.

Although no obvious differences in extraction or re-addition at either arm position were evident in *Tetrahymena* (Gibbons, 1965a), studies on other species show both positional differences on the outer fiber and differences in polymeric nature of extracted dyneins. Brief high salt extraction of sea urchin sperm flagella will result in the preferential removal of only the outer arm (Gibbons and Fronk, 1972), implying that dynein is bound less tightly to this site, whether for reasons inherent in the site or in the dynein. In the one case where ciliary and flagellar dynein were compared in the same species, that of scallop (*Aequipecten irradians*) gill cilia and sperm flagella (Linck, 1971, 1973b), low ionic strength dialysis of ciliary axonemes removed half of the ATPase as 14 S dynein, with the rest remaining tightly bound to the axoneme, removable as a normal 14 S component only after brief trypsinization. Removal of the initial 14 S dynein left an axoneme devoid of outer arms. In contrast, the sperm flagellar axoneme yielded all of its dynein to dialysis as a 14 S component, with no evidence of a polymeric or a bound form. These results, taken together, would suggest either that the inner and outer dynein binding sites *or arms* differ from one another in both cilia and flagella and that the dynein binding sites of cilia either differ from those in flagella or else binding is a function of a more tightly associated polymeric dynein existing in the former.

2. *Enzymatic properties*

In contrast to muscle myosin, perhaps the most characteristic enzymatic property of ciliary and flagellar dynein, whether 14 S, 30 S, axoneme-bound, or mechanochemically coupled during active beat, is its comparatively high degree of specificity for ATP as substrate. Gibbons (1966) reported that sucrose gradient-purified 14 S or 30 S dynein from *Tetrahymena* cilia hydrolyzed deoxy-ATP at a rate indiscernible from that of ATP, but split ITP, CTP, and inorganic tripolyphosphate at only 10% of the ATPase rate and hydrolyzed GTP and UTP at about 3% of the ATPase rate, all measured with magnesium as cofactor. For AMP, cyclic 3′, 5′-AMP, adenosine tetraphosphate, sodium pyrophosphate and *p*-nitrophenyl phosphate, hydrolysis was in all cases less than 1% of that for ATP. ADP hydrolysis at a rate nearly one-third that for ATP was attributed to the presence of an adenylate kinase. The flagellar 14 S ATPase from sea urchin sperm flagella, purified on Sepharose and hydroxylapatite, was found to hydrolyze GTP, CTP, and ADP at 5–8% of the ATPase rate while UTP, inorganic pyrophosphate, and *p*-nitrophenyl phosphate were not detectably hydrolyzed (Ogawa and Mohri, 1972).

Ciliary dynein bound to the axoneme but uncoupled from active motility split GTP, UTP, and ITP at less than 5% of the ATPase rate (Stephens and Levine, 1970); similar specificity was found in non-motile but enzymatically coupled axonemes of sea urchin flagella (Gibbons and Fronk, 1972). Using demembranated sea urchin sperm reactivated by exogenous nucleotide, Gibbons and Gibbons (1972) demonstrated clearly that ATP is the specific substrate during active flagellar beat; while other nucleotides were hydrolyzed at about 3% of the ATPase rate, no motility was detectable except in the case of ADP where feeble motility was observed after a brief lag period during which sufficient ATP was evidently generated by adenylate kinase. In all of these cases, whether free, bound, enzymatically coupled, or actively involved in motility, dynein showed an unquestionable specificity for ATP.

Divalent cations are required for dynein ATPase activity. The enzyme, regardless of source, is inactive in the presence of excess EDTA (Gibbons, 1966; Ogawa and Mohri, 1972), as are isolated ciliary axonemes (Stephens and Levine, 1970) and reactivatable glycerinated cilia (Gibbons, 1965b) or demembranated sperm flagella (Gibbons and Fronk, 1972). The substrate of dynein is $Mg\text{-}ATP^{2-}$, based on maximal specific activity of ciliary 14 S and 30 S dynein (Gibbons, 1966), flagellar 14 S dynein (Ogawa and Mohri, 1972), isolated ciliary and flagellar axonemes (Stephens and Levine, 1970; Gibbons and Fronk, 1972) and also based on optimal reactivation of demembranated sperm flagella (Gibbons and Gibbons, 1972). The ATPase activity of free or axoneme-bound dynein in the presence of calcium is typically one-half to two-thirds that of the magnesium-activated rate. Manganese at low concentration functions nearly as well as magnesium. Iron, cobalt, nickel, and comparatively high manganese concentrations (greater than 20 mM) activate more poorly than calcium, while beryllium, barium, strontium, zinc, cadmium, and mercuric ions have no appreciable activating effect (Gibbons, 1966; Mohri, 1968a; Stephens and Levine, 1970; Gibbons and Fronk, 1972; Ogawa and Mohri, 1972). The latter three ions, in fact, appear to be inhibitory.

Exactly what role calcium may play in ciliary motility is rather unclear. The binding of $Mg\text{-}ATP^{2-}$ and $Ca\text{-}ATP^{2-}$ to the enzymatic site of either free or axoneme-bound dynein appears to be nearly identical when judged by the Michaelis constant, and although the latter substrate is hydrolyzed moderately well, it induces only feeble motility in reactivated sperm (Gibbons and Fronk, 1972; Gibbons and Gibbons, 1972). Naitoh and Kaneko (1972) demonstrated that experimentally manipulated exogenous calcium concentrations can control the rate and direction of beat in reactivated *Paramecium*. The effect may be at a membrane or infraciliature level (Eckert, 1972), but the possibility remains that calcium may act at some regulatory enzymatic level, perhaps through competitive binding to the same site as magnesium-ATP, a possibility suggested by Gibbons and Gibbons (1972) from some preliminary data. It is significant, however, that both isolated *Tetrahymena* cilia (Gibbons, 1965b) and demembranated sea urchin sperm flagella (Gibbons and Gibbons, 1972) may be reactivated in the presence of EGTA, indicating no calcium requirement for motility *per se*.

The influence of ionic strength upon dynein ATPase appears to be dependent upon the physical state of the enzyme. In the case of dynein from *Tetrahymena* cilia, regardless of whether magnesium or calcium activated, 14 S dynein is markedly inhibited by increasing KCl concentrations, decreasing to about a quarter of its original value when an ionic strength of 0·8 is reached; the 30 S form, on the other hand, is enhanced about 3-fold over this same range (Gibbons, 1966). Using column-purified 14 S dynein from sea urchin sperm, Ogawa and Mohri (1972) observed a 20% enhancement of ATPase, maximal at a KCl concentration of about 0·3 M. Similarly, Gibbons and Fronk (1972) found a 6-fold enhancement in the activity of 14 S dynein obtained directly by high salt extraction of sea urchin sperm flagella, in this case having a maximum at 0·4 M. When bound to the axoneme, ciliary dynein appears to undergo a similar but substantially lesser activation with increasing salt concentration, maximizing at a KCl concentration of 0·3–0·4 M (Stephens and Levine, 1970). But using a pH-stat assay method to avoid low ionic strength complications from the presence of buffer, Gibbons and Fronk (1972) were able to demonstrate an initial 3-fold *decrease* in bound sperm flagellar ATPase, minimizing at about 0·15 M KCl, followed by a near-doubling in activity as the salt concentration reached 0·5 M. This is in contrast to the effect they found on free 14 S dynein, cited above, and gave the first clue to possible enzymatic differences between bound and free dynein. All of the above assays were performed in a pH range of 7·8–8·3, encompassing or approximating the pH optimum at low ionic strength of either 14 S, 30 S, or dynein bound to the axoneme.

Like ionic strength, pH appears to have a somewhat variable effect on dynein ATPase, depending upon source, state, or ionic strength. The marked differences in activity of free versus bound dynein at various pH and ionic strength values is related to dynein–axoneme interaction and will be discussed in the following sections. The behavior of isolated dynein and axoneme-bound but enzymatically and mechanochemically uncoupled dynein show a number of consistencies and will be discussed together. In the case of *Tetrahymena* ciliary dynein (Gibbons, 1966), the ATPase of the monomeric form is maximal at pH 9, with a shoulder at about pH 7; the 30 S polymeric form, on the other hand, has a major optimum at about pH 6 and another lesser optimum at pH 8·5. Thus there appear to be two optima on either side of neutrality with the 14 S form being favored under alkaline conditions and the 30 S under acidic ones. Gibbons (1966) attributed both these pH differences and the markedly different ionic strength behavior cited above to possible allosteric effects in the polymeric form. Bound dynein in non-motile ciliary axonemes shows an optimum at about pH 8 and a lesser one at about pH 6 (Stephens and Levine, 1970); the pH curves for this bound dynein appeared to be composites of the 14 S and 30 S curves of Gibbons (1966). The column-purified 14 S dynein from sea urchin sperm showed almost equal optima at pH values of 7 and 9 (Ogawa and Mohri, 1972) while salt-extracted dynein exhibited a broad optimum ranging from pH 8 to 9·5 (Gibbons and Fronk, 1972). All of these assays were performed in the presence of 1–2 mM magnesium-ATP and 10–30 mM Tris-HCl buffer, with the exception of the last-cited study where 0·1 M KCl and no buffer were used.

When measured under conditions of low salt, magnesium-activation, and at a pH of 8, the Michaelis constant, K_m, for 14S *Tetrahymena* dynein was found to be $3{\cdot}5 \times 10^{-5}$ M while that for the 30 S form was $1{\cdot}1 \times 10^{-5}$ M; the V_{max} values were found to be 3·5 and 1·3 μM P_i per min mg respectively (Gibbons, 1966). Column-purified 14 S sea urchin sperm dynein had a K_m of 3×10^{-5} M and a V_{max} of 2·7 μM P_i per min mg (Ogawa and Mohri, 1972), while the salt-extracted preparation had K_m values ranging from $2{\cdot}2 \times 10^{-5}$ to $8{\cdot}8 \times 10^{-5}$ M and V_{max} values of 1–4 μM P_i per min mg, depending upon salt concentration (Gibbons and Fronk, 1972). Axonemes which were non-motile but enzymatically coupled (i.e. whose dynein ATPase properties were modified by association with axonemal constituents) gave complex Michaelis–Menton kinetics. Reactivated sperm show apparent K_m values of $13\text{–}21 \times 10^{-5}$ M, depending upon whether maximal total ATPase, mechanochemically coupled ATPase, or beat frequency was employed as a rate (Gibbons and Gibbons, 1972).

Dynein, regardless of form or state of binding, consistently shows the same high degree of specificity for Mg-ATP^{2-} and (with the apparent exception of *Tetrahymena* ciliary 14 S dynein) is activated by relatively high K^+ concentrations at its alkaline pH optimum. It is its complex salt–pH behavior that sets axonemally bound, enzymatically coupled dynein apart from its free or bound but uncoupled counterparts.

3. *Dynein–axoneme interactions*

Because the nucleotide specificity, divalent cation requirements, and ionic strength behavior of isolated ciliary axonemes so closely resembled a composite of the properties of 14 S and 30 S ciliary dynein (albeit from a different species) and because the specific activity of dynein did not appear to change appreciably when released from the axoneme, Stephens and Levine (1970) concluded that there was little enzymatic evidence for interaction of dynein with other axonemal proteins, as compared with the case of muscle myosin where actin stimulates the magnesium-activated ATPase and the troponin–tropomyosin complex prevents actin–myosin interaction except in the presence of calcium. However, when the question was re-investigated using non-motile sperm flagellar axonemes (wherein the dynein apparently was still coupled enzymatically to the axoneme) rather significant differences between bound and free dynein were detected (Gibbons and Fronk, 1972). As already discussed above, when measured at its alkaline pH optimum, free dynein shows a marked increase in activity with increasing salt concentration while the bound form first decreases and then increases in activity, but it does so to a lesser extent than the free dynein. When pH versus activity profiles were determined for the free versus bound dynein, Gibbons and Fronk (1972) found that, in the absence of salt, free dynein had low activity and was little influenced by pH; in the presence of 0·1 M KCl the activity tripled with the appearance of an optimum at pH 8·5. In 0·5 M KCl the activity increased even further, nearly doubling again, and having a broad "optimum" ranging from about pH 7 to 9·5. The bound dynein, on the other hand, showed two nearly equal optima in the absence of salt, one at pH 8 and the other at pH 10; addition of 0·1 M KCl left the activity at pH 10 unchanged but halved the pH 8 activity. Increasing the salt to 0·5 M only slightly increased the pH 8 activity over that at 0·1 M but caused the appearance of two new optima, one at pH 5·5 and the other at pH 9·5. The latter optimum had nearly twice the ATPase activity as at the pH 10 optimum characteristic of the lower salt concentration. Using axonemes prepared under less optimal conditions, e.g. at a higher pH or lacking —SH protecting agents, Gibbons and Fronk (1972) went on to demonstrate that physically bound dynein could be enzymatically "uncoupled" from the axoneme, behaving in a manner characteristic of free dynein. Similar results had been obtained previously by Brokaw and Benedict (1971) by thiourea treatment of sea urchin flagella. Enzymatic uncoupling, resulting in a marked enhancement of ATPase activity and an elimination of characteristic ionic strength–pH behavior, would thus explain why some ciliary axoneme preparations appeared to have properties indiscernible from those of free dynein (cf. Stephens and Levine, 1970). When bound, coupled dynein was assayed under conditions which would induce motility in intact axonemes, Michaelis–Menton kinetics were not obeyed, but when assayed under several other sets of conditions, proper fit could be obtained. The complex kinetic behavior of the bound dynein fraction and the marked differences in pH and salt behavior of the free versus the bound dynein were interpreted by Gibbons and Fronk (1972) as evidence for interaction of dynein with other axonemal components.

Actively swimming reactivated sperm have a pH optimum for ATPase activity and maximum beat centered at pH 8·3, while broken tail fragments—non-motile but still having coupled dynein ATPase activity—show an optimum at about pH 9·5, as mentioned above. These fragments, and also potentially motile flagella in a highly viscous medium, show about a 3-fold *lower* ATPase activity than the actively swimming sperm, the difference being related to the amount of ATPase activity *directly coupled to motility* (Brokaw and Benedict, 1968, 1971; Gibbons and Gibbons, 1972). The actively swimming reactivated sperm also show optimum ATPase activity at a KCl concentration of 0·15–0·25 M while the non-motile fragments, as discussed above, show a minimum of activity at about 0·1 M KCl and a slight but steady increase at higher KCl concentrations. Unlike coupled dynein in the non-motile axonemes, the motile reactivated sperm show normal Michaelis–Menton kinetics. Such differences in pH and ionic strength behavior between actively motile and non-motile axonemes, plus the markedly higher specific activity of the former, lend still more enzymatic evidence for the interaction of dynein and axonemal components, in this case resulting directly in motility.

In summary, axonemal dynein is maximally active during mechanochemically coupled beat, shows a single physiological pH optimum, and is enhanced by K^+ (Gibbons and Gibbons, 1972); dynein *decreases* substantially in ATPase activity and has complex pH and ionic strength behavior when axonemes are rendered non-motile by mechanical disruption or high viscosity media. Dynein shows a marked *increase* in activity when either freed from the axonemes or enzymatically uncoupled from the axonemal complex (Gibbons and Fronk, 1972).

A different line of evidence for dynein–axoneme interaction, not directly dependent upon enzymology, was presented earlier by Gibbons (1965c). Addition of ATP to suspensions of *Tetrahymena* cilia or ciliary axonemes was found to cause a 20% decrease in turbidity through an apparent hydration of the organelle. Tibbs (1962) observed a similar phenomenon with perch sperm tails, while Raff and Blum (1966) found that the pellet height of sedimented cilia was a function of ATP concentration, with low concentrations causing a significant increase in volume through hydration and higher concentrations causing dissolution of dynein (cf. Raff and Blum, 1969). This hydration effect is in contrast to syneresis, the apparent contraction or dehydration seen in actomyosin gels subjected to ATP. Through reconstitution experiments, Gibbons (1965c) further demonstrated that the ATP response resided in some interaction between fractionated 30 S dynein and the outer fiber fraction and suggested that the direct effect of ATP is the relative expansion of the "active elements responsible for motility". The most recent evidence, to be discussed in detail later, appears to confirm this contention, for Summers and Gibbons

(1971) have clearly demonstrated an active sliding between outer fiber doublets.

When γ-^{32}P-ATP was used to reactivate glycerinated sperm flagella, Yanagisawa *et al.* (1968) found that label appeared in the bound GDP and GTP of the outer fibers, implying that the terminal phosphate of ATP is transferred to tubule-bound GTP during active beat. This transphosphorylation is presumably under the mediation of dynein since the amount of ^{32}P transferred was directly proportional to the amount of ATP split by the active flagella. This potentially important observation has yet to be repeated under more recent conditions of controlled reactivation, however.

B. "9+2" Microtubules: backbone of the axoneme

1. Isolation and physical–chemical properties

The insoluble fraction of the Gibbons (1963, 1965a) dialysis procedure for the preparation of dynein consists of the outer fiber doublet microtubules, separable by simple sedimentation, while the soluble dynein fraction contains also the central pair microtubule components.

Renaud *et al.* (1966, 1968) prepared acetone powder extracts of whole *Tetrahymena* cilia and also outer fiber fractions produced by exhaustive dialysis or high-salt extraction, demonstrating that the microtubule protein of the outer fibers could exist under physiological conditions as a 6 S particle with a molecular weight of about 103,000 daltons, consisting in turn of two monomers of 55,000 daltons. On alkaline disc electrophoresis the protein migrated as a very closely spaced doublet of equally dense bands. Stephens (1968a) used differential solubilization by dialysis to prepare outer fibers from sea urchin flagella and similarly found a monomer molecular weight of 56,000–62,000 daltons using denaturing solvents, and a 6 S dimer of 130,000 daltons when the organic mercurial Salyrgan was used to dissolve the outer fibers. Both this study and that of Renaud *et al.* (1968) indicated that the protein had a characteristic amino acid composition quite reminiscent of muscle actin. Stephens *et al.* (1967) demonstrated that both the *Tetrahymena* outer fibers and those from sea urchin flagella contained bound guanine derivatives in a ratio of one mole per mole of protein monomer; ciliary outer fibers contained the nucleoside and free base while those from sea urchin contained bound guanine nucleotides. The constancy in total ratio but the variability in specific bound compound indicated that the phosphoribosyl moiety was susceptible to degradation during protein preparation. In three other sea urchins, Yanagisawa *et al.* (1968) likewise found GMP, GDP, and GTP bound to the outer fiber in a total ratio of one mole of nucleotide per protein monomer. A careful re-examination of the stoichiometry of nucleotide binding in flagellar outer fibres indicated that only one nucleotide per dimer was tightly bound while the other was exchangeable with the medium (Stephens, 1969); a similar conclusion was drawn earlier from studies of the 6 S dimer of porcine brain microtubules (Weisenberg *et al.*, 1968). Freshly prepared sea urchin flagellar outer fibers typically contain roughly equimolar amounts of GDP and GTP per mole of dimeric protein (Stephens, 1969, 1971).

Shelanski and Taylor (1967, 1968) found that short-term (3 h) dialysis of sea urchin flagellar axonemes in the presence of low EDTA concentrations selectively solubilized only the central pair microtubules, thus allowing these workers to compare directly the protein components of both central pair and outer fibers. The central pair protein yielded monomer and dimer molecular weights of 61,000–62,000 and 122,000 daltons, respectively, while the comparable values for outer fiber protein were 60,000–61,000 and 118,000. Like the molecular weights, the amino acid compositions for these fractions were nearly identical and compared strikingly well with those from *Tetrahymena* cilia and a different sea urchin species, already discussed above. In addition to the obvious solubility differences, these two fractions also differed in respect to bound nucleotide; the soluble 6 S central pair protein readily lost bound nucleotide unless GTP was present in the preparation medium, whereas the intact outer fibers retained nucleotide even after extensive dialysis. The former fraction bound colchicine, the latter did not.

Mohri (1968b) investigated the amino acid composition of outer fibers from flagella of three additional species of sea urchin and found relatively uniform agreement among species and with compositions published by other workers. On the basis of uniqueness of properties, Mohri concluded that these proteins represented a separate class, quite apart from, for example, muscle actin or bacterial flagellin. He named the protein *tubulin* after its source, the microtubule.

A comparison of tubulin with muscle actin at a gross level reveals some interesting similarities: both proteins contain equimolar amounts of nucleotide di- or triphosphates, both are rich in glutamic and aspartic acids and have a relatively high content of hydrophobic amino acid and free —SH groups, both form polymers, and both interact with characteristic ATPases. However, when one examines these similarities more closely, the differences become indisputably significant. The polypeptide chain of the monomeric units are clearly dissimilar: actin has a minimal molecular weight of about 46,000 while tubulin has a value of about 59,000, both judged from co-electrophoresis on SDS-polyacrylamide gels (Stephens, 1970a). When actin and tubulin from the same species were compared on the basis of amino acid composition (Stephens and Linck, 1969) or tryptic or chymotryptic peptide mapping, little if any homology could be proved (Stephens, 1970a). Globular actin binds one mole of ATP per monomer weight of 46,000 daltons; on polymerization the terminal phosphate is split, yielding fibrous actin containing one mole of ADP per 46,000 daltons. The tubulin dimer contains equimolar amounts of GDP and GTP per dimeric molecular weight of approximately 118,000, at least in the polymeric form. No concrete evidence has yet to be mounted showing any interaction between myosin and tubulin or dynein and actin. Antibodies against outer fiber tubulin, though able to cross-react with tubulin from other sources and species, will not cross-react with actin from the same species (Fulton *et al.*, 1971), indicating no apparent homology in terms of even local antigenic sites. Finally, and probably most importantly, actin is a double helix of 55 Å globular units, having a repeat period of about 700 Å, while the microtubule is a hollow cylinder with a mean diameter of 220 Å, composed of 12–13 parallel protofilaments of linearly arranged 40 Å monomeric subunits. So, on the basis of primary structure, molecular weight, nucleotide specificity and binding ratio, plus the morphological form of the native polymer, actin and tubulin are clearly quite different proteins, but they do show some intriguing and puzzling parallels of possible evolutionary significance.

On a gross biochemical level, tubulins from central pair and outer fibers of cilia and flagella from various sources have nearly identical properties (cf. compilation in Stephens, 1971). But the fact that they constitute morphologically different sets of microtubules, coupled with the fact that solubility differences during preparation were employed to distinguish among them in the first place, would intuitively indicate that some chemical differences must exist.

2. *Subunit subtleties*

In addition to the clear-cut differential solubility of the central pair *singlet* microtubules versus the outer fiber *doublets* during biochemical fractionation—a phenomenon perhaps explicable on purely morphological or steric grounds alone—other distinctions exist which point to some type of built-in chemical specificity. Dynein arms attach only to the A-subfiber of the doublet microtubule and they, like the B-subfiber, do so only at two specific circumferential points (Gibbons and Grimstone, 1960). Long-term dialysis tends to solubilize the B-subfiber in *Tetrahymena* cilia (Gibbons, 1965a), as will dialysis of cilia but not flagella of *Aequipecten* (Linck, 1970, 1971). One central pair member can be differentially extracted in *Aequipecten* cilia (Linck, 1970, 1971) and *Chlamydomonas* flagella (Jacobs *et al.*, 1968) and, in fact, only one member of the central pair occurs in some mutants of *Chlamydomonas* (Warr, *et al.* 1966). Furthermore, the chemically more stable central tubule (C_1) bears two rows of radial projections as contrasted to one row on the other member (Hopkins, 1970). Thus, operationally at least, A-tubules differ from B-subfibers in being more stable and apparently possessing specific dynein and B-subfiber binding sites; B-subfibers are more stable than central pair singlets, and appear to vary in stability from organism to organism and from cilium to flagellum in the same organism; and, finally, the two members of the central pair appear to differ in terms of extractability, accessory structures, and possibly genetic determination.

Using electron microscopic observation after temperature or enzymic treatment of sperm tail material, Behnke and Forer (1966) classified microtubules into four stability categories: the A-tubule, B-subfiber, and central pair cases as above, and a fourth, the cytoplasmic microtubule of the spindle, characteristically sensitive to hypothermic or colchicine treatment. In addition, they noted differential sensitivity *within* the B-microtubules themselves and postulated that the material immediately adjacent to the A-tubule was different from that making up the remainder of the B-subfiber. The relatively labile nature of the B-subfiber, as originally observed by Behnke and Forer (1966), gave rise to four major studies wherein attempts were made to fractionate outer fiber doublets and to determine the nature of the subunits making up the two apparently dissimilar microtubules.

Stephens (1969) attempted to fractionate sea urchin outer fiber flagellar doublets by treatment with low concentrations of the detergent Sarkosyl, the philosophy being that perhaps the B-subfiber protein was preferentially soluble since only singlet microtubules could be reconstituted from a dilute Sarkosyl solution of outer fiber doublets (Stephens, 1968b, 1971). This proved to be the case; detergent treatment selectively removed portions of the B-subfiber wall but did not do so with any consistency since the amount of B-material removed was variable at various points along the doublet and also both tubules appeared to dissolve from the ends, producing a solution of components from both tubules in varying amount (Stephens, 1969, 1970b, and unpublished).

Later, based primarily on Behnke and Forer's (1966) observation of temperature-sensitivity of the B-subfiber, Stephens (1970b) demonstrated that incubation of sea urchin doublets for 2 min at 37–40°C in a low ionic strength medium resulted in complete dissolution of the B-subfiber, with little effect upon either the length or integrity of the A-subfiber. The two fractions thus obtained showed marked differences in solubility and in behavior on acidic, though not on basic, polyacrylamide gels. The protein components were named A- and B-tubulin, reflecting their subfiber sources. Amino acid analysis revealed that the A-subfiber component had no disulfides and eight free cysteines, while the B-tubule counterpart contained one disulfide bond and seven free cysteines. The A-tubulin had about seven more lysines plus arginines than the B-tubulin while the latter contained a comparable amount more of glycine plus valine residues. In line with this difference in trypsin-sensitive basic amino acids, tryptic peptide maps of each tubule component were unique with respect to seven peptides, but the remainder of the primary structure was indistinguishable on this basis. Assuming a molecular weight of 60,000 daltons for the two components, the number of tryptic peptides obtained corresponded quite well with the amino acid composition. But considering both the uncertainty in molecular weight and the large number of resultant peptides, it was not possible to eliminate unequivocally the possibility that the two tubulin fractions were not each, in turn, composed of very similar polypeptide chains (Stephens, 1970b). Later, it was pointed out that sulfonated derivatives of the A- and B-tubulin fractions *each* gave rise to distinct doublets upon isoelectric focusing (Stephens, 1971), implying perhaps four potentially different polypeptide chains, with 2 forming the A-tubulin dimer and yet two more constituting the B-tubulin.

Almost simultaneously, Jacobs and McVittie (1970) reported the fractionation of *Chlamydomonas* doublets into intact singlet A-tubules and soluble B-subfiber protein by long-term dialysis at both low and high ionic strength. The A-tubule component (like that from sea urchin) moved as a single band on standard basic urea-polyacrylamide gels, having a slightly higher mobility than that of the predominant B-subfiber protein. The latter fraction also showed small amounts of the A-component and, since central pair protein fractionated with the B-tubule protein, these workers concluded that the central pair probably contained both A- and B-proteins while the B-subfiber contained principally the B-electrophoretic component. The principal A-subfiber protein was named α while that from the B-tubule was β-tubulin. Jacobs and McVittie (1970) pointed out that there were apparently somewhat greater electrophoretic mobility differences between the α- and β-tubulin components of *Chlamydomonas* than existed between the A- and B-tubulin counterparts in sea urchin (Stephens, 1970b), indicating differences in A- and B-tubule proteins in flagella of different organisms. As in the case of the sea urchin study, the possibility that the α- and β-tubulin fractions, derived respectively from A- and B-tubules, were each in turn composed of several closely related polypeptide chains could not be eliminated.

In contrast to these two studies, wherein it was concluded that each tubule in the doublet was composed principally of one type of tubulin, Olmsted *et al.* (1971) and Witman *et al.* (1972b) have demonstrated that both doublets and intact singlet A-tubules from *Chlamydomonas* contain equal amounts of *two* tubulins which differ slightly in charge on basic urea-acrylamide gels and in apparent molecular weight when subjected to urea-SDS-polyacrylamide gel electrophoresis. Furthermore, Sarkosyl fractionation of the B-subfiber seems to indicate that the two tubulins are removed in a manner corresponding to pairs of protofilaments, giving rise to a model for microtubule substructure wherein the doublets are made up of alternating pairs of protofilaments composed of either of the two tubulins (Witman *et al.*, 1972b). Nearly complete solubilization of the doublet resulted in the production of ribbons consisting of *three* protofilaments, corresponding to the shared wall between the two subfibers; these were composed primarily of only one of the two tubulin fractions (Witman *et al.*, 1972b). The tubulins separable on urea-SDS gels were named tubulin-1 and tubulin-2, corresponding to apparent 56,000 and 53,000 dalton species respectively (Olmsted *et al.*, 1971). When either tubulin-1 or tubulin-2 was subjected to isoelectric focusing, it appeared as if each was in turn composed of two or three discrete bands with the mixed tubulins showing five distinct components. Such microheterogeneity was interpreted to mean that each class of tubulin (i.e. 1 or 2) may be made up of several sets of closely related monomers forming possible heterodimers of either tubulin-1 or tubulin-2 (Witman *et al.*, 1972b).

Thus the studies of Stephens (1970b) and of Jacobs and McVittie (1970) would indicate that A- and B-subfibers are each composed of a distinct tubulin while the work of Olmsted *et al.* (1971) and Witman *et al.* (1972b) would indicate that both subfibers are each composed of the same two tubulins. This latter claim is rather firmly based on the fact that each subfiber component can be resolved into two equally staining protein bands on either basic urea-acrylamide or urea-SDS-acrylamide gel electrophoresis after reduction and alkylation. Additional evidence for two tubulin types occurring in equal amount in outer fiber doublets comes from the original studies of Renaud *et al.* (1968) and later from Everhart (1971), where *Tetrahymena* outer fibers were shown to yield a doublet upon electrophoresis. In the latter study, the two components were separated and shown to have different amino acid compositions.

These apparent discrepancies have been rationalized by Feit *et al.* (1971) when they were able to resolve A- and B-tubulin from sea urchin and neurotubule protein from brain *each* into two equally dense protein bands by means of urea-SDS-acrylamide gel electrophoresis. These components, corresponding to the tubulins-1 and 2 discussed above, were interpreted here as being *two non-identical subunits of a heterodimer*, i.e. A-tubulin would be composed of two different subunits (A_x and A_y) and B-tubulin would be made up of a different pair of subunits (B_x and B_y). Thus if size and charge are conserved throughout evolution, all tubulins regardless of source would be expected to yield, upon electrophoresis, apparently identical protein doublets representing in fact the two distinct subunits of a specific tubulin heterodimer rather than two distinct "tubulins". Feit *et al.* (1971) demonstrated that both porcine and mouse brain tubulins could be resolved into essentially identical protein doublets on urea-SDS-acrylamide electrophoresis, but that the former protein appeared as two components on isoelectric focusing while the latter appeared to be four, further supporting the contention that corresponding heterodimeric subunits have essentially identical electrophoretic mobilities.

The concept of tubulin being a heterodimer was originally advanced by Bryan and Wilson (1971). These workers used charge separation of chick brain tubulin on urea-acrylamide gel electrophoresis to demonstrate that the two resulting equimolar components had distinctly different amino acid compositions. Later, Bryan (1972) found that the tubulin which can be purified from sea urchin eggs via vinblastine sulfate co-crystallization was likewise composed of identical amounts of two significantly different tubulin subunits. By varying electrophoretic gel composition, Bryan (1972) was able to show convincingly that the tubulin subunits had identical molecular weights, implying that the urea-SDS-acrylamide method used by various other workers was not separating on the basis of molecular weight but rather by charge. In this regard, it might be pointed out that Lee *et al.* (1972) have contended that the urea-SDS methodology is sensitive to differential urea binding and Weber and Kuter (1971) have employed urea to remove SDS from proteins: perhaps in such cases charge differences become unmasked. The consistent observations that all tubulin dimers thus far studied bind one GTP tightly and one loosely, plus the evidence that there is one colchicine and one vinblastine binding site per dimer of cytoplasmic tubulin (cf. Bryan, 1972), likewise, though circumstantially, would point toward the heterodimer concept.

As an argument against the heterodimer, Witman *et al.* (1972b) presented evidence, discussed briefly above, that the three protofilaments making up the shared wall of the flagellar outer fiber doublet in *Chlamydomonas* yielded only one tubulin band on urea-SDS-acrylamide gel electrophoresis and one band on isoelectric focusing, in contrast with five for the whole doublet, indicating that a dimer must be composed of this apparently single polypeptide chain as a homodimer at least in the shared wall of the doublet. On the other hand, exactly the opposite result has been obtained from the comparable structure of sea urchin doublets by Meza *et al.* (1971). The differences between these two groups of workers must be methodological and are presently unresolved. Witman *et al.* (1972b) further contend that the relative amounts of tubulin-1 and 2 extracted vary with the degree of disintegration of the B-tubule during Sarkosyl fractionation, and can be best explained if the tubule is composed of alternating pairs of protofilaments made up of either tubulin-1 or tubulin-2 but concede that, on the basis of multiple components in isoelectric focusing, tubulins-1 and 2 could be themselves heterodimers at least within the A- or B-tubule proper. Meza *et al.* (1972) find no such sequential extraction of tubulin components in the case of sea urchin sperm outer fiber doublets and argue that this indicates that each protofilament is, in fact, heterodimeric in composition; again, the reasons for these differences appear to be methodological or perhaps species-dependent.

Antibodies prepared against sea urchin outer fiber doublets were found to cross-react with tubulins prepared from flagellar outer fiber A- and B-fractions and the central pair fraction, from blastula cilia, and from the isolated mitotic apparatus of various other echinoid (though not asteroid) species (Fulton *et al.*, 1971), indicating that common antigenic determinants were present in all. At a quantitative level, however, significant differences in cross-reactivity to this antibody could be discerned between flagellar A- and B-subfiber tubulin fractions and also between mixed flagellar and ciliary tubulins from the same organism, implying that some determinants were present in the A-tubulin that were not present in the B-tubulin and, similarly, that flagellar tubulins contained determinants not present in the morphologically identical cilia. These studies would thus further indicate that various tubulin dimers are extremely similar but not quite equivalent at a primary structural level.

The bulk of the fractionation, antibody, and stability data support the idea that there are, operationally at least, several classes of tubulin, while the bulk of the chemical studies indicate that tubulin molecules are very likely composed of two non-identical but very closely related subunits. Mounting evidence points to the fact that the two apparent heterodimeric subunits appear to differ very little with regard to source or phylogeny, conserving their characteristic charge and size parameters. If, in fact, there are several heterodimeric tubulin classes, it will be of considerable interest to determine whether any of these dimers hold any polypeptide chains in common, with various combinations being used to form A-, B-, central pair, and cytoplasmic microtubules or parts thereof.

3. *Surface lattices and subunit arrangements*

André and Thiéry (1963) and Pease (1963), using negative staining, first noted that mammalian sperm tail doublets could be resolved into parallel protofilaments with diameters of about 40 Å and a longitudinal repeat of 80–88 Å. In cross-section, *Chlamydomonas* flagellar outer fibers show 13 protofilaments of about 45 Å diameter in each member of the doublet with three or four protofilaments of the A-tubule being shared with the B-subfiber (Ringo, 1967). Numerous other studies have likewise indicated either 12 or 13 protofilaments composing the 200–240 Å cylindrical walls of microtubules from diverse sources.

Employing optical diffraction of negatively stained micrographs of protozoan flagellar outer fiber doublets and central pair singlets, Grimstone and Klug (1966) were able to demonstrate a basic 40 × 50 Å surface lattice wherein protofilaments with an axial repeat of 40 Å were separated by a 50 Å center-to-center spacing; the 40 Å globular subunits of which the protofilaments were composed appeared to be arrayed at a pitch angle of about 10–15°. The resultant surface lattice is illustrated in Fig. 2a.

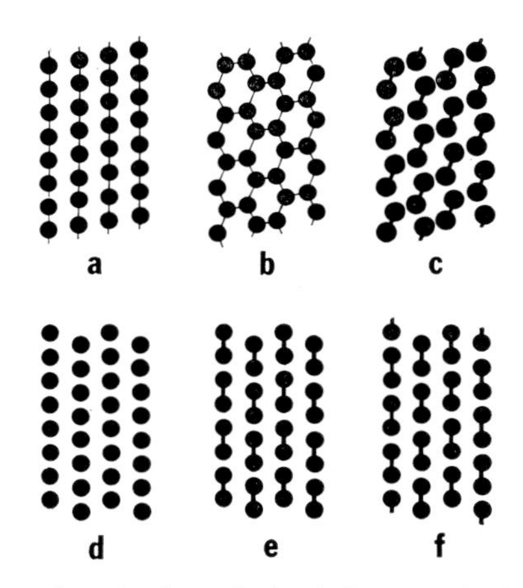

FIG. 2. Microtubule surface lattices derived from optical and X-ray diffraction. (a) protofilament lattice of Grimstone and Klug (1966) from optical diffraction; (b) perturbed protofilament lattice, giving rise to 80 Å periodicity (a 160 Å spacing could be generated by displacing every fourth subunit with respect to the surface of the page—Grimstone and Klug, 1966); (c) dimer-protofilament lattice derived from optical reconstruction by Chasey (1972); (d) half-staggered lattice of Cohen *et al.* (1971); (e) half-staggered dimer model (Stephens, 1971); (f) helical dimer model of Thomas and Henley (1971) with half-staggered subunits.

The diffraction patterns obtained by Grimstone and Klug (1966) also displayed prominent 80 Å, 160 Å, and 480 Å axial repeats; since these spacings were only observed on intact tubules and not those in which the cylindrical integrity was disrupted, they were explained in terms of perturbations occurring radially and in the plane of the basic lattice (Fig. 2b). Chasey (1972) performed optical diffraction and reconstruction on singlet tubules from *Tetrahymena* and arrived at the same basic surface lattice, but in this case the somewhat refined image appeared to indicate the position of dimers within the lattice structure (Fig. 2c). Other studies of negatively stained lung fluke and planarian sperm cortical singlets and outer fiber doublets revealed a prominent 80 Å banding running helically along the tubule surface (cf. Burton, 1970; Burton and Silveira, 1971) with little evidence for the existence of parallel protofilaments or 40 Å globular subunits.

When wet gels of sea urchin outer fibers were subjected to X-ray diffraction analysis (Cohen *et al.*, 1971), 50 Å equatorial, 40 Å off-meridional, and 20 Å meridional or near-meridional reflections were clearly observed. These spacings were consistent with a surface lattice wherein any point would lie near the center of an approximately orthogonal cell defined by its four neighbors, i.e. the 40 Å globular subunits are arrayed in half-staggered fashion from one protofilament to the next (Fig. 2d). It must be pointed out that this surface lattice is not at all consistent with that seen in most cases in negatively stained preparations. However, when the tubule preparation was dried prior to diffraction, the uniquely characteristic 20 Å meridional and 40 Å off-meridional reflections were not observed; instead a prominent 40 Å near-meridonal reflection appeared, yielding a pattern consistent with the surface lattice deduced from electron microscopy. From these data it would appear that drying during negative staining disrupts the coherence of the lateral bonding between protofilaments, resulting in diffraction from a simple linear array of subunits (Cohen *et al.*, 1971).

That negative staining or drying may cause changes in the apparent surface lattice of microtubules is implied from several studies. Henley (1970) showed that phosphotungstate "macerated" various ciliary and flagellar microtubules to quite a variable extent, either dissolving them or disrupting lateral coherence of the protofilament arrays. Thomas (1970) observed both an 80 Å helical and a 40 Å protofibrillar subunit configuration, within the same polyclad sperm cortical singlet microtubule, separated by a transition zone; she also observed that when doublet microtubules become singlets near their tips, a prominent helical configuration replaced the usual protofibrillar one. In platyhelminth sperm cortical singlets, Thomas and Henley (1971) observed the characteristic 80 Å helical form undergo a series of transitions to a 40 Å protofibrillar array through first a random lateral separation of protofilaments of 80 Å axial repeat period, followed by a splitting of the 80 Å units into 40 Å globular subunits. These data were interpreted to indicate that 80 Å dimers were associated end-to-end to form protofilaments, which in turn associated side-to-side with the dimers displaced relative to one another in such a manner that the characteristic 80 Å helical pattern was generated.

Physical disruption is clearly one way that a protofibrillar configuration can be generated from a helical one, but simple flattening or longitudinal displacement can result in an apparent subunit transition. Figure 3a is a model of the half-staggered subunit array of Cohen *et al.* (1971) built on the basis of 13 protofilaments per tubule (Fig. 1c in Cohen *et al.*, 1971), using strings of beads which may move longitudinally with respect to one another but whose "subunits" retain their nearest-neighbor relationships since all strings of beads are attached to the same plane. Tilting the attachment plane simulates flattening and slippage of the tubule, such as might be expected during the drying process in negative staining. It may be easily seen that this simple physical distortion will generate from

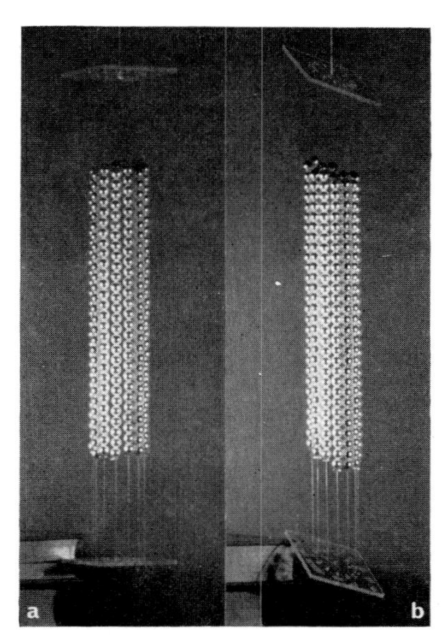

FIG. 3. Parallel bead-string model of the 13-fold half-staggered microtubule surface lattice of Cohen *et al.* (1971) in true cylindrical form (a) and in partially "flattened" or elliptical cross-sectional form (b) illustrating the illusion whereby a half-staggered array might appear protofibrillar upon drying during negative staining.

a half-staggered array (Fig. 3a) the protofilament configuration characteristic of electron microscopic observations (Fig. 3b). Furthermore, with appropriate manipulation, virtually any pattern seen in the electron microscope can be obtained.

The general structure of the microtubule as deduced from electron microscopy and X-ray diffraction (Cohen *et al.*, 1971), the biochemical concept of heterodimeric tubulin (Bryan and Wilson, 1971), and the observations of helical-protofibrillar subunit transitions (Thomas and Henley, 1971) can give rise to a relatively simple model for microtubule structure at a chemical level. This model has certain general properties which may have biological significance.

The following simplifying assumptions are made in establishing a model for a 13-fold radially symmetric singlet microtubule:

1. Tubulin dimers are heterodimers and associate head-to-tail to form protofilaments. The evidence for heterodimers is discussed above, while the longitudinal association of tubulin follows from observations of 40 Å diameter single protofilaments. Head-to-head association would require alternating tail-to-tail association, resulting in a protofilament lacking polarity.

2. Microtubules are composed of one dimeric tubulin type such that all protofilaments are equivalent; if protofilaments were of two types and arranged either alternately or in alternating pairs, the desired 13-fold symmetrical tubule would be excluded.

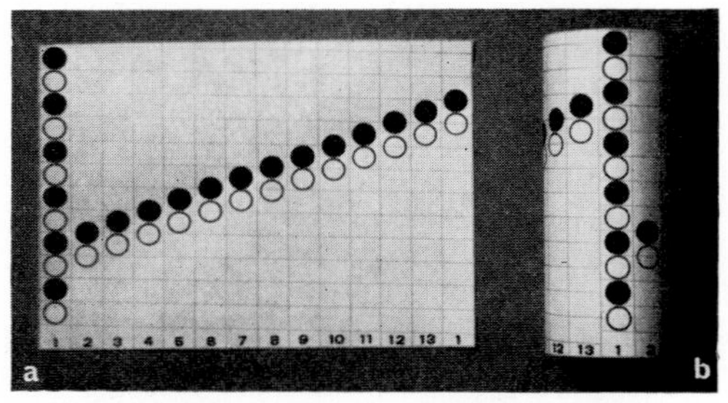

FIG. 4. Heterodimers arranged unidirectionally in helical array (Thomas and Henley, 1971) on the 13-fold half-staggered surface lattice derived from X-ray diffraction (Cohen *et al.*, 1971). (a) Surface lattice of 13 protofilaments on a sheet of paper; (b) cylinder formed from the lattice, showing a basic helical repeat of 240 Å, a dimer repeat of 80 Å, a pitch angle of approximately 20°, and a mean diameter of 205 Å with respect to the centers of the subunits.

3. All protofilaments of the tubule have the same polarity. If they did not and condition 2 was met, the 13-fold symmetrical tubule would again be excluded.

4. Subunits are half-staggered, as shown in Fig. 3a. This is based upon X-ray evidence (Cohen *et al.*, 1971).

5. Dimers within the half-staggered subunit array are helically disposed rather than half-staggered (Fig. 2f *vs.* 2e). This is based upon the helical 80 Å striping seen by Burton (1970), Burton and Silveira (1971) and Thomas and Henley (1971). Half-staggered dimers would have the same restrictions against 13-fold symmetry.

The subunit array of such a tubule is illustrated in Fig. 4a while the surface lattice generated from this array (conforming with the tubule of 13 units in two turns of the primitive helix, Cohen *et al.*, 1971) is shown in Fig. 4b. It should be immediately evident that such a tubule has inherent

40 Å (subunit), 80 Å (dimer), and 240 Å (repeat period) longitudinal periodicities in addition to having an overall polarity, handedness, and potential 80 Å helical striping. The same holds true for a 12-protofilament tubule generated in the same manner; in this case, however, alternating or paired protofilaments of a different composition or alternating protofilaments of opposite polarity could eliminate the properties of handedness in both cases and also polarity in the latter case. It should be emphasized that this is simply a model for the most simple tubule consistent with available data; production of a flagellar doublet in conjunction with a basal body would allow the tubule to be constructed in template fashion, with hetero- or homodimers, hetero- or homofilaments, alternating, specifically, or uniformly arrayed protofilaments. Until the potentially complex nature of outer fiber doublet association with dynein and with secondary axonemal proteins such as radial spokes and circumferential linkage is better understood, the manner by which radial and longitudinal bonding sites are specified on the microtubule surface lattice must remain a mystery.

C. *Secondary proteins: accessories for biological function*

Three other classes of protein are of vital importance in the functioning of the cilium or flagellum. The impervious membrane sheath is obviously necessary for retention of ATP, ions, and other small molecules within the axoneme. The amorphous matrix of the axoneme contains numerous minor proteins of unknown significance plus a major one of energetic importance, adenylate kinase, which catalyzes the reaction 2 ADP = AMP + ATP. Finally, and perhaps most importantly, there are the architectural proteins of the linkage-spoke complex which give symmetry to the axoneme and provide mechanical constraints relevant to motility.

1. *Membrane proteins*

At a point in history when the topic of membrane proteins is one of considerable controversy, it is perhaps presumptuous for someone not in the field even to mention the subject. But the relative clarity and consistency of the results thus far obtained independently by three sets of workers warrants discussion, for the data suggest a certain universality in ciliary and flagellar membranes which appear to correlate well with studies in other unrelated systems.

While investigating the microtubule components of *Chlamydomonas* flagella, Jacobs and McVittie (1970) prepared a membrane fraction by selectively solubilizing the axoneme with Sarkosyl and separating the membranes on a sucrose step gradient. Solubilization of the purified membrane fraction with mercaptoethanol-SDS and analysis by SDS-acrylamide gel electrophoresis revealed only a single high molecular weight component, with little trace of lower molecular weight material. From inspection of their gels, the molecular weight of this protein component appears to be in excess of 200,000 daltons.

Witman *et al.* (1972a) detached the membrane fraction from *Chlamydomonas* flagella by the Sarkosyl method of Jacobs and McVittie (1970) and also by simple shear, separating it from the axoneme or remnants by either differential or isopycnic centrifugation. On SDS-urea acrylamide gel electrophoresis, the SDS-solubilized membranes again revealed principally a single component of high molecular weight, estimated to be well in excess of the 170,000 dalton protein of the mastigoneme, a component found attached in lampbrush-fashion to the surface of the flagellum. The use of this different gel system by these workers gave appreciably better resolution of the single membrane protein and from the published photographs it appears as if this component comprises at least 90% of the membrane protein. Witman *et al.* (1972a) note that this protein may be comparable with the 200,000 dalton major human erythrocyte membrane protein analyzed by gel filtration by Gwynne and Tanford (1970) and the 240,000–255,000 dalton apparent counterpart reported by Lenard (1970) to comprise the major component of various mammalian erythrocyte membranes, as analyzed on SDS-acrylamide gels.

The argument has been made that the mechanical release of a membrane fraction by detergent treatment of an organelle isolated in the presence of EDTA may selectively remove numerous other membrane components prior to membrane analysis. This appears not to be the case, for results essentially identical to those above can be obtained in two other flagellar systems using a totally different methodology.

When membranes of hypertonically isolated molluscan gill cilia or sheared sea urchin or malluscan sperm tails are dispersed by treatment with Triton X-100 (Stephens and Linck, 1969) and the axonemes removed by high-speed centrifugation, the membrane proteins can be analyzed directly via SDS-acrylamide gel electrophoresis. In the case of sperm flagella, a major component of approximately 240,000 daltons is consistently obtained while gill cilia yield two components of approximately 115,000 and 60,000 daltons. In both cases, the only other proteins on the gel are those from the axoneme proper which are slightly soluble under these extraction conditions (e.g. adenylate kinase, central pair tubulin, and minor matrix proteins). The major membrane proteins constitute at least 85% of the total. It is not clear whether the two ciliary membrane proteins are related as monomer and dimer, whether they are distinct species, or whether they are a breakdown product of a 240,000 dalton protein (Stephens, unpublished).

2. *Adenylate kinase*

The presence of an adenylate kinase (ATP–AMP phosphotransferase) in flagella was first emphasized by Brokaw (1961). Gibbons (1966) noted that the apparent ability of gradient-purified 14 S and 30 S ciliary dynein to split ADP was due to a contaminating adenylate kinase, estimated to have a sedimentation coefficient of 10 S but aggregating further and distributing throughout the gradient. This point was proven conclusively by Ogawa and Mohri (1972) using chromatographically purified sea urchin sperm dynein; adenylate kinase activity in the dynein was negligible when compared with the crude extract.

Few specific data currently exist on ciliary or flagellar adenylate kinase. Like its muscle counterpart, it is stable to hot 0·1 N HCl and other harsh purification procedures. It may be ammonium sulfate-precipitated between 25 and 35% saturation and analysis of this cut by SDS-acrylamide gel electrophoresis reveals a major component with a molecular weight of about 35,000 daltons, as contrasted with muscle myokinase which migrates with an apparent molecular weight of 21,000, a number in agreement with accepted physico-chemical values (cf. Noda and Kuby, 1957). The protein typically purifies along with the axoneme proper, rather than the membrane or soluble matrix fractions, but is removable by low ionic strength dialysis or glycerination in many species; in others, about half is retained by the outer fiber fraction. As implied by Gibbons (1966), the enzyme aggregates badly, making a clean separation from tubulin and minor axonemal proteins quite difficult (Stephens and Levine, unpublished).

3. *Architectural proteins*

Dynein provides the enzymatic energy for axoneme function, evidently through direct interaction with outer fiber doublets, but it is the complex of architectural proteins of the axoneme that serves as a mechanical constraint on this interaction and which ultimately gives rise to the wave patterns so characteristic of cilia and flagella. At a biochemical level we know the least about these components, gaining what we do know principally from inference.

At a morphological level, in addition to the "9+2" microtubules and the paired dynein arms, at least six other fairly distinct general features are evident in the axoneme (Fig. 1). The outer fibers are connected together by means of *circumferential links*, sometimes running between A-tubules of adjacent doublets (Gibbons, 1965a) and at other times clearly connecting the A-tubule with the B-subfiber of the next doublet (Witman *et al.*, 1972a). Radially, the outer fibers are associated with the central pair by means of three distinct structures: a pair of *radial links* or *spokes* spaced longitudinally on the A-tubule 320 Å apart with 560–640 Å between pairs; a *radial link head* (originally interpreted as a "secondary fiber" in cross-section) at the end of each radial link; and the *transitional link* which attaches the radial link head to the central pair complex. This latter connection abuts the *central sheath*, a helical array encompassing the central pair and having a diameter of about 700 Å and a pitch of 160 Å (Warner, 1970; Hopkins, 1970). One member of the central pair (C_1 or the more chemically stable tubule) bears a pair of 180 Å *projections* 120° apart and spaced at 160 Å intervals along the length of the tubule (Chasey, 1969; Hopkins, 1970); the remaining member (C_2) appears to have only one row of such projections (Hopkins, 1970).

The thermal fractionation procedure for separating A-tubule and B-subfiber components (Stephens, 1970b) can be applied to the whole axoneme to remove selectively dynein, the B-subfibers, and central pair complex, leaving only A-tubules connected via circumferential links. These strap-like connections are typically about 300 Å wide, occurring at 900–1000 Å intervals, and giving a "bosun's ladder" appearance to the parallel array of singlet tubules. When the A-tubules were solubilized by mild acid treatment, a fairly pure protein component, evidently derived from the circumferential linkages, remained. Electrophoretically, this protein ran on urea-acrylamide gels as a single component with somewhat lesser mobility that tubulin and likewise ran on SDS-acrylamide gels as a single band with an apparent molecular weight of 150,000–165,000 daltons. This protein was named *nexin*, from the Latin *nexus*, meaning a tie binding together members of a group, in this case the outer fibers (Stephens, 1970c, 1971). When this procedure was applied to other organisms (e.g. *Asterias* or *Aequipecten*), another component of 80,000–90,000 daltons fractionated along with the nexin. It is not clear whether this is a monomeric form of nexin or another A-tubule-associated protein such as that of the radial link or spoke; the latter interpretation is currently the favored one since axonemes in which this occurs typically retain some of their radial structures after fractionation.

Considering the spacing of nexin along the outer fibers, there is probably no more nexin present in an axoneme than there is radial link, link head, or transitional link material. On this basis of quantity, three prominent bands consistently found on SDS-acrylamide gels of whole ciliary or flagellar axonemes probably correspond to the latter structures. These have approximate relative molecular weights of 70,000–75,000, 83,000, and 90,000 daltons and may be seen in Fig. 5 as a "triplet" particularly obvious in *Asterias* flagellar (gel *b*) and *Aequipecten* ciliary (gel *d*) axonemes. Both axonemes retain much of their radial structures (though no central pair in the former and only one central fiber in the latter) upon removal of dynein by dialysis; the three components in question remain principally with the resultant outer fiber fractions (gels *c* and *f*) rather than with the enriched dynein-central pair (gel *a*) or dynein and B- and C_1 tubule fractions (gel *e*). The nexin (band at 160,000 daltons) fractionates cleanly

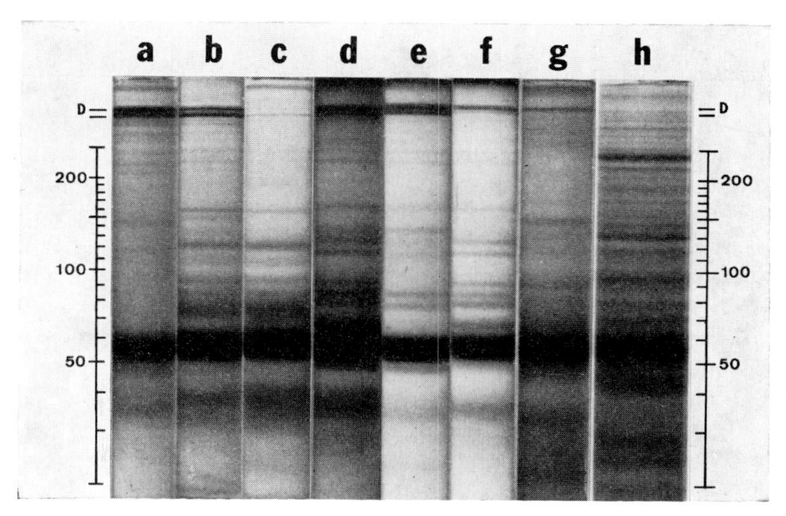

FIG. 5. Comparative SDS-acrylamide gel electrophoresis of fractionated flagellar axonemes, contractile axostyles, and mitotic apparatus. Scale = daltons $\times 10^{-3}$.

(a) supernatant containing doubly banded dynein (D =) and central pair tubulin (55,000) is obtained by dialysis of *Asterias* sperm flagellar axonemes (b) against Tris-EDTA; outer fibers, devoid of any dynein, remain insoluble (c). Axonemes (d) of *Aequipecten* cilia, when similarly dialyzed, yield a supernatant (e) containing one-half of the total dynein as a two-banded species, one member of the central pair, and all of the B-subfibers; the A-tubule-containing outer fiber fraction (f) still contains one-half of the dynein as a single-banded species (based on Linck, 1971.)

The flagellate contractile axostyle (g) contains a single dynein band, tubulin, and the linkage and/or spoke components at approximately 90,000 and 150,000 daltons (cf. Mooseker and Tilney, 1973).

The isolated sea urchin mitotic apparatus (h) shows little evidence for any dynein component (in spite of a relatively high concentration of a dynein-like component in the whole cell) and is dominated by tubulin (55,000) and the subunits of the 22 S matrix protein (240,000, 130,000, and 95,000). The fine band at 46,000 daltons could be actin, but, if so, it is present in much smaller amount in the isolated mitotic apparatus than in the whole cell. The general background material is interstitial ribosomal and cytoplasmic material (Stephens, in preparation).

with the outer fibers in the case of *Asterias* (gel *a* versus gel *c*), while in *Aequipecten* it appears to be distributed equally between the soluble and bound dynein-containing fractions (gels *e* and *f*). For reference, the fairly prominent band at 120,000 daltons so obvious on some gels (*b*, *c*, *d*, and *f*) is a tubulin dimer and that at 115,000 daltons on the ciliary axoneme and outer fiber gels (*d* and *f*) is likely to be a membrane component; these vary considerably from preparation to preparation. As discussed above, the 35,000 dalton component has been tentatively identified as adenylate kinase.

Sequential pulse-labeling of developing sea urchin embryos prior to and during ciliogenesis indicates that the major proteins of the linkage–spoke complex are synthesized *de novo* during the process of ciliogenesis itself, as contrasted to the synthesis of tubulin, adenylate kinase, a dynein structural component, and various minor constituents, which is uniform throughout post-fertilization, and further contrasted to the synthesis of the enzymatic component of dynein which is synthesized entirely during oogenesis (Stephens, 1972b). These observations would suggest that it is perhaps the programmed synthesis of the architectural elements which is responsible for the precisely timed assembly of the organelle.

As discussed above, dynein may be totally removed as a 14 S component, accompanied by central pair tubulin, by dialysis of echinoderm and other flagellar axonemes at low ionic strength; on SDS-acrylamide gel electrophoresis the dynein appears as a closely spaced doublet (Fig. 5, gel *a*) with essentially none remaining in the outer fiber fraction (gel *c*). The two dynein components have estimated molecular weights of 500,000 and 460,000 daltons, determined from SDS-acrylamide (3%) gels using such molecular weight standards as unreduced myosin, thyroglobulin, and hemocyanin (Linck, 1971, 1973b). In the case of *Aequipecten* cilia, however, comparable dialysis will remove only one-half the dynein ATPase, along with one central fiber and all B-subfibers. This dynein fraction also migrates as a characteristic doublet (gel *e*), but the outer fiber fraction retains about half of the activity and shows the upper member of the doublet (gel *f*), equal in intensity to the higher molecular weight component in the soluble dynein fraction. This remaining dynein is removable as a 14 S component, basically identical to the first, through brief trypsinization (Linck, 1971, 1973b). Salt extraction of flagellar axonemes removes almost all of the dynein as a 14 S particle, but this fraction, like the bound ciliary dynein, migrates principally as a single band on SDS-acrylamide gel electrophoresis (Stephens, unpublished; Gibbons, personal communication). A 13 S dynein-like ATPase has been isolated from sea urchin eggs and from isolated mitotic apparatus preparations (Weisenberg and Taylor, 1968); it was presumed to be either a mitotic ATPase or a store of ciliary dynein. This component was demonstrated to be, at least partly, a ciliary precursor by protein synthetic studies (Stephens, 1972b), and it, too, was found to migrate as a single-banded species (Stephens, 1972a). All of these data, taken together, would indicate that the true enzymatic component of dynein must correspond to the higher molecular weight component of the "dynein" doublet. Exactly what the lower molecular weight component (typically obtained when dynein is prepared by dialysis) might be is purely speculative at this point. It may be a material holding the enzymatic portion to the outer fiber (e.g. the inner lobe of the outer dynein arm seen by Allen, 1968), or a component of the linkage–spoke complex. It is highly unlikely that it is merely a breakdown product of the higher molecular weight dynein band since the two components are synthesized differentially during evelopment in the sea urchin; the high molecular weight dynein component pre-exists in the unfertilized egg while the lower molecular weight material is synthesized at a constant rate after fertilization (Stephens, 1972b).

Further evidence for the identity of various secondary components comes from the studies of Mooseker and Tilney (1973) on the contractile axostyle of the flagellate *Saccinobaculus*. This organelle is a "simplified flagellum", occurring intracellularly, and having an array of singlet tubules bearing a short arm much like the inner dynein arm of the axoneme; these tubules are interconnected with short linkages comparable in size to the circumferential (nexin) and radial links of the axoneme. The axostyle can be reactivated under conditions identical to those used by Gibbons and Gibbons (1972) for sperm flagella and has ATPase activity and specificity quite comparable to the reactivated sperm flagella. Analysis by SDS-acrylamide gel electrophoresis reveals several interesting features (Fig. 5, gel *g*). Morphologically and biochemically, tubules and tubulin dominate the organelle as the major band at 55,000 daltons indicates. Only a single band is evident in the dynein region, corresponding to the higher molecular weight dynein component of either cilia, flagella, or the developing egg. Bands at about 150,000 and 90,000 daltons most likely correspond either to nexin or to nexin and radial linkage components of the axoneme. A 35,000 dalton component is evident, as is adenylate kinase activity in the isolated organelle.

Finally, for comparison or perhaps controversy, another microtubular system—the mitotic apparatus—is illustrated along with the axoneme and axostyle (Fig. 5, gel *h*). Prepared by the Kane hexylene glycol method (Kane, 1962; Stephens, 1972a), this organelle contains tubulin as the single major component at 55,000 daltons plus subunits of the 22 S matrix protein with molecular weights of 240,000, 130,000, and 95,000. The numerous fine background components are most likely ribosomal subunits and general cytoplasmic proteins. In spite of the relatively high concentration of dynein-like component in the whole cell cytoplasm, no obvious dynein bands are apparent, indicating that dynein is not selectively concentrated in the isolated mitotic apparatus. Two components appear at 43,000 and 46,000 daltons; the latter corresponds in molecular weight to, and co-migrates with, actin, an alleged component of both the mitotic apparatus and the sperm flagellum (Behnke *et al.*, 1971). Such a component is never detected electrophoretically in ciliary or flagellar axonemes, within a limit corresponding to less than one actin filament per entire "9+2" complex. Tilney *et al.* (1972) have recently demonstrated that actin does, indeed, occur in several types of sperm, but does so in the *acrosomal filament*; this probably explains the repeated reports of sperm actin. The situation in the egg is less clear-cut, although it can be asserted that if, in fact, this band at 46,000 is actin, then it may still be artifactual. The whole egg contains an inordinate amount of 46,000 dalton material and this material is no more selectively concentrated in the mitotic apparatus than are the apparent ribosomal components (Stephens, in preparation).

It should be evident from the foregoing discussion of protein fractionations and properties that dynein is a unique ATPase, that the tubulins are a distinct class of protein, and that these two major contractile elements are assembled together as a functional motile unit—the axoneme—through circumferential and radial linkage proteins, the nature of which we are only just now beginning to decipher.

IV. Mechanisms and motility

Two general molecular mechanisms can be postulated for ciliary and flagellar bending. One involves the coordinated, sequential, active contraction of the outer fiber doublets, an idea originally suggested by Astbury *et al.* (1955) and later expanded upon by others (Machin, 1958; Brokaw, 1966; Lubliner and Blum, 1971). The other is a sliding filament mechanism wherein outer doublets would remain of constant length but move longitudinally relative to one another, perhaps by means of dynein crossbridges analogous to the myosin crossbridges of muscle (cf. Brokaw, 1971). Satir (1968) has mounted evidence which strongly favors this type of sliding mechanism by demonstrating, through serial sectioning of cilia tips fixed during active beat, that the outer fibers do remain of constant length. However, no bridging between adjacent doublets has ever been demonstrated, although it is quite plausible that, as in muscle, the active process is not preserved by fixation. On the other hand, the same contention can be made with regard to the possible contraction of the outer fiber doublets where the bend might be preserved but not the active contraction, thus leaving the situation somewhat ambiguous at the electron microscopic level. Complications from strong membrane diffraction (Cohen *et al.*, 1971) would appear to eliminate X-ray diffraction analysis as a more objective tool, at least for the time being.

The problem has recently been approached by Summers and Gibbons (1971) from an interesting direction. Flagellar axonemes were very gently trypsinized in order to partially disrupt the radial links and nexin bridges, thus removing from the doublets any mechanical constraint that these elements might endow, but otherwise leaving the axoneme morphologically intact and the dynein arms apparently unaffected. When ATP was added and these axonemes were observed under intense dark-field illumination, the axonemes were seen to elongate, sometimes to over five times their original length, disintegrating into individual doublets and groups, often spiraling or fragmenting as they did so. Electron microscopic examination after such ATP-induced activity demonstrated that the outer doublets were no longer in regular array. These effects had the same nucleotide specificity and divalent cation requirements as sperm flagellum reactivation, implying that the mechanisms were closely related. Summers and Gibbons (1971) concluded that these data support the hypothesis that "propagated bending waves of live-sperm tails are the result of ATP-induced shearing forces between outer tubules which, when resisted by the native structure, lead to localized sliding and generate an active bending moment".

Acknowledgements

The author wishes to thank Drs Catherine Henley, D. P. Costello, and L. G. Tilney for several very useful discussions of subunit surface lattices, Dr. I. R. Gibbons for many helpful comments on dynein chemistry, and these and numerous other authors for providing information about research in progress or in press. Many of the fractions illustrated in Figure 5 were prepared and run by M. S. Mooseker. The author's work reported herein was supported by USPHS grant GM 15,500 and Research Career Development Award GM 70,164 from the National Institutes of Health, Institute of General Medical Sciences.

References

AFZELIUS, B. (1959). *J. biophys. biochem. Cytol.* **5**, 269–278.
ALLEN, R. D. (1968). *J. Cell Biol.* **37**, 825–831.
ANDRÉ, J. and THIÉRY, J. P. (1963). *J. Microscopie* **2**, 71–80.
ASTBURY, W. T., BEIGHTON, E. and WEIBULL, C. (1955). *Symp. Soc. exp. Biol.* **9**, 282–305.
AUCLAIR, W. and SIEGEL, B. W. (1966). *Science, Wash.* **154**, 913–915.
BEHNKE, O. and FORER, A. (1966). *J. Cell Sci.* **2**, 169–192.
BEHNKE, O., FORER, A. and EMMERSEN, J. (1971). *Nature, Lond.* **234**, 408–410.
BLUM, J. J. (1971). *J. theor. Biol.* **33**, 257–263.
BROKAW, C. J. (1961). *Expl Cell Res.* **22**, 151–162.

Brokaw, C. J. (1966). *Nature, Lond.* **209**, 161–163.
Brokaw, C. J. (1971). *J. exp. Biol.* **55**, 289–304.
Brokaw, C. J. and Benedict, B. (1968). *Archs Biochem. Biophys.* **125**, 770–778.
Brokaw, C. J. and Benedict, B. (1971). *Archs Biochem. Biophys.* **142**, 91–100.
Bryan, J. (1972). *J. molec. Biol.* **66**, 157–168.
Bryan, J. and Wilson, L. (1971). *Proc. natn. Acad. Sci. U.S.A.* **68**, 1762–1766.
Burton, P. R. (1970). *J. Cell Biol.* **44**, 693–699.
Burton, P. R. and Silveira, M. (1971). *J. Ultrastruct. Res.* **36**, 757–767.
Chasey, D. (1969). *J. Cell Sci.* **5**, 453–458.
Chasey, D. (1972). *Expl Cell Res.* **74**, 140–146.
Child, F. M. (1959). *Expl Cell Res.* **18**, 258–267.
Cohen, C., Harrison, S. C. and Stephens, R. E. (1971). *J. molec. Biol.* **59**, 375–380.
Eckert, R. (1972). *Science, N.Y.* **176**, 473–481.
Everhart, L. P. (1971). *J. molec. Biol.* **61**, 745–748.
Feit, H., Slusarek, L. and Shelanski, M. L. (1971). *Proc. natn. Acad. Sci. U.S.A.* **68**, 2028–2031.
Fulton, C. M., Kane, R. E. and Stephens, R. E. (1971). *J. Cell Biol.* **50**, 762–773.
Gibbons, B. H. and Gibbons, I. R. (1972). *J. Cell Biol.* **54**, 75–97.
Gibbons, I. R. (1963). *Proc. natn. Acad. Sci. U.S.A.* **50**, 1002–1010.
Gibbons, I. R. (1965a). *Archs Biol., Liège* **76**, 317–352.
Gibbons, I. R. (1965b). *J. Cell Biol.* **25**, 400–402.
Gibbons, I. R. (1965c). *J. Cell Biol.* **26**, 707–712.
Gibbons, I. R. (1966). *J. biol. Chem.* **241**, 5590–5596.
Gibbons, I. R. (1967). Abstracts, 7th Internat. Congr. Biochem., Tokyo, p. 333.
Gibbons, I. R. and Fronk, E. (1972). *J. Cell Biol.* **54**, 365–381.
Gibbons, I. R. and Grimstone, A. V. (1960). *J. biophys. biochem. Cytol.* **7**, 697–712.
Gibbons, I. R and Rowe, A. J. (1965). *Science, Wash.* **149**, 424–426.
Grimstone, A. V. and Klug, A. (1966). *J. Cell Sci.* **1**, 351–362.
Gwynne, J. T. and Tanford, C. (1970). *J. biol. Chem.* **245**, 3269–3271.
Henley, C. (1970). *Biol. Bull.* **139**, 265–276.
Hopkins, J. M. (1970). *J. Cell Sci.* **7**, 823–839.
Iwaikawa, Y. (1967). *Embryologia* **9**, 287–294.
Jacobs, M. and McVittie, A. (1970). *Expl Cell Res.* **63**, 53–61.
Jacobs, M., Hopkins, J. M. and Randall, J. T. (1968). *J. Cell Biol.* **39**, 66 A.
Kane, R. E. (1962). *J. Cell Biol.* **12**, 47–55.
Lee, J. C., Frigon, R. P. and Timasheff, S. N. (1972). Abstracts of Papers, Sixteenth Annual Meeting of the Biophysical Society, p. 225a.
Lenard, J. (1970). *Biochemistry* **9**, 5037–5040.
Linck, R. W. (1970). *Biol. Bull.* **139**, 429.
Linck, R. W. (1971). Ph.D. Thesis, Brandeis University.
Linck, R. W. (1973a). *J. Cell. Sci.* **12**, 345–367.
Linck, R. W. (1973b). *J. Cell Sci.* **12**, 951–981.
Lubliner, J. and Blum, J. J. (1971). *J. theor. Biol.* **31**, 1–24.
Machin, K. E. (1958). *J. exp. Biol.* **35**, 796–806.

MEZA, I., HUANG, B. and BRYAN, J. (1971). Abstracts of Papers, Eleventh Annual Meeting of the American Society for Cell Biology, p. 192.
MEZA, I., HUANG, B. and BRYAN, J. (1972). *Expl Cell Res.* **74**, 535–540.
MOHRI, H. (1964). *Biol. Bull.* **127**, 381.
MOHRI, H. (1968a). *J. Fac. Sci. Tokyo Univ.* **8**, 307–315.
MOHRI, H. (1968b). *Nature, Lond.* **217**, 1053–1054.
MOOSEKER, M. S. and TILNEY, L. G. (1973). *J. Cell Biol.* **56**, 13–26.
NAITOH, Y. and KANEKO, H. (1972). *Science, N.Y.* **176**, 523–524.
NODA, L. and KUBY, S. A. (1957). *J. biol. Chem.* **226**, 551–558.
OGAWA, K. and MOHRI, H. (1972). *Biochim. biophys. Acta* **256**, 142–155.
OLMSTED, J. B., WITMAN, G. B., CARLSON, K. and ROSENBAUM, J. L. (1971). *Proc. natn. Acad. Sci.* **68**, 2273–2277.
PEASE, D. C. (1963). *J. Cell Biol.* **18**, 313–326.
RAFF, E. C. and BLUM, J. J. (1966). *J. Cell Biol.* **31**, 445–453.
RAFF, E. C. and BLUM, J. J. (1969). *J. biol. Chem.* **244**, 366–376.
RENAUD, F. L., ROWE, A. J. and GIBBONS, I. R. (1966). *J. Cell Biol.* **31**, 92A–93A.
RENAUD, F. L., ROWE, A. J. and GIBBONS, I. R. (1968). *J. Cell Biol.* **36**, 79–90.
RINGO, D. L. (1967). *J. Ultrastruct. Res.* **17**, 266–277.
SATIR, P. (1968). *J. Cell Biol.* **39**, 77–94.
SHELANSKI, M. L. and TAYLOR, E. W. (1967). *J. Cell Biol.* **34**, 549–554.
SHELANSKI, M. L. and TAYLOR, E. W. (1968). *J. Cell Biol.* **38**, 304–315.
STEPHENS, R. E. (1968a). *J. molec. Biol.* **32**, 277–283.
STEPHENS, R. E. (1968b). *J. molec. Biol.* **33**, 517–519.
STEPHENS, R. E. (1969). *Q. Rev. Biophys.* **1**, 377–390.
STEPHENS, R. E. (1970a). *Science, Wash.* **168**, 845–847.
STEPHENS, R. E. (1970b). *J. molec. Biol.* **47**, 353–363.
STEPHENS, R. E. (1970c). *Biol. Bull.* **139**, 438.
STEPHENS, R. E. (1971). *In* "Biological Macromolecules" (G. D. Fasman and S. N. Timasheff, eds), Vol. 5, pp. 355–391. Marcel Dekker, New York.
STEPHENS, R. E. (1972a). *Biol. Bull.* **142**, 145–159.
STEPHENS, R. E. (1972b). *Biol. Bull.* **142**, 489–504.
STEPHENS, R. E. (1974). *J. Cell Biol.* (in press).
STEPHENS, R. E. and LEVINE, E. E. (1970). *J. Cell Biol.* **46**, 416–421.
STEPHENS, R. E. and LINCK, R. W. (1969). *J. molec. Biol.* **40**, 497–501.
STEPHENS, R. E., RENAUD, F. L. and GIBBONS, I. R. (1967). *Science, N.Y.* **156**, 1606–1608.
SUMMERS, K. E. and GIBBONS, I. R. (1971). *Proc. natn. Acad. Sci. U.S.A.* **68**, 3092–3096.
TIBBS, J. (1962). *In* "Spermatozoan Motility" (D. W. Bishop, ed.), pp. 233–250. AAAS, Washington.
THOMAS, M. B. (1970). *Biol. Bull.* **138**, 219–234.
THOMAS, M. B. and HENLEY, C. (1971). *Biol. Bull.* **141**, 592–601.
TILNEY, L. G., HATANO, S. and MOOSEKER, M. S. (1972). *J. Cell Biol.* **55**, 261a.
WARNER, F. D. (1970). *J. Cell Biol.* **47**, 159–182.
WARR, J. R., MCVITTIE, A., RANDALL, J. and HOPKINS, J. M. (1966). *Genet. Res.* **1**, 335–351.
WATSON, M. R. and HOPKINS, J. M. (1962). *Expl Cell Res.* **28**, 280–295.

WEBER, K. and KUTER, D. J. (1971). *J. biol. Chem.* **246**, 4505–4509.

WEISENBERG, R. C. and TAYLOR, E. W. (1968). *Expl Cell Res.* **53**, 372–384.

WEISENBERG, R. C., BORISY, G. G. and TAYLOR, E. W. (1968). *Biochemistry* **7**, 4446–4479.

WINICUR, S. (1967). *J. Cell Biol.* **35**, c7–c9.

WITMAN, G. B., CARLSON, K., BERLINER, J. and ROSENBAUM, J. L. (1972a). *J. Cell Biol.* **54**, 507–539.

WITMAN, G. B., CARLSON, K. and ROSENBAUM, J. L. (1972b). *J. Cell Biol.* **54**, 540–555.

YANAGISAWA, T., HASEGAWA, S. and MOHRI, H. (1968). *Expl Cell Res.* **52**, 86–100.

Ciliary movement, evidence for its mechanism and forces exerted in ciliary bending

Chapter 4

Patterns of movement of cilia and flagella

MICHAEL A. SLEIGH

Department of Zoology, University of Bristol, England

I. Introduction

Seravin (1971) has aptly remarked "It is no exaggeration to state that all cilia and flagella beat in different ways". The aim of this chapter is to indicate the main forms of variation that occur in ciliary and flagellar movement and to indicate the order of size recorded for some of the main parameters of beating. Because some "flagella" show a ciliary beat, and some "cilia" have been observed to perform flagellar undulations, the convention of terminology mentioned in the introductory chapter (p. 2) will be followed.

The generalized pattern of movement of both forms of organelle involves the passage of waves of bending along the axis from one end to the other. The sequence of shapes assumed by the organelle during the passage of the bending wave depends upon such features as the length of the organelle, the frequency, direction of propagation and propagation velocity of the waves of bending and the symmetry of the bending movements—i.e. whether bending results in planar or three-dimensional waves and whether the amount of bending in one direction is greater than in any other direction. In very few cases is the pattern of movement known with any adequate degree of certainty, principally because of small size, rapid movement, crowding together of organelles and the occurrence of bending in three dimensions. Earlier reviews presented by Bishop (1962) on spermatozoa, by Holwill (1966a) on flagella, by Sleigh (1968) on cilia and by Jahn and Bovee (1967) on protozoan organelles give a wider range of examples than is included here; in this chapter I will concentrate on the best known patterns, particularly those which have been the subject of recent quantitative studies or which have been described since the earlier reviews.

The movement of many flagella is confined to a single plane, while in other cases the movement is helical, or at least has a substantial three-dimensional component. Similarly, in the case of ciliary movement, some cilia certainly complete the beat cycle within the same plane, but there have also been recent clear descriptions of ciliary beat cycles in which the movement is not planar. It is proposed therefore to discuss examples of planar and helical flagellar movement and planar and three-dimensional ciliary beating, in that order. Jahn and Bovee (1967) strongly emphasized that there are many examples of organelles which fall between these descriptions of "typical" cilia or flagella, and Phillips (1972) has described intermediate beat patterns for mammalian spermatozoa, but in the absence of complete information or quantitative data these will not be considered here. Clearly, there is a wide spectrum of patterns of movement in which the functioning of the basic axoneme mechanism is adapted to fit the needs of the organism.

Earlier descriptions of patterns of beating were based on slowly moving organelles studied by rather low-speed ciné photography, or on the use of stroboscopes and flash photography of phase contrast images. The use of high-speed ciné cameras with improved light sources, and particularly the development of Nomarski interference contrast optics have made possible a considerable improvement in the accuracy of information available.

II. Planar movement of flagella

The first accurate account of flagellar movement by Gray (1955) concerned the beating of the sperm tail of the sea urchin, along which approximately planar waves travel from base to tip. This pattern of movement has been shown to occur in sperm flagella in a number of echinoderms, annelids, coelenterates and a tunicate, as described in this volume by Brokaw (Chapter 5). Similar planar waves have been reported for flagellate Protozoa of several groups, notably by Brokaw (1963) for *Polytoma*, by Jahn *et al.* (1963) and Brokaw and Wright (1963) for the longitudinal

TABLE I. Some examples of parameters of flagellar beating[a]

	Length (μm)	Frequency (Hz)	Wavelength (μm)	Amplitude (μm)	Temperature (°C)	Shape of beat	Reference
Bull (sperm)	70	22	35	9·5	37	Helical	Rikmenspoel (1966)
Psammechinus miliaris (sperm)	40	35	24	4	17	Planar	Gray (1955)
Strigomonas oncopelti	17	16·8	14·4	2·4	22	Planar[b]	Holwill (1965)
Poteriodendron sp.	35	40	4	~2	~20	Planar	Sleigh (1964)
Euglena viridis	100	12	35	6	?	Helical	Holwill (1966b)

[a] For further examples see Holwill (1966a).

[b] Waves propagated basally or distally.

flagellum of the dinoflagellate *Ceratium* and by Sleigh (1964) for several other flagellates. In all of these cases the waves were propagated distally, but in the trypanosomid flagellate *Strigomonas* (= *Crithidia*) *oncopelti*, Holwill (1965) reported that planar waves may be propagated in either direction along the flagellum, and the movement of bending waves from the tip of the flagellum towards the base has also been reported for other trypanosomid flagellates.

The shape of the waves seen in these cases usually appears to be approximately sinusoidal, but Silvester and Holwill (1972) came to the conclusion that it is not yet possible to decide whether the waveform is best represented by a sine wave, a meander wave or a wave composed of circular arcs and straight lines (Brokaw and Wright, 1963). Commonly the length of the flagellum includes little more than one complete wave, as described for *Lytechinus*, *Ciona* and *Chaetopterus* by Brokaw (1965), while in the flagellate *Poteriodendron* 8 or 9 waves may be seen within the flagellar length (Sleigh, 1964). The amplitude and wavelength of flagellar waves in different species are quite variable, and bear no particular relation to the length of the flagellum (Table I). In some cases the wavelength and amplitude of the flagellar waves remain approximately constant as the waves pass along the flagellum; more frequently, however, both amplitude and wavelength vary along the flagellum, and commonly both of these parameters increase during the propagation of a wave. The influence of variation in wavelength and amplitude upon the hydrodynamic effects of flagellar activity has been discussed by Holwill and Miles (1971), and is summarized in this volume in the chapter on Hydrodynamics by Holwill (Chapter 8). The effects of changes in the physical and chemical environment upon the waveform of invertebrate sperm flagella are considered by Brokaw (p. 93, this volume); one striking change is that the sperm tails of *Ciona*, which show a planar beat in sea water, perform a helical movement when placed in viscous solutions with a low concentration of thiourea (p. 106).

III. Helical movement of flagella

Many flagella have been described as showing helical undulations, but rarely is the form of the wave anything like a true helix in the sense that a cross-section of the wave is circular and the envelope of the movement of the flagellum is cylindrical. Probably the nearest approach to a genuine helical waveform among descriptions of flagella is that by Jahn *et al.* (1963) for the transverse flagellum of *Ceratium*, which moves in a semicircular groove at the surface of the body (Fig. 1), forming a long helix with many turns. However, the form of movement of the basal part of the flagellum is not known, and it may be that the waveform changes close to the implantation. In the case of *Euglena*, for example, Holwill (1966b) found that the wave that originated at the flagellar base was almost planar, but became helical as the undulations progressed towards the tip.

The vast majority of flagella that show movements in three dimensions propagate waves that are elliptical in cross section; it seems likely that

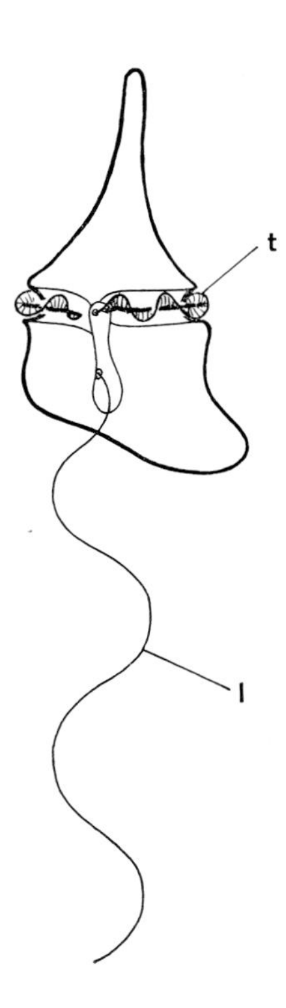

FIG. 1. Diagram to illustrate the two principal patterns of movement of flagella as shown by the two flagella carried by a cell of *Ceratium*. The longitudinal flagellum (l) performs a planar beat, while helical waves travel along the transverse flagellum (t), which lies in an equatorial groove and has a lateral fin. Information from Jahn *et al.* (1963).

these waves are more or less planar at the flagellar base, but become more nearly helical towards the tip, so that a true helix may be seen distally in flagella that are long enough to contain several or many wavelengths. Almost all of the flagella described in this category are sperm tails; in motion the sperm normally rotate about their long axis, so that accurate information on their movement is difficult to obtain. In fact, it is only in cases where the head shows a marked asymmetry, permitting an estimate of the rate of rotation, that accurate information is available. In the case of the bull sperm Rikmenspoel (1966) found that the amplitude of bending in the plane of the flattened head was about three times that perpendicular to the head flattening, that the principal amplitude increased towards the tip, and that the forward velocity increased with the amplitude and was proportional to the rate of rotation; an average bull sperm swimming at a velocity of about 100 $\mu m\ s^{-1}$ and propagating waves at a frequency of about 22 Hz rotated about 7 times a second. Almost all sperm, including many with an apparently planar tail waveform, show some rotation, but the rotational frequency is normally very much lower than the wave frequency. In most sperm it is assumed that the rotation occurs because the tail movement has a greater or lesser helical component, while, according to Thompson (1966), sperm of the mollusc *Archidoris* have a predominantly planar wave but rotate because of a prominent spiral keel running the full 200 μm length of the sperm. In the majority of known cases the orientation of the rotation is anticlockwise when viewed from the flagellar tip, but clockwise motion has been observed (Bishop, 1958), and in some organisms rotation in both directions has been reported (Thompson and Blum, 1967). The simultaneous propagation along insect sperm tails of two series of waves, differing in frequency and amplitude, is described by Phillips in this volume, p. 387. Although helical waves normally travel towards the flagellar tip, the propagation of helical waves from the tip towards the base has been described by Bovee *et al.* (1963). Comparative calculations upon flagellar propulsion of micro-organisms by planar waves, uniform helical waves and non-uniform helical waves have been carried out by Coakley and Holwill (1972).

IV. Planar movement of cilia

It is now clear that while nearly all cilia perform movements where the effective stroke occurs more or less in one plane, in many cases the cilium moves out of this plane during the recovery stroke. Confirmation that a ciliary beat is planar is normally more difficult than confirmation that it is three-dimensional. However, it is clear that the beat of the giant compound cilia that form the comb plates of ctenophores is planar, and that the movement of the compound abfrontal cilia of *Mytilus* gills is at least approximately planar—the latter cilia remain in sharp focus throughout the beat in the very clear ciné film by Baba and Hiramoto (1970) (Fig. 2).

The movement of the compound cilia that circulates water around the segmental gills of the worm *Sabellaria*, first described by Sleigh (1968), is close to a typical ciliary cycle (Fig. 3); although the recovery stroke may take place a little to one side of the effective stroke, the beat is not markedly three-dimensional. The characteristic features are an effective stroke of shorter duration than the recovery stroke, and differences in shape of the cilium in the two phases of beat, such that the extended cilium moves through a large arc in the effective stroke, while it is bent and "unrolls"

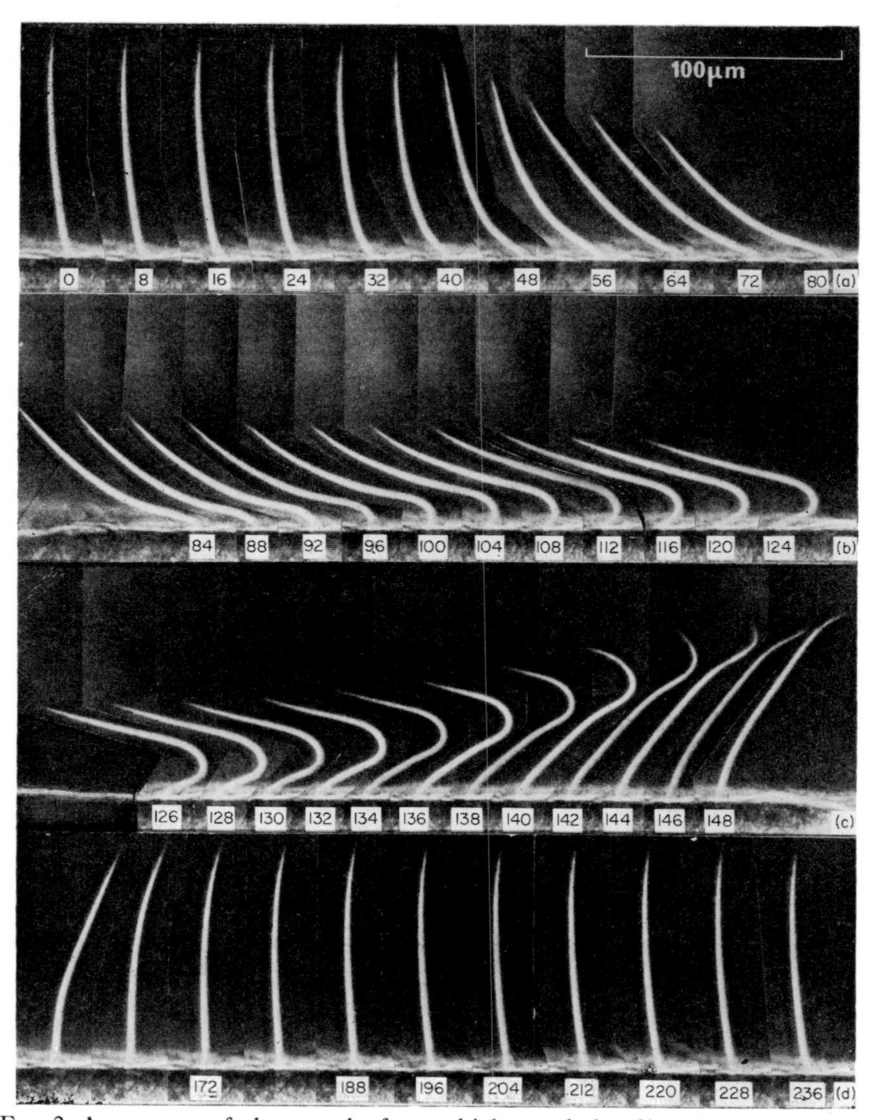

FIG. 2. A sequence of photographs from a high-speed ciné film of the movement of an abfrontal cilium of *Mytilus* during one cycle of beating. The numbers indicate the frames printed from the sequence; the film was taken at 450 frames per second. Reproduced from Baba and Hiramoto (1970), with permission.

through a small volume of water in the recovery stroke. The movement of a ctenophore comb plate is very similar to this, although it is more than an order of magnitude larger (Fig. 4).

In the movement of a *Mytilus* abfrontal cilium (Fig. 2), the extended and bent phases of the movement are still evident, but the unrolling "recovery" stroke occupies a considerably smaller part of the whole cycle than the extended "effective" stroke; in fact, the cilium may move exceedingly

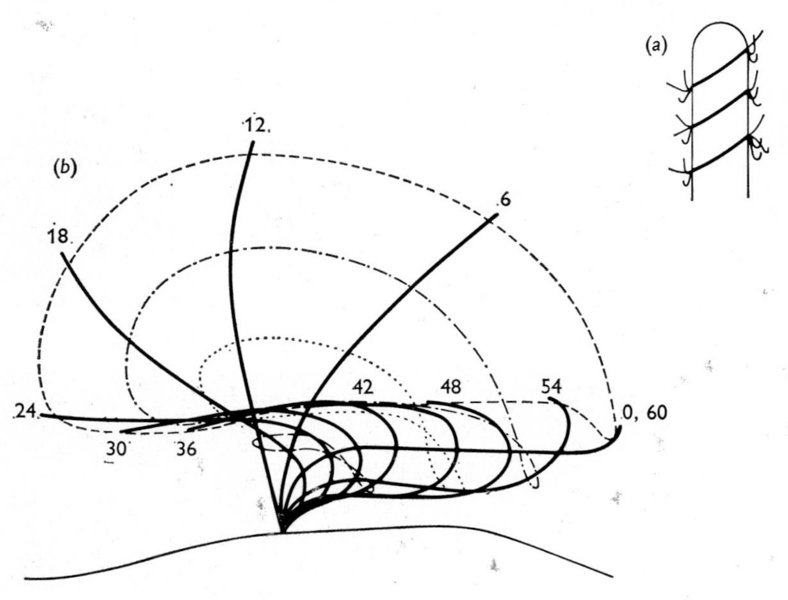

FIG. 3. The movement of a cilium from a dorsal gill of the worm *Sabellaria*. The position of the ciliary rows on a cylindrical gill is indicated in (a); the effective stroke of the beat is towards the tip of the gill. The positions of a cilium at 6 ms intervals throughout the cycle of beating are shown in (b). Water is propelled towards the left by the effective stroke, which is complete within about 24 ms. The basal part of the cilium starts to move back towards the right before the 18 ms profile, so that the recovery movement begins before the effective stroke has ended. From Sleigh and Holwill (1969), with permission.

slowly or even stop during the effective stroke of this unusual pattern of beating, so that the usual terminology for the two phases of the cycle does not seem very appropriate (Sleigh and Holwill, 1969; see also Holwill, p. 161, this volume).

Among the variables that may be used to describe the movement of cilia, the most important are the angular velocity of the effective stroke and the rate of propagation, radius of curvature and length of arc of the bending wave that travels up the cilium in the recovery stroke (Table II).

TABLE II. Some examples of parameters of ciliary beating

	Length (μm)	Frequency (Hz)	Effective stroke angular velocity (degrees ms^{-1})	Recovery stroke bend		Temperature (°C)
				Propagation rate (mm s^{-1})	Radius of curvature (μm)	
Sabellaria gill cilium	32	16·6	7·5	0·75	3·5	20
Mytilus abfrontal cilium	90	~2	0·3	1·5	10	20
Paramecium body cilium	12	30	12	0·35	~2	18
Pleurobrachia comb plate a	600	5	4	9	70	21
Pleurobrachia comb plate b	600	20	10	17	85	21

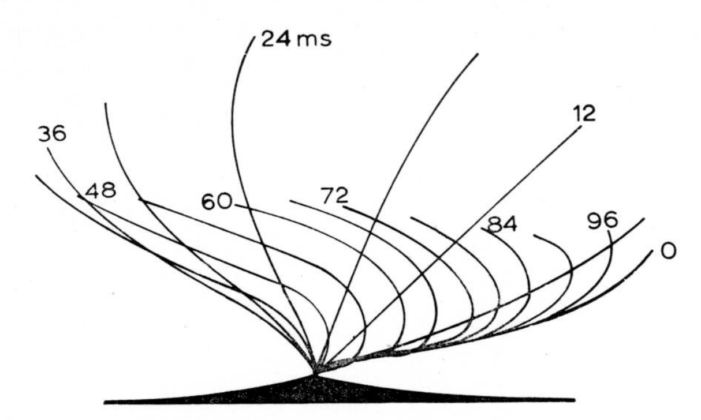

FIG. 4. A sequence of profiles to illustrate the movement of a comb plate of *Pleurobrachia* when beating at about 10 c s^{-1}. The compound cilium moves to the left for about 40 ms in the effective stroke and the recovery stroke occupies about 60 ms. From Sleigh and Jarman (1973), with permission.

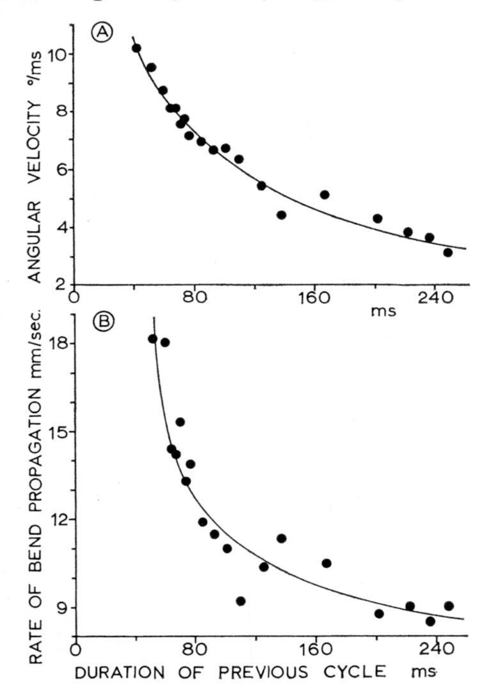

FIG. 5. Changes in parameters of movement of a comb plate of *Pleurobrachia* when the frequency of beat is varied. Mechanical stimuli were used to control the frequency of metachronal excitation of the observed comb plate. Because of slight irregularities of the rate of beating, individual beats were analysed and parameters were related with the duration of the previous beat. Changes in the angular velocity of the effective stroke are shown above (A), and changes in the rate of propagation of the bending wave in the recovery stroke are shown below (B). From Sleigh and Jarman (1973), with permission.

Although these parameters cover a fairly wide range, it is interesting that the figures mostly represent rather similar rates of sliding between the fibrils of the axoneme, and indeed rates that correspond with those reported in comparable systems of sliding fibrils, including muscle (Sleigh, 1973). It is interesting that in ctenophore comb plates, where the rate of beat may be varied under mechanical control at constant temperature, both the angular velocity of bending in the effective stroke and the velocity of bend propagation in the recovery stroke increase as the frequency of beating increases (Fig. 5), so that the rate of sliding of fibrils must be variable—apparently it depends upon the interval since the previous beat (Sleigh and Jarman, 1973).

V. Three-dimensional movement of cilia

A further indication of the diversity of ciliary beating is the fact that some cilia show an almost helical beat, especially under some experimental conditions. Even when beating under supposedly normal conditions some types of cilia bend sharply to the left in the recovery stroke,

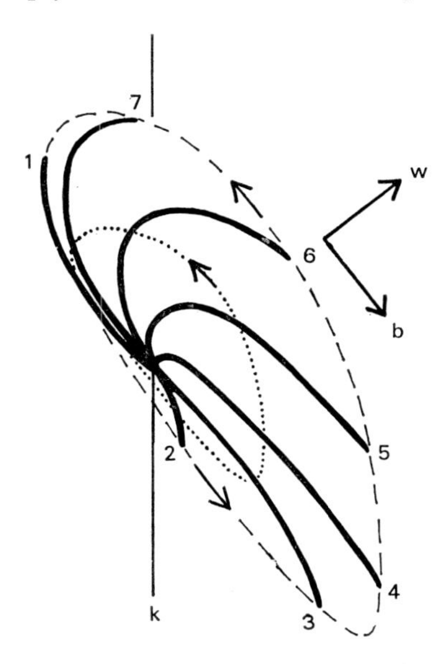

FIG. 6. The cycle of beat of a cilium of *Paramecium* as seen by an observer looking down on the cell surface. In the effective stroke (1–3) the cilium moves in a nearly vertical plane, but it moves to the side in the recovery stroke (4–7) with the cilium curved close to the cell surface, so that the ciliary tip traces out an anticlockwise loop. k indicates the orientation of ciliary kineties, b the direction of the effective stroke and w the direction of propagation of metachronal waves. Drawn from data by Machemer (1972).

the tip performing an anticlockwise movement as seen from above, while cilia in other species bend to the right in the recovery stroke, so that an observer looking down on the ciliated surface sees the ciliary tip trace out a clockwise path. The occurrence of three-dimensional beating of cilia has been suspected for some years on the evidence of rapid-fixed metachronal waves of such organisms as *Paramecium*, particularly in studies by Párducz (reviewed in English, Párducz, 1967), and some ciné films also indicated that the beating of some cilia was not planar. However, it is difficult to establish the exact form of beat, especially where many cilia lie close together.

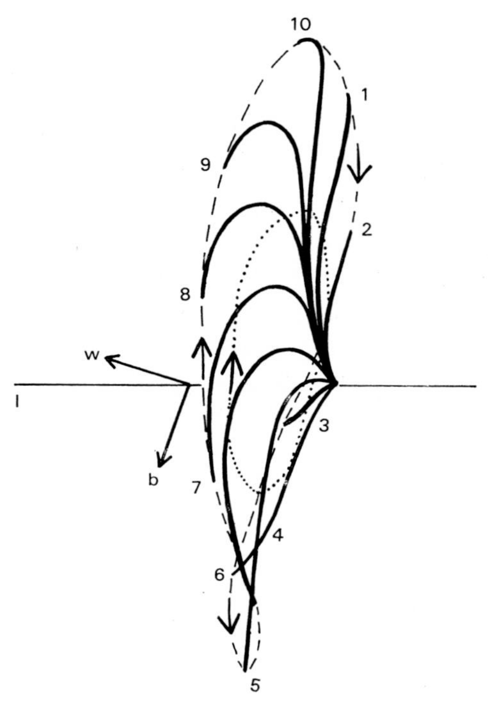

FIG. 7. The cycle of a beat of a cilium from the lateral epithelium of a gill filament of *Mytilus*, as seen by an observer looking down on the lateral epithelium. The effective stroke (1–4) takes place in a near-vertical plane, and in the recovery stroke (5–10) the cilium bends to one side so that the ciliary tip traces out a clockwise loop. l indicates the edge of the line of lateral cells, b the direction of the effective stroke, and w the direction of propagation of metachronal waves. Drawn from observations of Aiello and Sleigh (1972).

Proof that the cilia of *Paramecium* move in a manner similar to that described by Párducz has recently been obtained by Machemer (1973), by careful study of flash photographs of living ciliates taken with Nomarski interference contrast optics, and comparison of these photographs with models of metachronal waves. *Paramecium* cilia are about 12 μm long and perform an effective stroke in which the cilium moves almost in one plane through an angle of about 150°. At the end of this stroke the cilium swings to one side, so that an observer looking in the direction of the effective stroke would see the cilium move to his left, with the main part of the cilium being drawn around the ciliary base by the unrolling of the cilium, and tracing out a path close to and almost parallel with the cell surface (Fig. 6, see also Machemer, p. 206, this volume). Apparently, the effective stroke takes place principally in the vertical plane and the recovery stroke principally in the horizontal plane, bending movements in the basal region providing transitions between the two planar phases. The movement therefore can scarcely be regarded as helical, although in a viscous medium the beat is modified to a more nearly helical form (Machemer, 1972).

Recently it has been shown by Aiello and Sleigh (1972) that the movement of the lateral cilia of *Mytilus* gill resembles a mirror image of that described for *Paramecium*. In the effective stroke the extended cilium swings down across the face of the gill, and then moves to the right of the direction of the effective stroke, to unroll close to the gill surface on its way back to the starting position (Fig. 7).

Other systems where the cilia are known to move to the left are in *Opalina* (see Machemer, p. 221, this volume) and branchial cilia of the ascidian *Corella* (Sleigh, unpublished), while in the velar cilia of the nudibranch veliger larva the cilia in their recovery stroke move to the right of the effective stroke (Sleigh, unpublished). It seems likely that in at least the majority of cases where simple cilia show diaplectic metachronism, dexioplectic metachronal waves are associated with an anticlockwise motion of the ciliary tip, while laeoplectic metachronism is associated with a clockwise movement of the tip of the cilium. Compound cilia may show a three-dimensional beat in a number of cases, e.g. in *Stentor*, where the fanned-out tips of the plate-like membranelles are twisted to present a maximum surface to the water in the effective stroke (Sleigh, 1972b), providing an excellent example of the conclusion that the sliding fibril mechanism of cilia and flagella is capable of functional adaptation allowing a wide diversity of patterns of movement.

References

AIELLO, E. and SLEIGH, M. A. (1972). *J. Cell Biol.* **54**, 493–506.
BABA, S. A. and HIRAMOTO, Y. (1970). *J. exp. Biol.* **52**, 675–690.
BISHOP, D. W. (1958). *Nature, Lond.* **182**, 1638–1640.
BISHOP, D. W. (1962). *Physiol. Rev.* **42**, 1–59.
BOVEE, E. C., JAHN, T. L., FONSECA, J. and LANDMAN, M. (1963). *Absts. 7th Ann. Meet. Biophys. Soc. New York*, 1963. MD2.

Brokaw, C. J. (1963). *J. exp. Biol.* **40**, 149–156.
Brokaw, C. J. (1965). *J. exp. Biol.* **43**, 155–169.
Brokaw, C. J. and Wright, L. (1963). *Science, N.Y.* **142**, 1169–1170.
Coakley, C. J. and Holwill, M. E. J. (1972). *J. theor. Biol.* **35**, 525–542.
Gray, J. (1955). *J. exp. Biol.* **32**, 775–801.
Holwill, M. E. J. (1965). *J. exp. Biol.* **42**, 125–137.
Holwill, M. E. J. (1966a). *Physiol. Rev.* **46**, 696–785.
Holwill, M. E. J. (1966b). *J. exp. Biol.* **44**, 579–588.
Holwill, M. E. J. and Miles, C. A. (1971). *J. theor. Biol.* **31**, 25–42.
Jahn, T. L. and Bovee, E. C. (1967). *In* "Research in Protozoology" (T.-T. Chen, ed.), Vol. 1, pp. 41–200. Pergamon Press, Oxford.
Jahn, T. L., Harmon, W. M. and Landman, M. (1963). *J. Protozool.* **10**, 358–363.
Machemer, H. (1972). *J. exp. Biol.* **57**, 239–259.
Machemer, H. (1973). *Acta Protozool.* **11**, 295–300.
Párducz, B. (1967). *Int. Rev. Cytol.* **21**, 91–128.
Phillips, D. M. (1972). *J. Cell Biol.* **53**, 561–573.
Rikmenspoel, R. (1966). *In* "Dynamics of Fluids and Plasmas" (S. Pai, ed.), pp. 9–34. Academic Press, New York.
Seravin, L. N. (1971). *Adv. comp. Physiol. Biochem.* **4**, 37–111.
Silvester, N. R. and Holwill, M. E. J. (1972). *J. theor. Biol.* **35**, 505–523.
Sleigh, M. A. (1964). *Quart. J. microsc. Sci.* **105**, 405–414.
Sleigh, M. A. (1968). *Symp. Soc. exp. Biol.* **22**, 131–150.
Sleigh, M. A. (1972a). *In* "Essays in Hydrobiology" (R. B. Clark and R. Wootton, eds), pp. 119–136. Exeter University.
Sleigh, M. A. (1972b). *J. Protozool.* **19** Suppl. 52.
Sleigh, M. A. (1973). *In* "Cell Biology in Medicine" (E. E. Bittar, ed.), pp. 525–569. Wiley, New York.
Sleigh, M. A. and Holwill, M. E. J. (1969). *J. exp. Biol.* **50**, 733–743.
Sleigh, M. A. and Jarman, M. (1973). *J. Mechanochem. Cell Motility* **2**, 61–68.
Thompson, T. E. (1966). *Phil. Trans. R. Soc. Lond.* B **250**, 343–375.
Thompson, T. E. and Blum, M. (1967). *Ann. ent. Soc. Am.* **60**, 632–642.

Chapter 5

Movement of the flagellum of some marine invertebrate spermatozoa*

C. J. BROKAW

Division of Biology, California Institute of Technology, Pasadena, California, U.S.A.

I. Introduction

Many marine invertebrates reproduce by liberating large numbers of eggs and spermatozoa into the surrounding sea. The spermatozoa from these animals typically have a simple "9+2" flagellum about 40 μm long, a small head, and are vigorously motile even in dilute suspensions in sea water. These qualities make them one of the most favorable subjects for the experimental study of flagellar motility. The motility of these spermatozoa has been reviewed recently in two more general reviews of sperm motility, by Bishop (1962) and by Nelson (1967).

* Preparation of this manuscript has been assisted by support from the National Science Foundation (GB 32035) and the United States Public Health Service (GM 18711).

Sir James Gray initiated the modern study of the movement of invertebrate sperm flagella with his paper on the movement of sea urchin spermatozoa (Gray, 1955). Using dark-field microscopy he made observations with both continuous and stroboscopic illumination, and obtained photographs of the sea urchin sperm flagellum with exposures of approximately 1/500 s (Fig. 1). His work provided a wealth of new information about the movements of these sperm flagella, including the following conclusions:

1. Bending waves pass along the sperm flagellum with sustained amplitude, indicating that active elements are present throughout the flagellum.

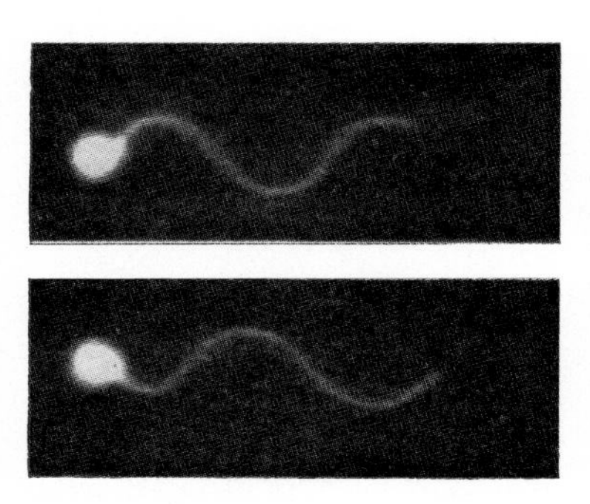

FIG. 1. Photograph showing the flagellum of a motile spermatozoon from the sea urchin, *Psammechinus miliaris*. Dark-field photomicrographs taken with exposures of approximately 1/500 s. (Reprinted from Gray, 1955, with permission.)

The observed waveforms cannot be duplicated by a passive elastic filament which is oscillated at one end in a viscous medium. This conclusion has been further substantiated by a theoretical treatment of bend propagation on a passive filament by Machin (1958) and by observations of nearly normal swimming movements by isolated sperm flagella (Brokaw, 1965, 1970).

2. Bending of the flagellum is predominantly in one plane. Subsequent observation has shown this to be typical of marine invertebrate spermatozoa, although spermatozoa from *Ciona* can be induced to generate helical bending patterns under abnormal conditions (Brokaw, 1966).

3. Unless constrained by a surface, spermatozoa rotate slowly (about 1 rotation per 10 beats of the flagellum) about a longitudinal axis. The forces responsible for rotation are still unknown.

4. The flagellar beat pattern is usually asymmetrical, so that as it swims, the spermatozoon also rotates around an axis perpendicular to the beat plane of the flagellum. The combination of this rotation and the rotation about a longitudinal axis causes the spermatozoon to swim in a helical path with a straight axis, when unconstrained, and also to become trapped at any impenetrable surface. The tendency of these spermatozoa to swim in circles next to the surface of a microscope slide greatly facilitates observation and photography.

5. Accurate measurements of beat frequency (f = 34·5 beats per s at 18°C), wave amplitude (b = 4 μm) and wavelength (λ = 24 μm) were obtained for *Psammechinus miliaris* spermatozoa. These measurements were later used to calculate a swimming velocity, based on some simple hydrodynamic assumptions, which agreed well with observed values (Gray and Hancock, 1955).

6. A thin terminal piece is observed at the end of the flagellum.

7. Spermatozoa sometimes slow down gradually without any noticeable change in wavelength or amplitude.

Gray discussed his observations in terms of a model containing contractile elements distributed along the sides of the flagellum, and suggested that elastic properties of the flagellum might be significant in controlling the propagation of bending waves.

II. Advances in methodology for observation and measurement of sperm motility

Only a small number of species of marine invertebrates has been used for quantitative studies of sperm motility. Sea urchin spermatozoa have been used more extensively than any others, but the few comparative studies which have been made suggest that further study of spermatozoa from other groups would be valuable. Such studies require a dependable source of spermatozoa in mature condition and a method for liberating the spermatozoa. The book by Costello *et al.* (1957) is a basic reference for some of this information. Once spermatozoa have been obtained, activation of their motility is sometimes a problem, and pure sea water may not always be the optimum environment for normal motility (cf. Tyler, 1953). Much more basic work is needed to develop methods for handling spermatozoa from other species of marine invertebrates. Appendix I attempts to provide a bibliographic guide to some of the research studies which describe techniques for various species.

The use of electronic flash tubes as illuminators for stroboscopic observation and photomicrography has led to a significant improvement in resolution, because adequate exposure for photomicrography can be obtained with flash durations at least an order of magnitude less than the 2 ms exposures used by Gray (1955). Specially constructed illuminators were used at first (Brown and Popple, 1955; Gray, 1958; Brokaw, 1963, 1965; Goldstein, 1969) but more recently, commercial units of superior capabilities have become available from the Chadwick-Helmuth Company, Monrovia, California, U.S.A. These illuminators make it relatively easy to obtain high-quality photomicrographs showing the configuration of a moving sperm flagellum. With cinematography or multiple-flash exposures, information about rates of bending of the flagellum can also be obtained (cf. Brokaw, 1970, and Fig. 2). However, photographs can only be obtained of a few selected spermatozoa. These are usually spermatozoa which are swimming at an interface, such as the surface of a microscope slide, and do not represent a randomly selected sample of the entire sperm suspension. Unless the movement of all the spermatozoa in the suspension is very similar, photography and stroboscopic measurement of the beat frequencies of individual spermatozoa are not certain to be reliable measures of the average motility in the suspension. Considerable caution is thus

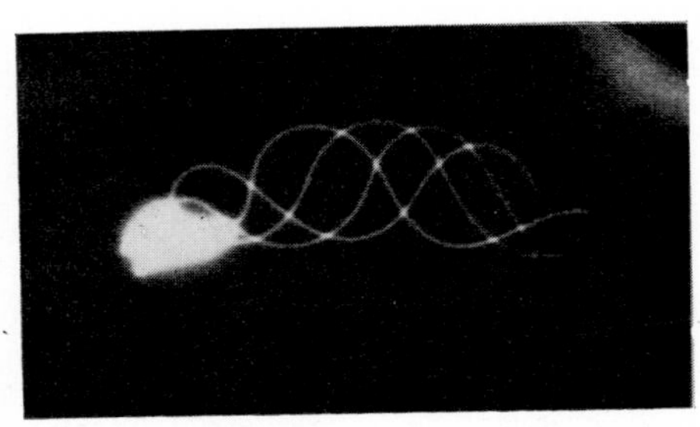

FIG. 2. Photograph showing the flagellum of a motile spermatozoon from the sea urchin *Lytechinus pictus*. Multiple exposure dark-field photomicrograph taken with flashes at intervals of 1/100 s.

required in interpreting results from experiments where photography or stroboscopic measurement of beat frequencies are used to obtain information about the effects of experimental variables on parameters of movement.

Various methods have been proposed for measuring the average motility of spermatozoa in a suspension, and some have been used with marine invertebrate spermatozoa. The "dark field track method" (Rothschild and Swann, 1949) using dark-field illumination and exposure durations of 0·5–1·0 s to produce an image of the path of the sperm head is a convenient method for providing photographic information about swimming velocities of individual spermatozoa, as it allows a relatively large sample of spermatozoa to be measured. Gibbons and Gibbons (1972) have recently used this technique to compare the motility of live and reactivated sea urchin spermatozoa, with respect to swimming velocity and the degree of asymmetry of the path of the sperm head.

Branham (1966) and Nelson (1972) measured sea urchin sperm motility by comparing the centrifugal sedimentation rates of live and immobilized spermatozoa. This method depends on the orientation of spermatozoa in a centrifugal field, so that the swimming movements of live spermatozoa carry them more rapidly to the bottom of the centrifuge tube. Optical density measurements were used to detect the altered distribution of spermatozoa in the centrifuge tube, and the method therefore gives an average motility for the entire suspension.

Timourian and Watchmaker (1970) oriented sea urchin spermatozoa in a flow-through spectrophotometer cuvette and found that the sperm suspension had minimum optical density when the spermatozoa were oriented by flow parallel to the light path. After stoppage of the flow, the swimming movements of the spermatozoa randomized their orientation, causing the optical density of the suspension to increase. The optical density change observed during the first 30 s after flow stoppage was used as a measure of the average motility of the spermatozoa in the suspension.

These methods are simple enough to offer considerable potential for purposes such as screening of chemical inhibitors of sperm motility, and for other applications where an average measure of sperm motility is more important than detailed information about parameters of the flagellar bending waves.

III. Observations on normal motility

Spermatozoa are normally non-motile and show little or no metabolic activity during storage within the male reproductive tract, and become active when diluted into an appropriate medium. The mechanism by which motility is suppressed during sperm storage has been studied in sea urchin and starfish spermatozoa (cf. Mohri and Yasumasu, 1963). Factors which may contribute to this effect include low oxygen partial pressure and elevated carbon dioxide partial pressure, as a result of metabolic activity in very dense aggregates of spermatozoa; physical obstruction of movement; and effects of ions such as potassium or zinc. In some cases, activation of optimum motility appears to require the addition of chelating agents or other substances to sea water (Tyler, 1953; Brokaw, 1965; Miller and Brokaw, 1970). Consequently, it is not always a straightforward task to describe the "normal" motility of spermatozoa, and to distinguish between "activation of normal motility" by a substance such as ethylene diamine tetraacetate and "enhancement of motility" by a substance such as dimethyl sulfoxide.

A representative example of the movement of a marine invertebrate spermatozoon is shown by the photomicrograph of a spermatozoon from the sea urchin *Lytechinus pictus* in Fig. 2. Other photomicrographs of the "normal" movements of the flagellum of sea urchin spermatozoa have been published by Brokaw (1965) (*Lytechinus pictus*); Brokaw *et al.* (1970)

(*Strongylocentrotus purpuratus*); and Gibbons and Gibbons (1972) (*Colobocentrotus atratus*). Spermatozoa from an annelid, *Chaetopterus variopedatus*, and a tunicate, *Ciona intestinalis* were shown by Brokaw (1965). Photographs showing the movements of spermatozoa from two species of coelenterates, the hydroids *Tubularia crocea* and *Campanularia calceolifera*, were shown by Miller and Brokaw (1970) and Brokaw *et al.* (1970). The normal flagellar movements of all these species are remarkably similar in pattern.

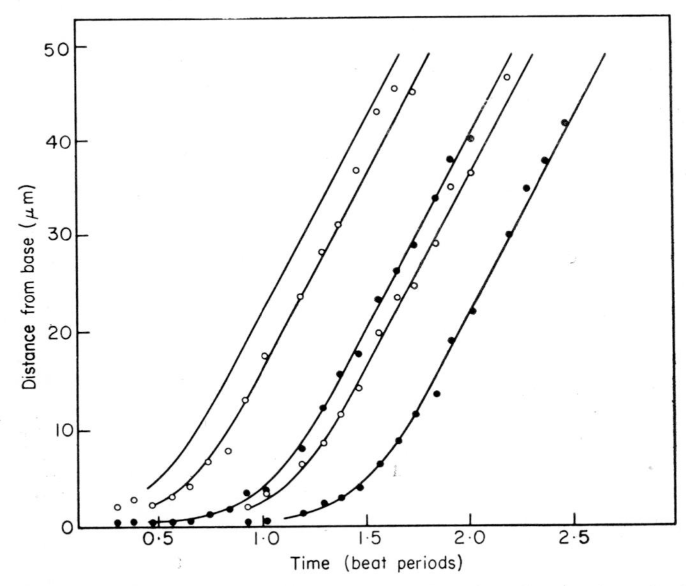

FIG. 3. Progression of bending points (○) and unbending points (●) along a flagellum, as determined from multiple-flash photomicrographs of a *Lytechinus* sperm flagellum. The points, derived from several bending cycles, have been repositioned to show the passage of one pair of bends along the flagellum. The curve on the left is a one-cycle repeat of the line through the points at the right of the figure, to show the extent of the straight region preceding the first bent region. (Reprinted from Brokaw, 1970, with permission.)

Significant variations among these species are found with respect to the normal beat frequency, which can be measured from photographs or determined directly by stroboscopic measurements. At 16°C, *Ciona* spermatozoa have a mean beat frequency of 35 beats per second (Brokaw, 1965) while *Chaetopterus* spermatozoa have a mean beat frequency of 26 beats per second. *Lytechinus* and *Tubularia* spermatozoa fall within this range. The difference in beat frequency between *Ciona* and *Chaetopterus* spermatozoa, and a small difference in size, indicate that *Ciona* spermatozoa are normally performing work against the viscous resistances of the medium at a rate which is two to three times greater than for *Chaetopterus* spermatozoa (Brokaw, 1965). It has not yet been determined whether these differences between species reflect differences in the ATP concentrations maintained within the flagellum, in the spatial density of active sites, or in the kinetic parameters of the active sites.

The flagellar bending waves on these spermatozoa do not have the form of an ideal wave, which can be fully described by two spatial parameters, such as wavelength and amplitude, in addition to the beat frequency, and the bending waves are not adequately approximated by a sinusoidal wave, which is defined with respect to an imaginary x-axis. A more

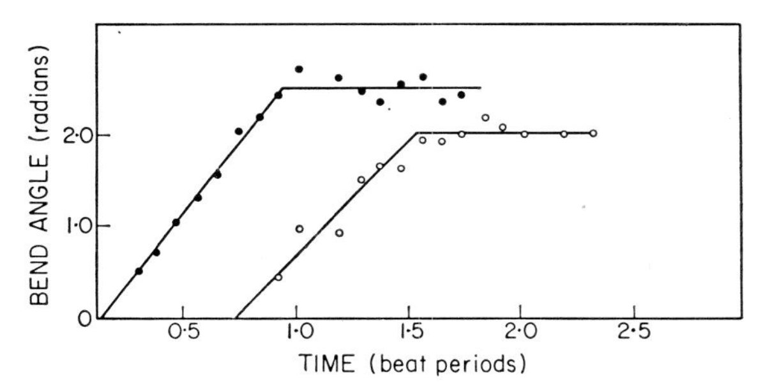

FIG. 4. Measured values of the total angle of bend in the bent regions corresponding to those in Fig. 3. The angles for the two bends should actually have opposite signs, but have been plotted as shown, for compactness. (Reprinted from Brokaw, 1970, with permission.)

adequate description of the wave should be made with respect to a coordinate which measures distance along the flagellum itself. A function which gives the curvature of the flagellum as a function of distance along the length of the flagellum will completely describe the shape of the wave. However, there is no accurate procedure yet available for making direct measurements of the curvature of the image of a flagellum on a photograph. An extensive discussion of this problem has been presented by Silvester and Holwill (1972).

Flagellar bending waves can be closely approximated by a "sine-generated" wave, in which the curvature is a sinusoidal function of distance along the flagellum, but a fairly elaborate procedure is required to derive a sine-generated waveform which accurately fits a particular image. An accurate, and considerably more convenient curve-fitting procedure has been suggested by Brokaw and Wright (1963). They approximated the flagellar waveform by a series of circular arcs and straight lines. The most complete analysis of flagellar bending waves has been made on two photographs of headless sea urchin sperm flagella by Brokaw (1970). Some of the results of this analysis are shown in Figs 3 and 4. A bend develops near the base of the flagellum with a nearly linear rate of increase of bend angle until the normal bend angle is reached; this remains fairly constant as the bend propagates along the flagellum. The velocity of propagation of the ends of the bent region, or the inflection point between bends, accelerates slowly to a relatively constant value. As a result, the bends are shorter and have a smaller radius of curvature while they are being initiated at the base of the flagellum. These results obtained with headless sperm flagella appear to be representative of the bending behavior of the flagella on normal spermatozoa, shown by Brokaw *et al.* (1970) and by Brokaw (1965). Analysis by this method provides sufficient information for calculation of bending moments and energy expenditure, even though it cannot tell us the precise form of the curvature as a function of length along the flagellum.

IV. Effects of the physical environment on movement

Observations of the effects of changes in viscosity of the medium on flagellar movement are the simplest means for controlled variation of the mechanical impedance encountered by the mechanochemical system in flagella, and are therefore a particularly important source of information about this system. An extensive series of experiments on the effects of viscosity on the movement of spermatozoa from three species of marine invertebrates (Brokaw, 1966) has provided data for comparison with predictions of a number of theoretical models for flagellar movement.

These viscosity experiments were carried out with solutions containing a high molecular weight methyl cellulose. With this material, high viscosities can be obtained at very low concentrations (around 1% by weight), so that osmotic or chemical effects of the methyl cellulose are negligible. However, these methyl cellulose solutions show clearly non-Newtonian viscosity behavior, which complicates the interpretation of the results. It would be well to have additional experimental observations using other molecules for increasing the viscosity of the solutions.

Two distinct patterns of response to increased viscosity were found in the spermatozoa which were examined. One pattern, found with spermatozoa of *Lytechinus* and *Ciona*, was characterized by the fact that the peak curvature in the bent regions propagated along the flagella was relatively unchanged at increased viscosities. As a consequence, the amplitude of the bending waves decreased as the wavelength decreased with increasing viscosity. Effects of viscosity on beat frequency, wavelength and bend propagation velocity are illustrated in Fig. 5. The propagation velocity, equal to the product of wavelength times frequency, was nearly inversely proportional to the square root of the viscosity. The results shown in Fig. 5 appear to be most simply explained by a theory which suggests that the wavelength of the movement is determined by an internal viscous resistance to bending of the flagellum (Brokaw, 1972b), and may be considered to provide strong experimental support for that theory.

Calculations based on these data suggest that the total energy expenditure against the external viscous resistance decreases with increasing viscosity (Brokaw, 1966). The data for *Ciona* spermatozoa are good enough

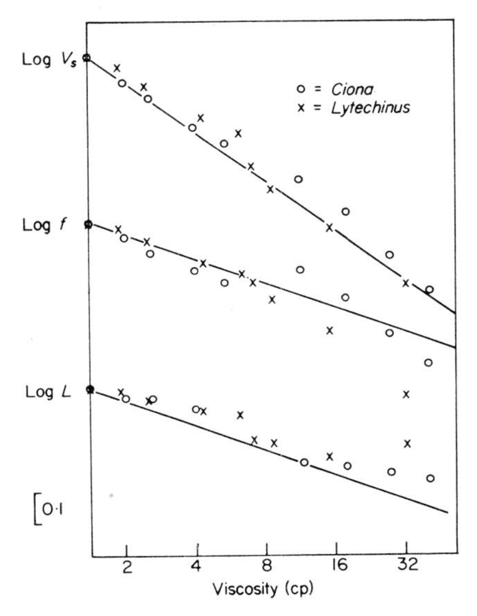

FIG. 5. Experimental data from Brokaw (1966) which show the decrease in bend propagation velocity, V_s, beat frequency, f, and wavelength measured along the flagellum, L, as the external viscosity of the medium is increased. The straight lines with slopes of $-0{\cdot}50$ (for V_s) and $-0{\cdot}25$ (for f and L) indicate the results predicted by a theoretical analysis of small-amplitude wave propagation along a flagellum of infinite length, when the wavelength is regulated by the internal viscous bending resistance (Brokaw, 1972b).

to suggest that this decrease is approximately proportional to beat frequency, so that the energy expenditure per beat remains approximately constant (Brokaw and Benedict, 1968b). In a series of similar experiments, the metabolism of *Ciona* spermatozoa was measured as a function of viscosity, by using pH-stats to monitor the acid production associated with metabolism (Brokaw and Benedict, 1968b). These experiments demonstrated that the metabolism decreased at high viscosities, indicating that 80% or more of the sperm metabolism was tightly coupled to motility, and that the amount of movement-coupled metabolism per beat was approximately constant at about $0{\cdot}27 \times 10^{-19}$ mol of O_2 per beat per spermatozoon.

Spermatozoa of *Chaetopterus* show a different pattern of response to increased viscosity, characterized by an increase in curvature in the bent regions, so that the wave amplitude remains nearly constant as wavelength and frequency decline. There is a greater than 10-fold increase in the calculated energy expenditure per beat against external viscous resistances (Brokaw, 1966). Preliminary results from attempts to measure the metabolism of *Chaetopterus* sperm as a function of viscosity indicate that it does not decrease in proportion to beat frequency, as in *Ciona* spermatozoa, but that the increase in metabolism per beat is much less than the increase in viscous work per beat (Brokaw and Benedict, unpublished measurements).

The behavior of *Chaetopterus* spermatozoa at high viscosities is not unique to this species, as a similar response to increased viscosity is shown by both *Lytechinus* spermatozoa (Brokaw, 1966) and *Strongylocentrotus purpuratus* spermatozoa (Brokaw, unpublished) when the beat frequency is decreased by thiourea (see below). This type of movement is an important observation, as it reveals that the control mechanism for flagellar bending can generate bends over a substantial range of curvatures. No satisfactory explanation for the control of the amplitude of bending, i.e. the curvature, has yet appeared.

The mechanical environment of the flagellum can be modified also by changing the boundary conditions at the ends of the flagellum. The presence of a head introduces additional viscous resistance to linear and rotational movement of the flagellum, but calculations of this effect suggest that it will be small (Gray and Hancock, 1955; Brokaw, 1965) and there are no experimental measurements of its effect. A stronger effect is produced by attachment of the sperm head to the microscope slide or coverglass, as frequently occurs. In some cases, linear movement of the flagellum is restricted by attachment, and in other cases both linear and rotational movements are restricted. Attached spermatozoa typically beat with a slightly lower frequency (10–20%) and with a "compressed" wave pattern showing a reduced radius of curvature of bends near the base, and a continually increasing velocity of bend propagation (V_s) (Brokaw, 1965 and Fig. 6). This effect is an additional source of information about the control of curvature, but it has not yet received detailed analysis.

Spermatozoa are also often observed with both the head and the tip of the flagellum attached to the slide. In these cases, a large variety of unusual bending patterns can be observed, depending on the positions of the attachment points. These effects also remain to be analyzed in detail.

Holwill (1969) made detailed measurements of the beat frequency of spermatozoa from the sea urchins *Lytechinus pictus* and *Strongylocentrotus purpuratus* as a function of temperature. In the range from 5 to 21°C, the results showed an approximately linear relationship between log f/T and $1/T$, as expected for a simple kinetic model in which the beat frequency

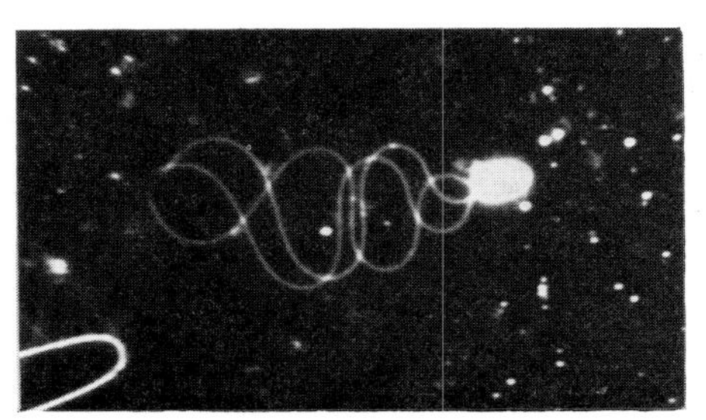

FIG. 6. Multiple exposure dark-field photomicrograph of a spermatozoon from the sea urchin *Centrostephanus coronatus* which has its head attached to the microscope slide. The flash rate was 36 Hz.

was equivalent to a chemical reaction rate. Values for activation enthalpy and activation entropy were calculated which were in agreement with measurements on other systems.

Temperature also influences other parameters of the movement of these spermatozoa, but these effects have not been systematically studied.

MacGinitie (1938) noticed an activation of the motility of *Urechis* spermatozoa by light, and I have seen indications of similar effects with other species, as well as inhibition of the motility of spermatozoa when exposed to high levels of illumination under the microscope. Effects of light on the respiratory enzymes of sea urchin spermatozoa were observed by Spikes (1949). It seems most likely that effects of light on the motility of these spermatozoa are mediated by photosensitive molecules within the mitochondrial metabolic apparatus, and do not indicate a direct effect of light on the mechanochemical system within flagella.

V. Effects of the chemical environment on movement

In contrast to freshwater animals, and a great many animals in all environments which practise internal fertilization, fertilization in the marine invertebrates whose spermatozoa are discussed in this chapter occurs in sea water, and optimum motility would be expected to be found in sea water. However, as already noted, full activation of normal motility, and maintenance of motility over extended periods is often found only after addition of various substances—chelating agents, egg extracts, and miscellaneous organic compounds—to sea water. On the other hand, marine spermatozoa appear to be relatively tolerant of variations in their ionic environment. Simple artificial sea water solutions containing Na, K, Mg, Ca, Cl and SO_4 ions appear to support motility as well as or better than natural sea waters.

Spermatozoa from the mussel *Mytilus* are motile in isotonic NaCl (Brokaw, unpublished) and in isotonic KCl, even after repeated washing (Nelson, 1955). Spermatozoa from *Ciona* show nearly normal motility in 0·55 M NaCl containing 1 mM thioglycollate, but may be dependent on traces of divalent cations carried over with the sperm sample, as they do

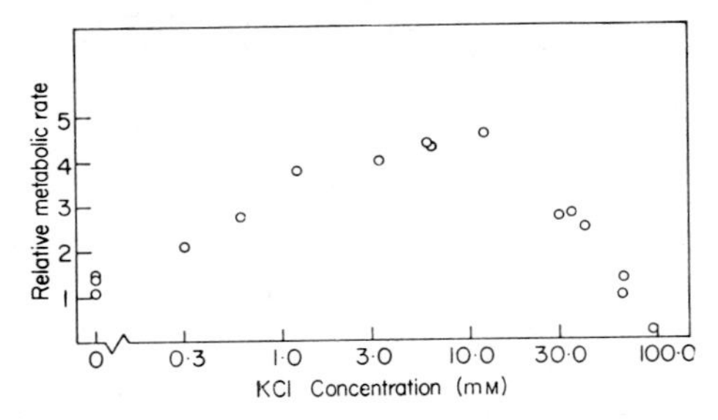

FIG. 7. Metabolic activity of suspensions of *Chaetopterus* spermatozoa as a function of KCl concentration in an artificial sea water solution. Metabolic activity was measured by the pH-stat method developed for *Ciona* spermatozoa (Brokaw and Benedict, 1968b). Observations of motility gave results which roughly parallel the measurements of metabolic activity.

not swim in 0·55 M NaCl containing 1 mM EDTA (Brokaw, unpublished). However, the motility of spermatozoa of some species is sensitive to the potassium ion concentration. Sea urchin spermatozoa swim more slowly in solutions containing greater than the normal (9–10 mM) concentration of potassium ions (Nelson, 1969, 1972; Baker and Presley, 1969); a 50% reduction in swimming speed was measured when the potassium concentration was elevated to 20–30 mM. To some extent this reduction in swimming velocity appears to be the result of an increase in the asymmetry of the beat pattern of the sperm flagellum (Brokaw, 1966). *Chaetopterus* spermatozoa show optimum motility and maximum metabolic activity at a KCl concentration close to that normally found in sea water (Fig. 7). Increases in KCl concentration do not inhibit the normal swimming movements of *Tubularia* spermatozoa, but interfere with their ability to turn in response to chemotactic stimuli (Brokaw and Benedict, unpublished observations).

These scattered observations indicate that there are significant differences between the effects of the ionic environment on motility of spermatozoa from different species of marine invertebrates, and suggest that more extensive comparative studies might be valuable. However, in these studies, as in a great many similar studies of the effects of the chemical environment on motility, there is no evidence that the ionic environment is directly influencing the mechanochemical system within the flagella. Many of the effects of the chemical environment may be mediated by effects on the integrity of the metabolic processes which supply ATP to the sperm flagellum, and may not therefore be directly informative about the mechanisms of flagellar movement. Rothschild (1951) has provided an extensive listing of substances which are reported to influence the metabolic activity and/or the motility of sea urchin spermatozoa, and additional information of this type has been gathered by Steinbach (1966) and by Nelson (1972).

More useful information about chemical effects on the flagellum itself can be obtained from studies with reactivated spermatozoa (see Chapter 6) and from studies in which several parameters of movement are measured, rather than just a single measure of motility. The latter studies permit analysis of the interrelationships between movement parameters, and may provide evidence for more direct effects of a chemical agent on flagella which are not mediated through alterations in the ATP production. However, very little work of this nature has been reported. The most complete study of this type involves the effects of thiourea.

Thiourea is known to be an inhibitor of the muscle actomyosin system (Ruegg *et al.*, 1963), an inhibitor of the movement of glycerinated sea urchin spermatozoa (Brokaw and Benedict, 1968a) and both an uncoupling agent and an inhibitor of the ATPase activity of sea urchin sperm flagella (Brokaw and Benedict, 1971). In the same concentration range, it inhibits the motility of live spermatozoa (Brokaw, 1965, 1966), but this inhibition may reflect effects on the metabolic apparatus of the spermatozoa as well as the mechanochemical system of the flagella. The effects of thiourea on sea urchin and *Ciona* spermatozoa are observed within 1–2 min, so that it appears to penetrate the sperm membrane readily. However, it does not have similar effects on *Chaetopterus* spermatozoa, which may be much less permeable to thiourea.

Most extensive measurements have been made with *Ciona* spermatozoa. At concentrations in the range of 0–0·18 M thiourea, there is a nearly linear decrease in beat frequency from 35 to 12 beats per second with increasing thiourea concentration. The wavelength and radius of curvature of the bending waves remain almost constant, so that the calculated energy expenditure per beat against external viscous resistances should decrease in proportion to the beat frequency. Measurement of sperm metabolism as a function of thiourea concentration indicated that the metabolic rate decreased in proportion to the beat frequency, so that the energy available per beat should be constant (Brokaw and Benedict, 1968b). These spermatozoa appear to maintain a constant stoichiometry of ATP usage per beat, rather than a constant energy efficiency, as they are inhibited by thiourea. However, these results may require re-evaluation in the light of evidence that thiourea acts as an uncoupler of flagellar ATPase activity.

Two other effects have been observed with thiourea-inhibited spermatozoa from *Ciona*. At concentrations of 0·12–0·2 M thiourea, some of the spermatozoa are observed swimming with longer wavelengths at about half the usual beat frequency, and others with shorter wavelengths at about twice the usual beat frequency. These appear to be discrete classes of movement, and abrupt transitions between classes are observed with individual spermatozoa. These phenomena may be related to the wavelength-mode preferences predicted from computer simulation studies of a sliding filament model for flagellar movement (Brokaw, 1972c), but further study is needed.

In the presence of 0·1–0·2 M thiourea, and methyl cellulose concentrations which raise the viscosity to 8–10 centipoise or greater, *Ciona* spermatozoa change their movements to a helical pattern (Brokaw, 1966). This observation is a particularly clear demonstration of the ability of a flagellum to generate either planar or helical bending waves with small modifications of the environmental conditions.

Sea urchin spermatozoa (*Lytechinus pictus* and *Strongylocentrotus purpuratus*) show a similar decrease in beat frequency without much change in wave shape or size, as the thiourea concentration is increased. They do not show the multiple mode phenomena observed with *Ciona* spermatozoa, and at increased viscosities in the presence of thiourea, they do not convert to helical movements, but instead show planar movements similar to those observed with *Chaetopterus* spermatozoa at high viscosities. The effects of thiourea on sea urchin spermatozoa have not been studied quantitatively, because the movements are not very stable with time.

VI. Interpretation

In many cases, an analysis of the hydrodynamic interactions between a spermatozoon and its environment is required before observations and measurements of parameters of sperm motility can be interpreted in terms of internal mechanisms. These hydrodynamic interactions are the subject of Chapter 8 in this volume (p. 143). A computer program for numerical integration of the bending moments resulting from interaction between a moving spermatozoon and its viscous environment has recently been incorporated into a program that will simulate the movement of a flagellum (Brokaw, 1972a). This computer simulation method is a powerful technique for examining the results of assumptions about internal mechanisms of flagella, and has been an important adjunct to reaching the following conclusions about the mechanisms of movement of the simple flagella exemplified by marine invertebrate spermatozoa:

1. The active mechanochemical transduction process in simple 9+2 flagella is probably a sliding filament process similar to that postulated for muscle. The evidence for this conclusion is presented elsewhere in this volume (Chapter 7).

2. Spontaneous initiation and propagation of bending waves by flagella will occur if the active sliding process is locally controlled by the curvature of the flagellum (Brokaw, 1971, 1972a). This control mechanism is an application to a sliding filament model for flagella of the simple feedback control originally suggested by Machin (1958, 1963). It appears to be an adequate explanation for bend initiation and propagation in sperm flagella, but as yet there is no direct evidence to support the assumption that the active process is controlled by curvature.

3. Observations on the effects of altered external viscosity on the parameters of sperm movement are consistent with the suggestion that these spermatozoa use what seems to be the simplest available mechanism for stabilizing the wavelength of their bending waves—the presence of a relatively high internal viscous bending resistance within the flagellum (Brokaw, 1972a, b and c). However, we cannot exclude the possibility that they have evolved a more efficient mechanism for wavelength regulation, with properties which mimic the behavior expected when wavelength is regulated by internal viscosity.

4. The amplitude of the bending waves of sperm flagella does not behave as expected if the parallel elastic properties of structures within the flagellum are important in regulating the amplitude. In particular, there are many observations similar to Gray's original observation that spermatozoa can sometimes swim very slowly with bending waves which have a normal wavelength and amplitude. Further study is required to determine what method is used for amplitude regulation, and to obtain information about elastic properties of flagellar structures. Studies of starting and stopping transients might be particularly instructive.

5. Although the movements of sperm flagella from only a very few species of marine invertebrates have been examined critically, and the results have contributed a great deal to our current understanding of mechanisms of flagellar movement, many phenomena have been observed which remain unexplained. We have only begun to exploit the advantages of these spermatozoa as experimental material which can provide important information about the internal mechanisms of flagellar motility.

Appendix I

A taxonomic guide to references containing information on methods for study of the motility of marine invertebrate spermatozoa

Sea urchin spermatozoa

General references: Rothschild (1951); Tyler and Tyler (1966)
Motility of live spermatozoa: Tyler (1953); Brokaw (1965, 1966); Branham (1966)
Reactivated spermatozoa: Kinoshita (1958, 1959); Brokaw (1966); Gibbons and Gibbons (1972); Brokaw and Benedict (1968a)
Isolation of flagella for biochemical experiments: Brokaw and Benedict (1971); Gibbons and Fronk (1972)

Other groups

Coelenterates: Miller and Brokaw (1970)
Molluscs:
Bivalves: Nelson (1955)
Cephalopods: Bishop (1958); Austin *et al.* (1964); Martin *et al.* (1970)
Annelids: Brokaw (1965, 1966)
Echiuroids: MacGinitie (1938)
Tunicates: Brokaw (1965, 1966); Brokaw and Benedict (1968b); Ursprung and Schattbach (1965)

References

Austin, C. R., Lutwak-Mann, C. and Mann, T. (1964). *Proc. R. Soc.* B **161**, 143–152.
Baker, P. F. and Presley, R. (1969). *Nature, Lond.* **221**, 488–490.
Bishop, D. W. (1958). *Nature, Lond.* **182**, 1638–1640.
Bishop, D. W. (1962). *Physiol. Rev.* **42**, 1–59.
Branham, J. M. (1966). *Biol. Bull.* **131**, 251–260.
Brokaw, C. J. (1963). *J. exp. Biol.* **40**, 149–156.
Brokaw, C. J. (1965). *J. exp. Biol.* **43**, 155–169.
Brokaw, C. J. (1966). *J. exp. Biol.* **45**, 113–139.
Brokaw, C. J. (1970). *J. exp. Biol.* **53**, 445–464.
Brokaw, C. J. (1971). *J. exp. Biol.* **55**, 289–304.
Brokaw, C. J. (1972a). *Biophys. J.* **12**, 564–586.
Brokaw, C. J. (1972b). *J. Mechanochem. Cell Motility* **1**, 151–155.
Brokaw, C. J. (1972c). *J. Mechanochem. Cell Motility* **1**, 203–211.
Brokaw, C. J. and Benedict, B. (1968a). *Archs Biochem. Biophys.* **125**, 770–778.
Brokaw, C. J. and Benedict, B. (1968b). *J. gen. Physiol.* **52**, 283–299.
Brokaw, C. J. and Benedict, B. (1971). *Archs Biochem. Biophys.* **142**, 91–100.

Brokaw, C. J., Goldstein, S. F. and Miller, R. L. (1970). *In* "Comparative Spermatology" (B. Baccetti, ed.), pp. 475–486. Academic Press, New York.
Brokaw, C. J. and Wright, L. (1963). *Science, N.Y.* **142**, 1169–1170.
Brown, R. H. J. and Popple, J. A. (1955). *Med. Biol. Illus.* **5**, 23–28.
Costello, D. P., Davidson, M. E., Eggers, A., Fox, M. H. and Henley, C. (1957). "Methods for obtaining and handling marine eggs and embryos." Marine Biol. Lab., Woods Hole, Mass.
Gibbons, B. H. and Gibbons, I. R. (1972). *J. Cell Biol.* **54**, 75–97.
Gibbons, I. R. and Fronk, E. (1972). *J. Cell Biol.* **54**, 365–381.
Goldstein, S. F. (1969). *J. exp. Biol.* **51**, 431–441.
Gray, J. (1955). *J. exp. Biol.* **32**, 775–801.
Gray, J. (1958). *J. exp. Biol.* **35**, 96–108.
Gray, J. and Hancock, G. J. (1955). *J. exp. Biol.* **32**, 802–814.
Holwill, M. E. J. (1969). *J. exp. Biol.* **50**, 203–222.
Kinoshita, S. (1958). *J. Fac. Sci., Univ. Tokyo* IV **8**, 219–228.
Kinoshita, S. (1959). *J. Fac. Sci., Univ. Tokyo* IV **8**, 427–437.
MacGinitie, G. (1938). *J. exp. Zool.* **79**, 237–242.
Machin, K. E. (1958). *J. exp. Biol.* **35**, 796–806.
Machin, K. E. (1963). *Proc. R. Soc.* B **158**, 88–104.
Martin, A. W., Jr., Thiersch, J. B., Dott, H. M., Harrison, R. A. P. and Mann, T. (1970). *Proc. R. Soc.* B **175**, 63–68.
Miller, R. L. and Brokaw, C. J. (1970). *J. exp. Biol.* **52**, 699–706.
Mohri, H. and Yasumasu, I. (1963). *J. exp. Biol.* **40**, 573–586.
Nelson, L. (1955). *Biol. Bull.* **109**, 295–305.
Nelson, L. (1967). *In* "Fertilization" (C. Metz and A. Monroy, eds), pp. 27–97. Academic Press, New York.
Nelson, L. (1972). *Biol. Reprod.* **6**, 319–324.
Rothschild, Lord (1951). *Biol. Rev.* **26**, 1–27.
Rothschild, Lord, and Swann, M. M. (1949). *J. exp. Biol.* **26**, 164–176.
Ruegg, J. C., Straub, R. W. and Twarog, B. M. (1963). *Proc. R. Soc.* B **158**, 156–176.
Silvester, N. R. and Holwill, M. E. J. (1972). *J. theor. Biol.* **35**, 505–523.
Spikes, J. D. (1949). *Am. Nat.* **83**, 285–298.
Steinbach, H. B. (1966). *Biol. Bull.* **131**, 166–171.
Timourian, H. and Watchmaker, G. (1970). *Devl Biol.* **21**, 62–72.
Tyler, A. (1953). *Biol. Bull.* **104**, 224–239.
Tyler, A. and Tyler, B. S. (1966). *In* "Physiology of Echinodermata" (R. Boolootian, ed.), pp. 639–682, J. Wiley, New York.
Ursprung, H. and Schattbach, E. (1965). *J. exp. Zool.* **159**, 370–383.

Chapter 6

Isolated, reactivated and laser-irradiated cilia and flagella

STUART F. GOLDSTEIN

Department of Zoology, University of Minnesota, Minneapolis, Minnesota, U.S.A.

I. Introduction

Cilia and flagella are distinguished by their physical remoteness from the rest of the cell. The question immediately arises of these conspicuous motile cell extensions: is their source of movement contained within them, or are they moved by something within the cell body? This question has been answered directly by isolating them from the cell body. Under appropriate conditions they exhibit quite normal beating and swimming through their bathing solution, even when completely isolated from the rest of the cell.

Since there is no indication that the basic axonemal structure of motile cilia and flagella has any function other than movement, they appear to be completely self-contained organelles specialized for motility. Isolated cilia and flagella thus provide pure systems for studies of both biochemistry and motile behavior.

Although biochemical analyses generally require complete isolation, many biophysical studies can be performed with the organelle attached to the cell body. The surrounding membrane can be damaged or removed with glycerol or detergents if desired, exposing the axoneme to the external medium. In some—but not all—types of cells this treatment results in separation of the cilia or flagella from the cell body. Suspension of these organelles in solutions containing ATP often results in beating and swimming. These reactivated organelles have been useful for studies of the energy sources and ions necessary for motility, as well as for studies of variations in beating under different conditions.

The term "model" in practice refers to a cilium or a flagellum whose membrane has been disrupted or removed, whether or not it has been isolated from the rest of the cell. The term is perhaps somewhat unfortunate, as it suggests a synthetic construct; the essential components of cilia and flagella have not yet been completely enumerated, let alone recombined into motile organelles.

II. Anchorage of cilia and flagella to the cell

Before discussing the detachment of cilia and flagella from the cell body, it might be well to consider briefly some ways in which they are anchored. They terminate in a basal body, generally beneath the cell surface, and any connected structures are attached to the basal body. Specialized anchoring structures have not been described in some cells, such as a number of invertebrate spermatozoa.

In some cells, material extends from the distal (outermost) region of each triplet of the basal body toward the plasma membrane. These have been referred to as "transitional fibres" (Gibbons and Grimstone, 1960). They have been found in invertebrate spermatozoa (Afzelius, 1971), flagellate protozoa (Gibbons and Grimstone, 1960; Dingle and Fulton, 1966), vertebrate sensory cells (Flock and Duvall, 1965; Reese, 1965), epithelium (Thornhill, 1967) and vertebrate embryos (Doolin and Birge, 1966). These structures are curved sheets rather than fibres in basal bodies of rhesus monkey oviduct cilia, and may be sheets in some of the other cells as well (Anderson, 1972).

In many cells, fibres extend into the cell body from the proximal (innermost) region of the basal body; these rootlet fibers are generally striated. They have been seen in a variety of cells, including ciliated epithelium (Fawcett and Porter, 1954), vertebrate sensory cells (Schulte, 1972) and flagellate protozoa (Tamm, 1972).

Cells with a number of cilia or flagella commonly contain fibres running more or less parallel to the plasma membrane, connecting the basal bodies into a network. These are especially highly developed in ciliates (Allen, 1971).

There are many variations in the patterns of these accessory structures, and many cells contain more than one type. Whatever other functions they may perform, they give the impression of helping to orient cilia and flagella and of helping to hold them in place at their insertion into the cell. These anchorage structures are described in detail by Pitelka (Chapter 16, this volume).

The outer doublets usually appear to be lightly attached to the flagellar membrane. In the trypanosomes (Vickerman, 1969) and some trypanosomatid flagellates (Brooker, 1970), the flagellar membrane is in turn connected to the plasma membrane of the cell body by a series of small desmosome-like connections. These connections apparently help to anchor and orient the flagellum.

III. Site of separation of isolated cilia and flagella

In view of the fact that the basal body is very often anchored to the cell body, it is not surprising that the basal body usually remains in the cell when a cilium or flagellum is removed. Breakage generally occurs in or just above the transition region at the base of the organelle, whatever the method of isolation; Blum (1971) has reviewed this subject. This has been noted for a variety of cell types, including flagellates (Rosenbaum, *et al.*, 1969), ciliates (Child and Mazia, 1956), and metazoan cells (Iwaikawa, 1967). New organelles can often grow from the basal bodies, indicating that basal bodies are not permanently damaged by the isolation procedures.

Spermatozoa, however, often sever in the neighborhood of the mid-piece during fertilization, but not necessarily in the transition region (see Blum, 1971). Sperm flagella are often easily detached, but the fate of the basal body does not seem to have been studied in spermatozoa. Pease (1963) found that rat spermatozoa which were forced through a hypodermic needle after treatment with sodium phosphotungstate or formalin frayed at their tip rather than breaking. Presumably they were protected elsewhere by their heavy sheath. Brown (1954) found that human spermatozoa which were segmented by freezing broke most often between the head and mid-piece, but commonly broke at other points. Summers and Gibbons (1971) found that sea urchin sperm flagella isolated in solutions containing Triton X-100 often had their basal body attached.

Blum (1971) has suggested that the transition region is specialized for breakage; however, there are relatively few cases in which natural breakage is known to occur. Weakness or flexibility in this region might result from specialization for bend initiation. Perhaps the ciliary necklace (Gilula and Satir, 1972) is the dotted line on which they tear?

IV. Bulk isolation of cilia and flagella

Removal of these organelles is usually accomplished by agitating the cell suspension, so that they are sheared off. This is often accompanied by treatment of the cells with chemicals such as alcohol, glycerol or a mild detergent. Most of these procedures are performed in the cold. The cilia or flagella are separated from the more rapidly sedimenting cell bodies by centrifugation. Some procedures damage the organelles.

A number of workers have developed techniques for isolating cilia from *Tetrahymena pyriformis* for biochemical analysis. Child and Mazia (1956) developed a method based on the procedure of Mazia and Dan (1952), as modified by Mazia (1955): The cells were bathed in ethanol at −10°C for several hours and transferred to a 1% solution of digitonin. The cilia detached at their base, but the preparations were not clean. Child (1959) improved this technique by transferring the cells from ethanol to 0·1 M KCl, pH 7, and stirring vigorously. He also presented a technique in which the cells were suspended in 20% (v/v) cold glycerol for 5–10 min, with occasional stirring. Watson and Hopkins (1962) developed a technique involving ethanol, ethylenediaminetetraacetate (EDTA) and Ca^{++}, which has served as the basis for numerous subsequent procedures: the cells were suspended in cold 0·025 M sodium acetate, pH 7·0; 150 ml of this suspension were mixed with 750 ml of cold 12% (v/v) ethanol in 0·025 M sodium acetate containing 0·1% EDTA, pH 7·0; the cells were still motile at this point; 25 ml of 1 M $CaCl_2$ were added, and ciliary detachment began immediately, as the pH dropped to 5·8. Whole cilia were produced. A slightly modified procedure, in which the Ca^{++} was added at an earlier stage, produced segments 1 μm long as well as whole cilia (Hopkins and Watson, 1963; Watson *et al.*, 1964). The method of Watson and Hopkins (1962) was modified by Gibbons (1965a): the cells were suspended in a cold solution containing 30 mM NaCl and 0·2 M sucrose; 250 ml of this suspension were added to 1 l of a cold solution containing 11% ethanol, 2·5 mM EDTA and 15 mM tris-thioglycolate buffer, pH 8·3 (the buffer is a 15 mM solution of tris (hydroxymethyl) aminomethane whose pH has been adjusted to 8·3 with thioglycolate); the cells were motile at this point; 12 ml of 1 M $CaCl_2$ were quickly added while swirling; the cilia began to detach immediately, and the pH was about 7·5. Watson and Hynes (1966) used a variation of this procedure in which the Ca^{++} was added simultaneously with the ethanol. Gibbons (1965b) suspended the cells in 20 ml of 0·18% NaCl and 0·2 M sucrose and cooled to 0°C. This suspension was mixed with 100 ml of a solution containing 70% (v/v) glycerol, 5 mM KCl, 2·5 mM $MgSO_4$ and 20 mM tris-thioglycolate, pH 8·3, and stored at −20°C. The suspension was shaken vigorously in a Vortex mixer. These cilia could be reactivated (see below). Winicur (1967) showed that cilia isolated by an ethanol–EDTA–Ca^{++} procedure could also be reactivated if the membranes were disrupted afterwards. Nozawa and Thompson (1971) simply suspended the cells in a cold solution of 0·1 M NaCl, 3 mM EDTA and 0·2 M phosphate buffer, pH 7·2, and gently homogenized in a loose-fitting hand homogenizer.

Cilia have been removed from *Tetrahymena pyriformis* so that the growth of new cilia from the basal bodies could be studied. Child (1965a) briefly exposed the cells to 5 mM EDTA in acetate buffer, pH 6·0, followed by addition of 5 mM or greater $CaCl_2$ along with shearing forces. Rosenbaum and Carlson (1969) used a variation of this procedure.

A number of workers have isolated pellicles of *Tetrahymena pyriformis* or the associated basal bodies and rootlet structures. Seaman (1960) suspended the cells in cold 40% ethanol, and then in 1% digitonin in 0·4 M KCl. They were then suspended in 0·25 M sucrose and ground with powdered quartz in a tissue homogenizer. Most of the procedures of later workers have involved successive refinements of this method (Hoffman, 1965; Argetsinger, 1965; Satir and Rosenbaum, 1965; Wolfe, 1970; Flavell and Jones, 1971). Williams and Zeuthen (1966) isolated the oral apparatus by suspending the cells in unbuffered *t*-butyl alcohol and mixing well. At a concentration of 1·5 M, the cilia were removed from the oral apparatus. At higher concentrations, the cilia remained attached to the isolated apparatus. Procedures for removing components from the pellicle are rather harsh, and are not designed to preserve the cilia.

Cilia have been stripped from *Paramecium caudatum* by treatment with 5×10^{-3} M chloral hydrate (Grebecki and Kuznicki, 1961; Kennedy and Brittingham, 1968); this procedure allowed regrowth of the cilia. The cortical structures of *Paramecium* have been isolated by Metz *et al.*, (1953), who sonicated cells following fixation in formalin; by Smith-Sonneborn and Plaut (1967), who sheared the cells by passage through a Pasteur pipette after fixation in various solutions; and by Hufnagel (1969), who sheared the cells in a tissue homogenizer.

The large oral cilia of *Stentor coeruleus* are shed in response to a number of compounds. Treatment with several chlorides, sulfates, acetates, sugars, urea and albumen (generally 1% solution), as well as 25% sea water, can cause shedding of the entire membranellar band, including the underlying basal bodies (Tartar, 1957, 1961). Other treatments, including exposure to 20% (w/w) sucrose for about 2 s, cause shedding of the oral cilia without removal of the basal bodies (Tartar, 1968).

Flagella have been removed from flagellate protists, especially *Chlamydomonas* and the closely related alga, *Polytoma*, either to study the flagella themselves or to study development of new ones from the old basal bodies. Tibbs (1957) suspended cells of *P. uvella* in distilled water, added a few drops of chloroform or toluene, and shook by hand. Flagella of *P. caeca* and *Chlorogonium elongatum* could be freed by shaking in distilled water. These were precipitated out of suspension with acetic acid and alcohol. Better preparations were obtained after removal from suspension by centrifugation (Tibbs, 1958). Hagen-Seyfferth (1959) removed flagella from *Chlamydomonas eugametos* with high temperature, low or high pH, or ethanol. Dubnau (1961) removed flagella from *Ochromonas danica* by mechanical shearing. Brokaw isolated flagella from *C. moewusii* (Brokaw, 1960) and *P. uvella* (Brokaw, 1961) by suspension in cold glycerol. These extracted flagella could be reactivated (see below), as could *C. snowiae* isolated by the same technique (Chorin-Kirsh and Mayer, 1964). Rosenbaum and Child (1967) removed flagella from *Ochromonas* by vigorous mixing in a fluted tube, and from *Euglena* and *Astasia* either by vigorous mixing or by subjecting them to a rapid change of pH. The pH shock was obtained by adding 1 N acetic acid while stirring rapidly, to bring the pH from 6·8 to 4·7. After 2–2·5 min, the pH was returned to 6·8 with 1 N KOH. Rosenbaum *et al.* (1969) removed flagella from *C. reinhardii* by vigorous mixing in a fluted tube; Jacobs and McVittie (1970) removed flagella in this species by the method of Watson and Hynes (1966). Mintz and Lewin (1954) found that flagella were shed by *C. moewusii* in about 5 min after reducing the pH from 8·0 to 3·0 with dilute HCl. The flagella of *Chlamydomonas* are resorbed under some culture conditions (Lewin, 1953).

Stephens and Linck (1969) isolated flagella from the gill of a marine bivalve, *Pectin irradians*, by suspending pieces in double-strength sea water, prepared by adding 30 g l^{-1} of NaCl to sea water, and agitating gently for 10 min, in a manner similar to that described by Auclair and Siegel (1966).

Sperm tails have been removed from the head and mid-piece by numerous workers, for studies of flagellar biochemistry or preparation of reactivated models. Spermatozoa do not seem to be capable of regenerating an amputated flagellum. Tibbs (1957, 1958) isolated flagella from spermatozoa of char, perch and brown trout by shaking by hand in distilled water. He also isolated tails from perch spermatozoa by suspending them in a cold solution containing 0·5 M sucrose, 2·5 mM EDTA and 0·05 M tris, pH 7·6, and homogenizing in a high-speed homogenizer for 20 s (Tibbs, 1965). Flagella can be removed from spermatozoa of some marine invertebrates by drawing dilute suspensions through a fine-tipped Pasteur pipette, although this treatment causes breakage rather than removal in others (Brokaw, 1965). Nelson (1955) obtained a preparation of tails from spermatozoa of *Mytilus* either by freezing them to a slushy consistency and grinding in a mortar and pestle or by diluting them and mixing in a hand homogenizer. Mohri removed the tails from sea urchin spermatozoa either by suspending them in Ca^{++}- and Mg^{++}-free artificial sea water and homogenizing in a glass homogenizer (Mohri, 1964) or by subjecting them to mild sonication (Mohri, 1968). Stephens (1970) separated tails from sea urchin spermatozoa by suspending them in cold sea water containing 10^{-4} M EDTA and homogenizing them in a high-speed mixer. Ogawa and Mohri (1972) glycerinated sea urchin spermatozoa before sonicating them in cold Ca^{++}- and Mg^{++}-free sea water. Gibbons and Fronk (1972) obtained intact axonemes and axonemal fragments from sea urchin spermatozoa by extraction in a cold solution containing 1% (w/v) Triton X-100, 0·1 M KCl, 5 mM $MgSO_4$, 0·5 mM EDTA, 1 mM ATP, 1 mM dithiothreitol and 10 mM tris-phosphate buffer, pH 7·0. The basal body often remained attached to its axoneme (Summers and Gibbons, 1971).

Some methods of separation have been used with mammalian spermatozoa. Bishop (1950) separated tails from bull and rabbit spermatozoa by sonication. Nelson (1954) obtained tail fragments from bull spermatozoa by freezing them to a slushy consistency and grinding them in a mortar and pestle. Englehardt and Burnasheva (1957) removed flagella from bull spermatozoa with a homogenizer. Pernot (1956) deflagellated guinea pig spermatozoa by sonication. Brown (1954) obtained tails and fragments from human spermatozoa by rapid freezing. Tails of spermatozoa of some mutant bulls are naturally detached from the head (Nelson, 1967). Work on spermatozoa before 1962 has been reviewed by Bishop (1962).

Cilia have been removed from sea urchin embryos by brief exposure to sea water whose osmolarity was doubled by adding 29·2 g l^{-1} of NaCl (Auclair and Siegel, 1966). Iwaikawa (1967) deciliated sea urchin embryos by exposure to either hypertonic sea water or 1 M sodium acetate. Cytochalasin B apparently causes loss of some of the cilia in embryos of the echinoderms *Lytechinus* and *Dendraster* (Wessells *et al.*, 1971).

V. Injury and removal of individual cilia and flagella

Microknives and microbeams have been used to remove preselected cilia and flagella from single cells, in studies of both isolated organelles and growth of new organelles from the cell body. In some cases, organelles can be severed at any desired point along their length. Ultraviolet (u.v.) microbeams can be used to produce localized lesions along an organelle or to cause the entire organelle to be sloughed off. Laser microbeams have been used to produce localized lesions along flagella and, at higher power, to sever them completely. The main shortcoming of the u.v. microbeams used is the length of time required to inflict damage (of the order of 1 min). This can be shortened to well under a millisecond by the use of a pulsed laser microbeam. A laser can be synchronized with photographic recording equipment and, under suitable conditions, to a beating organelle. It is therefore a useful tool for producing instantaneous damage on an actively beating cilium or flagellum.

Terni (1933) sliced the long undulating membrane of urodele spermatozoa with a needle, and could divide them into three or more pieces. Tamm (1967) used a tungsten knife to sever the anterior flagellum of *Peranema* at various points, in studies of regeneration; he could do this to cells enucleated at various stages of the cell cycle (Tamm, 1969). Lindemann and Rikmenspoel (1972) used a glass microknife to sever tails of bull and *Drosophila* spermatozoa. Several workers have used microdissection to rearrange fields of cilia (see Sleigh, 1962a).

Terni (1933) used a u.v. microbeam of 8 μm diameter to produce lesions in urodele spermatozoa. Walker (1961) used a u.v. microbeam of 4 μm diameter to irradiate regions of the flagellum of *Trypanosoma*. Wise (1965) used u.v. microbeams of 4, 12 and 25 μm diameter to irradiate cirri and membranelles of *Euplotes*. When the base of a cirrus was irradiated, disintegration proceeded in stages over a period of a few minutes: first beating stopped, then individual cilia splayed out and became limp, and finally individual cilia disintegrated, starting with the tip; neighboring cilia outside the beam were unaffected—even when they were in the same cirrus. Hanson (1955, 1962) used a u.v. microbeam of 5 μm diameter to irradiate portions of one of the gullets of double-gullet cells of *Paramecium aurelia*. Frankel (1960) used a 5 $\mu m \times 5$ μm u.v. microbeam to irradiate oral areas of a ciliate, *Glaucoma chattoni*. Jerka-Dziadosz (1972) used a u.v. microbeam of 5 μm^2 area to irradiate cirri of the hypotrichous ciliate, *Urostyla weissei*. The basal bodies appeared to be sloughed off and regenerated. Wise, Hanson, Frankel and Jerka-Dziadosz studied regrowth and reorganization in irradiated regions, and used a microbeam technique described by Uretz *et al.* (1954); Jenkins and Sawyer (1970) have used a laser for this purpose. Microbeam techniques have been reviewed by Smith (1964).

A pulsed ruby laser microbeam has been used to produce lesions approximately 2 μm long in flagella and reactivated models of echinoderm spermatozoa (Goldstein, 1969). The same technique has been used to injure locally or completely sever the flagellum of a trypanosomatid flagellate, *Crithidia oncopelti* (Goldstein *et al.*, 1970a, b). The laser beam was reduced and focused by aiming it backwards through the microscope, so that it emerged from the objective lens as a spot of about 2 μm diameter. The 794·3 nm radiation was absorbed by a blue dye (brilliant blue FCF) dissolved in the water. Cells were selected in which the cell body adhered to the glass while the tail beat freely. The laser could be synchronized with a regularly beating flagellum: The specimen was illuminated with a stroboscopic light which flashed at its frequency of beating, so that the flagellum was in the same position each time the light flashed. Cross hairs which located the beam position were placed over the desired spot on the image of the flagellum. The laser was then fired in synchrony with the stroboscope, so that it flashed when the flagellum was under the cross hairs. Results were recorded photographically (see section VIII below).

VI. Preparation and reactivation of models

Flagellar and ciliary models which can be reactivated to beat when supplied with a source of chemical energy have been prepared by several workers. The goal is disruption of the membrane to allow entry of ATP and other molecules, without any degradation of the internal structures which are necessary for beating. There is no single procedure which can be applied without change to all types of cells, but there are some rules of thumb. Until recently, glycerol was generally used for "extraction", as preparation of the models is often called. A non-ionic detergent, Triton X-100, now appears to be the best available agent for dissolving the membrane. Glycerol causes partial disruption of the membrane (Satir and Child, 1963). It is effective in strongly hypertonic media (Gibbons, 1965b), indicating that disruption is not simply an osmotic effect. Triton X-100 completely dissolves the membrane (Gibbons and Gibbons, 1972). The reactivating medium contains an energy source, plus other substances which hopefully provide an optimum environment for motility. ATP is the only nucleoside triphosphate which induces appreciable motility, although Bishop and Hoffmann-Berling (1959) report slight activity in ITP. ADP is sometimes effective, because of the presence of adenylate kinase in the organelles. Frequency of beating increases with ATP concentration. Above about 0·1–1 mM, the frequency continues to rise but the amplitude decreases rapidly. Optimum frequency and amplitude are generally less than in the live cells. Mg^{++} is necessary, and cannot be replaced by Ca^{++}. KCl is required; Na^{+} generally cannot replace K^{+}, and attempts to replace Cl^{-} are not usually reported—although Bishop and Hoffmann-Berling (1959) found that NaCl, but not KI, could substitute for KCl in reactivation of mammalian spermatozoa. A chelator—generally EDTA—and antioxidants, such as thioglycolate, cysteine or dithiothreitol, are often used. High molecular weight polymers, such as polyvinylpyrrolidone (PVP), dextran, bovine serum albumen or polyethylene glycol, often improve motility. The reason for this is not clear:

they might simply coat the denuded organelle and prevent flow of fluid through the axoneme as it beats (Gibbons and Gibbons, 1972). Extraction and storage in glycerol are usually carried out below 0°C. Organelles are ready for reactivation within minutes.

Hoffmann-Berling (1954, 1955) pioneered the development of flagellar and ciliary models when he adapted techniques for the glycerination of muscle to grasshopper spermatozoa and trypanosomes, as well as to several other types of cell motile systems. Flagella were extracted in a cold 50% glycerol solution containing 2×10^{-3} M EDTA, 10^{-2} M phosphate buffer, pH 7·2, and KCl to adjust the ionic strength to about 0·18μ. They could be reactivated in a solution containing 10^{-3} M ATP, 2×10^{-3} M cysteine, 5–10×10^{-2} M $MgCl_2$, 10^{-2} M phosphate buffer, pH 7·2 and KCl to adjust the ionic strength to about 0·18μ. Beating increased with ATP concentration between 10^{-5} and 10^{-3} M, and increased with temperature between 0 and 30°C. Beating could occur in sperm tails separated from the head and mid-piece. Since then numerous other types of cilia and flagella have been extracted and reactivated with variations of this procedure. Bishop (1958) reactivated squid spermatozoa following glycerination. Kinoshita (1958, 1959) reactivated glycerinated sea urchin and starfish spermatozoa. He found that a metal-chelating agent was required in the reactivating solution, and suggested that flagella use a chelator identical to muscle relaxing factor. This idea stimulated serious discussion at the time, and was characteristic of attempts of that period to equate flagellar motility with muscular contraction. Later workers have not always found a chelator to be necessary (Brokaw, 1961; Gibbons, 1965b). Bishop and Hoffmann-Berling (1959) reactivated flagella of several types of mammalian spermatozoa in which the membrane had been disrupted by digitonin; 50% glycerol was less successful, presumably because it was not able to dissolve the heavy sheath of these spermatozoa. These flagella could be reactivated after isolation from the spermatozoa. PVP in the reactivation medium improved beating, as did chick serum or 1% agar. Alexandrov and Arronet (1956) reactivated glycerinated models of ciliated epithelium from frog palate and rat trachea. Groups of cilia sometimes beat synchronously, but did not exhibit normal metachrony.

All of these models failed to mimic live organelles in one important respect: although they bent rhythmically from side to side, they did not propagate bends down their length, and hence they did not swim through the water. Models of this period have been reviewed by Bishop (1962) and Sleigh (1962b).

Propagation of bends in reactivated models was first achieved by Brokaw (1961), and has been achieved by several subsequent workers. Brokaw extracted flagella of *Polytoma uvella* in a glycerol solution similar to that of Bishop and Hoffmann-Berling (1959): Cells were suspended in 3–4 volumes of a cold solution containing 70% (v/v) glycerol, 0·01 M $MgCl_2$ and 0·02 M tris-thioglycolate buffer, pH 7·8, for about 30 min. They were then stirred vigorously. The cell bodies were removed by centrifugation. The flagella were reactivated by dilution of the glycerol suspension 5–10 times with a solution of ATP, 0·05 M KCl, 0·004 M $MgCl_2$ and 0·02 M tris-thioglycolate buffer, pH 7·8. Best results were obtained with 10^{-5}–10^{-4} M ATP. Beat frequency rose to about one-sixth of normal at this concentration. At greater concentrations the frequency continued to rise, but the amplitude fell sharply. Bends were propagated along the flagella, and they swam through the medium. The addition of 2·5% PVP decreased sticking to the glass and improved motility (Brokaw, 1963). Essentially the same procedure, using 0·25 M KCl, has been applied to glycerination of spermatozoa of marine invertebrates (Brokaw, 1966a, 1967). Motility was observed when the ATP was replaced with ADP. These flagella did not usually detach. Gibbons (1965b) adapted this technique to the cilia of *Tetrahymena pyriformis*. Motility was not appreciably affected by the chelating agents EDTA or EGTA. He found—as have others—that diluting the glycerol suspension of cilia with too much reactivating medium caused a decrease in motility. Winicur (1967) reactivated cilia which had been removed from *T. pyriformis* by the ethanol–EDTA–Ca^{++} method of Gibbons (1965a). After removal, the cilia were treated with glycerol, ethylene glycol, or digitonin plus sucrose. Some reactivation was obtained with ADP. Motility could be improved by addition of 1 mM xylose or dextrose (Winicur, 1971). Child (1965b) reactivated lateral cilia on the gill of a mussel, *Modiolus demissus*. The tissue was placed in a cold solution containing 40% (v/v) ethylene glycol, 10% (v/v) glycerol, 0·1 M KCl and 0·01 M phosphate buffer, pH 7·5, for at least 1·5 h. The tissue was then washed in a cold solution containing 0·1 M KCl, 0·005 M $MgCl_2$, 0·01 M histidine and 0·01 M phosphate buffer, pH 7·5 and reactivated in the same solution containing 3 mM ATP. They appeared to beat with a normal waveform, and showed various forms of coordinated activity, including normal metachrony.

The main shortcomings of the glycerol procedures are a low percentage of reactivation—typically 25–50% (Brokaw and Benedict, 1968)—and an often somewhat abnormal flagellar waveform. Gibbons and Gibbons (1972) introduced a procedure using Triton X-100 which is largely free of these shortcomings. One volume of sea urchin semen was diluted with 1–2 volumes of sea water; this diluted semen was added to 20 volumes of a solution containing 0·04% (w/v) Triton X-100, 0·15 M KCl, 4 mM $MgSO_4$, 0·5 mM EDTA, 0·5 mM mercaptoethanol and 2 mM tris-HCl buffer, pH 8·0 and gently swirled for 30 s. Flagella remained attached to the head, and could be reactivated by diluting the suspension into a large volume of solution containing 0·15 M KCl, 2 mM $MgSO_4$, 0·5 mM EDTA, 5 mM dithiothreitol, 1 mM ATP, 2% polyethylene glycol and 20 mM tris-HCl buffer, pH 8·0. Preparations showed 95–100% motility, with fairly normal waveforms. Beat frequency and forward swimming speed were about two-thirds of normal. Motility was not appreciably affected by chelators. Very slight motility was observed with ADP.

Models of cells of *Paramecium caudatum* have been prepared by extraction with glycerol (Naitoh, 1969) and Triton X-100 (Naitoh and Kaneko, 1972) for studies of ciliary control and metachronal coordination. See Naitoh and Eckert (this volume, Chapter 12) for a detailed discussion of these studies.

Lindemann and Rikmenspoel (1972) disrupted the membrane of bull spermatozoa by puncturing the head with a glass microprobe.

VII. Observations on models and isolated organelles

The ability of isolated organelles to beat normally when provided with a source of chemical energy indicates quite directly that they are self-contained, and do not require any structures within the cell body for movement. The specificity for the requirement of ATP confirms its use as the primary energy source. The ability of some models to use ADP confirms the presence of adenylate kinase, allowing more efficient use of energy sources (Raff and Blum, 1968).

Measurements on dephosphorylation of ATP by live spermatozoa and reactivated models have extended *in vitro* measurements of enzyme activity. Flagella appear to use about 1 ATP molecule per axonemal ATPase molecule for each beat, relatively independently of the work performed per beat (Brokaw, 1967). A large fraction of the ATPase activity appears to be coupled to movement: the activity can be greatly reduced by rendering the spermatozoa immotile through mechanical breakage (Brokaw and Benedict, 1968), and this reduction can be appreciably relieved by treating the flagella with uncoupling agents (Brokaw and Benedict, 1971). When intact reactivated models are prevented from beating by increasing the viscosity of the medium, their ATPase activity decreases to approximately that of mechanically homogenized spermatozoa (Gibbons and Gibbons, 1972).

The role of ions in the relationship between ATPase activity and motility may also be studied with models; for example, Gibbons and Gibbons (1972) found that $MgATP^{2-}$, $MnATP^{2-}$ and $CaATP^{2-}$ were all hydrolyzed by sperm tail model ATPase, but motility was active with Mg^{2+}, reduced with Mn^{2+} and very feeble with Ca^{2+}.

Frequency and amplitude of beating vary with ATP concentration, and the most normal beating in models occurs at reasonably physiological concentrations. Since most cilia and flagella are supplied with ATP from the cell body, diffusion of ATP along the organelle would result in a steep concentration gradient, with the concentration at the tip probably being quite low (Brokaw, 1966b; Nevo and Rikmenspoel, 1970). This suggests that the frequency of beating is a function of the ATP concentration at the base, and that the speed of bend propagation (which is normally constant for most of the length of a flagellum) is relatively independent of ATP concentration (Brokaw and Goldstein, 1965).

Gibbons (1965a) removed the axonemal ATPase from ciliary models of *Tetrahymena pyriformis* by extraction with EDTA at low ionic strength. This ATPase protein—called "dynein"—was in two forms, with sedimentation coefficients of 14 S and 30 S. The 30 S form may be a polymer of the 14 S form (Gibbons and Rowe, 1965). The 30 S dynein could recombine with the outer axonemal doublets, with an accompanying return of ATPase activity. The removal and recombination of 30 S dynein and ATPase activity was correlated with the disappearance and return of the arms on the outer doublets. The arms thus appear to contain a significant portion of the axonemal ATPase activity.

Tibbs (1962) measured the decrease in light scattered by perch sperm flagella on addition of ATP. Gibbons (1965c) made similar measurements on models of *Tetrahymena* cilia. The decrease in turbidity was apparently due to an increased hydration of the axonemal proteins accompanying ATP hydrolysis. The effect increased with increased binding of dynein. Raff and Blum (1966, 1969) found that the pellet formed when such models were centrifuged had a decreased density when ATP was added at less than 5 mM concentration, also suggesting increased hydration; above 5 mM, partial solubilization of the protein resulted. Tibbs (1965) isolated flagella from perch spermatozoa by mixing in a high-speed homogenizer in a solution containing 0·5 M sucrose, 2·5 M EDTA and tris buffer, pH 7·6. Pellets of flagella sedimented in this solution were up to 40% heavier in suspensions to which ATP and Mg^{++} were added. This appeared to be due to sucrose uptake associated with ATPase activity.

Summers and Gibbons (1971) subjected models of sea urchin sperm flagella to light digestion with trypsin. Addition of ATP caused the microtubules to slide past one another, providing strong evidence that bends are developed by active sliding of microtubular structures.

Arronet and Konstantinova (1964) compared the effects of high pressure on intact tissue and glycerinated models of ciliated palate epithelium from two species of frog; intact tissue from *Rana temporaria* and *Rana ridibunda* beat at pressures up to 1592 and 1694 atm, respectively, while the corresponding values for glycerinated tissues were 1500 and 1656 atm. The close parallel between results on intact and glycerinated tissues was interpreted as indicating that high pressure acted directly on the proteins involved in ciliary beating.

VIII. Observations on laser-irradiated flagella

The pulsed laser microbeam provides a method of producing an almost instantaneous lesion or break in a flagellum. It can therefore be used to study the effect of damage on bends already established at the time of irradiation, as well as on subsequent bend formation and propagation.

The laser appears to cause intense heating of the irradiated region. A lesion can be seen under the light microscope as a break in the flagellum. The effects of laser irradiation on lateral gill cilia of a freshwater mussel, *Elliptio*, have been examined in the electron microscope (Goldstein and Satir, 1971; Goldstein, 1972): Lesions of about 30 μm diameter were made in actively beating tissue. Partially damaged cilia could be found in the periphery of these lesions. Cilia appeared coagulated within an irradiated region, and the boundary between damaged and intact regions of a cilium appeared quite sharp. The matrix, which is difficult to preserve even under the best conditions, was very easily damaged. Both the central pair of microtubules and the outer doublets were affected, but the central pair was noticeably more sensitive. The doublets sometimes appeared to fuse to one another. The membrane apparently sealed over the stump of a severed cilium.

In echinoderm sperm flagella bends originate at the base and travel to the tip, as they do in most flagella. When these flagella were irradiated, the region between the irradiated point and the head could continue to beat for at least a few cycles if that region was greater than about 10 μm long (Fig. 1). In the region beyond the irradiated point, bends which were already established at the time of irradiation continued to travel to the tip (Fig. 2). The speed of travel along the flagellum decreased, and the size and shape of a bend often changed. However, no new bends ever formed beyond an irradiated region, so that this region was straight after the old bends had traveled off the tip (Fig. 1). Irradiation in the center of a bent region apparently caused that region to straighten (Fig. 3); that is, the bend did not appear on subsequent exposures (Goldstein, 1969).

In these flagella, then, the base seems somehow specialized to initiate bends, in the sense that bends could originate only in a region which contained the base. A similar conclusion has been reached from experiments involving bulk irradiation of bull spermatozoa (Rikmenspoel and van Herpen, 1969). The inability to initiate bends beyond the irradiated point was not due to depletion of ATP in that region. The most direct indication of this came from irradiation of glycerinated models, which produced results similar to those on live spermatozoa. These results are

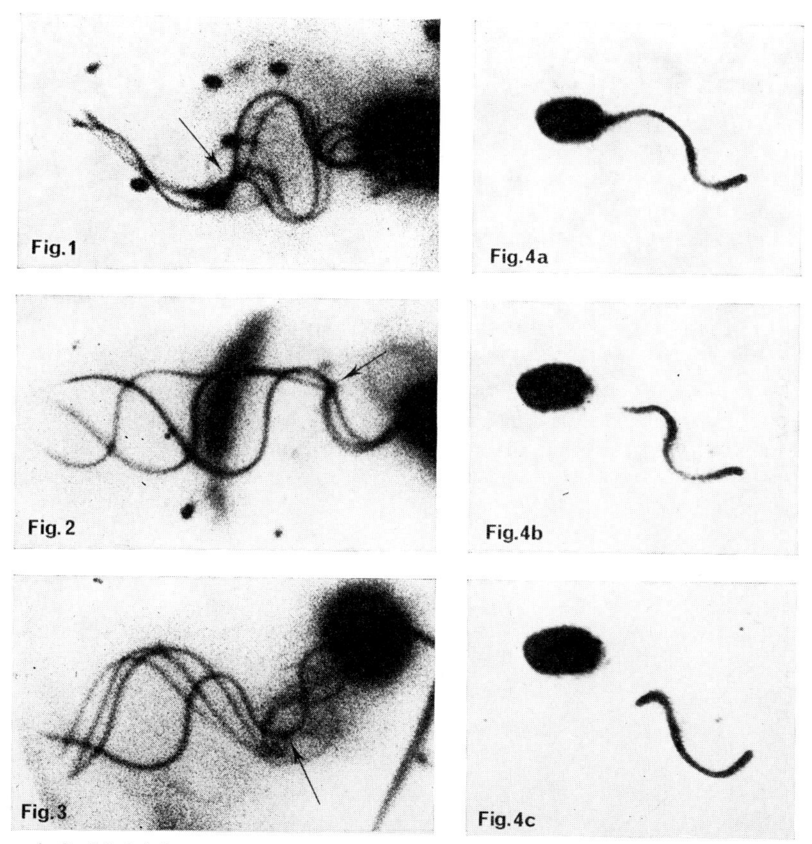

FIGS. 1–3. Multiple-exposure photographs of laser-irradiated sea urchin sperm flagella. Arrows indicate irradiated point.

FIG. 4. Frames from a film of a cell of *Crithidia oncopelti* whose flagellum was severed near the base. (a) Cell just before irradiation; bends are travelling toward the cell body. (b) Cell just after irradiation; bends are still travelling toward the cell body. (c) Cell about 1 s later; bends are now travelling toward the tip.

in agreement with the finding of Brokaw and Benedict (1968) that mechanically broken sea urchin sperm flagella are not motile. The elimination of a bend by irradiation of points within it suggests that a bend is in some sense organized as a unit of activity.

The flagellum of *Crithidia oncopelti* has the unusual ability to originate bends at either its base or its tip. Bends usually travel from tip to base, and the direction often changes (Holwill, 1965). This flagellum has a simple 9+2 ultrastructure (Burnasheva *et al.*, 1968); when it was irradiated at one end, bends could continue to originate at the other end. Occasionally, however, bends would originate at or near the irradiated region and travel toward the tip. This could happen even in fragments which were completely severed from the cell body (Fig. 4). These isolated fragments could beat for up to 10 cycles after irradiation and occasionally the direction of bend propagation was reversed (Goldstein *et al.*, 1970b).

The ability to initiate bends at any point along the flagellum of *C. oncopelti* indicates that the complex basal structures of cilia and flagella are not necessary for bend initiation. The difference between points along this flagellum and those along most flagella might be usefully thought of as a difference in activation threshold: points along most flagella bend when activated by events in an adjacent region (Brokaw, 1966c); points along this flagellum are capable of initiating bends without any activation, at least when they are on the end of a segment. This notion receives some support from the finding of Lindemann and Rikmenspoel (1972) that fragments of severed bull sperm flagella produced propagated bends only when they were held in a bent position. There have been reports of beating in fragments of sperm tails of urodeles (Terni, 1933), *Drosophila* (Lindemann and Rikmenspoel, 1972) and humans (Brown, 1954), but these tails contain structures additional to the 9+2 axoneme which may alter their mechanical characteristics.

The ability of fragments to beat for up to 10 cycles indicates that some chemical energy is stored within flagella, and is in good agreement with the notion that ATP diffuses down a flagellum from the cell body (Brokaw, 1966b; Nevo and Rikmenspoel, 1970). Mechanically severed sea urchin sperm flagella can continue to beat for 1 or 2 min (Brokaw, 1965).

Walker (1961) irradiated small regions along the flagellum of *Trypanosoma* with a u.v. microbeam, and found that this did not impair motility in neighboring regions. This was interpreted as evidence for transmission of a signal through the damaged, non-motile region. However, these flagella apparently beat in a manner similar to that of *Crithidia*, so beating might have been occurring independently in different regions. The possibility of passage of beating through an immobilized region has not been studied in detail, and the nature of a possible signal remains unsolved.

A beginning has thus been made in locating the activities along a flagellum that are associated with the development and movement of a bend. Further direct experiments on localized regions should yield increasingly detailed information.

References

AFZELIUS, B. A. (1971). *J. Ultrastruct. Res.* **37**, 679–689.

ALEXANDROV, V. Y. and ARRONET, N. I. (1956). *Dokl. Acad. Nauk SSSR* **110**, 457–460.

ALLEN, R. D. (1971). *J. Cell Biol.* **49**, 1–20.

ANDERSON, R. G. W. (1972). *J. Cell Biol.* **54**, 246–265.

ARGETSINGER, J. (1965). *J. Cell Biol.* **24**, 154–157.

ARRONET, N. I. and KONSTANTINOVA, M. F. (1964). *Tsitologiya* **6**, 743–746.

AUCLAIR, W. and SIEGEL, B. W. (1966). *Science, Wash.* **154**, 913–915.

BISHOP, D. W. (1950) *Anat. Rec.* **108**, 574.

BISHOP, D. W. (1958). *Anat. Rec.* **132**, 414.

BISHOP, D. W. (1962). *Physiol. Rev.* **42**, 1–59.

BISHOP, D. W. and HOFFMANN-BERLING, H. (1959). *J. cell. comp. Physiol.* **53**, 445–466.

BLUM, J. (1971). *J. theor. Biol.* **33**, 257–263.

BROKAW, C. J. (1960). *Expl Cell Res.* **19**, 430–432.

BROKAW, C. J. (1961). *Expl Cell Res.* **22**, 151–162.

BROKAW, C. J. (1963). *J. exp. Biol.* **40**, 149–156.

BROKAW, C. J. (1965). *J. exp. Biol.* **43**, 155–169.

BROKAW, C. J. (1966a). *J. exp. Biol.* **45**, 113–139.

BROKAW, C. J. (1966b). *Am. Rev. resp. Dis.* **93**, Suppl. 32–40.

BROKAW, C. J. (1966c). *Nature, Lond.* **209**, 161–163.

BROKLAW, C. J. (1967). *Science, Wash.* **156**, 76–78.

BROKAW, C. J. and BENEDICT, B. (1968). *Archs Biochem. Biophys.* **125**, 770–778.

BROKAW, C. J. and BENEDICT, B. (1971). *Archs Biochem. Biophys.* **142**, 91–100.

BROKAW, C. J. and GOLDSTEIN, S. (1965). *J. Cell Biol.* **27**, 15A.

BROOKER, B. E. (1970). *Z. Zellforsch. mikrosk. Anat.* **105**, 155–166.

BROWN, R. L. (1954). *J. Urol.* **71**, 503–509.

BURNASHEVA, S. A. OSTROVSKAYA, M. V. and YURZINA, G. A. (1968). *Acta Protozool.* **6**, 357–364.

CHILD, F. M. (1959). *Expl Cell Res.* **18**, 258–267.

CHILD, F. M. (1965a). *J. Cell Biol.* **27**, 18A.

CHILD, F. M. (1965b). *In* "Progress in Protozoology", *2nd Intern. Conf. Protozool., London*, Exerpta Med., p. 110.

CHILD, F. M. and MAZIA, D. (1956). *Experientia* **12**, 161–162.

CHORIN-KIRSH, I. and MAYER, A. M. (1964). *Pl. Cell Physiol., Tokyo* **5**, 441–445.

DINGLE, A. D. and FULTON, C. (1966). *J. Cell Biol.* **31**, 43–54.

DOOLIN, P. and BIRGE, W. (1966). *J. Cell Biol.* **29**, 333–345.

DUBNAU, D. A. (1961). Ph.D. thesis, Columbia Univ., University Microfilms No. 61-3427.

ENGELHARDT, V. A. and BURNASHEVA, S. A. (1957). *Biokhimiya* **22**, 554–560.

FAWCETT, D. W. and PORTER, K. R. (1954). *J. Morph.* **94**, 221–281.

FLAVELL, R. A. and JONES, I. G. (1971). *J. Cell Sci.* **9**, 719–726.

FLOCK, A. and DUVALL, A. J. (1965). *J. Cell Biol.* **25**, 1–8.

FRANKEL, J. (1960). *J. exp. Zool.* **143**, 175–194.

GIBBONS, B. H. and GIBBONS, I. R. (1972). *J. Cell Biol.* **54**, 75–97.
GIBBONS, I. R. (1965a). *Archs Biol., Liège* **76**, 317–352.
GIBBONS, I. R. (1965b). *J. Cell Biol.* **25**, 401–403.
GIBBONS, I. R. (1965c). *J. Cell Biol.* **26**, 707–712.
GIBBONS, I. R. and FRONK, E. (1972). *J. Cell Biol.* **54**, 365–381.
GIBBONS, I. R. and GRIMSTONE, A. V. (1960). *J. biophys. biochem. Cytol.* **7**, 697–715.
GIBBONS, I. R. and ROWE, A. J. (1965). *Science, Wash.* **149**, 424–426.
GILULA, N. B. and SATIR, P. (1972). *J. Cell Biol.* **53**, 494–509.
GOLDSTEIN, S. F. (1969). *J. exp. Biol.* **51**, 431–441.
GOLDSTEIN, S. F. (1972). *Acta Protozool.* **11**, 259–262.
GOLDSTEIN, S. F., HOLWILL, M. E. J. and SILVESTER, N. R. (1970a). *Visual* **8**, 45–48.
GOLDSTEIN, S. F., HOLWILL, M. E. J. and SILVESTER, N. R. (1970b). *J. exp. Biol.* **53**, 401–409.
GOLDSTEIN, S. F. and SATIR, P. (1971). *Biophys. J.* **11**, 249a.
GREBECKI, A. and KUZNICKI, L. (1961). *Bull. Acad. pol. Sci. Cl. II Sér. Sci. biol.* **9**, 459–462.
HAGEN-SEYFFERTH, M. (1959). *Planta* **53**, 376–401.
HANSON, E. D. (1955). *Proc. natn. Acad. Sci. U.S.A.* **41**, 783–786.
HANSON, E. D. (1962). *J. exp. Zool.* **150**, 45–68.
HOFFMAN, E. J. (1965). *J. Cell Biol.* **25**, 217–228.
HOFFMANN-BERLING, H. (1954). *Biochim. biophys. Acta* **14**, 182–194.
HOFFMANN-BERLING, H. (1955). *Biochim. biophys. Acta* **16**, 146–154.
HOLWILL, M. E. J. (1965). *J. exp. Biol* **42**, 125–137.
HOPKINS, J. M. and WATSON, M. R. (1963). *Expl Cell Res.* **32**, 187–189.
HUFNAGEL, L. A. (1969). *J. Cell Biol.* **40**, 779–801.
IWAIKAWA, Y. (1967). *Embryologiya* **9**, 287–294.
JACOBS, M. and MCVITTIE, A. (1970). *Expl Cell Res.* **63**, 53–61.
JENKINS, R. A. and SAWYER, H. R. (1970). *Expl Cell Res.* **63**, 192–195.
JERKA-DZIADOSZ, M. (1972). *J. exp. Zool.* **179**, 81–96.
KENNEDY, J. R. and BRITTINGHAM, E. (1968). *J. Ultrastruct. Res.* **22**, 530–545.
KINOSHITA, S. (1958). *J. Fac. Sci. Tokyo Univ.* (Sect. IV) **8**, 219–228.
KINOSHITA, S. (1959). *J. Fac. Sci. Tokyo Univ.* (Sect. IV) **8**, 427–437.
LEWIN, R. A. (1953). *Ann. N.Y. Acad. Sci.* **56**, 1091–1093.
LINDEMANN, C. B. and RIKMENSPOEL, R. (1972). *Science, Wash.* **175**, 337–338.
MAZIA, D. (1955). *Symp. Soc. exp. Biol.* **9**, 335–357.
MAZIA, D. and DAN, K. (1952). *Proc. natn. Acad. Sci. U.S.A.* **38**, 826–838.
METZ, C. B., PITELKA, D. R. and WESTFALL, J. A. (1953). *Biol. Bull. mar. biol. Lab., Woods Hole* **104**, 408–425.
MINTZ, R. and LEWIN, R. A. (1954). *Can. J. Microbiol.* **1**, 65–67.
MOHRI, H. (1964). *Biol. Bull. mar. biol. Lab., Woods Hole* **127**, 381.
MOHRI, H. (1968). *Nature, Lond.* **217**, 1053–1054.
NAITOH, Y. (1969). *J. gen. Physiol.* **53**, 517–529.
NAITOH, Y. and KANEKO, H. (1972). *Science, Wash.* **176**, 523–524.
NELSON, L. (1954). *Biochim. biophys. Acta* **14**, 312–320.
NELSON, L. (1955). *Biol. Bull. mar. biol. Lab., Woods Hole* **109**, 295–305.

NELSON, L. (1967). *In* "Fertilization" (C. B. Metz and A. Monroy, eds), Vol. 1, pp. 27–97. Academic Press, New York.
NEVO, A. C. and RIKMENSPOEL, R. (1970). *J. theor. Biol.* **26**, 11–18.
NOZAWA, Y. and THOMPSON, G. A., JR. (1971). *J. Cell Biol.* **49**, 712–721.
OGAWA, K. and MOHRI, H. (1972). *Biochim. biophys. Acta* **256**, 142–155.
PEASE, D. C. (1963). *J. Cell Biol.* **18**, 313–326.
PERNOT, E. (1956). *Bull. Soc. Chim. biol.* **38**, 1041–1054.
RAFF, E. C. and BLUM, J. J. (1966). *J. Cell Biol.* **31**, 445–453.
RAFF, E. C. and BLUM, J. J. (1968). *J. theor. Biol.* **18**, 53–71.
RAFF, E. C. and BLUM, J. J. (1969). *J. biol. Chem.* **244**, 366–376.
REESE, T. S. (1965). *J. Cell Biol.* **25**, 209–230.
RIKMENSPOEL, R. and VAN HERPEN, G. (1969). *Biophys. J.* **9**, 833–844.
ROSENBAUM, J. L. and CARLSON, K. (1969). *J. Cell Biol.* **40**, 415–425.
ROSENBAUM, J. L. and CHILD, F. M. (1967). *J. Cell Biol.* **34**, 345–364.
ROSENBAUM, J. L., MOULDER, J. E. and RINGO, D. L. (1969). *J. Cell Biol.*, **41**, 600–619.
SATIR, B. and ROSENBAUM, J. L. (1965). *J. Protozool.* **12**, 397–405.
SATIR, P. and CHILD, F. M. (1963). *Biol. Bull. mar. biol. Lab., Woods Hole* **125**, 390.
SCHULTE, E. (1972). *Z. Zellforsch. mikrosk. Anat.* **125**, 210–228.
SEAMAN, G. R. (1960). *Expl Cell Res.* **21**, 292–302.
SLEIGH, M. A. (1962a). "The Biology of Cilia and Flagella", pp. 125–126. Pergamon Press, Oxford.
SLEIGH, M. A. (1962b). "The Biology of Cilia and Flagella", pp. 109–110. Pergamon Press, Oxford.
SMITH, C. L. (1964). *Int. Rev. Cytol.* **16**, 133–153.
SMITH-SONNEBORN, J. and PLAUT, W. (1967). *J. Cell Sci.* **2**, 225–234.
STEPHENS, R. E. (1970). *J. molec. Biol.* **47**, 353–363.
STEPHENS, R. E. and LINCK, R. W. (1969). *J. molec. Biol.* **40**, 497–501.
SUMMERS, K. E. and GIBBONS, I. R. (1971). *Proc. natn. Acad. Sci. U.S.A.*, **68**, 3092–3096.
TAMM, S. L. (1967). *J. exp. Zool.* **164**, 163–186.
TAMM, S. L. (1969). *J. Cell Sci.* **4**, 171–178.
TAMM, S. L. (1972). *J. Cell Biol.* **54**, 39–55.
TARTAR, V. (1957). *Expl Cell Res.* **13**, 317–332.
TARTAR, V. (1961). "The Biology of Stentor", pp. 129; 152 *et seq*. Pergamon Press, Oxford.
TARTAR, V. (1968). *Trans. Am. microsc. Soc.* **87**, 297–306.
TERNI, T. (1933). *C. r. Ass. Anat.* **24**, 651–654.
THORNHILL, R. A. (1967). *J. Cell Sci.* **2**, 591–602.
TIBBS, J. (1957). *Biochim. biophys. Acta* **23**, 275–288.
TIBBS, J. (1958). *Biochim. biophys. Acta* **28**, 636–637.
TIBBS, J. (1962). *In* "Spermatozoan Motility" (D. W. Bishop, ed.), pp. 233–250. American Association for the Advancement of Science, Washington, D.C.
TIBBS, J. (1965). *Biochem. J.* **96**, 340–346.
URETZ, R. B., BLOOM, W. and ZIRKLE, R. (1954). *Science, Wash.* **120**, 197–199.
VICKERMAN, K. (1969). *J. Cell Sci*, **5**, 163–193.

WALKER, P. J. (1961). *Nature, Lond.* **189**, 1017–1018.
WATSON, M. R., ALEXANDER, J. B. and SILVESTER, N. R. (1964). *Expl Cell Res.* **33**, 112–129.
WATSON, M. R. and HOPKINS, J. M. (1962). *Expl Cell Res.* **28**, 280–295.
WATSON, M. A. and HYNES, R. D. (1966). *Expl Cell Res.* **42**, 348–356.
WESSELLS, N. K., SPOONER, B. S., ASH, J. F., BRADLEY, M. O., LUDUENA, M. A., TAYLOR, E. L., WRENN, J. T. and YAMADA, K. M. (1971). *Science, Wash.* **171**, 135–143, footnote 46.
WILLIAMS, N. E. and ZEUTHEN, E. (1966). *C. r. Lab. Trav. Lab. Carlsberg* **35**, 101–118.
WINICUR, S. (1967). *J. Cell Biol.* **35**, C7–C9.
WINICUR, S. (1971). Ph.D. thesis (part I), California Institute of Technology.
WISE, B. N. (1965). *J. exp. Zool.* **159**, 241–268.
WOLFE, J. (1970). *J. Cell Sci.* **6**, 679–700.

Chapter 7

The present status of the sliding microtubule model of ciliary motion

PETER SATIR

Department of Physiology–Anatomy, University of California, Berkeley, California, U.S.A.

I. Origins

The sliding microtubule model of ciliary motion owes a heavy intellectual debt to the sliding filament theory of muscle contraction (Huxley and Hanson, 1954; Huxley and Niedergerke, 1954). The sliding filament theory demonstrated that gross changes in shape within muscle cell cytoplasm could be brought about by cyclical interactions of structural elements that in themselves remained of fixed length. This was in dramatic contrast to the then generally accepted dogma, based on model fibers; for example, work on macroscopic collagen fibers had been used to suggest that generalized protoplasmic contractility was brought about by phase-transitions that caused sudden massive shortening of serially arranged fibrous proteins. This hypothesis is designated the contractile fiber hypothesis or more simply, the contractile hypothesis.

In the early 1950s, when Manton (1952), Fawcett and Porter (1954) and others firmly established the 9+2 pattern of ciliary axonemal microtubules, Bradfield (1955) and Gray (1955) presented detailed but entirely speculative models of ciliary motion based on the sequential shortening of the microtubules. By 1959, however, Afzelius and others were drawn to a sliding microtubule model of ciliary motion as a viable, and possibly preferable, alternative to the contractile model. By simple geometry, one can easily estimate that contraction of only a few per cent of total length of a doublet microtubule on one side of the cilium would be enough to account for the observed bending. It is still problematical to measure overall microtubule length with enough precision to decide between the alternative hypotheses on the basis of direct measurement.

II. Experimental basis

The sliding microtubule model, however, makes several predictions, and is testable in special circumstances. The structure and biochemistry of axonemal microtubules are reviewed in this volume by Warner (Chapter 2) and Stephens (Chapter 3), but it is important to recall here that the axonemal microtubules provide regular attachment sites for dynein ATPase arms, radial spokes and perhaps other protein complexes as well.

The cilium is an active organelle, and the microtubules together with these attached moieties are the only structures necessary to reinitiate the beat form in isolated detergent-treated organelles. Such ciliary models are in effect naked axonemes (Satir and Child, 1963; Gibbons and Gibbons, 1972a) that beat and swim normally when placed in ATP solutions containing the proper cations. The microtubules are the main continuous longitudinal structures of such models. They provide flexible rods of considerable length and rigidity within the axoneme. The sliding hypothesis states that the force responsible for ciliary motility is produced when the axonemal microtubules, which do not change in length, tend to slide with respect to one another. Accordingly, the hypothesis predicts that in different stroke positions the morphological relationships of the microtubules will change in systematic fashion so that both qualitatively and quantitatively the geometry of the bent cilium will be reflected in the displacement of the microtubules. For the lateral cilia of the freshwater mussel, it has proved possible to demonstrate that this is so (Satir, 1965, 1968).

One simple illustration of the qualitative relationship is shown in Fig. 1. In the lateral cilium basal body, the positions of the doublet microtubules remain fixed, so that the sliding displacement is seen as changes in the microtubule configuration at the tip of the cilium. Once the tip configuration of one known stroke stage is defined, the tip pattern for all other stages

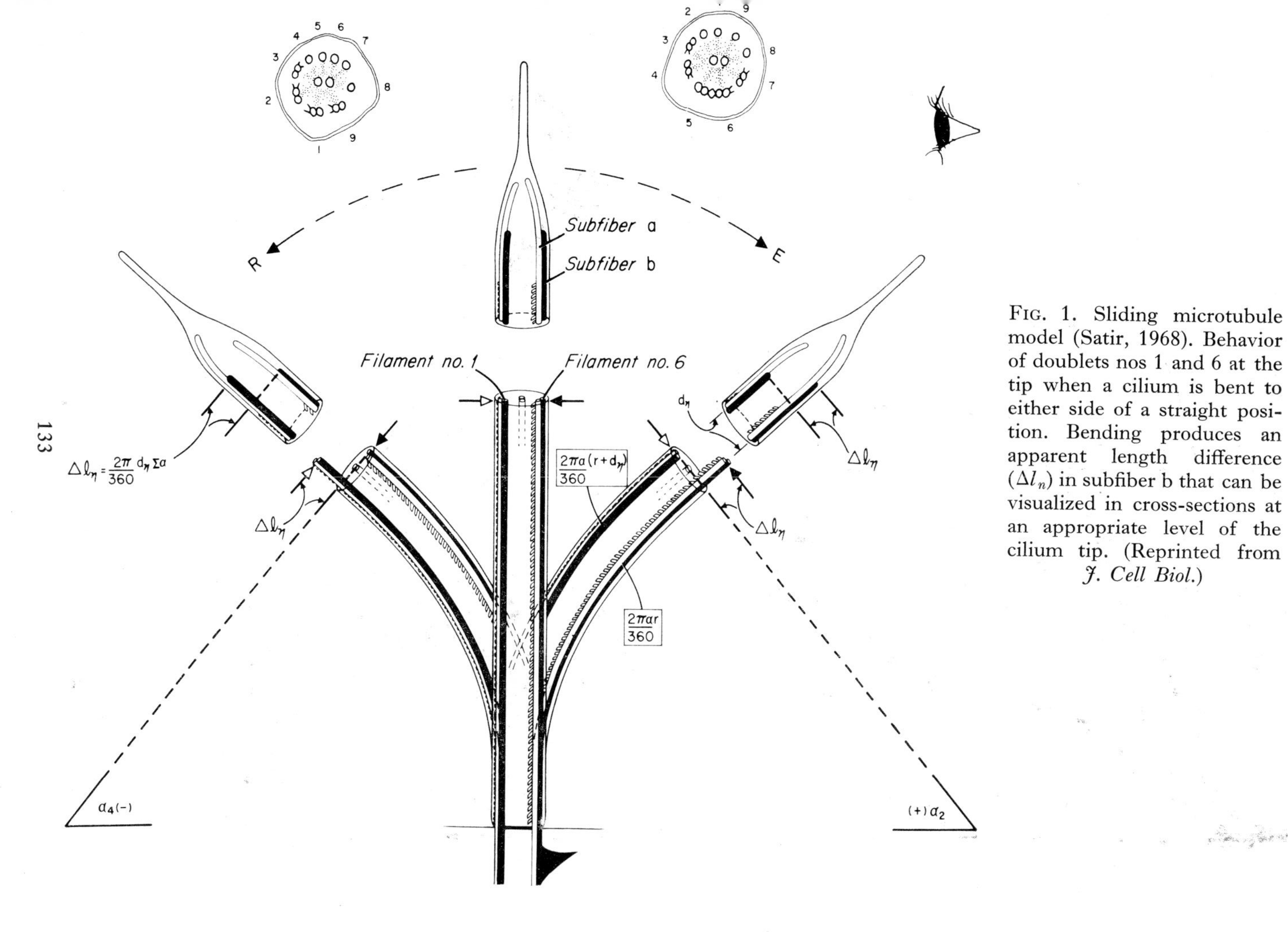

FIG. 1. Sliding microtubule model (Satir, 1968). Behavior of doublets nos 1 and 6 at the tip when a cilium is bent to either side of a straight position. Bending produces an apparent length difference (Δl_n) in subfiber b that can be visualized in cross-sections at an appropriate level of the cilium tip. (Reprinted from *J. Cell Biol.*)

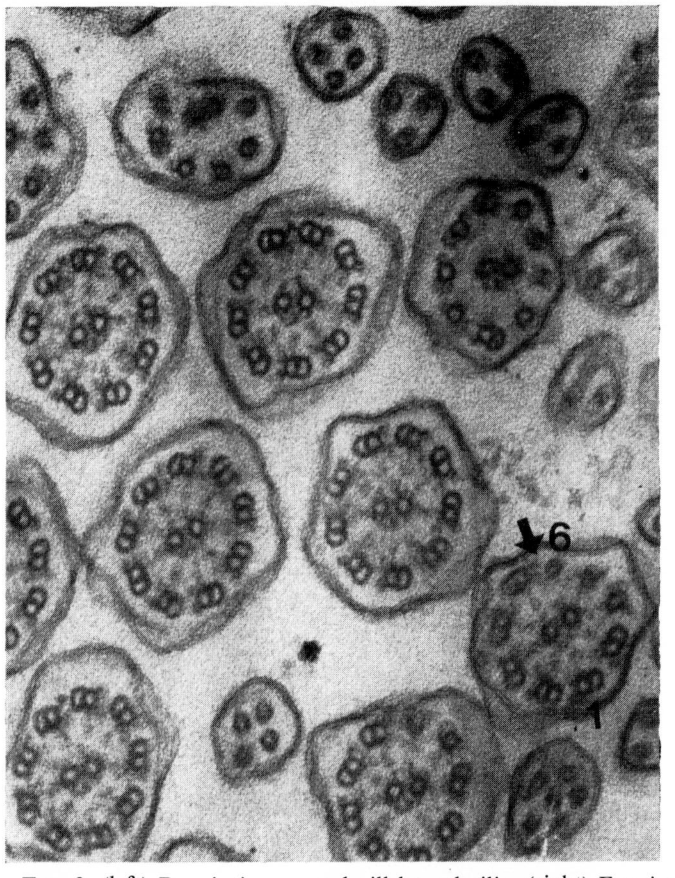

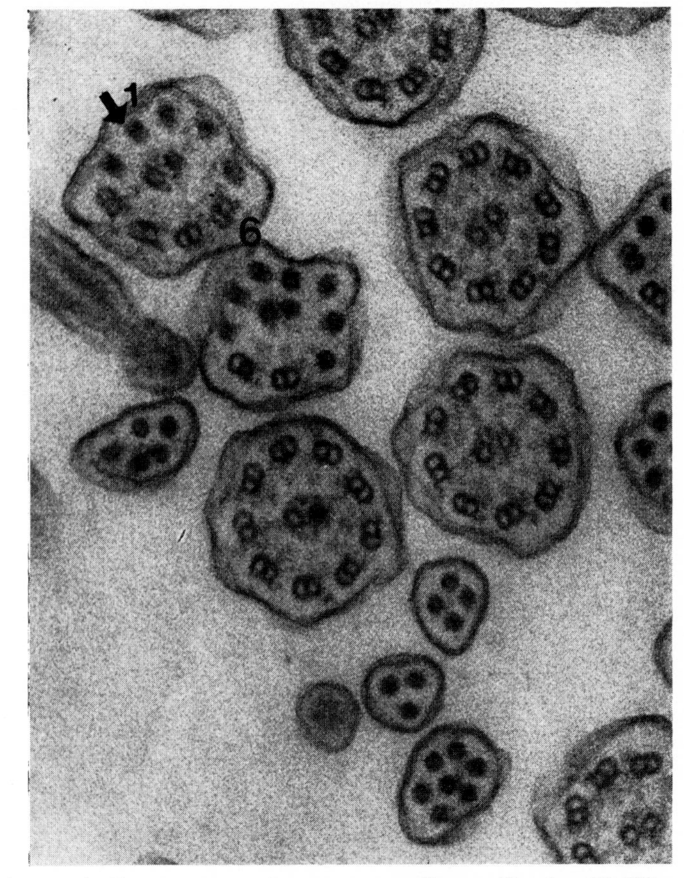

FIG. 2. (left) R-pointing mussel gill lateral cilia, (right) E-pointing. Arrows indicate cross-sections corresponding to Fig. 1. × 95,000.

can be predicted. For example, subfibers b of doublets nos 1 and 6 behave as if they are of equal length in a straight cilium. To bend the cilium in the direction of its effective stroke (E-pointing), subfiber b of no. 6 would slide out past subfiber b of doublet no. 1, while to bend the cilium in the opposite direction (R-pointing), no. 1 would slide out past no. 6. This can be visualized in cilia fixed in the appropriate stroke stages in cross-sections where some of the subfibers b have terminated (Fig. 2).

For a cilium composed of straight regions and circular arcs (Brokaw, 1965; Satir, 1967), the total sliding of any doublet microtubule relative to doublet no. 1 will be given by the equation:

$$\Delta l_n = d_n \Sigma \alpha \tag{1}$$

where Δl_n is the sliding of doublet no. n in μm; $\Sigma\alpha$ is the sum in radians of the angles subtended by the circular arcs, and d_n is the distance in

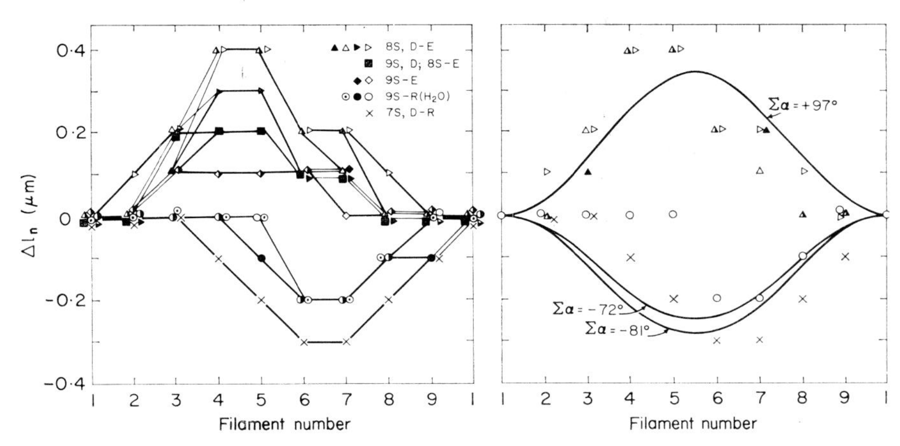

FIG. 3. Comparisons of prediction of sliding model to reconstructed ciliary tips. For complete method and notation, see Satir (1968). (Left) Measured points in E *vs.* R cilia. (Right) For maximum bends in E and R directions, predicted values (curves) match measured points. (Reprinted from *J. Cell Biol.*)

μm between doublets nos n and 1 projected on the axis of the cilium. For doublets 6 *vs.* 1, d_n is approximately the axonemal diameter (0·2 μm). In E-pointing cilia with a single bend, α may be maximally about 100°, arbitrarily designated +100°. In R-pointing cilia, α may be minimally −80°. Therefore, Δl_n varies from about +0·35 to −0·28 μm, where a negative Δl_n means that doublet no. 1 is sliding outward. Where this has been examined in detail, the quantitative correspondence between predicted and experimentally determined values for Δl_n is good (Fig. 3). The doublets move as integral units.

From these data, conclusions may be drawn that (1) the ciliary microtubules do not appreciably shorten as the organelle beats and (2) that the microtubules slide with respect to one another, at least at the ciliary tip.

III. Verification

When tritonated axonemes of sea urchin sperm are briefly treated with low concentrations of trypsin, the circumferential and radial links holding the axoneme together are interrupted. Addition of ATP then causes active

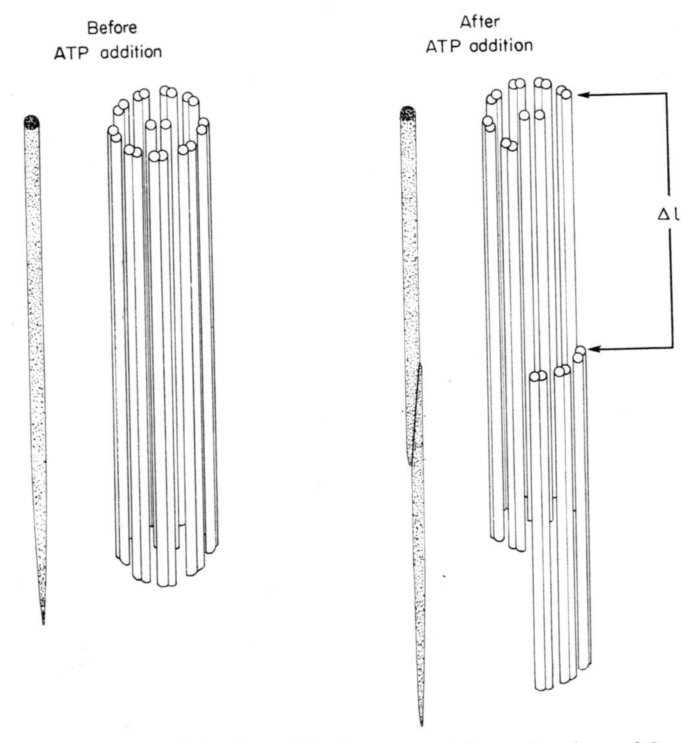

FIG. 4. Sliding in trypsinized models, interpreted from the data of Summers and Gibbons (1971). In each pair of diagrams, the figure to the left represents a dark-field light microscope image of the axoneme. This is interpreted at the electron microscope level in the adjacent diagram.

sliding of the microtubules (Summers and Gibbons, 1971) with or without bending. The net result is that the axoneme grows longer and thinner (Fig. 4) as groups of doublets crawl over one another. Finally all doublets are separated. It thus seems clear that ATP induces microtubule sliding in axonemes as a primary event in ciliary motility. This sliding is normally converted into bending by the series of intermicrotubule connections within the axoneme that are digested by the trypsin. These connections provide mechanical feedback for bend amplitude and propagation.

Brokaw (1971, 1972) has been able to develop a computer model based on the sliding microtubule process that predicts the detailed form of wave propagation and swimming of real isolated flagella with startling accuracy. A feedback loop is devised in the model by assuming that the magnitude and direction of the local shearing force produced by sliding are directly proportional to the local curvature of the axoneme. In this way, active sliding does not occur simultaneously throughout the axoneme, but propagates with a fixed relationship to the propagating bends. This model is referred to in Chapter 5 by Brokaw. One important consequence of the model is that starting from any initial position, the flagellum will develop a stable movement pattern within a short time. After two or three beat cycles, the transient effects of the initial position have disappeared, to be replaced by a stable, periodic movement that is independent of the initial state. Therefore, in this version of the sliding hypothesis, no special initiator is required for bending at the base of the axoneme, but a control mechanism is required to maintain phase differences between the active elements along the axonemal length.*

IV. Consequences

The sliding microtubule model lends strong support to the suggestion that there is a constant basic motility mechanism for all 9+2 cilia. Regardless of the detailed feedback loop between microtubule sliding and bend propagation, the gross amount of observable sliding is seen to depend on d_n, the axonemal diameter. The model thus accounts for the constancy of this parameter among various 9+2 cilia. The precise form of beat of a particular cilium seems, however, to depend upon modulations of the basic mechanism caused by specific environmental or morphological parameters.

During sliding, any connections extending from the microtubules would become stretched were they firmly anchored at their projecting ends. Such connections include not only links between adjacent microtubules or links to the central sheath (the radial spokes) but also microtubule–membrane linkages, such as the ciliary necklace (Gilula and Satir, 1972). For the radial spokes, evidence of stretch can be seen best in bent regions along the axoneme, but not in the straight regions distal to a bend where the spokes are, usually, perpendicular to the microtubules (Satir, 1973).

* Lastly, further work with gill cilia (Satir and Warner, 1973; Warner and Satir, 1974) uses radial spoke attachment sites as markers of length along axonemal doublets. By direct measurement and photographic translation, it can be shown that minimal local contractions of the microtubules are absent in both straight and bent regions of the axoneme, and that relative displacements of microtubules where Δl_n accumulates occur only in the bent regions, as indicated in Fig. 1.

Therefore, a particular spoke head must be able to attach and detach periodically from a passing row of central attachment sites (Warner and Satir, 1974). This could provide part of the physical basis of variations of shear-resistance within the axoneme.

Similarly, sliding would produce directional distortion of the permanent membrane–microtubule complex at the necklace region of somatic cilia. Such distortion could be coupled to membrane depolarization in such a way as to make the cilium an effective mechanoreceptor. A mechanosensitive mechanism based on sliding microtubules could be retained in cilia-based sensory organelles, such as the insect campaniform sensillum, or the hair cell of the vertebrate ear.

V. Difficulties

The major difficulty encountered by the sliding microtubule model is to account for the simultaneous propagation of bends of opposite sense along one axoneme. At first glance, this seems to require simultaneous sliding of a single doublet in opposite directions at the same time. Actually, during such propagation no overall change in Δl_n occurs, so that gross tip configuration changes are absent, and there can be simply compensating local adjustments of shear and bend. On the other hand, to satisfy eqn (1) it would seem that when a new bend is initiated at the base of an axoneme, some doublets must slide in a specified direction along their entire length to produce the required tip change. It is difficult to explain why this new sliding does not produce observable perturbation of the bends propagating further along the axoneme. Brokaw's computer equations give at least one successful version of the physical adjustments within the axoneme that satisfy these conditions.

The elegance of Brokaw's model does not ensure that it is a unique solution; nor is it claimed to be. Bend initiation and propagation could be separable phenomena, for example. This may be particularly important in ciliary motion, where the bends in opposite directions on the axoneme may be highly asymmetric (Satir, 1967) or even monophasic (Sleigh, 1969), so that an effective-recovery stroke cycle is generated. In such cases, there may need to be special ways of timing stroke phase and directional indicators that would introduce considerable complexity into a mechanical feedback loop. There is no reason to suppose that directional, timing or phase, and ionic controls are not part of the fundamental sliding mechanism, but they are not yet successfully incorporated into the model. Equation (1) is only correct if there is no sliding at the base of the cilium, as seems to be the case for mussel gill lateral cilia for which the equation was derived. The initiation problem is less difficult if sliding can occur at the base, as it may in mammalian cilia.

VI. Problems of mechanism

It is not too difficult to envision that ciliary microtubules slide by a ratchet mechanism analogous to that presumed to act in actomyosin systems, including striated muscle. The difficulty is that hard evidence for such a mechanism in ciliary motion is still elusive. Because their trypsinized models lengthen to over five times their original length, Summers and Gibbons (1971) conclude that the basic interaction in sliding occurs between adjacent doublet microtubules. If this is so, it seems evident that the dynein arms form one component of the sliding mechanism, especially since the arms are trypsin resistant, and so are present in the treated models while other possible linkages are disrupted (Summers and Gibbons, 1973). Also, dynein is the main ciliary ATPase, and the conditions for activating the models are precisely those for activating dynein.

The basic mechanism is likely to remain obscure until the additional components, primarily those responsible for attachment to the adjacent doublet, are identified. There have been repeated attempts to identify actin itself in the axoneme. Hitherto, these claims have proved incorrect, and it remains to be seen whether the most recent claims of Behnke *et al.* (1971), for example, will stand the test of time.

A second possibility is that tubulin substitutes for actin in the microtubule system. This is attractive because of the similar characteristics of tubulin and actin. There is no known repeat of microtubule substructure that stands out as a probable attachment site period, but perhaps this is still to be discovered, particularly as more information becomes available on the assembly of subfiber b. Of course, dynein does attach at one end to unknown periodic sites along subfiber a; all we now know of this interaction is that it is sensitive to Mg^{++} concentration. A final possibility is that one or another of the elusive links interacts with dynein in the sliding mechanism.

One would hope that electron microscopy would prove useful in studying the intimate sliding cycle, just as it has in striated muscle (Reedy, 1968). Unfortunately, no one has yet been able to read the details of the mechanism from cross-sections, and good longitudinal sections of correct orientation and thickness are virtually unknown. Most of the dynein arms are seemingly not attached to anything in normal preparations, and their kinks tell us nothing yet. However, it is now possible to prepare axonemes in rigor by quickly diluting out the ATP from swimming tritonated sperm models (Gibbons, personal communication). Under such conditions, the arms are all attached to the adjacent doublets. These preparations should prove valuable in an analysis of arm action. One good question presented by the normal morphology is: why are there two rows of arms whose structures differ slightly? Gibbons and Gibbons (1972b) have removed the outer row of dynein arms in tritonated models where the radial spokes are intact; in these preparations, beat form is maintained but frequently falls to half normal.

If one assumes that ATP is dephosphorylated at every active interaction site along a doublet every time the doublet slides by one dynein-arm period ($\sim$175 Å), we can calculate the maximum distance between successive interaction sites. We arrive at a rough estimate of the maximum total sliding during a single ciliary beat by adding the absolute values of Δl_n for every doublet in E-plus R-pointing cilia (Fig. 3). The sum is about 2·7 μm or about 0·3 μm per doublet. Values of ATP utilization (Brokaw, 1967) or the specific activity of dynein (Gibbons, 1966) suggest that approximately 1 ATP molecule is dephosphorylated per dynein per beat. Thus, active interaction sites must be spaced at average distances of no more than about a third of a micron apart along a doublet. At such a distance, the interaction site would handle a cycle of 20 or so consecutive arms as sliding progressed, while each arm would see an interaction site only once per beat. Of course, the distance between interaction sites might be considerably smaller than this maximum, particularly if the active sliding region propagates along a doublet once per beat. In this case, an interaction site will see only a single dynein, and the mechanochemical mechanism will be cyclic with a periodicity corresponding to beat frequency, which is much longer than the corresponding cycle in striated muscle.

The ciliary waveform could be produced either by microtubules on opposite sides of the cilium actively sliding in the same direction at different phases of beat or microtubules on one side actively sliding in opposite directions at different beat stages. Subfiber a is highly polarized (Warner and Satir, 1973) and all subfibers a around the axoneme have equivalent structural polarities. Such polarity is usually reflected mechanistically. In Summers and Gibbons' (1971) models, sliding maintains a constant polarity along the length of each doublet and although the analysis is not yet conclusive, it seems probable that the polarity of sliding is the same on all nine doublets. In this case, the first alternative would seem the correct one. However, a variation of the second alternative—that microtubules at both sides of the cilium actively slide in one direction during the effective stroke and in the opposite direction during the recovery stroke—would also fit the available structural data. Conceivably, one row of arms could be used for sliding one way, and the other row for sliding the other way (Sleigh, personal communication), but this possibility now seems unlikely because of the finding by Gibbons and Gibbons (1972b) discussed above.

VII. Conclusions

It can be seen from the foregoing that the sliding microtubule hypothesis has been a valuable one in clarifying the mechanism of ciliary motion. The following points may be regarded as settled:

1. The cilium is an active organelle, with the axoneme alone responsible for the production of bending.
2. Microtubule sliding accompanies ATP dephosphorylation as a fundamental event in ciliary motility.

The following may be regarded as highly probable:

1. There is no local longitudinal contraction of the microtubule doublets during motility.
2. The basic sliding interaction takes place between the dynein arms on one doublet and some attachment site on the adjacent doublet.
3. Sliding is normally converted into bending by a feedback mechanism involving connecting links within the axoneme that are stretched as sliding proceeds.
4. The sliding mechanism is common to all motile cilia and flagella.

The following may be regarded as speculative or unsettled at this time:

1. All doublets actively slide in the same direction.
2. Active sliding does not occur simultaneously along the length of a doublet.
3. No special initiator is required for sliding and bending since initiation is a consequence of a mechanical feedback loop.
4. Cellular and environmental controls change the form of wave propagation and initiation by altering the relationship between sliding and bend propagation.

Acknowledgements

This article was written while the author was a guest of the Zoological Institute, Faculty of Science, University of Tokyo, Japan, under the auspices of the NSF US-Japan Cooperative Science Program and the John Simon Guggenheim Memorial Foundation. The original work reported herein was supported by a grant from USPHS (HL 13849).

References

Afzelius, B. (1959). *J. biophys. biochem. Cytol.* **5**, 269–278.
Behnke, O., Forer, A. and Emmensen, J. (1971). *Nature, Lond.* **234**, 408–410.

Bradfield, J. R. G. (1955). *Symp. Soc. exp. Biol.* **9**, 306–334.
Brokaw, C. J. (1965). *J. exp. Biol.* **43**, 155–169.
Brokaw, C. J. (1967). *Science, Wash.* **156**, 76–78.
Brokaw, C. J. (1971). *J. exp. Biol.* **55**, 289–304.
Brokaw, C. J. (1972). *Biophys. J.* **12**, 564–586.
Fawcett, D. W. and Porter, K. R. (1954). *J. Morph.* **94**, 221–281.
Gibbons, B. H. and Gibbons, I. R. (1972a). *J. Cell Biol.* **54**, 75–97.
Gibbons, B. H. and Gibbons, I. R. (1972b). *J. Cell Biol.* **55**, 84a.
Gibbons, I. R. (1966). *J. biol. Chem.* **241**, 5590–5596.
Gilula, N. B. and Satir, P. (1972). *J. Cell Biol.* **53**, 494–509.
Gray, J. (1955). *J. exp. Biol.* **32**, 775–801.
Huxley, A. F. and Niedergerke, R. (1954). *Nature, Lond.* **173**, 147–149.
Huxley, H. and Hanson, J. (1954). *Nature, Lond.* **173**, 149–152.
Manton, I. (1952). *Symp. Soc. exp. Biol.* **6**, 306–319.
Reedy, M. (1968). *J. molec. Biol.* **31**, 155–176.
Satir, P. (1965). *J. Cell Biol.* **26**, 805–834.
Satir, P. (1967). *J. gen. Physiol.* **50** (2), 241–258.
Satir, P. (1968). *J. Cell Biol.* **39**, 77–94.
Satir, P. (1973). *In* "Behaviour of Microorganisms" (A. Pérez-Miravete, ed.), pp. 214–228. Plenum press, London.
Satir, P. and Child, F. M. (1963). *Biol. Bull.* **125**, 390.
Satir, P. and Warner, F. D. (1973). *Abstr. 4th Intern. Congr. Protozool., Univ. Clermont*, p. 363.
Sleigh, M. A. (1969). *In* "Handbook of Molecular Cytology" (A. Lima-de-Faria, ed.), pp. 1244–1258. North-Holland, Amsterdam.
Summers, K. E. and Gibbons, I. R. (1971). *Proc. natn. Acad. Sci. U.S.A.* **68**, 3092–3096.
Summers, K. E. and Gibbons, I. R. (1973). *J. Cell Biol.* **58**, 618–629.
Warner, F. D. and Satir, P. (1973). *J. Cell Sci.* **12**, 313–326.
Warner, F. D. and Satir, P. (1974). In preparation.

Chapter 8

Hydrodynamic aspects of ciliary and flagellar movement

M. E. J. HOLWILL

Department of Physics, Queen Elizabeth College, London, England

I. Introduction

From the hydrodynamic point of view cilia and flagella are essentially motile organelles which are used either to propel an organism through its liquid environment or to induce a flow of liquid over a stationary cell surface. Important quantities associated with such a system are the force communicated to the liquid and the energy dissipated during movement. Although a rigorous evaluation of these quantities is difficult, correct solutions are important, for not only are the answers relevant to basic hydrodynamics but they reflect also on the mechanochemical system which bends cilia and flagella. The energy dissipated against hydrodynamic forces represents the minimum energy which the cell must provide, in chemical form, for the activity of the cilium or flagellum. Demonstrably correct theoretical calculations of the hydrodynamic energy dissipation are of great value to the biochemist or biophysicist who is concerned with the molecular interactions that are responsible for flagellar undulation or ciliary beating.

This chapter will be concerned solely with a consideration of the interaction between organelles and their surrounding medium. Discussions of the internal structure and function of cilia and flagella will be found in other chapters of this book.

II. General hydrodynamic principles

1. *Origin of forces*

When a body moves through a liquid the forces acting on it are, in general, of two types. The first type, referred to as *inertial*, arise because the movement of the body imparts momentum to the surrounding liquid; the magnitude of the force is related to the rate of change of the momentum. The other type of force originates from frictional effects at the interface between the body and fluid. There is considerable evidence that no relative motion exists between the layer of liquid in contact with the body surface and the body surface itself, so that tangential stresses, giving rise to *viscous* forces, are set up in the liquid. By the principle of action and reaction, these viscous forces will also act on the body.

It is possible to set up general equations of motion which include the effects of both inertial and viscous forces, but the solution of such equations, except in trivial cases, is extremely difficult and often analytically impossible. In many cases of practical importance it is found that one type of force dominates the other, which can be neglected by comparison. An order of magnitude ratio between inertial and viscous forces acting on a system is given by Reynolds number, Re, which may be expressed

$$Re = \frac{lv\rho}{\eta} \tag{1}$$

In this equation l is a characteristic length of the body and v is its velocity, while η and ρ are the viscosity and density of the liquid in which the body is moving. Equation (1) relates to a body moving steadily with a velocity v through the liquid. Cilia and flagella elicit forces from the medium by virtue of their oscillatory movement, and they may induce a steady propulsive speed in the cell to which they are attached, although the velocity of each organelle varies with time. In considering the movement of cilia and flagella the velocity to be used in eqn (1) should be characteristic of the oscillatory motion, and may be considerably greater than the velocity of the cell itself. For an undulating flagellum beating with frequency f and amplitude a, an order of magnitude value for v is fa; a cilium of length L executing an oar-like movement with frequency f gives rise to a value of v of the order fL. In evaluating the stress set up in the fluid by a motile cilium or flagellum, the length of the organelle is less critical than its diameter, which is therefore chosen as the characteristic length l in eqn (1). Typical order of magnitude values for the parameters associated with the movement of cilia and flagella are given in Table I. Reynolds number

TABLE I. Order of magnitude values relevant to an evaluation of Reynolds number for ciliary and flagellar movement

Parameter	Order of magnitude
Frequency of beat f	10–100 Hz
Amplitude of flagellar wave a	1–10 μm
Length of cilium L	10–100 μm
Diameter of cilium or flagellum	0·1 μm
Viscosity of medium η	$\sim$1 mN s m^{-2}
Density of medium ρ	10^{3} kg m^{-3}

characteristic of this movement thus lies in the range 10^{-3}–10^{-6}, so that the forces which act on the system are predominantly viscous, and inertial forces may be neglected by comparison (see also Lighthill, 1969).

2. *Force coefficients*

A rigorous theoretical treatment of the movement of flagellated micro-organisms requires solution of the Stokes equations with the appropriate boundary conditions. These equations, which relate to viscous flow in an incompressible fluid, are

$$\nabla . \mathbf{q} = 0 \tag{2}$$

and

$$\nabla p = \eta \nabla^2 \mathbf{q}, \tag{3}$$

where $\mathbf{q}$ is the velocity and p the pressure in the fluid. It has not been possible to obtain analytical solutions to eqns (2) and (3) for conditions which correspond to the movement of real cilia and flagella. Useful information has been obtained, however, by making certain approximations relating to the movement itself or to the forces elicited thereby.

A very useful approximation is that made by Gray and Hancock (1955) of expressing the force acting on a straight cylinder in terms of surface coefficients of force. If a cylinder of length L moves with component velocities V_L and V_N along and normal to its axes the forces acting on the cylinder in these directions are, respectively,

$$F_L = -C_L V_L L \tag{4}$$

and

$$F_N = -C_N V_N L \tag{5}$$

where C_L and C_N are the appropriate force coefficients. For a long thin cylinder

$$C_N = 2C_L \tag{6}$$

and

$$C_L = \frac{2\pi\eta}{\ln\left(\frac{2L}{r}\right) - 0{\cdot}5} \tag{7}$$

where r is the radius of the cylinder (Hancock, 1953; Gray and Hancock, 1955). Most of the theoretical calculations relating to flagellar movement have made use of eqns (4)–(7). In the following sections the results of these calculations will be described and estimates made of the errors introduced by the approximations made.

III. Undulating flagella

A. Smooth organelles

Hydrodynamic treatments of undulating systems under conditions where viscous forces predominate over inertial ones were first undertaken by Taylor (1951, 1952) who showed that such a system could propel itself. Since the first of his papers was concerned with an undulating sheet and the second with an undulating cylinder having a small ratio of amplitude to radius the equations he derives are of limited value for quantitative prediction of the behaviour of flagellated organisms. The limitation on amplitude was removed in an analysis by Hancock (1953), who obtained equations from which various hydrodynamic characteristics of the system could be calculated. The form of the equations was such that computation was not simple, a factor which led Gray and Hancock (1955) to develop the concept of force coefficients described in the previous section. The application of Gray and Hancock's technique to a number of systems will now be discussed.

1. Plane waves

Consider the movement of a cell propelled by a single cylindrical flagellum which undulates in a plane as shown diagrammatically in Fig. 1. Microscopic observation of a real cell of this type would allow measurement of amplitude, wavelength, frequency of beat and velocity of propulsion. A theoretical analysis of the system predicts the propulsion velocity in terms of the wave parameters for comparison with experimental results. The thrust developed by the flagellum is first calculated and then equated to the drag on the cell body, making the assumption that an equilibrium condition has been established.

If, as in Fig. 1, the system is propelled along the x-axis of a Cartesian co-ordinate system, the components of force relevant to propulsion are

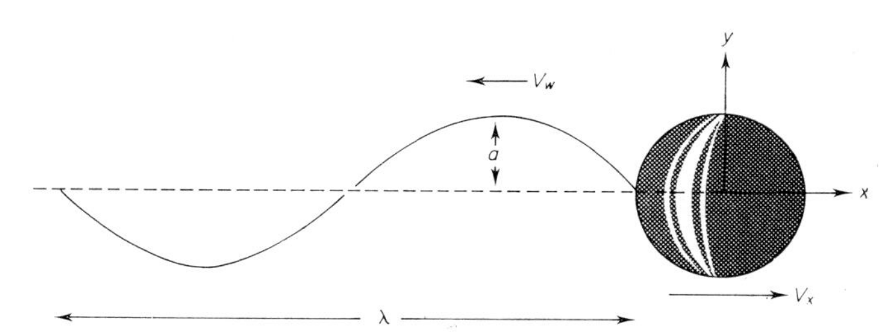

FIG. 1. Parameters associated with propulsion by planar waves of amplitude a and wavelength λ. V_w, V_x are the wave and propulsion velocities respectively.

those in the x-direction. Assuming that the displacement, y, is an arbitrary, single-valued function of distance x and time t, the force acting on a small element of the flagellum at any instant is

$$\mathrm{d}F = \left\{ \frac{(C_N - C_L)\dfrac{\mathrm{d}y}{\mathrm{d}t}\dfrac{\mathrm{d}y}{\mathrm{d}x} - V_x\left[C_N + C_L\left(\dfrac{\mathrm{d}y}{\mathrm{d}x}\right)^2\right]}{\left[1+\left(\dfrac{\mathrm{d}y}{\mathrm{d}x}\right)^2\right]^{\frac{1}{2}}} \right\} \mathrm{d}x, \tag{8}$$

where V_x is the velocity of the element in the x-direction and dx is the projected length of the element on the x-axis. To evaluate the force developed in the x-direction by the entire flagellum eqn (8) must be integrated over the flagellar length. The drag experienced by the cell body can be expressed as $C_B V_x R$ where C_B is a drag coefficient and R is a linear dimension of the body. For a spherical body $C_B = 6\pi\eta$ while R is the radius of the sphere. The velocity of the system at any instant is thus given by

$$C_B V_x R = \int_{x=0}^{L} \mathrm{d}F. \tag{9}$$

The velocity of an isolated flagellum, such as those obtained by glycerol extraction procedures (e.g. Brokaw, 1961; Douglas and Holwill, 1972), is obtained by setting C_B equal to zero in eqn (9).

Both the experimental and theoretical propulsion velocities are functions of time. Experimental values are usually quoted as time averages and it is convenient therefore to express the theoretical values in this way. The time averaged value, $\bar{V}_x$, is given by

$$\bar{V}_x = \frac{1}{T}\int_0^T (V_x)_t \mathrm{d}t \tag{10}$$

where $(V_x)_t$ is the value of the propulsion velocity at time t and T is a suitable time period. For an undulating flagellum T is conveniently taken as the period of the oscillation.

Once the propulsive velocity is known, the energy dissipated by the flagellum in overcoming viscous forces can be calculated from an equation derived by Carlson (1959). The mean power $\bar{P}$ is given by

$$\bar{P} = C_B R \bar{V}_x^2 + \frac{1}{T}\int_0^T \int_0^L (C_N V_N^2 + C_L V_L^2)\left[1+\left(\frac{dy}{dx}\right)^2\right]^{\frac{1}{2}} dx\, dt \qquad (11)$$

where V_N, V_L are velocities normal to and parallel to an element of the flagellum at the position x.

Flagellar waveforms vary considerably among species and even among individual organisms of the same species (Holwill, 1966a). Ideally, therefore, the application of the hydrodynamic equations should be made on an individual basis and the various derivatives determined from the actual flagellar movement. Since the observed waves are seldom regular or uniform this type of analysis requires the specification of the flagellar shape at short time intervals. Such information could, for example, be obtained by high-speed ciné-photographic methods. Expedient calculation of propulsive velocities and power expenditures would, in this situation, require a computer, using a program similar to that suggested by Brokaw (1970).

To obtain order of magnitude estimates of propulsive velocities and power expenditures, and to predict the variation of these quantities as the wave parameters change, it is useful to investigate the behaviour of flagella which propagate uniform, regular waves. Many authors (e.g. Gray and Hancock, 1955; Carlson, 1959) have chosen the sine wave for analytical purposes, although some reports (Brokaw and Wright, 1963; Brokaw, 1965) suggest that some flagella bear uniform waves which consist of circular regions, joined by straight lines. A further possibility is that the waves are meander-like, as discussed by Silvester and Holwill (1972), but neither this type of wave nor the one containing circular arcs is as amenable to mathematical analysis as the sine wave. Although the actual shape of the flagellum may provide important information about the internal mechanism which bends it, the hydrodynamic forces developed by the three wave shapes differ by only a few per cent, provided the wave parameters are similar (Brokaw, 1965; Silvester and Holwill, 1972).

The ways in which the propulsion velocity and power dissipation vary with the ratio of amplitude to wavelength for an isolated flagellum propagating uniform waves are shown in Figs 2 and 3. As can be seen, the value for $\bar{V}_x$ is always a small fraction of the wave velocity, V_w, and, according to the equations of Gray and Hancock, can never exceed one-half. The direction of propulsion is opposite to that in which the waves are propagated. A measure of the hydrodynamic efficiency of an isolated flagellum propagating planar waves is given by the ratio $\frac{\bar{V}_x^2}{P}$ where P is the total power dissipated against viscous forces to maintain a propulsion velocity $\bar{V}_x$. The efficiency is found to vary as the amplitude/wavelength ratio is altered, and passes through a broad maximum in the region $a/\lambda = 0{\cdot}16$ (Holwill and Burge, 1963). Measurements made on a number of organisms

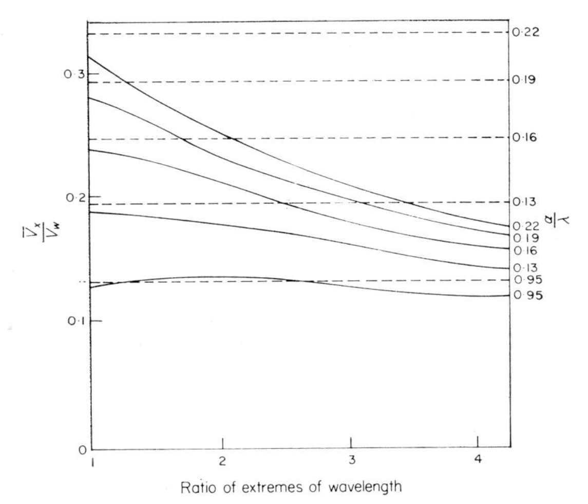

FIG. 2. The solid curves show how the ratio of the propulsion velocity $\bar{V}_x$ to the wave velocity V_w for an isolated filament undulating with constant amplitude but with varying wavelength depends on the magnitude of the non-uniformity, expressed as the ratio of the wavelengths at either end of the filament. The interrupted curves are obtained by inserting mean wave parameters in the equations of Gray and Hancock (1955). a is the amplitude while λ is the mean wavelength. (Reproduced with permission from Holwill and Miles, 1971.)

which sustain uniform, planar waves give values for a/λ close to that corresponding to the maximum hydrodynamic efficiency.

The non-uniformities in observed flagellar movements often take the form of increases in amplitude or wavelength as a wave moves along the organelle. It is of interest to investigate the theoretical behaviour of flagella which bear waves of this type since predictions are often made on the basis of average values for the amplitude and wavelength. The results of such calculations (Holwill and Miles, 1971) for an isolated flagellum bearing a wave which has an exponential increase in wavelength at constant amplitude are summarized in Figs 2 and 3. It is generally found that for corresponding ratios of the extreme values of amplitude and wavelength, the propulsive velocity for the varying wavelength case deviates further from curves obtained using the mean value of the parameters in Gray and Hancock's (1955) equations than does that for an isolated flagellum bearing waves of varying amplitude. The analysis by Holwill and Miles (1971) shows that for isolated flagella the use of average values in the equations of Gray and Hancock (1955) and of Carlson (1959) predicts the propulsive velocity and power expenditure satisfactorily when the amplitude varies alone, but not when the wavelength varies, either alone or with the amplitude.

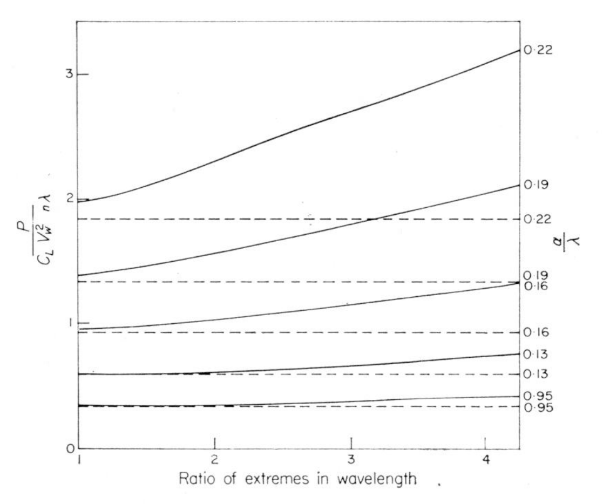

FIG. 3. The solid curves show the dependence of the external power expenditure P of an isolated filament undulating with constant amplitude but with varying wavelength on the magnitude of the non-uniformity expressed as the ratio of the wavelengths at either end of the filament. The interrupted curves are obtained by inserting the mean wave parameters in the equations of Gray and Hancock (1955) and of Carlson (1959). n is the number of wavelengths sustained by the filament; V_w is the wave velocity; λ is the mean wavelength; a is the amplitude; C_L is the longitudinal coefficient of force for the filament. (Reproduced with permission from Holwill and Miles, 1971.)

Most experimental observations concerning flagellar motility have been made on intact organisms, in which a flagellum propels an inert body. In this case the effects on propulsive velocity of non-uniform wave parameters are not so marked as in the case of an isolated organelle. This is illustrated in Fig. 4 for the particular case of a 2 : 1 ratio for the extremes of wavelength and a constant amplitude. The deviation from the curve (interrupted in Fig. 4) obtained from Gray and Hancock's (1955) equations becomes generally smaller as the body size increases. For an organism with a relatively large cell body, such as a protozoan, the value of the abscissa

$\left(\frac{C_B}{C_L} \cdot \frac{R}{n\bar{\lambda}}\right)$ is roughly 3 so that the difference between the two curves is only about 3%. Spermatozoa, on the other hand, have rather small heads and a value for $\left(\frac{C_B}{C_L} \cdot \frac{R}{n\bar{\lambda}}\right)$ as low as 0·5, and the discrepancy between the curves in Fig. 4 at this value of the abscissa is greater than 10%. When the power dissipation is considered as a function of body size, however, the

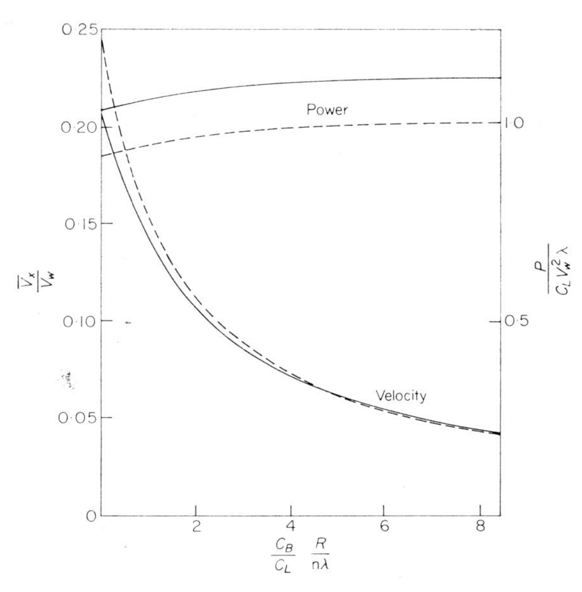

FIG. 4. The solid curves show how the propulsion velocity, $\bar{V}_x$, and the external power expenditure, P, of a filament propelling an inert body vary with the body size for a value of $a/\lambda = 0{\cdot}16$ where λ and a are the mean wavelength and amplitude of the oscillation. The results of the corresponding computations made by inserting the mean wave parameters in equations derived by Gray and Hancock (1955) and by Carlson (1959) are shown by the interrupted curves. V_w is the wave velocity; n is the number of waves sustained by the filament; C_B, C_L are force coefficients associated with the body and filament, R is the hydrodynamically equivalent radius of the head. (Reproduced with permission from Holwill and Miles, 1971.)

discrepancy between the curve calculated from Carlson's (1959) equation using mean values for the wave parameters and that calculated more rigorously does not decrease significantly as the body size increases. Thus, although the propulsive velocities of many organisms may be calculated using mean wave parameters, the energy dissipated against viscous forces may be significantly in error if this approximation is used, and the methods described by Holwill and Miles (1971) are preferable.

2. *Three-dimensional waves*

It is probably a rather rare occurrence for a flagellum to propagate absolutely planar waves. More often than not the movement has a three-dimensional component. The waveform in this category which has received most attention from the theoretical viewpoint is the circular helix, although this also is seldom observed on flagella attached to eukaryotic cells but exists on the flagella of many bacterial cells. It is, however, useful to discuss this waveform in an account concerning eukaryotic flagella, since it represents the limit to which a uniform three-dimensional wave can extend. Waves with shapes between a plane sinusoid and a circular helix, and thus more representative of waves observed on real flagella, have also been investigated from a hydrodynamic viewpoint, and the results will be referred to later.

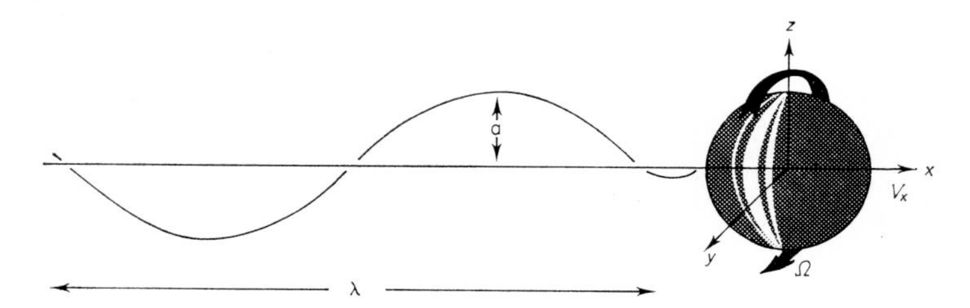

FIG. 5. Parameters associated with propulsion by three-dimensional waves of wavelength λ and amplitude a. V_x is the propulsion velocity while Ω is the angular velocity of the body.

Figure 5 shows, diagrammatically, a helical flagellum propelling a cell body. As for the case of two-dimensional waves, the thrust generated by the flagellum in the x-direction can be evaluated and equated to the hydrodynamic drag experienced by the body. The three-dimensional character of the waves also produces a torque which tends to rotate the entire flagellum about its axis of progression. In the absence of a cell body to oppose this rotation, the rate at which the flagellum rotates about the axis just compensates the rate of wave propagation, the flagellum remains stationary, and no propulsive forces are generated (Taylor, 1952; Chwang and Wu, 1971). With a cell body attached to the flagellum the torque generated by the propagated waves is communicated to the body, which is forced to rotate, but in so doing sets up an opposing torque. This opposing torque prevents free rotation of the system and allows a propulsive thrust to be generated along the axis of the helix. When the steady state is reached, the couple generated by the flagellum is balanced by that induced by the body rotation, and an equation describing this equilibrium can be written.

Hence, in discussing the hydrodynamics of a flagellum which propagates helical waves two equations can be set up, one to describe equilibrium of forces in the direction of propulsion, the other to describe rotational equilibrium. Both equations involve terms in the propulsion velocity, V_x, and the angular velocity, Ω, of the body about the axis of propulsion. The two simultaneous equations can be solved for V_x and Ω separately and yield equations for these velocities in terms of the flagellar wave parameters and the dimensions of the head (Chwang and Wu, 1971). These equations predict that the propulsion velocity is zero both when the body is absent, as already mentioned, and when the body is very large. The propulsion velocity thus passes through maximum value for a particular body size, provided the wave parameters remain unchanged (Fig. 6). Although a number of eukaryotic cells sustain three-dimensional waves

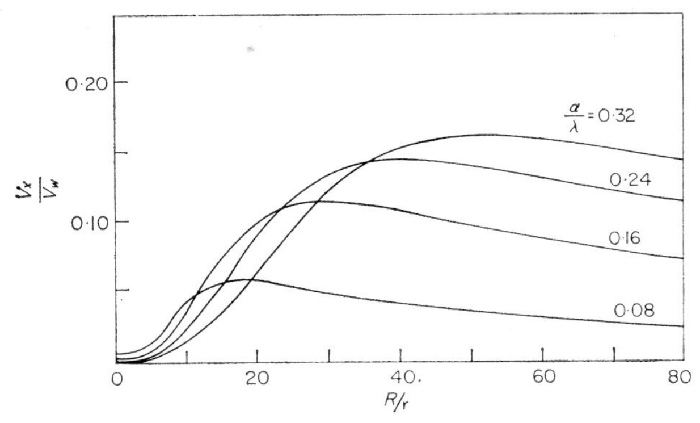

FIG. 6. Variation of propulsion velocity V_x with body size R (expressed in relation to the flagellar radius r) for an inert body propelled by a filament executing uniform helical waves of amplitude a and wavelength λ. (Reproduced with permission from Chwang and Wu, 1971.)

on their flagella, very few observations have been made of the magnitude of the wave parameters. For those organisms on which observations have been made (e.g. Holwill, 1966b), the equations of Chwang and Wu (1971) predict that the optimum body size is in the range of 15–40 times the flagellar radius. Since most smooth flagella have a radius of about 0·1 μm, the optimum body size is the hydrodynamic equivalent of a sphere with a radius between 1·5 and 4 μm. While many protozoa do have cell bodies which lie within this range of sizes, a number which are known to propagate three-dimensional flagellar waves are considerably larger. The body sizes of such organisms, if they are propelled by smooth flagella, appear to be determined, perhaps not surprisingly, by factors other than optimum hydrodynamic performance. In this regard, however, it is interesting to consider the organism *Euglena viridis*. This organism has, on average, a body hydrodynamically equivalent to a sphere of radius 10 μm (Holwill, 1966b) and hence is larger than the optimum expected if the organism is propelled by a smooth flagellum of radius 0·1 μm. Electron microscope studies (Pitelka and Schooley, 1955) have shown that the flagellum of *E. viridis* carries hair-like appendages, which have no obvious role and which may be wrapped round the flagellum during movement. The effective radius of the flagellum could thus be increased to 0·2 μm or more, so that the body size is close to the optimum value. The flagella of other organisms bear scales of various shapes and sizes, again for no obvious reason (e.g. Pitelka, 1963), and hence have radii greater than that associated with a smooth organelle. It is possible that these flagellar appendages exist to improve the hydrodynamic performance of the motile cell.

The hydrodynamic efficiency of propulsion by helical waves is proportional to V_x^2/P, as for planar waves, where P is the total power dissipated against viscous forces in maintaining a propulsion velocity V_x. For a given body size, the efficiency is a function of the amplitude/wavelength ratio of the flagellar undulations. The efficiency passes through a broad maximum in the region of $a/\lambda = 0{\cdot}16$ (Holwill and Burge, 1963; Chwang and Wu, 1970; Schreiner, 1971). The ratio a/λ for *E. viridis* lies close to this value (Holwill, 1966b), but no other records of sufficient detail relating to three-dimensional waves on eukaryotic cells are available. It is interesting to note that the a/λ ratios for flagellar bundles on a large number of prokaryotic cells are in the region of 0·16 (e.g. Liefson, 1960).

Propulsion of micro-organisms by flagellar waves of elliptical cross-section and by non-uniform three-dimensional flagellar waves has been investigated theoretically by Coakley and Holwill (1972). For given wave parameters and body size the propulsion velocity of a uniform wave with elliptical cross-section is found to be intermediate between that for a planar wave and that for a circular helix. The calculations relating to non-uniform three-dimensional waves lead to conclusions similar to those reached for non-uniform planar waves, i.e. propulsion velocities can reasonably be predicted by inserting mean wave parameters in equations derived for uniform waves (Holwill and Burge, 1963; Chwang and Wu, 1971; Schreiner, 1971), but the power dissipation will be considerably in error if mean wave parameters are used. Reasonable estimates of power expenditure by non-uniform waves require the use of equations similar to those of Coakley and Holwill (1972).

B. *Hispid flagella*

As mentioned earlier, the force elicited from the medium by a smooth, undulating flagellum is such as to propel the organism in a direction opposite to that of wave propagation. Flagella carrying hairs, or mastigonemes, can produce a force which induces propulsion in the direction of wave propagation (Jahn *et al.*, 1964; Holwill and Sleigh, 1967). A qualitative description of this phenomenon has been given by Jahn *et al.* (1964), who consider the mastigonemes to be rigid and to remain perpendicular to the flagellar surface during movement (Fig. 7). The appendages are therefore assumed to have no motile power of their own. The movement of mastigonemes at the crest of a wave is such as to produce a propulsive

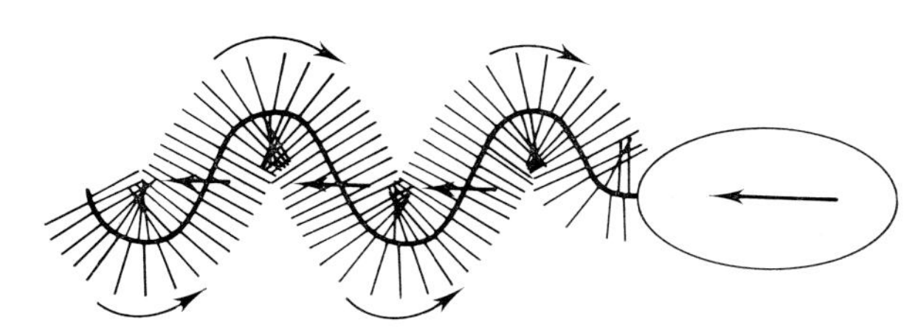

FIG. 7. Movement of mastigonemes (assumed to remain perpendicular to surface of flagellum). A wave moving in the direction of the arrows on the axis will cause mastigonemes to move in the direction of the peripheral arrows. If the organism is free to move it will do so in the direction of the arrow within the "cell body". (Reproduced with permission from Jahn *et al.*, 1964.)

force in the direction of the wave. The flagellar shaft elicits a force in the opposite direction. If the force due to the mastigonemes is greater than that produced by the shaft, a net force in the direction of the flagellar wave will result.

A quantitative analysis of propulsion by hispid flagella has been performed by Holwill and Sleigh (1967) who used the method of force coefficients discussed earlier. For a filament bearing a sinusoidal wave and propelling an inert spherical body eqns (8) and (9) yield,

$$\frac{V_x}{V_w} = \frac{a^2k^2[1-(C_L/C_N)]}{(2RC_B/n\lambda C_N)(1+\frac{1}{2}a^2k^2)^{\frac{1}{2}}+(2C_L/C_N)+a^2k^2} \tag{12}$$

where $k = 2\pi/\lambda$ and n is the number of wavelengths on the filament. From eqn (12) it is clear that the sign of $\frac{V_x}{V_w}$ depends on the magnitude of the ratio C_L/C_N. A negative value for V_x/V_w indicates that the two velocities are in the same direction while a positive value indicates that they are opposite in direction. For a smooth flagellum eqn (6) holds and the numerator of eqn (12) has the value $+\frac{1}{2}a^2k^2$. V_x/V_w is therefore positive and the propulsive force is in the opposite direction to that of wave propagation. The numerator of eqn (12), and hence V_x/V_w, can only become negative if the ratio C_L/C_N is greater than 1. If hispid flagella function in the manner described the ratio of their surface coefficients should have an appropriate value.

To calculate C_L/C_N, Holwill and Sleigh (1967) assumed the flagellum to be a cylindrical shaft with rod-like filaments projecting normal to its surface. If the cylinder carries two rows of q filaments per unit length (Fig. 8), and the flagellum moves such that the projections lie in the plane of beating, approximate values for the surface coefficients of the rough cylinders are

$$C_N = C_N^C + 2qLC_L^P \tag{13}$$

and

$$C_L = C_L^C + 2qLC_N^P, \tag{14}$$

where L is the length of a projection. The superscripts C and P refer to the cylinder and projections respectively. To obtain eqns (13) and (14)

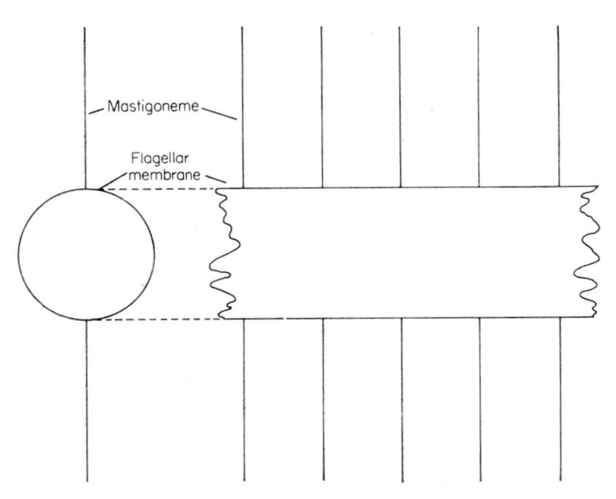

FIG. 8. Arrangement of mastigonemes assumed for calculation of force coefficients.

the influences on the fluid flow around one projection produced by movement of neighbouring projections and the main cylinder have been neglected. If the projections are not too close together and the fluid flow around the system is streamlined the errors arising from this approximation are not likely to be large. Using data obtained from electron microscopy together with eqns (13) and (14), the ratio C_L/C_N was found to be 1·8 for the hispid flagella of two chrysomonad flagellates, and hence eqn (12) predicts that the propulsive force is in the direction of wave propagation, as experimentally observed. As for smooth flagella the propulsive efficiency of hispid flagella passes through a maximum value for an amplitude/wavelength ratio of about 0·16. The experimental values lie in this region.

Although the agreement between theory and experiment provides some justification for the assumption that mastigonemes function in the manner described earlier, recent electron microscope evidence (Bouck, 1971) suggests that the mastigonemes may remain perpendicular to the plane of flagellar beating during movement. In this configuration no propulsive force would be exerted if the mastigonemes remain passive (Holwill and Sleigh, 1967), and alternative explanations for the phenomenon of reversal would need to be sought. Preliminary observations of fluid flow about hispid flagella (Miss P. D. Peters, personal communication) strongly suggest that the mastigonemes lie in the plane of beat during movement, but further experiments are needed to confirm this.

C. *Predicted values*

Wave parameters and propulsive velocities have been recorded by a number of observers using photographic and stroboscopic techniques. In most cases the observations relate to planar waves although a limited

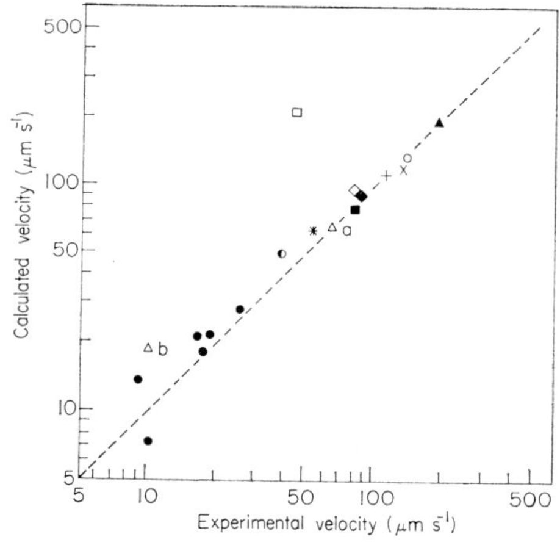

FIG. 9. Plot of observed velocity against theoretical velocity for a number of flagellated micro-organisms. ● Trypanosomatid flagellates (Holwill, 1964, 1965). ×, □ Bull sperm (Gray, 1958; Rikmenspoel *et al.*, 1960). ▲ Sea urchin sperm (Gray, 1955). ■ *Euglena viridis* (Holwill, 1966b). ◐ *Chlamydomonas* sp., ○ *Bodo saltans* (Holwill, 1964). Δ*a* *Polytoma uvella*; Δ*b* isolated flagella from *P. uvella* (Brokaw, 1961). + Rabbit sperm (Shack *et al.*, personal communication). ◇ Spore of *Blastocladiella emersonii*; ◆ *Allomyces* (Miles, 1969). * *Ochromonas malhamensis* (Holwill and Sleigh, 1967).

amount of information has been obtained for three-dimensional undulations. Table II contains a number of parameters relating to a variety of micro-organisms. From values such as those in Table II, the equations

TABLE II

Organism	Beat frequency (Hz)	Amplitude (μm)	Wave-length (μm)	Body size (μm) (hydrodynamically equivalent sphere)	Observed velocity (μm s^{-1})	Calculated velocity (μm s^{-1})	Reference
Crithidia oncopelti	17	2·4	14·4	3·8	17·0	21	Holwill (1965)
Euglena viridis	12	6	35	9·7	80	95	Holwill (1966b)
Psammechinus miliaris sperm	35	4	24	0·5	191·4	191	Gray and Hancock (1955)
Lytechinus pictus sperm	30	4·6	22·6	0·75	158	158	Brokaw (1965)
Ochromonas malhamensis	68·4	1·0	7	3	55–60	61	Holwill and Sleigh (1967)

discussed earlier can be used to predict the propulsive velocity of the organism. Figure 9 is a graph of theoretically predicted velocities against those observed experimentally. Good agreement can be seen between the two sets of values, thereby providing some confidence in the hydrodynamic assumptions made in the analysis. The validity of these assumptions will be discussed in a later section.

IV. Motion of cilia

Tonsate (or oar-like) movement is that characteristic of most cilia and some flagella. The tonsate motion comprises only one part of the ciliary beat cycle (usually that part known as the effective stroke). During the other part of the cycle (the recovery stroke) a bend is initiated at the base of the cilium and propagates along the organelle to restore it to a position at the beginning of the effective stroke. Figure 10 shows, in diagrammatic form, an idealized ciliary beat cycle. Cilia usually occur in large numbers

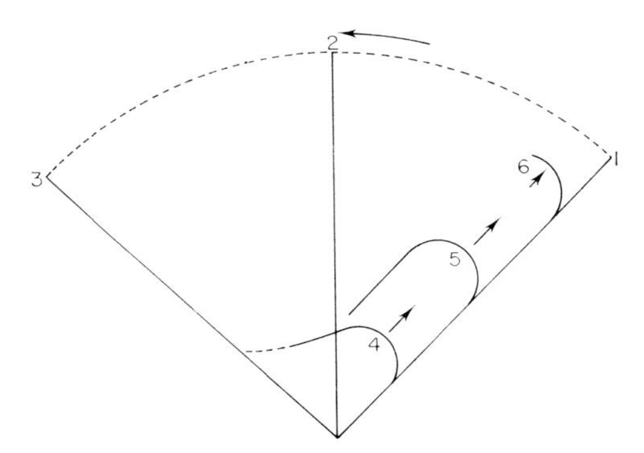

FIG. 10. Idealized form of ciliary beat: 1, 2, 3 effective stroke; 4, 5, 6, recovery stroke.

on the cell surface, a factor which can pose considerable difficulty in hydrodynamic analysis. The simplest hydrodynamic situation arises when the cilia form an isolated clump on a cell surface and the cilia within the clump beat synchronously. The system can then be treated as a single unit executing the movement of Fig. 10 and lends itself to analysis by the use of force coefficients. More frequently cilia are distributed fairly evenly over the cell surface and beat asynchronously, so that adjacent cilia are slightly out of phase, and the movement gives rise to metachronal waves. According to whether the metachronal wave propagates in the direction of the effective stroke or against it, the metachronal wave is described as symplectic or antiplectic (Fig. 11). Movement of the effective stroke perpendicular to the direction of wave propagation is also found in ciliates. According to whether the effective stroke is to the right or left of the wave direction, the movement is known as dexioplectic or

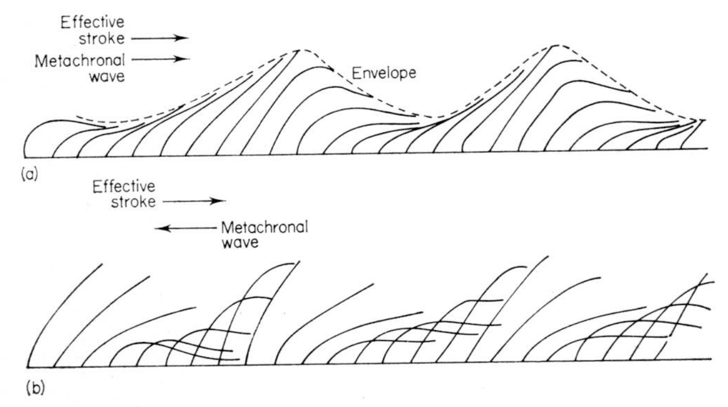

FIG. 11. Cilia executing (a) symplectic, and (b) antiplectic metachrony.

laeoplectic metachrony (Knight-Jones, 1954). This type of motion is not amenable to analysis by the use of force coefficients described earlier, but requires consideration of the overall movement of the entire field of cilia.

A. *Isolated groups of cilia*

The forces exerted by a group of cilia in which the individual members beat synchronously, so that the group can be considered as a single filament, can be estimated by the method of force coefficients described earlier. To estimate the forces with reasonable accuracy it is necessary to know how the shape of the cilium varies during its beat cycle. In principle, this information could be used in a computer program to work out the forces involved. It is convenient, however, to consider the forces which act on a cilium executing an idealized beat cycle (Fig. 10). During the effective stroke the cylinder which represents the cilium rotates about one end with an angular frequency ω, so that the (normal) force acting on an element $\mathrm{d}y$ at a distance y from the hinged end of the cylinder is

$$\mathrm{d}F = C_N \omega y \mathrm{d}y, \tag{15}$$

while the rate at which work is done by the element is

$$\mathrm{d}P = \omega y \mathrm{d}F = C_N \omega^2 y^2 \mathrm{d}y. \tag{16}$$

By integrating over the length of the cylinder, the total force generated is seen to be proportional to the square of the length of the cilium while the power dissipation varies as the cube of this length (Sleigh and Holwill, 1969). Expressions for the force exerted and the power dissipated during the recovery stroke can also be obtained using this type of analysis.

Viscous forces generated by the large abfrontal cilium of *Mytilus* have been calculated by integrating eqn (15) and a similar expression for the tangential force acting on an element (Baba and Hiramoto, 1970). These authors found that the maximum force exerted by the cilium was in the region of $2–3 \times 10^{-8}$ N. Sleigh and Holwill (1969) used the hydrodynamic equations to estimate the power dissipated against viscous forces by gill cilia of *Sabellaria* and the solitary abfrontal cilia of *Mytilus*. The viscous

TABLE III. Viscous work (in $J \times 10^{-15}$) done by a single cilium. (Modified with permission from Sleigh and Holwill, 1969)

	Sabellaria	*Mytilus*
Effective stroke	0·4	0·1
Recovery stroke	0·1	0·2
Total	0·5	0·3

work done by the two types of cilia is summarized in Table III. For *Sabellaria* the power dissipated against viscous forces is less during the recovery stroke than during the effective stroke, a result which is to be expected, since the observed resultant movement of water as a consequence of ciliary movement is in the direction of the effective stroke. In the case of *Mytilus* on the other hand, the cilium performs more viscous work in the recovery phase than in the effective stroke, and in this case there is little resultant water movement. This cilium probably has some function other than the propulsion of water and may be used to keep the gill surface clean.

B. *Extensive fields of cilia*

Some aspects of the complex hydrodynamic problem posed by the metachrony of large fields of cilia have been discussed by Blake (1971a, b, c, 1972). In one approach to the problem, Blake represents the ciliary motion by a surface envelope, which contains the tips of the individual cilia. Metachronal waves are thus modelled as undulations of this surface envelope. In applying the relevant Stokes flow equations (2) and (3), it is assumed that no slip occurs between the envelope and the fluid, and hence no account is taken of fluid squeezed out or sucked in by the ciliary movement. The envelope model is therefore satisfactory only for symplectic metachronism (Fig. 11a), in which fluid movement across the envelope is minimal. Considerable fluid flow is likely to occur across the envelope during the propagation of antiplectic metachronal waves (Fig. 11b), so the model described is probably not a realistic one for this case.

An alternative approach, applicable to both symplectic and antiplectic metachrony, is described by Blake (1972) and considers the movement of and interactions between individual cilia within a field. The field of cilia is represented by an infinite array of long, flexible cylinders each attached at one end to a plane surface. The latter model is a better physical representation of a ciliated surface than the envelope model which, however, allows discussion of propulsion of finite, as well as infinite, systems, and may thus better represent a motile organism. Both models will therefore be considered here.

The tips of individual cilia which participate in metachronal wave motion have components of movement normal and parallel to the cell surface. Accordingly, in the surface envelope approach, the undulations which best model the ciliary movement contain both transverse and longitudinal components. The shape of the surface envelope will not, in general, be sinusoidal and hence several analyses concerned with infinitely large, undulating sheets at low Reynolds number (e.g. Taylor, 1951; Reynolds, 1965; Tuck, 1968) cannot be applied to ciliary metachronism. The predictions of these authors are in accord with those of Blake for limiting cases of his models.

To represent the undulations of an extensible, undulating sheet Blake (1971b) uses the equations

$$x_0 = x+\beta \cos (kx+\sigma t)+\gamma \sin (kx+\sigma t) \tag{17}$$

$$y_0 = \alpha+\delta \sin (kx+\sigma t) \tag{18}$$

in which x_0, y_0 is the position of a particle (representing a ciliary tip) at time t, $k = 2\pi/\lambda$ where λ is the length of the metachronal wave and σ is the angular frequency of the wave. α, β, γ and δ are constants which characterize the shape of the wave. If β and γ are both zero, the wave is entirely transverse (Fig. 12f, Table IV) whereas if δ and γ (or β) are zero the wave is wholly longitudinal (Fig. 12g, Table IV).

The propulsion velocity, V_x, for an infinite sheet undulating in the manner described by eqns (17) and (18) is, to second order

$$V_x = \tfrac{1}{2}\sigma k(\delta^2+2\delta\beta-\beta^2-\gamma^2) \tag{19}$$

while the power expended is, to first order,

$$P = \eta\sigma^2 k(\beta^2+\gamma^2+\delta^2). \tag{20}$$

From eqn (19) it is clear that, for a transverse wave ($\beta = \gamma = 0$) V_x is positive while for a longitudinal wave ($\delta = \gamma$ (or β) $= 0$) V_x is negative.

Thus, longitudinal undulations alone propel a sheet in the direction of wave travel, whereas transverse oscillations alone cause a sheet to move in the direction opposite to that of wave propagation (Fig. 12). The net

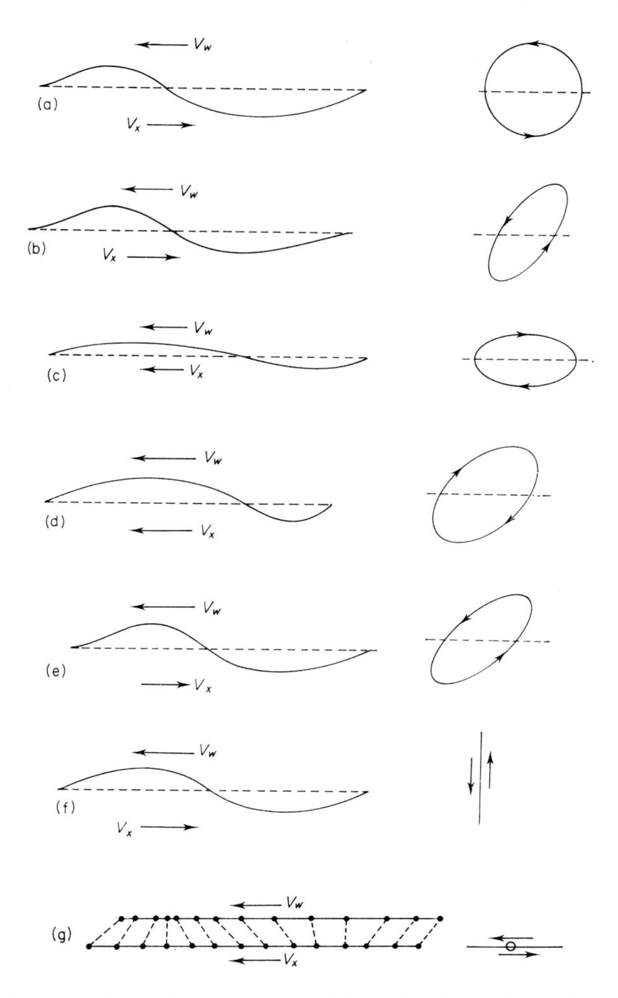

FIG. 12. Surface envelope shapes and particle paths for varying β, γ and δ (eqns (17) and (18)). See also Table IV. (Reproduced with permission from Blake, 1971b.)

force produced by a sheet containing both types of oscillation will be in a direction determined by the dominant component. Figure 12 illustrates the propulsive effect of an undulating sheet for a number of different wave shapes.

As mentioned earlier, the concept of an undulating envelope as a model for ciliary motion is reasonable for symplectic metachronism but is probably not satisfactory for antiplectic metachronism. In terms of the model, symplectic metachronism (Fig. 11a) corresponds to a predominantly transverse wave motion, so that organisms such as *Opalina* with cilia co-ordinated in this manner would be expected, and are observed, to move in the direction opposite to that of wave propagation. Antiplectic metachronism (Fig. 11b) corresponds to a predominantly longitudinal wave

TABLE IV. Characteristics associated with the wave shapes shown in Fig. 12. (Reproduced with permission from Blake, 1971b)

Wave	β	γ	δ	V_x/V_w	E (%)
a	0·5	0	0·5	0·25	25
b	0·25	0·25	0·5	0·19	19
c	−0·5	0	0·25	−0·22	30
d	−0·4	0·3	0·5	−0·2	16
e	0·3	0·4	0·5	0·15	9
f	0	0	0·5	0·125	12·5
g	0·5	0	0	−0·125	12·5

β, γ, δ are coefficients in eqs (17) and (18).
V_x is the propulsion velocity, V_w is the wave speed while
E is the efficiency (eqn 21).

and hence, according to this model, organisms which exhibit this type of metachrony are expected to move in the same direction as that of wave propagation. The metachronal waves of *Paramecium* appear antiplectoid and move forward over the body surface during locomotion, which is observed to occur in the direction of the metachronal waves. Thus although the physical representation of antiplectic metachronism by the model is limited, there is qualitative agreement between prediction and experimental observation.

The ratios of propulsive velocity, V_x, to wave speed, V_w, for the various wave shapes shown in Fig. 12 are given in Table IV. A measure of the efficiency of propulsion E is given by V_xT/P, where T is a characteristic thrust per unit area. For simplicity T is taken to be $2\,\eta V_x k$, which gives

$$E = \tfrac{1}{2}\,\frac{(\delta^2+2\delta\beta-\beta^2-\gamma^2)^2}{\delta^2+\beta^2+\gamma^2} \tag{21}$$

The maximum efficiency occurs for $\gamma = 0$ and $\beta = 0{\cdot}6\,\delta$ (or $\delta = -0{\cdot}6\beta$).

This corresponds roughly to the wave shape given by Fig. 12c. The maximum velocity occurs when $\gamma = 0$ and $\delta = |\beta|$(Fig. 12a) and is achieved with an efficiency of 25%.

Blake (1971a) has also considered the propulsive effects of undulations of an initially spherical envelope. As expected from the case of an infinite undulating sheet, movement of the sphere is in the direction of wave movement for predominantly longitudinal waves (Fig. 13a) and opposite to it for predominantly transverse oscillations (Fig. 13b). For transverse

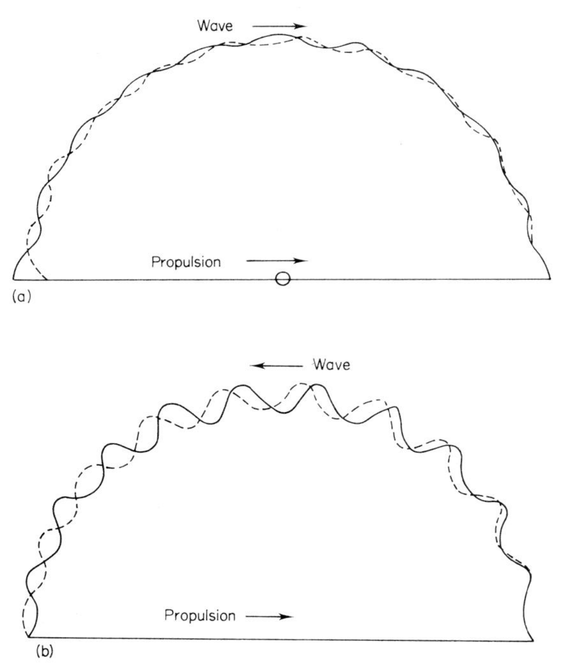

FIG. 13. Predominantly longitudinal (a) and transverse (b) waves on a hemispherical surface. The broken lines represent the wave positions one-quarter of a period later than the full lines. (Reproduced with permission from Blake, 1971a.)

waves alone, Blake (1971b) has shown that the propulsion velocity for the spherical case lies between 40 and 45% of that of an infinite flat sheet with similar wave parameters. The lower propulsion velocity for the spherical organism might be expected intuitively, since the cilia at the front and rear are not beating in a direction favourable for propulsion. Many ciliated micro-organisms have flat, disc-like shapes, so the propulsion velocities are likely to be intermediate between those predicted by the flat sheet and spherical sheet models.

With wave parameters similar to those observed on real organisms, both models predict propulsion velocities of the observed order of magnitude. For example, a wave with angular frequency 25 Hz and a shape similar to that of Fig. 13b, will propel a sphere of radius 100 μm at a speed of about 100 μm s^{-1}. These magnitudes are similar to those of the ciliate *Opalina* (Sleigh, 1968). An undulating sheet with characteristics close to those of the ciliary movement of *Opalina* also produces a propulsion velocity of order 100 μm s^{-1}.

A surface envelope model with characteristics similar to those of the antiplectoid metachronal waves of *Paramecium* predicts a propulsive

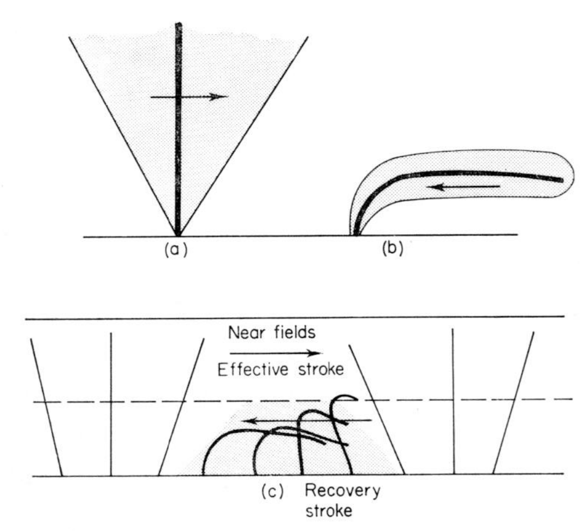

FIG. 14. Volume of fluid influenced by near-field effects in (a) the effective and (b) the recovery strokes of an individual cilium. In (c) the influence of the near-field effects on flow in the inter-ciliary fluid of a metachronal wave is indicated. (Reproduced with permission from Blake, 1972.)

velocity which is significantly lower than that generally observed. The discrepancy is attributed to the rather poor physical representation by the model of antiplectic metachrony discussed earlier.

In considering a model which retains the individuality of cilia within a field, Blake (1972) considers the influence on fluid flow both near to and far from the cilia. The "near-field" effects give information about fluid flow within the field of cilia while "far-field" effects are also involved when the propulsion velocity of the system is considered. For an element of a single cilium beating on an infinite plane sheet, the effective near field takes the form of a sphere of radius $\frac{1}{2}h$, where h is the height of the element above the plane. Hence, during the effective stroke, the near-field of the cilium influences a relatively large volume of fluid while in the recovery stroke the affected volume is rather small (Fig. 14a, b). A qualitative assessment of fluid flow within the ciliary field can be obtained by considering the combined near-field effects of the cilia. From Fig. 14c it can be seen that the layer of inter-ciliary fluid near the plane is influenced about equally by the effective and recovery strokes while that region remote from the plane is influenced only by the effective strokes. Thus one would expect oscillatory fluid motion in the regions close to the plane and a predominantly unidirectional flow in the outer layers. Little experimental evidence is available concerning fluid flow in the regions discussed, but Sleigh and Aiello (1973) report that the oscillatory component of fluid flow is small in the outer regions of inter-ciliary fluid moved by the comb plates of *Pleurobrachia*.

In a quantitative consideration of the system Blake defines an interaction velocity, U^*, to take account of the effects of neighbouring cilia. If V is the contribution of a particular cilium to the local mean velocity field while the overall mean velocity is U then

$$U^* = U - V + v' \tag{22}$$

where v' is the oscillatory variation of velocity about the mean and has a time-averaged value of zero. To evaluate the mean velocity field in the inter-ciliary fluid Blake neglects v' to set up integral equations from which U can be calculated by numerical analysis. In a review article of this type it is not appropriate to discuss further the complex analysis needed to obtain the mean velocity field, but a summary of the results will be given. Using an analytical representation of the movement of cilia on three organisms *Opalina*, *Paramecium* and *Pleurobrachia*, Blake obtained the mean velocity profiles shown in Fig. 15. The value of U at the tip of the cilia represents the propulsive velocity of the system and gives values of the order of magnitude observed experimentally. The ratio U/c, where c is the metachronal wave velocity, is found to be less than one for *Opalina* and to lie between 1 and 3 for *Paramecium*. The predicted values of U are thus of order 100 μm s^{-1} and 1000 μm s^{-1} respectively for the two organisms.

From Fig. 15b and c it is clear that a mean backflow of fluid is expected in the lower inter-ciliary region of fluid propelled by cilia in antiplectic metachrony although this may be an artefact resulting from the approximations made in the analysis. Such backflow has not been reported for *Pleurobrachia*, the only organism for which fluid flow has been critically examined (Sleigh and Aiello, 1973), but this organism may well violate the low Reynolds number assumption on which the analysis depends.

The force exerted by an individual cilium in a metachronal wave varies with the phase of the beating, as might be expected, and is different for symplectic and antiplectic rhythms. Thus for antiplectic metachrony, the force exerted parallel to the plane of attachment is larger during the effective stroke than during the recovery stroke. In symplectic metachrony the maximum force parallel to the plane of attachment occurs during the recovery stroke immediately before the effective stroke begins. Such a different distribution of forces is not unexpected when the movement of individual cilia in the two types of metachronal wave is considered (Fig. 11). Thus cilia in symplectic metachrony are expected to have a co-operative effect during the effective stroke but not during the recovery stroke while the reverse is true for cilia executing antiplectic metachrony.

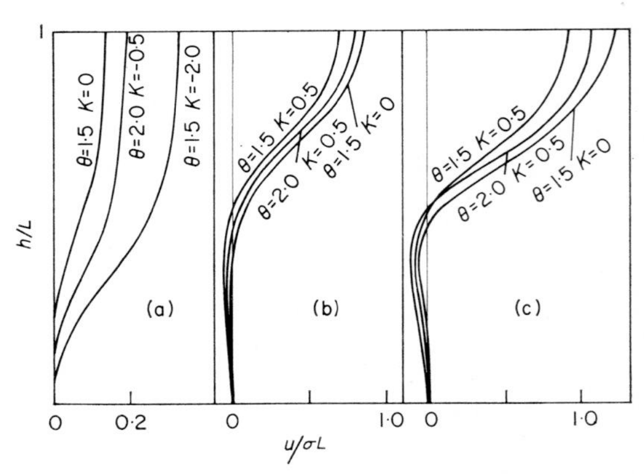

FIG. 15. Variation of mean fluid velocity U in the inter-ciliary fluid with height h above the surface of attachment for metachronal wave patterns similar to those on (a) *Opalina*, (b) *Paramecium*, and (c) *Pleurobrachia*. L is the total length of a cilium while σ is the beat frequency. θ is the ratio of force coefficients while $K=\sigma L/C$ where C is the metachronal wave velocity. (Reproduced with permission from Blake, 1972.)

The rate of working for individual cilia is also found to be different in the two types of metachrony, again in the manner to be expected from the co-operative effects mentioned above. For an individual cilium the power dissipated is largest during the effective stroke for antiplectic metachrony and during the last part of the recovery stroke in symplectic metachrony. For a given beat frequency and ciliary length the results indicate that antiplectic metachrony is the more efficient means of propulsion.

V. Model micro-organisms

To investigate the physical behaviour of a living system it is frequently convenient to make a scale model which can be manipulated under controlled experimental conditions. An examination of the hydrodynamic behaviour of a micro-organism using this approach is best made with models which are considerably larger than the micro-organism. For predictions made on the basis of the behaviour of the model system to be applicable to the real organism, it is important that a scaling factor be considered in the construction of the model. Of paramount importance in the scaling factor is the Reynolds number. Conclusions derived from a study of the model (Reynolds number *Rm*) will obtain for the full scale system (Reynolds number *Re*) only if *Rm* is of the same order as *Re* (e.g. Holwill, 1970). As mentioned earlier, the Reynolds number for flagellar movement of eukaryotic micro-organisms lies in the range 10^{-6}–10^{-3}, so that for correct simulation, the Reynolds number of a model micro-organism should be significantly less than one.

Several workers have used scaled-up models of micro-organisms in attempts to understand the hydrodynamic forces involved in propulsion

TABLE V. Reynolds numbers (*Rm*) associated with various model micro-organisms. (Reproduced with permission from Holwill, 1970)

Model	*Rm*	Reference
Wires rotating in water	> 100	Metzner (1920)
Cylinder rotating in air	> 400	Lowndes (1944a)
or water	> 500	Lowndes (1944a)
Leather thong undulating in water	about 250	Lowndes (1944b)
Model cell body rotating in water	about 4000	Brown (1945)
Human body with arm rotated	about 27,000	Brown (1945)
Model spermatozoon in glycerine	about 0·2	Taylor (1952)

of the living system. Table V contains descriptions of the models used, together with estimates of the respective Reynolds numbers. Taylor's (1952) model is the only one in which viscous forces dominate inertial ones and his main conclusion, that undulating filaments can propel a system at low Reynolds number, can reasonably be applied to the living system.

The other models described in Table V were used by their authors to support the idea that many organisms propel themselves by body oscillation, which is induced by flagellar activity. According to these authors the movement of the flagellum elicits no propulsive force from the medium directly, but only indirectly by causing the body movement. This conclusion is supported solely by the behaviour of models in which inertial forces dominate viscous ones and is therefore untenable. A theoretical hydrodynamic analysis of the system (Holwill, 1966b) shows that body gyration of the type seen on such organisms as *Euglena* provides insufficient force to propel the organism at the speeds observed. Flagellar activity, on the other hand, can provide sufficient force to account for the observed speeds of propulsion.

In a recent paper Jahn and Votta (1972) support the original conclusions of the authors reporting the model experiments on the grounds that "the fact that the Reynolds numbers are different does not necessarily prove that the principles of the mechanisms are different". While this may be true if the Reynolds numbers of the two systems are, say, 100 and 1000, so that the forces involved are predominantly inertial in both cases, it is certainly not so when the Reynolds number for the model system is greater than 100 and that for the actual system is 10^{-3} or less. In this last situation, which corresponds to some of the model micro-organism experiments (Table V), the forces acting on the model system are largely inertial, but those acting on the real system are predominantly viscous (see section II.1 for a discussion of these two types of force). The forces acting on the model system are, as mentioned earlier, not representative of those acting on the living organism and conclusions based on the behaviour of the former cannot be correctly applied to the latter.

VI. Validity of hydrodynamic analysis

1. *Individual, isolated cilia and flagella*

The validity of the hydrodynamic analyses as applied to a living system depends on the accuracy of the model chosen to represent the system and on the nature of any approximations made during the analysis. In considering the movement of individual, isolated organelles, the system is modelled by a motile cylindrical filament. The cylindrical shape is justified by the results of electron microscope examination of numerous cilia and flagella (see e.g. Holwill, 1966a, for references). The movement of the filament can often be matched by a suitable analytical equation, but if this is not possible the movement can be simulated in a computer. The model therefore appears to represent the living system reasonably well.

The main hydrodynamic approximation made concerns the calculation of the surface coefficients C_N and C_L which are used in equations based on Gray and Hancock's (1955) analysis. The approximations made by Gray and Hancock, and by a number of other authors, are given in eqns (6) and (7), and relate to a long, thin straight cylinder. In applying these equations to motile, undulating systems some assumptions are clearly necessary. One of these involves the choice of a length to use in eqn (7) for the calculation of an absolute value of C_L. Usually, and arbitrarily, the length is taken as the wavelength of the undulating system. Since the length appears within a logarithm, a large error in its value will not dramatically affect the magnitude of the surface coefficient. A second approximation relating to the surface coefficients concerns the ratio C_N/C_L. The value of 2 for this ratio is only valid as a first approximation and may be as low as 1·8 for flagella (Brokaw, 1965) or 1·6 for cilia (Blake, 1972).

Recent work (Cox, 1970; Batchelor, 1970; Tillett, 1970) on Stokes' flow about slender bodies of length l and radius R shows that $C_N/C_L = 2+0(\varepsilon^{-1})$ where $\varepsilon = \ln(l/R)$ and $C_L = 2\pi\eta/(\varepsilon - 0{\cdot}807)$, for cylinders with straight centre lines. Since ε is not large, even for thin flagella, accurate values of C_N/C_L require higher approximations than those generally used. If the centre line of the body is curved the effect of the curvature on the flow will be small when compared with the flow about a straight cylinder, provided the effective wavelength and amplitude of the curve are much larger than the radius of the cylinder, as they are for flagella. Some indication of the effects of incorrect choice of the ratio C_N/C_L has been given by Coakley and Holwill (1972). Using Gray and Hancock's (1955) equations a change of C_N/C_T from 2·0 to 1·8 produces a 20% reduction in the calculated propulsion velocity of an isolated filament executing sinusoidal oscillations with an amplitude/wavelength ratio of 0·16. The calculated power dissipation is, however, little affected by this change.

In the analyses described earlier, the forces which act on each element of a cilium or flagellum are summed over the length of the organelle. No account is taken of the effect on a particular element of the movement of fluid due to other parts of the organelle. In an accurate analysis of the system, the filament should be considered as a unit rather than as a series of infinitesimal elements. This, however, is a considerable hydrodynamic problem, the solution of which will require a sophisticated extension of the type of analysis undertaken by, for example, Cox (1970), in respect of inert filaments.

In the hydrodynamic analysis it is usually assumed that the filament is moving in a fluid of infinite extent. This is usually far from the situation of a thin fluid film confined by microscope slide and cover slip under which cilia and flagella are normally examined. Little attention has been paid to this approximation although the effects could be significant.

The good agreement between the predicted and observed velocities for a number of organisms (Fig. 9) suggest that the approximations made in the analysis are not unreasonable. More direct confirmation of this comes from a comparison of the experimentally measured force exerted by a cilium (Yoneda, 1962) and the calculated value (Yoneda, 1962; Sleigh and Holwill, 1969). The values obtained give order of magnitude agreement. Better agreement would probably be achieved if an accurate model (rather than an idealized one) were used to simulate the ciliary movement.

A difficulty associated with the use of force coefficients is that the pattern of fluid flow about an organelle cannot be predicted. This pattern can be determined experimentally and would provide additional data against which to test appropriate hydrodynamic equations. Additional information would be very useful in an examination of those cases, such as hispid flagella, where the approximations made are clearly extreme.

2. *Metachronal waves*

The first hydrodynamic analyses specifically concerned with metachronism appeared quite recently (Blake, 1971a, b, c, 1972) and it is most encouraging that the predicted velocities are of the same order as those observed experimentally. Accurate observations of an organism which exhibits symplectic metachrony, and therefore is reasonably modelled by the surface envelope approach, are confined to *Opalina* (Sleigh, 1962) so that comparison of theory with experiment is necessarily limited. The models discussed specifically include a spherical envelope and a flat sheet of infinite extent, both of which are assumed to undulate in the simulation of metachronal waves. *Opalina* is a flat, disc-shaped organism of finite extent, and hence does not match either of the models described. As Blake (1971b) suggests, the velocity of *Opalina* is expected to fall between the two theoretical extremes; these extremes are both of the same order as the experimentally observed velocity of *Opalina* so that the model appears to be satisfactory in this case. Clearly, many more experimental studies of ciliated micro-organisms are needed for comparison with the theoretical equations. The surface envelope model is reasonable only if no fluid is squeezed out or sucked in by the cilia. Careful examination of the fluid flow in the neighbourhood of metachronal waves could ascertain whether fluid movement across the "envelope" occurs.

The more sophisticated model of large fields of cilia (in which the individuality of cilia is preserved (Blake, 1972)) is physically more satisfactory than the surface envelope approach and provides data about fluid flow which can be subjected to experimental investigation. The amount of experimental data of this type available is very limited and, as mentioned earlier, refers to an organism for which the assumption that viscous forces are significantly greater than inertial ones may not be valid. The observed propulsion velocities are of the same order as those predicted theoretically, but many more experimental results are required to test the theory and to suggest possible improvements to the model.

VII. Conclusion

Although considerable attention has been focused on the hydrodynamic aspects of the movement of individual undulating flagella, and the velocity of organisms with such appendages can be predicted with reasonable accuracy, a complete solution to this problem has yet to be formulated. Nevertheless, it seems likely that the power dissipation estimated using the method of surface coefficients is close to that obtaining in practice and may therefore be used for comparison with the energy available from biochemical sources within the cell. The analysis of metachronism initiated by Blake (1971a, b, c, 1972) appears to be very promising and it is to be hoped that further theoretical and experimental investigations of this important and intriguing movement will be forthcoming.

Having considered in reasonable detail those aspects of ciliary and flagellar movement that have been subject to hydrodynamic study let us finally examine briefly two problems which have been neglected from a theoretical viewpoint, in the hope that this will shortly be rectified. Metachronal wave movement is induced, at least in part, by viscous coupling between adjacent cilia (e.g. Sleigh, 1969; Machemer, 1972). So far few attempts have been made to analyse the nature of this coupling, which could provide a physical explanation of the wavelength and velocity. This problem is closely related to that of flagellar synchronization, which occurs when two flagella approach closely. Machin (1963) has analysed this phenomenon but the equations he used are not satisfactory for ciliary movement, and no quantitative predictions concerning metachrony can be made from them.

The second problem concerns the movement of ciliary comb plates on ctenophores such as *Beroe* which occurs at Reynolds numbers in the region of unity. For this situation both inertial and viscous forces would appear in the equations of motion, the solution of which would be complex. It would, for instance, be interesting to compare the dependence of propulsion velocity on beat frequency for this case with that of a ciliated organism which functions at a much lower Reynolds number.

It should be clear from this review that while much has been accomplished in the field of ciliary and flagellar hydrodynamics, many problems yet remain to be solved. It is the author's hope that this review will stimulate other workers to investigate these problems.

References

BABA, S. A. and HIRAMOTO, Y. (1970). *J. exp. Biol.* **52**, 675–690.
BATCHELOR, G. K. (1970). *J. Fluid Mech.* **44**, 419–440.
BLAKE, J. R. (1971a). *J. Fluid Mech.* **46**, 199–208.
BLAKE, J. R. (1971b). *J. Fluid Mech.* **49**, 209–222.
BLAKE, J. R. (1971c). *Bull. Aust. Math. Soc.* **5**, 255–264.

Blake, J. R. (1972). *J. Fluid Mech.* **55**, 1–23.
Bouck, G. B. (1971). *J. Cell Biol.* **50**, 362–384.
Brokaw, C. J. (1961). *Expl Cell Res.* **22**, 151–162.
Brokaw, C. J. (1965). *J. exp. Biol.* **43**, 155–169.
Brokaw, C. J. (1970). *J. exp. Biol.* **53**, 445–464.
Brokaw, C. J. and Wright, L. (1963). *Science, Wash.* **142**, 1169–1170.
Brown, H. P. (1945). *Ohio J. Sci.* **45**, 243–301.
Carlson, F. D. (1959). *Proc. natn. Biophys. Conf.* **1**, 443–449.
Chwang, A. T. and Wu, T. Y. (1971). *Proc. R. Soc.* B **178**, 327–346.
Coakley, C. J. and Holwill, M. E. J. (1972). *J. theor. Biol.* **35**, 525–542.
Cox, R. G. (1970). *J. Fluid Mech.* **44**, 791–810.
Douglas, G. J. and Holwill, M. E. J. (1972). *J. Mechanochem. Cell Motility.* **1**, 213–223.
Gray, J. (1955). *J. exp. Biol.* **32**, 775–801.
Gray, J. (1958). *J. exp. Biol.* **35**, 96–108.
Gray, J. and Hancock, G. J. (1955). *J. exp. Biol.* **32**, 802–814.
Hancock, G. J. (1953). *Proc. R. Soc.* A **217**, 96–121.
Holwill, M. E. J. (1964). Ph.D. Thesis, University of London.
Holwill, M. E. J. (1965). *J. exp. Biol.* **42**, 125–137.
Holwill, M. E. J. (1966a). *Physiol. Rev.* **46**, 696–785.
Holwill, M. E. J. (1966b). *J. exp. Biol.* **44**, 579–588.
Holwill, M. E. J. (1970). *Nature, Lond.* **226**, 1046–1047.
Holwill, M. E. J. and Burge, R. E. (1963). *Archs Biochem. Biophys.* **101**, 249–260.
Holwill, M. E. J. and Miles, C. A. (1971). *J. theor. Biol.* **31**, 25–42.
Holwill, M. E. J. and Sleigh, M. A. (1967). *J. exp. Biol.* **47**, 267–276.
Jahn, T. L. and Votta, J. J. (1972). *A. Rev. Fluid Mech.* **4**, 93–116.
Jahn, T. L., Landman, M. D. and Fonseca, J. R. (1964). *J. Protozool.* **11**, 291–296.
Knight-Jones, E. W. (1954). *Q. Jl microsc. Sci.* **95**, 503–521.
Liefson, E. (1960). "Atlas of Bacterial Flagellation." Academic Press, New York.
Lighthill, M. J. (1969). *A. Rev. Fluid Mech.* **1**, 413–446.
Lowndes, A. G. (1944a). *Proc. zool. Soc. Lond.* **113**, 99–107.
Lowndes, A. G. (1944b). *Proc. zool. Soc. Lond.* **114**, 325–338.
Machemer, H. (1972). *J. exp. Biol.* **57**, 239–259.
Machin, K. E. (1963). *Proc. R. Soc.* B **158**, 88–104.
Metzner, P. (1920). *Biol. Zbl.* **40**, 49–87.
Miles, C. A. (1969). Ph.D. Thesis, University of London.
Pitelka, D. R. (1963). "Electron-microscopic Structure of Protozoa." Pergamon Press, New York.
Pitelka, D. R. and Schooley, C. N. (1955). *Univ. Calif. Publs Zool.* **61**, 79–128.
Reynolds, A. J. (1965). *J. Fluid Mech.* **23**, 241–260.
Rikmenspoel, R., van Herpen, G. and Eijkhout, P. (1960). *Physics Med. Biol.* **5**, 167–181.
Schreiner, K. E. (1971). *J. Biomechanics* **4**, 73–83.

SILVESTER, N. R. and HOLWILL, M. E. J. (1972). *J. theor. Biol.* **35**, 505–523.
SLEIGH, M. A. (1962). "The Biology of Cilia and Flagella." Pergamon Press, Oxford.
SLEIGH, M. A. (1968). *Symp. Soc. exp. Biol.* **22**, 131–150.
SLEIGH, M. A. (1969). *Int. Rev. Cytol.* **25**, 31–54.
SLEIGH, M. A. and AIELLO, E. (1973). *Acta Protozool.* **11**, 265–277.
SLEIGH, M. A. and HOLWILL, M. E. J. (1969). *J. exp. Biol.* **50**, 733–743.
TAYLOR, G. I. (1951). *Proc. R. Soc.* A **209**, 447–461.
TAYLOR, G. I. (1952). *Proc. R. Soc.* A **211**, 225–239.
TILLETT, J. P. K. (1970). *J. Fluid Mech.* **44**, 401–417
TUCK, E. O. (1968). *J. Fluid Mech.* **31**, 305–308.
YONEDA, M. (1962). *J. exp. Biol.* **39**, 307–317.

Chapter 9

Mechanics of ciliary movement

YUKIO HIRAMOTO

Biological Laboratory, Tokyo Institute of Technology, Tokyo, Japan

I. Introduction

Although there are some distinct differences between cilia and flagella from the standpoint of morphology and the beating pattern, both have a common 9+2 pattern at the ultrastructural level and it is believed that the basic mechanisms of beating are the same.

There is a considerable number of reviews on the mechanics of ciliary and flagellar movements (e.g. Gray, 1928; Harris, 1961; Sleigh, 1962; Brokaw, 1966b; Holwill, 1966; Yoneda, 1967). In this chapter, mechanics of the movement of cilia, especially quantitative analyses, will be dealt with.

II. Force and bending moment

A. Viscous resistance

The first step in the study of the mechanics of ciliary movement is to record the movement in a quantitative fashion. If the movement is planar, the movement of every part of the cilium can satisfactorily be described by a set of photographs of the cilium taken at proper intervals from the direction normal to the beating plane after an appropriate magnification (cf. Sleigh, 1968, and this volume, p. 79). The shape of the cilium beating with a regular frequency can be observed through a microscope by stroboscopic illumination (e.g. Gray, 1930). Complete description of non-planar movements of the cilium is difficult.

The shapes of a large abfrontal cilium of *Mytilus* at various phases of the beating cycle, obtained from a high-speed cinematographic record (Baba and Hiramoto, 1970), are shown in Fig. 1. During movement, every part of the cilium exerts a force on the surrounding medium. The amount of the force depends on the shape of the beating cilium, the beating

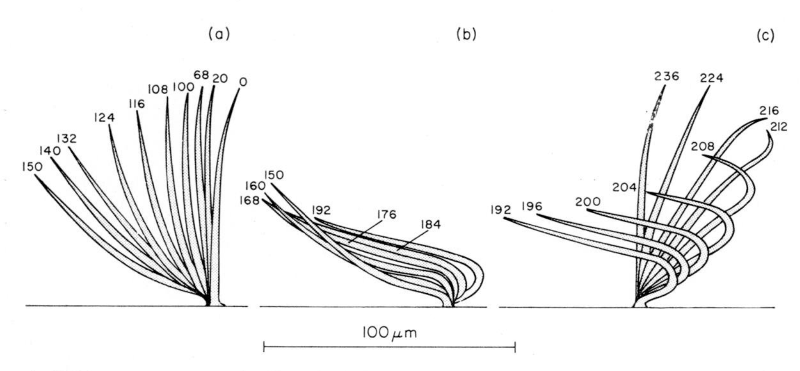

FIG. 1. The movement of a large abfrontal cilium of *Mytilus edulis*.

The shapes of the cilium at various phases of beating were obtained by tracing pictures from high-speed microcinematographs (450 frames per second). Numbers indicate frame numbers. (a) Effective phase. (b) Recovery phase. (c) Preparatory phase.

velocity and the viscosity of the medium. Harris (1961) estimated the torque around the base of the cilium during its effective stroke. He assumed that the torque is equal to one-half of the couple required to overcome the resistance of the medium to a prolate spheroid, having a major radius equal to the length L of the cilium and a minor radius equal to the radius a of the cilium, which rotates about its minor axis with an angular velocity (ω) equal to that of the beating cilium. The torque (T) is given by

$$T = \frac{4\pi\eta\omega L^3}{3\{\ln(2L/a) - 0{\cdot}5\}} \tag{1}$$

where η is the viscosity of the medium. The large abfrontal cilium of *Mytilus* is regarded as a half of a prolate spheroid with an axial ratio (L/a) of $50 \sim 60$ (Baba and Hiramoto, 1970). Substituting 55 for L/a, eqn (1) gives

$$T = 1{\cdot}0\eta\omega L^3 \tag{2}$$

Yoneda (1962) obtained an equation (eqn 3) similar to eqn (2) from experiments using a model similar in figure to the cilium:

$$T = 1{\cdot}1\eta\omega L^3. \qquad (3)$$

As shown by eqn (1), the value of the torque varies rather little for a considerable change in the axial ratio (L/a), e.g.

$$T = 0{\cdot}76\eta\omega L^3 \qquad (4)$$

when $L/a = 200$.

The cilium is not always straight and the direction of the movement of each part of the cilium is not always normal to the ciliary axis, especially during the recovery stroke (cf. Fig. 2). According to Gray and Hancock (1955), the force exerted on a short cylindrical element (of length ds) of

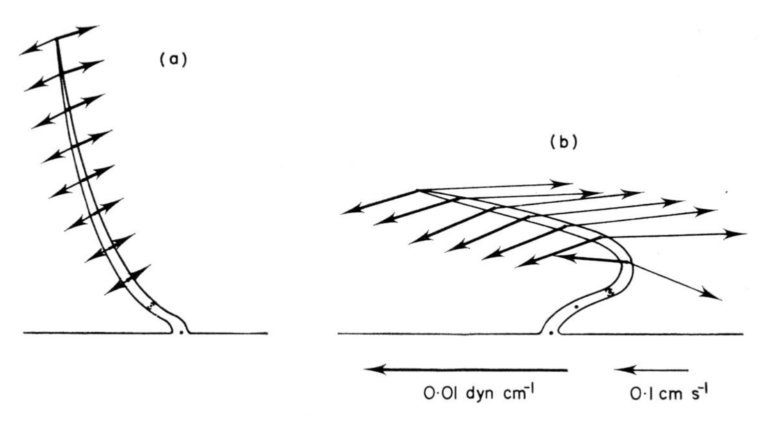

FIG. 2. Velocities and forces exerted on a beating cilium. Thin vectors indicate velocities and thick vectors indicate forces per unit length. *Mytilus* large abfrontal cilium.

the flagellum by its movement in a viscous medium consists of the component (dF_N) normal to the flagellar axis and the component (dF_T) tangential to the axis, which are given by

$$\mathrm{d}F_N = C_N V_N \mathrm{d}s \qquad (5)$$

$$\mathrm{d}F_T = C_T V_T \mathrm{d}s \qquad (6)$$

where V_N and V_T are the component of the velocity of the element normal to the flagellar axis and that tangential to the axis respectively, and C_N and C_T are the respective drag coefficients. In the case of thin cylindrical filaments, C_T is given approximately by

$$C_T = \frac{-2\eta}{\ln(a/2\lambda)+0{\cdot}5} \qquad (7)$$

where λ is the wavelength in the case of undulatory movement, and C_N is twice C_T for long filaments (Gray and Hancock, 1955).

The resistance of the medium to an ellipsoid translating along its axis is given by Oberbeck's formulas (cf. Lamb, 1932; Happel and Brenner, 1965). If it is assumed that the drag coefficients per unit length (C_N, C_T) are constant along the axis of the spheroid, they are given by

$$C_T = \frac{2\eta}{\ln(2L/a) - 0{\cdot}5} \tag{8}$$

$$C_N = \frac{4\eta}{\ln(2L/a) + 0{\cdot}5}. \tag{9}$$

Equation (8) is the same as eqn (7) if $\lambda = L$. Substituting 55 for L/a in eqns (8) and (9) gives $C_T = 1{\cdot}5\eta$ and $C_N = 2{\cdot}4\eta$. If it is assumed that the drag coefficient (C_N) is constant along the length of the model (and the cilium) in Yoneda's experiment (1962), $C_N = 3{\cdot}3\eta$ from eqn (3).

Machin (1958) used eqn (10) instead of eqn (7).

$$C_N = \frac{4\eta}{2{\cdot}0 - \ln R} \tag{10}$$

where R is the Reynolds number. Because the Reynolds number is in the order of $10^{-2} \sim 10^{-4}$ in ciliary movement, $C_N = 1{\cdot}7\eta \sim 4{\cdot}8\eta$.

Sleigh and Holwill (1969), Baba and Hiramoto (1970) and Rikmenspoel and Sleigh (1970) estimated resistances of the medium during ciliary

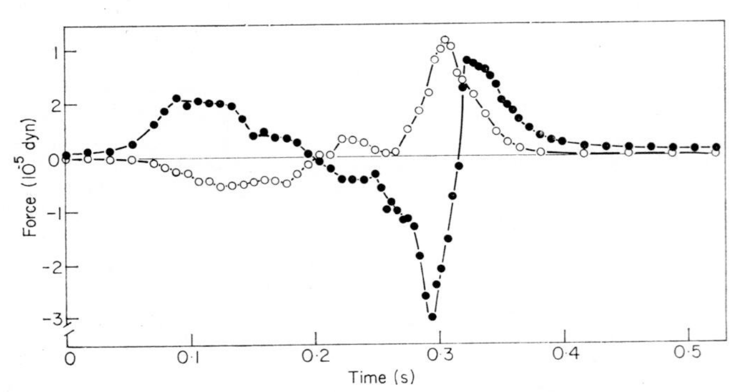

Fig. 3. The forces exerted on the cell through a beating cilium.

●: the component of the force tangential to the cell surface (positive in the direction of the recovery stroke). ○: the component of the force normal to the cell surface (positive in the direction from outside to inside of the cell). *Mytilus* large abfrontal cilium.

movement following the above principle. Thick vectors shown in Fig. 2 indicate forces exerted on various parts of a *Mytilus* large abfrontal cilium determined from velocities shown by thin vectors in this figure assuming that $C_N = 3{\cdot}3\eta$, $C_T = 1{\cdot}65\eta$. Because inertia is practically negligible in ciliary movement, the only external forces exerted on the cilium during movement are the viscous resistances mentioned above.

The forces exerted on various parts of the cilium are transmitted through the cilium to the cell body. When the cell (or tissue) is fixed, a current is generated in the medium. In Fig. 3, the component of the force exerted on the cell parallel to the cell surface (solid circles) and the component normal to the cell surface (open circles) during the movement of a *Mytilus* large abfrontal cilium (cf. Baba and Hiramoto, 1970) are shown. In reaction to the forces mentioned above, forces equal in magnitude and opposite in direction are exerted on the medium by the beating of the cilium. Because the velocity of the current generated by the ciliary movement is proportional to the applied force, the amount of the medium carried by the ciliary movement during a fixed period is determined by integrating the force shown in Fig. 3 over that period. Since the force is proportional to the velocity and the time required to move a definite distance is inversely proportional to the velocity, the amount of the medium carried by the cilium during one beat cycle is independent of the velocity of beating, provided that the form of the cilium during movement is the same.

B. Bending moment due to viscous resistance

A bending moment develops around every part of the cilium due to the viscous resistances of the medium acting on the moving cilium. If it is assumed that the cilium is straight and rotates around its base, the bending moment $[M(x)]$ around the point a distance x away from the base of the cilium during the effective stroke, is given by

$$M(x) = C_N \left(\frac{L^3}{3} - \frac{L^2 x}{2} + \frac{x^3}{6}\right)\omega, \tag{11}$$

where L is the length of the cilium and ω is the angular velocity.

When the form of the cilium changes during movement and/or when the direction of the velocity of the movement is not normal to the ciliary axis, the bending moment is determined from the forces acting on the cilium at various regions as follows (cf. Brokaw, 1970). The normal component of force acting on any small element, ds, of the length of the cilium, resulting from its movement through a viscous medium is represented by eqn (5). The normal component of shear force, $F_N(s)$, and bending moment, $M(s)$, at point s on the co-ordinate along the ciliary axis are represented by

$$F_N(s) = F_N(0) + \int_{s=0}^{s=s} \mathrm{d}F_N \tag{12}$$

$$M(s) = M(0) + \int_0^s F_N \, \mathrm{d}s. \tag{13}$$

Because both $F_N(s)$ and $M(s)$ are zero at the free end of the cilium, the bending moment around any point on the cilium is determined from velocities of various parts of the cilium during beating using eqns (5), (12) and (13). The bending moment around a point on the cilium is also determined by summing the moments caused by the resistances acting on all the elements of the cilium distal to the point in question.

Following the principles mentioned above, Baba and Hiramoto (1970) and Rikmenspoel and Sleigh (1970) determined viscous bending moments at various regions of the cilium during its beating. More complete results determined from a microcinematographic record are shown in Fig. 4, indicating changes in viscous bending moment around five points (basal point and the points 10 μm, 30 μm, 50 μm and 70 μm away from the base) on a *Mytilus* large abfrontal cilium (90 μm in length). As shown in this figure, the magnitude of the bending moment (solid circles) decreases as the distance from the base increases while the variation in bending moments occupies a similar duration of time at all points along the cilium. Although the change in bending moment occurs slightly later in more distal regions, the phase difference in bending moment between the proximal region and a distal one is smaller than the phase difference of change in curvature (open circles).

C. Bending moment due to elastic deformation

When the cilium bends, an elastic bending moment (M_e) is set up owing to the passive deformation of the structure in the cilium. The active bending moment (M_a) responsible for active bending and unbending balances the sum of the viscous bending moment (M_v) and the elastic bending moment (M_e).

$$M_a + M_v + M_e = 0 \tag{14}$$

M_e is the sum of the elastic bending moments due to the deformation of all the elements passively bent in the cilium.

$$M_e = \sum_{i=1}^{n} q_i A_i k_i^2 / \rho \tag{15}$$

Where q_i is Young's modulus of the ith element, n is the number of elements, $A_i k_i^2$ is the second moment of area of the cross-section of that element and $1/\rho$ is its curvature.

Rikmenspoel and Sleigh (1970) concluded that the elastic bending moment is negligibly small as compared with the viscous bending moment in the movements of *Sabellaria* and *Modiolus* cilia. However, because the value of flexural rigidity of the cilium used by them in the calculation

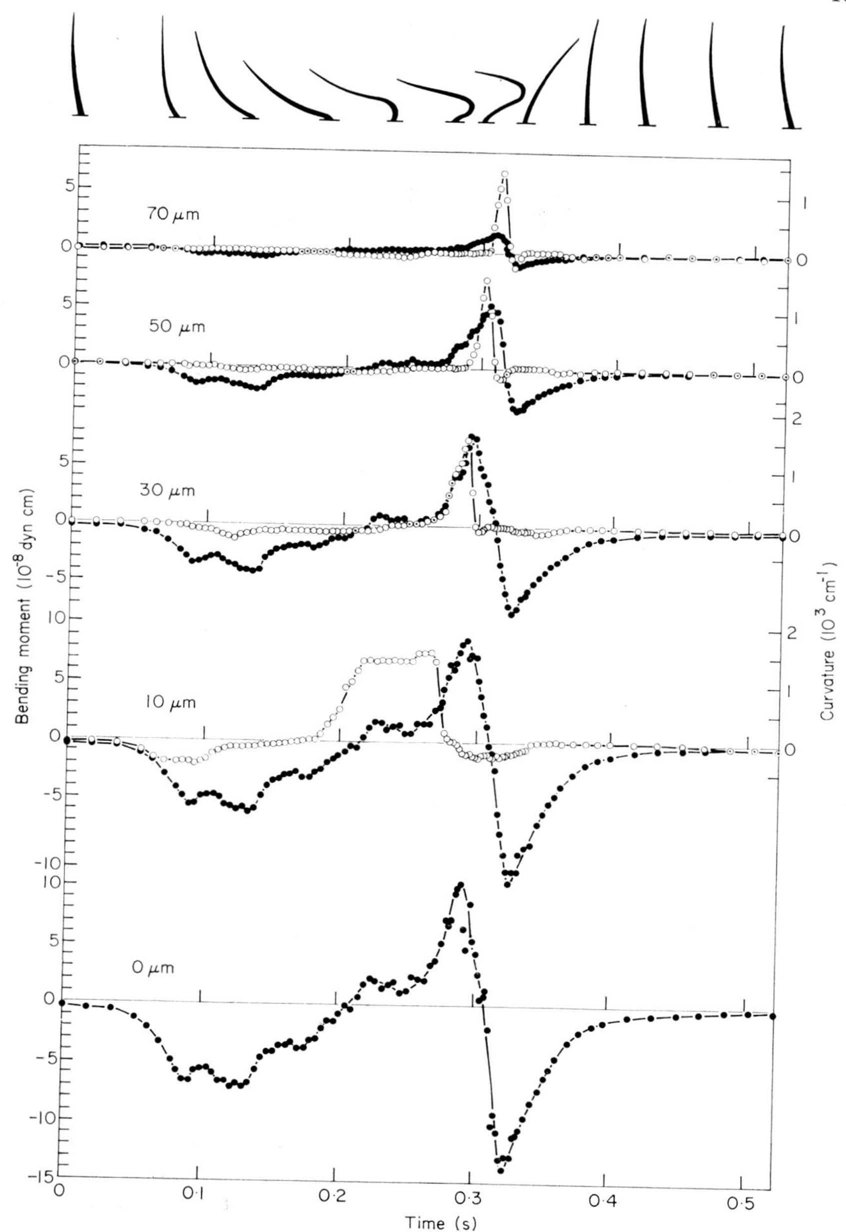

FIG. 4. Changes in the curvature and in the bending moment at various regions of a cilium during a single beat.

Curvature (open circles) and bending moment (solid circles) at 70 μm, 50 μm, 30 μm and 10 μm from the base and bending moment around the base are shown. *Mytilus* large abfrontal cilium (length: 90 μm).

of the bending moment is very small (10^{-14} dyn cm^{-2} per single cilium), a value which is not reliable, as mentioned later, it is possible that the elastic bending moment is similar to or larger than the viscous bending moment. In the above discussion, it is assumed that the passively bent elements in the cilium are perfectly elastic, but it is possible that they are visco-elastic as in many other biological materials. If this is the case, the magnitude of the bending moment due to passive deformation depends not only on the amount of deformation (curvature) but also on the rate of deformation.

D. *Relationship between the beating velocity and the bending moment*

When a muscle shortens, extra energy (indicated by the shortening heat plus mechanical work) is liberated as well as energy (maintenance heat) liberated during isometric contraction (Hill, 1938). The rate of extra energy liberation depends on the load and the velocity of shortening, and the relationship between the load and the velocity of shortening of the muscle is represented by the well known Hill equation (1938):

$$(P+a)(V+b) = \text{constant.} \tag{16}$$

It is important to know whether or not such a relationship exists in ciliary and flagellar movements in studies of the mechanism of these movements. The forces applied to cilia as well as the velocity of ciliary movement are affected by the viscosity of the medium. Although there is a considerable literature on the effects of viscosity of the medium on the ciliary activity (cf. Sleigh, 1962), little work has been done to determine quantitative force–velocity relationships in ciliary or flagellar movement.

Brokaw (1966a) reported a relationship between the viscous bending moment (M_0) and the bending velocity (V) in sperm flagella as follows:

$$M_0(V+100\ \mu\text{m s}^{-1}) = \text{constant.} \tag{17}$$

Yoneda (1962) determined the beating velocities and the torques around the base of the cilium in media of various viscosities for *Mytilus* large abfrontal cilia. The relationship between the beating velocity and the torque obtained from his data is shown in Fig. 5. Using the same material Yoneda (1960) measured the force exerted by the cilium when the beating was stopped with a microneedle during the course of the effective stroke. The magnitude of the force, which is different in different positions of arrest, is practically inversely proportional to the distance of the position of arrest from the base of the cilium, i.e. the torque at the base is constant irrespective of the position of arrest. This torque, which is regarded as the torque in the case of null beating velocity is indicated in Fig. 5. As shown in Fig. 5, the moment at the base increases as the beating velocity decreases, though the quantitative relationship is different from those of the muscle and the flagellum.

The active bending moment, which is responsible for the motive force of active bending and unbending of the cilium, is determined from eqn (14) if the viscous bending moment and the elastic bending moment are known. Because no reliable value of the flexural rigidity of the passively bent component of the cilium is known, an exact relationship between the

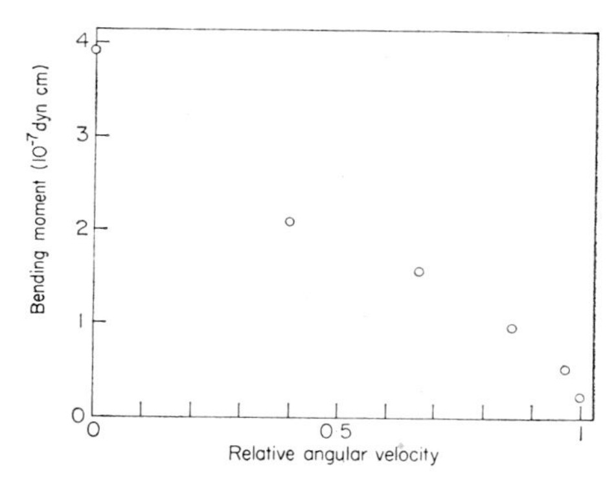

FIG. 5. Relationship between the beating velocity and the bending moment at the base of a cilium.

Constructed from Yoneda's data (1960, 1962) on *Mytilus* large abfrontal cilia.

active bending moment and bending velocity (ω) cannot be obtained. If it is assumed that the elastic bending moments are practically the same at various angular velocities, because the curvatures at the ciliary base are not much different at different velocities in Yoneda's experiment, the relationship shown in Fig. 5 may scarcely be affected by the magnitude of the elastic bending moment except by a shift of the curve parallel to the moment axis.

E. *Force exerted by a single cilium*

The *Mytilus* large abfrontal cilium is a compound one consisting of many component cilia. The number of components in a compound cilium is approximately proportional to the cross-sectional area of the compound cilium (Baba, 1972). The torques at the base per single component cilium calculated from data of Yoneda (1960, 1962) and Baba (1972) are $2{\cdot}7 \times 10^{-8}$ dyn cm when the beating of the cilium is stopped during the effective stroke and $1{\cdot}4 \times 10^{-9}$ dyn cm during the effective stroke in sea water.

Values similar to the latter are obtained from Rikmenspoel and Sleigh's data (1970) in *Sabellaria* and *Modiolus* cilia freely beating in sea water and from the data shown in Fig. 4 for *Mytilus* cilia.

If the above torque is due to the active contraction of contractile elements in a half of the cross-section of the cilium, the tension is $3M_cR^3$ where M_c is the torque and R is the radius of the cilium. Substituting the above values for M_c and 0·14 μm for the radius gives 3×10^7 dyn cm^{-2} for the maximal tension and $1{\cdot}5 \times 10^6$ dyn cm^{-2} for the tension during the effective stroke. The former is about 10 times the maximal tension generated by frog striated muscle (2·7 kg cm^{-2}, Gordon *et al.*, 1966).

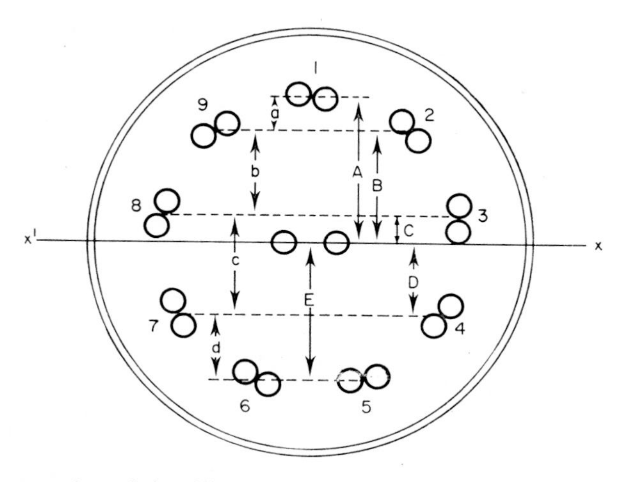

FIG. 6. Cross-section of the cilium.

$x'x$ indicates the line passing the centre of the cilium, perpendicular to the beating plane. The bending moment around $x'x$ caused by contraction of doublet microtubules 1, 2, 3, 8 and 9 is $F_c(A+2B+2C)$ where F_c is the contractile force of each doublet. The bending moment around $x'x$ caused by contraction of doublets 4, 5, 6 and 7 is $2F_c(D+E)$. The bending moment around $x'x$ caused by sliding forces (F_s) between adjacent doublets is $2F_s(a+b+c+d)$.

If it is assumed that doublet microtubules nos 4, 5, 6 and 7 shown in Fig. 6 or microtubules nos 8, 9, 1, 2 and 3 generate contractile forces, the bending moment around $x'x$ axis is represented as follows:

$$M_c = rF_c(2 \sin 30° + 2 \sin 80°) \approx 2{\cdot}9rF_c, \tag{18}$$

$$M'_c = rF_c(\sin 90° + 2 \sin 50° + 2 \sin 10°) \approx 2{\cdot}9rF_c, \tag{18'}$$

where r is the distance of each doublet from the centre of the cilium and F_c is the mean contractile force per single doublet. Substituting the above values for M_c or M'_c and assuming r to be 0·1 μm (10^{-5} cm) gives

$F_c = 9 \times 10^{-4}$ dyn and $F_c = 5 \times 10^{-5}$ dyn for the maximal force generated in each doublet and for the contractile force generated in each doublet during the effective stroke in sea water, respectively. The value of this maximal force is about 20 times the maximal force per thick filament ($4{\cdot}3 \times 10^{-5}$ dyn) generated in striated muscles, calculated from the maximal tension mentioned above and the number of thick filaments per cross-section (cf. Huxley, 1969).

On the other hand, if it is assumed that a sliding force (F_s) is generated between each pair of adjacent doublets in the direction so as to generate bending moment (M_c) around $x'x$ axis,

$$M_c = 2rF_s\{(\sin 90° - \sin 50°) + (\sin 50° - \sin 10°) + (\sin 10° + \sin 30°) + (\sin 70° - \sin 30°)\} \approx 3{\cdot}9rF_s. \quad (19)$$

Substituting the above values for M_c and assuming r to be 0·1 μm gives $F_s = 7 \times 10^{-4}$ dyn and $F_s = 4 \times 10^{-5}$ dyn for the maximal sliding force and for the sliding force generated during the effective stroke in sea water, respectively.

From each of the nine doublet microtubules in cilia, pairs of arms project with a periodicity of 170 Å (Gibbons and Rowe, 1965; Grimstone and Klug, 1966). If it is assumed that all the arms contained in a cilium 30 μm long contribute to the generation of sliding, the maximal sliding force per single arm is calculated to be 2×10^{-7} dyn. This value is about a half the maximal force per single bridge ($4{\cdot}3 \times 10^{-7}$ dyn) in striated muscles, calculated from the maximal tension per thick filament mentioned above and the number of bridges contained in a filament (cf. Gordon *et al.*, 1966; Huxley, 1969).

In conclusion, the results of the estimates of the active forces generated in cilia are favorable for the "sliding hypothesis" rather than the "contraction hypothesis", provided that the force generated by an arm in cilia is of the same order as the force generated by a bridge in striated muscles.

III. Work done during ciliary movement

A. *Work to overcome viscous forces*

The rate of work done by the cilium to overcome the viscous resistance of the medium is the total sum of the scalar products of the velocity and the force exerted on the medium by all component parts of the cilium. By integrating the rate of work over a definite period, total work done during that period is obtained.

Yoneda (1962) determined the rate of mechanical work done by the cilium in *Mytilus* large abfrontal cilia by the following equation, assuming that the cilium rotates around its base without bending of the ciliary shaft:

$$\begin{aligned}\text{Rate of mechanical work} &= \text{Torque} \times \text{Angular velocity}\\ &= 1{\cdot}1\eta\omega^2 L^3 \qquad (20)\end{aligned}$$

Sleigh and Holwill (1969) determined the work done to overcome the viscous resistance of the medium in *Sabellaria* and *Mytilus* cilia, assuming that the cilium consists of straight part(s) and circular part(s). Their results are shown in Table I. As shown in this table, the work done per beat by a single cilium is of order of 10^{-8} erg.

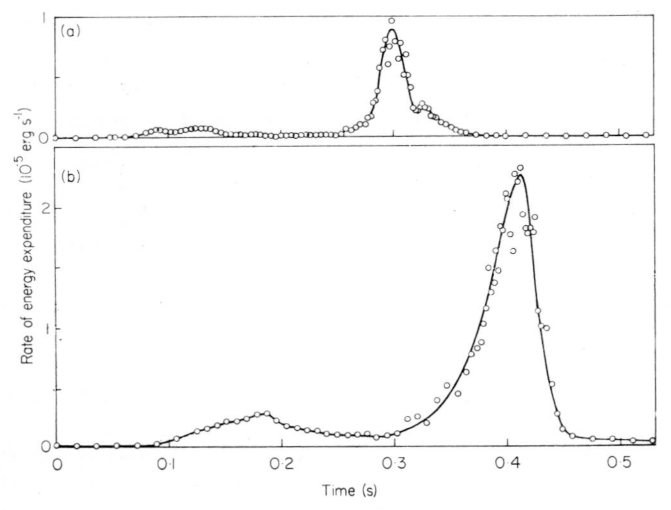

FIG. 7. The rate of work necessary to overcome viscous resistances of the medium. (a) The change in the rate of the viscous work in a *Mytilus* large abfrontal cilium during a single beat in normal sea water (viscosity: 0·995 centipoise). (b) The change in the rate of the same cilium during a single beat in sea water containing polyvinyl pyrrolidone (viscosity: 9·9 centipoise).

Figure 7 shows the rate of viscous work done by a *Mytilus* large abfrontal cilium calculated from the velocities of component parts of the cilium which were determined from a high-speed cinematographic record (Baba and Hiramoto, 1970). The work done during a beat cycle can be obtained by integrating the rate of work over the cycle. In the example shown in this figure, the work done by the cilium to overcome viscous forces is $3{\cdot}6 \times 10^{-7}$ erg beat^{-1} in normal sea water and $17{\cdot}1 \times 10^{-7}$ erg beat^{-1} in high-viscosity sea water. Because the number of component cilia contained in this cilium is about 36, judging from the relation between the thickness of the cilium and the number of component cilia (Baba, 1972), the work done by a single component cilium is $1 \cdot 0 \times 10^{-8}$ erg beat^{-1} in sea water and $4 \cdot 7 \times 10^{-8}$ erg beat^{-1} in high-viscosity sea water.

B. *Work done to overcome elastic forces*

The work done per unit length (E) to overcome elastic forces in the cilium is given by

$$E = S/(2\rho^2) \tag{21}$$

where S is the flexural rigidity of the passively bent component in the cilium and ρ is its radius of curvature. An equivalent amount of energy is stored in the bent cilium and is released at the time of unbending. If the passively bent components are not perfectly elastic, part of the work done at the time of bending changes into heat during bending and unbending.

Sleigh and Holwill (1969) calculated the work done to overcome elastic forces during beating in *Sabellaria* and *Mytilus* cilia, assuming that the

TABLE I. Work done (in 10^{-8} ergs) by a single cilium (Sleigh and Holwill, 1969)

	Sabellaria			*Mytilus*		
	viscous	elastic	total	viscous	elastic	total
Effective stroke	0·4	1·9	2·3	0·1	0·7	0·8
Recovery stroke	0·1	3·2	3·3	0·2	1·1	1·3
Total work for complete cycle	0·5	—	—	0·3	—	—

flexural rigidity of the component cilium is 2×10^{-12} dyn cm^{-2}. Their results are shown in Table I. The elastic work is larger than the viscous work in this table. However, the absolute value of the elastic work depends on the value of the flexural rigidity used in the calculation. For example, if a value of 10^{-14} dyn cm^{-2} (Rikmenspoel and Sleigh, 1970) is adopted, the elastic work becomes smaller than the viscous work.

C. *Total energy dissipated by the cilium*

Since the pioneer work of Hoffmann-Berling (1955), it has been shown by many investigators that ATP serves as the energy donor for ciliary and flagellar movements. Gibbons and Rowe (1965) isolated the ATP dephosphorylating enzyme (dynein) associated with the arms projecting from outer doublets in *Tetrahymena* cilia. According to Gibbons (1966), the number of ATP molecules dephosphorylated per second per dynein molecule coincides approximately with the frequency of beat of the cilium. The same conclusion was obtained by Brokaw (1967) by comparing the rate of ATP dephosphorylation, beating frequency and external mechanical work in glycerinated sperm flagella. Brokaw and Benedict (1968a, b) suggested that an actively motile sperm flagellum uses at least one, and possibly two, ATP molecules per dynein molecule during each flagellar beat cycle. Proportionality between the movement-coupled ATPase activity and beat frequency is observed in sperm extracted with Triton X-100 (Gibbons and Gibbons, 1972); the ratio of the number of ATP molecules dephosphorylated per sperm per beat by the movement-coupled ATPase to the number of dynein molecules contained in one sperm is calculated to be 1·5. Yanagisawa (1967) estimated the energy dissipated by sea urchin spermatozoa to be 3×10^{-7} erg s^{-1} $cell^{-1}$ from the rate of consumption of creatine phosphate. A similar value is obtained from the data of O_2-consumption determined by Mohri (1956). If the mean frequency of beat of sperm in the experimental conditions of Yanagisawa (1967) and Mohri (1956) is assumed to be 10 s^{-1}, energy consumption is 3×10^{-8} erg $beat^{-1}$, which is similar to the value calculated from ATP dephosphorylation in sperm models. Sleigh and Holwill obtained $2{\cdot}4 \times 10^{-8}$ erg $beat^{-1}$ and $3{\cdot}8 \times 10^{-8}$ erg $beat^{-1}$ for energy dissipation of cilia in *Sabellaria* and *Mytilus*, respectively, assuming that one ATP molecule is dephosphorylated per dynein per beat. These values are larger than the values of viscous work shown in Table I.

The number of arms contained in the cilium from which the data shown in Fig. 6 was obtained is estimated to be $2{\cdot}3 \times 10^6$, if it is assumed that the cilium contains 4·7 component cilia per μm^2 of its cross-section (cf. Baba, 1972) and that pairs of arms project from each outer doublet with a periodicity of 170 Å (Gibbons and Rowe, 1965; Grimstone and Klug, 1966). If one ATP molecule is dephosphorylated per beat by a single arm, the total energy liberated is $1{\cdot}6 \times 10^{-6}$ erg $beat^{-1}$. This value is sufficient to account for the viscous work of the cilium beating in sea water mentioned above ($3{\cdot}6 \times 10^{-7}$ erg $beat^{-1}$), but is insufficient for the viscous work of the cilium beating in a viscous medium ($1{\cdot}7 \times 10^{-6}$ erg $beat^{-1}$), if the efficiency of transducing chemical energy into mechanical work is considered.

As is well known, the total energy liberated during muscular contraction (the sum of heat and mechanical work) becomes larger when larger external work is performed by muscle (Fenn, 1923, 1924). It is probable that a similar mechanism for the liberation of energy is present in cilia and flagella and that more than one molecule of ATP is dephosphorylated per beat per dynein molecule when the external work is large.

IV. Stiffness of the cilium

The cilium behaves as a more or less rigid rod during the effective stroke whereas it is drawn back as a limp, non-elastic body during the recovery stroke (cf. Fig. 1). Carter (1924) reported that the beating can be stopped with a microneedle during the effective stroke while it is not easy to stop the cilium during the recovery stroke because it frequently slips below the needle. These facts appear to suggest that the cilium is stiff during the effective stroke and limp during the recovery stroke. It is also possible that the stiffness of the cilium for bending in one direction is different from that for bending in the opposite direction.

Rikmenspoel (1965, 1966) determined the flexural rigidity of sperm flagella from beating frequency and the propagation velocity of the bending wave, assuming that the propagation of the bending wave in flagella is governed by a principle fundamentally the same as that for the propagation of a mechanical wave along an elastic rod. Rikmenspoel and Sleigh (1970), applying this principle to the propagation of the bending wave during the recovery stroke in cilia, estimated the flexural rigidities in cilia of *Sabellaria* and *Modiolus* to be $(2 \pm 1) \times 10^{-13}$ dyn cm^2 and $(7 \pm 1) \times 10^{-13}$ dyn cm^2, respectively. These values correspond to flexural rigidities of a single component cilium of $(0{\cdot}8 \pm 0{\cdot}4) \times 10^{-14}$ dyn cm^2 and $(2{\cdot}8 \pm 0{\cdot}4) \times 10^{-14}$ dyn cm^2, respectively. Rikmenspoel (1965, 1966) assumed that the bending wave is amplified by an active process at every point along the flagellum to keep the amplitude constant, because the amplitude of the wave of mechanical vibration ought to decrease distally along the length in elastic rods if the motive force of the vibration is localized at the proximal region (cf. Machin, 1958). However, the bending moment and the amount of bending (curvature) do not change in parallel with each other, as shown in Fig. 4. If the bending of the ciliary shaft were a passive elastic or visco-elastic bending by the external force, the bending of the shaft would change *in parallel with* the bending moment applied to it, or the former would *lag behind* the latter in phase. The fact that sharp maxima and minima of the curvature *precede* those of bending moment may hardly be explained by assuming that the propagation of the bending wave is governed by a mechanism which is fundamentally the same as that of the propagation of flexural vibration in an elastic rod. Therefore, the estimates of flexural rigidity mentioned above are questionable.

Brokaw (1966b) estimated the flexural rigidity of the *Mytilus* large abfrontal cilium to be 5×10^{-12} dyn cm^2 from the bending moment during the effective stroke obtained by Yoneda (1962) and the curvature of the cilium. It is questionable whether this value indicates the actual stiffness of the cilium because there is no direct correlation between the curvature and bending moment of the cilium during the effective stroke (cf. Rikmenspoel and Sleigh, 1970).

Recently Baba (1972) determined the flexural rigidity of a *Mytilus* large abfrontal cilium from the change in its curvature when a force is applied to it with a microneedle. When a force is applied as shown in the inset of Fig. 8, a linear relationship exists between the curvature of the cilium and the bending moment as shown in Fig. 8. The slope of the line, which represents the flexural rigidity of the cilium, is almost constant in the same cilium irrespective of the direction of bending, the phase of ciliary beating, and the region of the cilium to which the needle is applied. The flexural rigidity per single component cilium is $2\text{–}3\times 10^{-10}$ dyn cm^2.

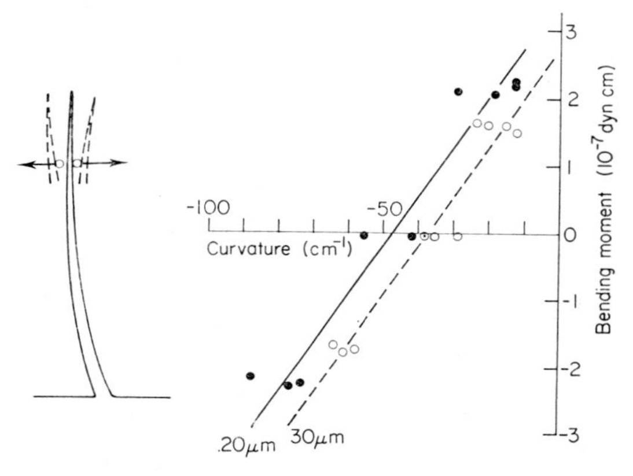

FIG. 8. Determination of the flexural rigidity of the cilium.

Responses of the curvature at 20 μm (solid circles) and at 30 μm (open circles) from the base of a *Mytilus* large abfrontal cilium to the bending moment due to the external force are shown. The slopes of the lines indicate the flexural rigidity (Baba, 1972).

If this value is substituted for S in eqn (21), the elastic work required to overcome elastic forces is calculated to be of the order of $10^{-6}\text{–}10^{-5}$ erg per beat per single cilium, which is very large in comparison with estimates of the viscous work and of the total work mentioned above. This fact may indicate that the flexural rigidity determined by Baba (1972) is mainly due to a property of the structural component responsible for active bending, rather than being due to a passively bent component. The outer doublet microtubules with arms projecting from them are the structural component which seems to be responsible for generating active bending forces. If it is assumed that the doublets are not connected with one another, the Young's modulus of the microtubule is calculated to be 2×10^{13} dyn cm^{-2}. On the other hand, if it is assumed that the microtubules are connected with one another, the Young's modulus of the microtubule is calculated to be $5 \sim 9 \times 10^{10}$ dyn cm^{-2}. The latter assumption is more probable, because Young's moduli similar to the latter are reported for various protein fibres; e.g. 10^{11} dyn cm^{-2} for flagellin of bacterial flagella (Fujime *et al.*, 1972); of the same order for F-actin of striated muscles (Fujime, 1970; Fujime *et al.*, 1972); and of the order of 10^{10} dyn cm^{-2} for keratin and collagen (Mason, 1965, 1967). From the preceding results, it is concluded that, in the case of active bending, the connections among the doublets are broken because a large amount of energy is necessary to overcome elastic forces of microtubules connected to one another. This conclusion implies that the active bending of the cilium cannot be explained by the activity of contractile elements at one side of the cilium (e.g. doublets constituting one side of the $x'x$ line in Fig. 6), unless the connections among the remaining passively bent doublets are broken during active bending of the cilium. If the connections among the doublets are not broken without producing large internal stresses during active bending of the cilium, it is necessary that all the doublets synchronously contract (and/or expand) to different degrees.

The above results are well explained by assuming that the active bending of the cilium results from *active sliding* between adjacent doublets as follows. The active sliding force is generated by exchange of attachments of arms to the binding sites of the adjacent microtubules which face the arms. The connections among the doublets are kept as a whole, while the work to overcome elastic deformation can be small because of detachment at the time of sliding. If the Young's modulus of the microtubule is $5\text{–}9 \times 10^{10}$ dyn cm^{-2} as mentioned above, the elastic work is of order of $10^{-8}\text{–}10^{-9}$ erg per beat per single cilium, which is the same order as that of the viscous work.

In conclusion, the results of the flexural rigidity estimates support the "sliding filament hypothesis" of ciliary movement (Satir, 1965, 1967, 1968; Sleigh, 1968; Brokaw, 1968, 1971, 1972a, b; Summers and Gibbons, 1971), rather than the "contraction hypothesis".

V. Active region

The degree of bending (curvature) of the cilium during movement is different in different regions of the cilium. It is important in understanding the mechanism of ciliary movement to question whether the motive force of active bending is localized at a special region of the cilium or whether the sites of the motive force exist over a wide extent in the cilium.

As shown in Fig. 4, the change in the curvature of the cilium during movement generally precedes the change in the bending moment at that point and there is no direct correlation in magnitude between the curvature and the bending moment. This fact indicates that the bending of the cilium during movement is not passive, in response to the external viscous force, but involves an active process in the cilium (cf. Baba and Hiramoto, 1970).

Machin (1958) concluded from his mathematical analysis that the movement of the flagellum is not explained by passive oscillation caused by active vibration at the proximal region and that the active bending forces are generated in the flagellum over the entire length. The same conclusion is obtained from the analysis of the relationship between the bending moment and the curvature of the flagellum in starfish spermatozoa (Hiramoto and Baba, unpublished).

Kinosita and Kamada (1939) found, as a result of an experiment using a microneedle to arrest the movements of *Mytilus* large abfrontal cilia,

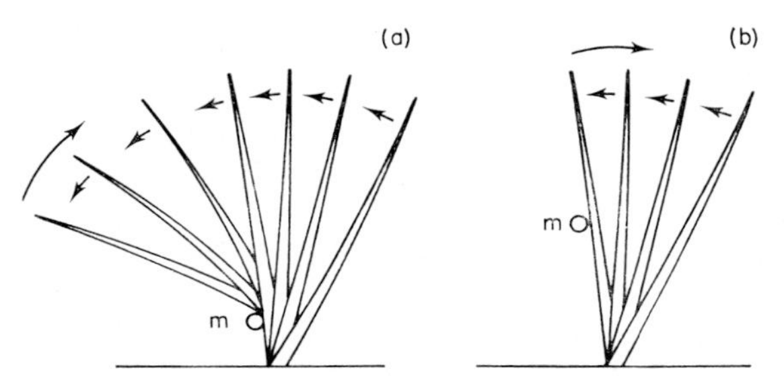

FIG. 9. Responses of a *Mytilus* large abfrontal cilium to the arrest of the effective stroke with a microneedle.

(a) Arrest with a microneedle (m) at proximal region of the cilium. (b) Arrest at distal region of the cilium (Kinosita and Kamada, 1939).

that the cilia bend over the needle to complete the stroke when the needle is applied in the region within 10–20% of the total length from the base of the cilium (Fig. 9a), whereas the cilia are arrested without further bending when the needle is applied to the distal region (Fig. 9b). The frequency of the ciliary beating increases when this "proximal region" (the region within 10–20% from the base) is pinched with a pair of microneedles (Kinosita and Kamada, 1939) or stimulated photodynamically (Tsuchiya, 1971), whereas no change in frequency is observed when these stimuli are applied to the distal region. These facts may indicate that a difference in excitability exists in different regions of the cilium, at least in this material.

Three distinct processes are involved in the beating activity of the cilium: (1) excitation, (2) bending, and (3) propagation of the bending wave. It is considered that the excitation starts from the basal body in normal cilia (cf. Sleigh, 1962), but it can start elsewhere in some cases. Because bending and propagation of the bending wave are observed in fragments of flagella severed from the cell body (Brokaw, 1965; Goldstein *et al.*, 1970) and in cilia and flagella in which the cell membranes are removed (e.g. Hoffmann-Berling, 1955; Brokaw, 1963, 1966a; Satir and

Child, 1963; Gibbons and Gibbons, 1972), it is clear that these processes can be initiated without the basal body or the cell membrane. It seems likely that the sites of motive force for bending and unbending are distributed over the full length of the outer doublet microtubules with arms projecting from them. Propagation of bending waves is observed in flagella extracted with glycerine or Triton X-100 (Brokaw, 1963, 1965, 1966a; Gibbons and Gibbons, 1972), though propagation was not observed in early experiments with glycerinated models (see Sleigh, 1962). It is therefore likely that the mechanism of bend propagation in flagella is built into structures other than the membrane.

Acknowledgements

I thank the following who have read the manuscript of this chapter and given helpful advice: Professor H. Kinosita, Professor P. Satir, Dr. H. Morhi, Dr. M. Yoneda and Dr. S. A. Baba.

References

BABA, S. A. (1972). *J. exp. Biol.* **56**, 459–467.
BABA, S. A. and HIRAMOTO, Y. (1970). *J. exp. Biol.* **52**, 675–690.
BROKAW, C. J. (1963). *J. exp. Biol.* **40**, 149–156.
BROKAW, C. J. (1965). *J. exp. Biol.* **43**, 155–169.
BROKAW, C. J. (1966a). *J. exp. Biol.* **45**, 113–139.
BROKAW, C. J. (1966b). *Am. Rev. Resp. Dis.* **93**, (suppl.) 32–40.
BROKAW, C. J. (1967). *Science, Wash.* **156**, 76–78.
BROKAW, C. J. (1968). *Symp. Soc. exp. Biol.* **22**, 101–116.
BROKAW, C. J. (1970). *J. exp. Biol.* **53**, 445–464.
BROKAW, C. J. (1971). *J. exp. Biol.* **55**, 289–304.
BROKAW, C. J. (1972a). *Biophys. J.* **12**, 564–586.
BROKAW, C. J. (1972b). *Science, Wash.* **178**, 455–462.
BROKAW, C. J. and BENEDICT, B. (1968a). *Archs Biochem. Biophys.* **125**, 770–778.
BROKAW, C. J. and BENEDICT, B. (1968b). *J. gen. Physiol.* **52**, 283–299.
CARTER, G. S. (1924). *Proc. R. Soc.* B **96**, 115–122.
FENN, W. O. (1923). *J. Physiol.* **58**, 175–203.
FENN, W. O. (1924). *J. Physiol.* **58**, 373–395.
FUJIME, S. (1970). *J. Phys. Soc. Japan* **29**, 751–759.
FUJIME, S., MARUYAMA, M. and ASAKURA, S. (1972). *J. molec. Biol.* **68**, 347–359.
GIBBONS, B. H. and GIBBONS, I. R. (1972). *J. Cell Biol.* **54**, 75–97.
GIBBONS, I. R. (1966). *J. biol. Chem.* **241**, 5590–5596.
GIBBONS, I. R. and ROWE, A. J. (1965). *Science, Wash.* **149**, 424–426.
GOLDSTEIN, S. F., HOLWILL, M. E. J. and SILVESTER, N. R. (1970). *J. exp. Biol.* **53**, 401–409.
GORDON, A. M., HUXLEY, A. F. and JULIAN, F. J. (1966). *J. Physiol.* **184**, 170–192.
GRAY, J. (1928). "Ciliary Movement." Cambridge University Press.
GRAY, J. (1930). *Proc. R. Soc.* B **107**, 313–332.

GRAY, J. and HANCOCK, G. J. (1955). *J. exp. Biol.* **32**, 802–814.
GRIMSTONE, A. V. and KLUG, A. (1966). *J. Cell Sci.* **1**, 351–362.
HAPPEL, J. and BRENNER, H. (1965). "Low Reynolds Number Hydrodynamics." Prentice-Hall, Engelwood Cliffs.
HARRIS, J. E. (1961). *In* "The Cell and the Organism" (J. A. Ramsey and V. B. Wigglesworth, eds), pp. 22–36. Cambridge University Press.
HILL, A. V. (1938). *Proc. R. Soc.* B **126**, 136–195.
HOFFMANN-BERLING, H. (1955). *Biochim. Biophys. Acta* **16**, 146–154.
HOLWILL, M. E. J. (1966). *Physiol. Rev.* **46**, 696–785.
HUXLEY, H. E. (1969). *Science, Wash.* **164**, 1356–1366.
KINOSITA, H. and KAMADA, T. (1939). *Jap. J. Zool.* **8**, 291–310.
LAMB, H. (1932). "Hydrodynamics." Cambridge University Press.
MACHIN, K. E. (1958). *J. exp. Biol.* **35**, 796–806.
MASON, P. (1965). *Kolloid-Z. Z. Polym.* **202**, 139–147.
MASON, P. (1967). *Kolloid-Z. Z. Polym.* **218**, 46–52.
MOHRI, H. (1956). *J. exp. Biol.* **33**, 73–81.
RIKMENSPOEL, R. (1965). *Biophys. J.* **5**, 365–392.
RIKMENSPOEL, R. (1966). *Biophys. J.* **6**, 471–479.
RIKMENSPOEL, R. and SLEIGH, M. A. (1970). *J. theor. Biol.* **28**, 81–100.
SATIR, P. (1965). *J. Cell Biol.* **26**, 805–834.
SATIR, P. (1967). *J. gen. Physiol.* **50** (6–2), 241–258.
SATIR, P. (1968). *J. Cell Biol.* **39**, 77–94.
SATIR, P. and CHILD, F. M. (1963). *Biol. Bull.* **125**, 390.
SLEIGH, M. A. (1962). "The Biology of Cilia and Flagella." Pergamon Press, Oxford.
SLEIGH, M. A. (1968). *Symp. Soc. exp. Biol.* **22**, 131–150.
SLEIGH, M. A. and HOLWILL, M. E. J. (1969). *J. exp. Biol.* **50**, 733–743.
SUMMERS, K. E. and GIBBONS, I. R. (1971). *Proc. natn. Acad. Sci. U.S.A.* **68**, 3092–3096.
TSUCHIYA, T. (1971). *Annotnes zool. jap.* **44**, 57–64.
YANAGISAWA, T. (1967). *Expl Cell Res.* **46**, 348–354.
YONEDA, M. (1960). *J. exp. Biol.* **37**, 461–468.
YONEDA, M. (1962). *J. exp. Biol.* **39**, 307–317.
YONEDA, M. (1967). *Protein, Nucleic Acid and Enzyme* **12**, 1179–1185 [in Japanese].

Note added in proof

Recently, Lindemann *et al.* (1973) determined flexural rigidity of motionless bull sperm flagella to be 5×10^{-11} dyn cm^2, which decreases to 4×10^{-12} dyn cm^2 by addition of ATP to the medium. Okuno and Hiramoto (1973) determined flexural rigidity of echinoderm sperm flagella immobilized by CO_2, and obtained about 10^{-12} dyn cm^2 when the flagella were bent slowly. Higher values were obtained when the speed of bending was increased. It seems likely that sliding occurs among the outer doublet microtubules when motionless flagella are bent by external forces.

LINDEMANN, C. B., RUDD, W. R. and RIKMENSPOEL, R. (1973). *Biophys. J.* **13**, 437–448.
OKUNO, M. and HIRAMOTO, Y. (1973). *Zool. Mag., Tokyo* **82** (in press).

Coordination and control of cilia

Chapter 10

Ciliary activity and metachronism in Protozoa

HANS MACHEMER

*Department of Biology, University of California, Los Angeles, U.S.A.**

I. Introduction

Groups of cilia and flagella often exhibit a conspicuous temporo-spatial organization of their movements called metachronism. This phenomenon is seen as a system of waves travelling over the ciliated or flagellated cell surface. For a long time ciliary metachronism has been compared to the transmission of rhythmic bending of ears in a cornfield moved by a breeze (Valentin, 1842). In both systems certain stages of the individual oscillation occur temporally shifted in one direction; perpendicular to this shift in phase the oscillating units are synchronized, thus creating travelling wave fronts. The well-known analogy between ears and cilia has an important limitation: cilia (or flagella) produce *active* bending motions which may or may not be accompanied by metachronism. The nature of the co-ordinating process leading to metachronism has been studied for more than 100 years (recent reviews by Kinosita and Murakami, 1967; Párducz, 1967; Jahn and Bovee, 1967; Sleigh, 1969; Seravin, 1971; Jahn and Votta,

* On leave of absence from: Zoologisches Institut, Fachbereich Biologie, Universität Tübingen, D-74 Tübingen, Federal Republic of Germany.

1972) and was discussed from the viewpoint of internal transmission of signals (Engelmann, 1868) or mechanical coupling between cilia (Verworn, 1891). In the last two decades progress in the study of ciliary metachronism was stimulated by at least two factors. Firstly, it was found that the direction of metachronal waves is related to the ciliary power stroke by certain fixed angles (Knight-Jones, 1954). Secondly, the development of new techniques such as rapid fixation (Párducz, 1952) and improvements in microscopic systems provided new means for observation and recording of ciliary motion at the lower limit of optical resolution. Many new data have now been accumulated, but it appears that we are still a long way from a general understanding of the mechanisms of metachronism. It is the purpose of this article to review these data on the basis of a concept of ciliary metachronism which has been critically reconsidered. Special attention will be given to the relations between metachronism and ciliary activity. Current hypotheses concerning the nature of metachronal coordination will be discussed in connection with ciliary systems to which they are related.

II. The concept of metachronism

Metachronism is a physical phenomenon whose properties occur in space and in time. An appreciation of some of its general properties, however simple as they may be, appears to be indispensible for the understanding of any specific case of metachronal coordination of cilia.

A. Metachrony and synchrony

In populations of cilia or flagella that are arranged in two dimensions the occurrence of metachronal waves will always be accompanied by ciliary synchronization. This is schematically illustrated by a "field" of numbers, as seen in surface view, each number representing a distinct stage of the ciliary cycle (Fig. 1). The metachronal field shown in Fig. 1a represents a momentary state of coordinated ciliary activity which may be seen on the cell surface of *Paramecium* (Fig. 6). In Fig. 1, line a is a unique coordinate parallel to which all cilia are in the same state of cyclic motion (same numbers). This is identical to the definition of *synchronization.* All other possible lines across the field, such as b, b′, b″, show different degrees of phase-shift (metachrony), the "wavelength" along these lines being expressed by the length of a complete sequence of ciliary stages (numbers 1–7 inclusive). It is obvious that metachrony and synchrony are inseparable aspects of any metachronal organization occurring in a field; this is true at any time in the metachronal process.

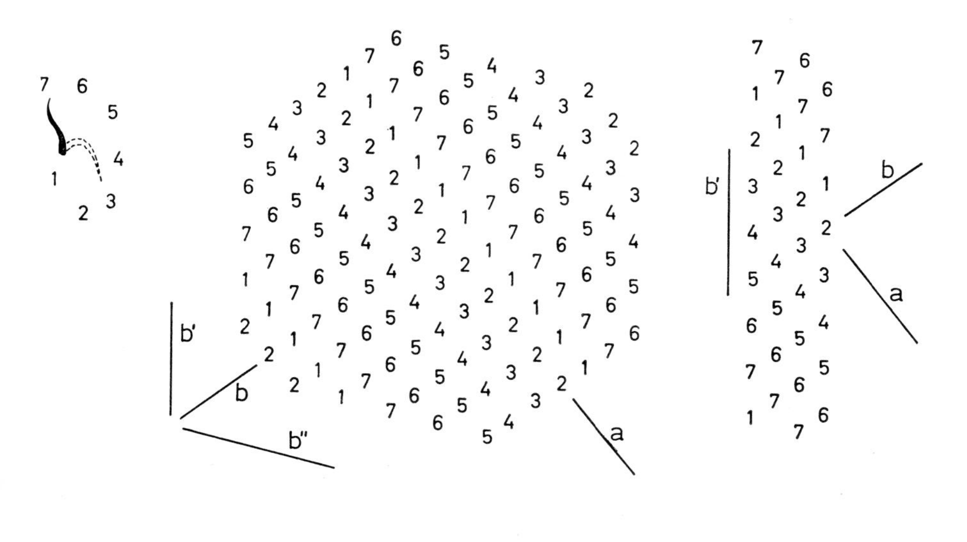

FIG. 1. Properties of a field of cilia during metachronal coordination, surface view. Different stages of the ciliary cycle (left) are represented by numbers 1–7. The two-dimensional system has two main coordinates: the line of ciliary synchronization (a) and the mainline of metachrony (b) in which phase-shift is maximal and which is normal to a. Direction and length of waves is determined along b. Other lines like b′ and b″, which are intermediate between a and b show not the true but an apparent wavelength and wave direction. These properties of the metachronal field are independent from morphological arrangements of cilia, which may occur in large surface areas (1a) or in narrow bands (1b). Modified from Machemer (1972c).

B. Properties of the metachronal field

Closer examination of the metachronal fields in Fig. 1 shows that the "wavelength" decreases if lines of reference are turned from line a towards line b, the latter forming a right angle to line a. Since there are no shorter "waves", only "waves" of equal length parallel to b, this line is another important co-ordinate, comparable with the line of synchrony, and may be called the *mainline of metachrony*. Along this line the true wavelength of a metachronal system, λ, is determined. The mainline of metachrony coincides with the direction in which metachronal waves travel. In fact, a metachronal field is specifically defined if the direction and amount of phase-shift along the mainline of metachrony are known.

Under certain conditions, observation of metachronism along lines other than the mainline of metachrony may provide correct information about wave direction and wavelength. As shown in Fig. 2a, the true wavelength (λ) is defined by the apparent wavelength (λ'),* seen in the focusing plane (b′) and the cosine of the angle β between the mainline of metachrony (b) and b′, according to

$$\lambda = \lambda' \cos \beta. \tag{1}$$

In practice, line b is determined as the normal to the line of synchronization (a), the latter being more easily observable. Equation (1) may, therefore, be modified by regarding the angular deviation α of the plane of focus (b′) from a (Fig. 2b). Since $\cos \beta = \sin \alpha$,

$$\lambda = \lambda' \sin \alpha. \tag{2}$$

For reasons of convention α is measured in a clockwise direction, if the observer looks down on the ciliated cell surface (Fig. 2c).

C. *Types of ciliary metachrony*

It was an important finding by Knight-Jones (1954) that the power strokes of cilia involved in a metachronal system occur at specific angles to the direction of metachronal transmission. This author did not discriminate between an observed propagation of waves along morphological rows of cilia (line b′ in Fig. 1b) and the true wave direction (along line b). His terminology, as briefly outlined in the following, is nonetheless appropriate to examine possible relations between ciliary activity and properties of metachronism (Fig. 3). In an *antiplectic* type of metachrony waves are travelling in a direction opposite to that of the ciliary power stroke. In *symplectic* metachrony, on the other hand, power stroke and wave proceed in the same direction. Both of these types are termed *orthoplectic*, since power stroke and wave transmission occur in the same plane. Metachrony is called *diaplectic*, if wave and ciliary power stroke occur at right angles to each other. This may be realized in a *dexioplectic* type of metachrony, where the power stroke occurs toward the right of an observer looking in the wave direction. Conversely, in *laeoplectic* metachrony the observer would see power strokes pointing toward the left.

All four types of metachrony are said to occur in Metazoa, although most groups show only one type of diaplectic metachrony. In Protozoa

* The difference between "true" and "apparent" direction or length of waves may be illustrated by the following example. An observer watching crests of water-waves rolling towards a shore at an acute angle, will see breaking crests travelling along the shoreline. The direction and length of these breaking waves does not correspond to the true wave direction and true wavelength on the water surface. Wave parameters determined along the shoreline may, therefore, be called "apparent".

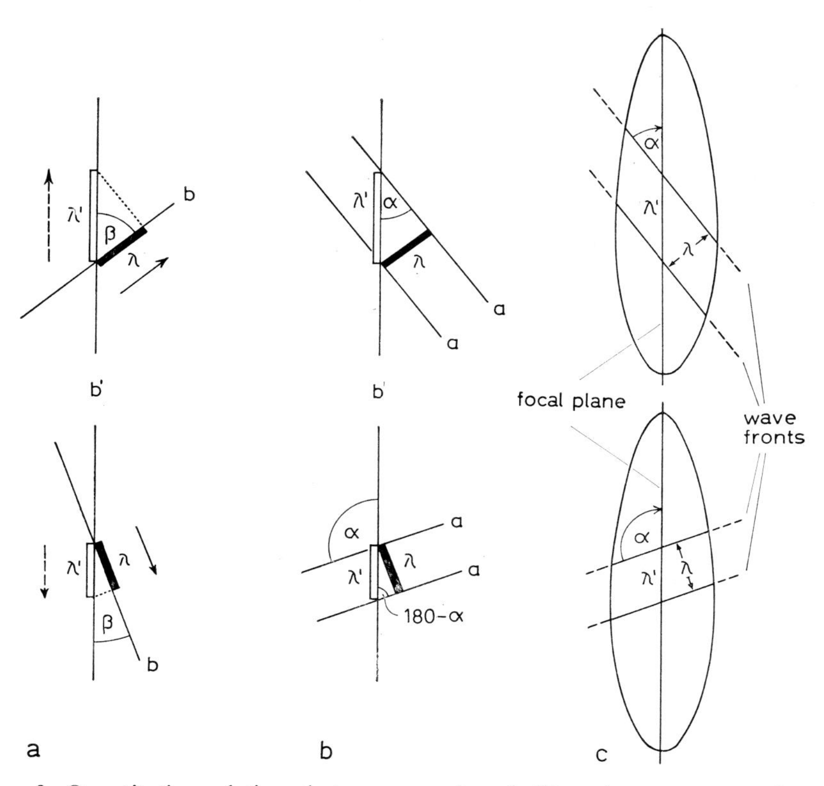

FIG. 2. Quantitative relations between wavelength (λ) and apparent wavelength (λ') appearing in any given line (e.g. b′) across a metachronal field in surface view (e.g. in *Paramecium* at normal viscosity (upper row) or high viscosity (40 cP, lower row)). 2a. Illustration of $\lambda = \lambda' \cos \beta$. β is the angle between main line of metachrony (b) and b′. 2b. Illustration of $\lambda = \lambda' \sin \alpha = \lambda' \sin (180 - \alpha)$. α is the angle between line of synchrony (a) and b′. 2c. Practically, α is measured as the angular deviation of the plane of focus (b′) from the wavefronts in *clockwise* direction. Dark bars and complete arrow: length and directions of waves; light bars and dashed arrow: length and direction of waves as they appear in the focal plane.

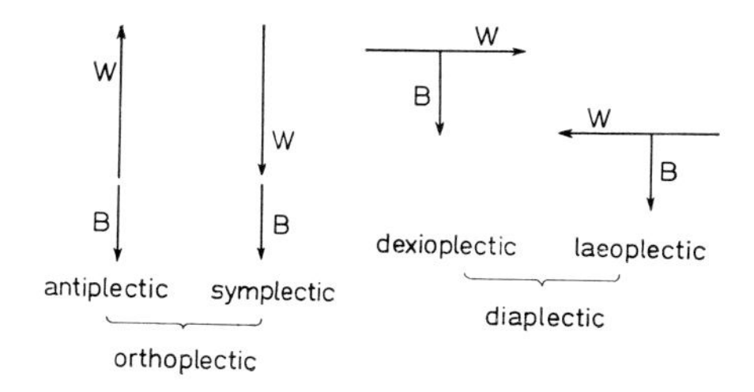

FIG. 3. Four basic types of metachronal coordination, defined by relations between directions of ciliary beating (B) and wave transmission (W) as seen in the plane of the metachronal field.

the occurrence of all of these types, except for laeoplectic metachrony, has been reported. However, some data may have to be re-examined, since the direction of wave transmission was not always separated from a given plane of optical focus or a pathway along ciliary rows. It appears

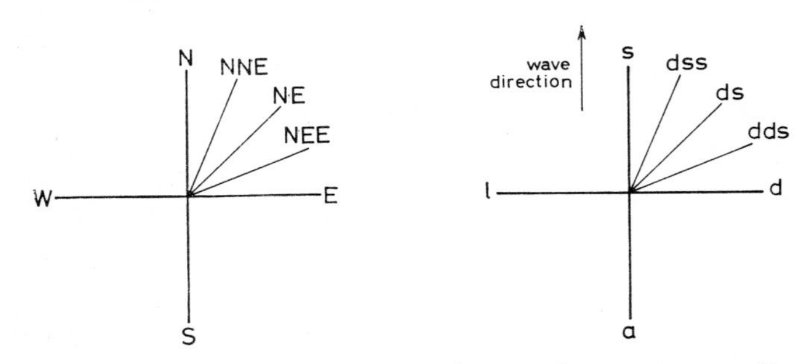

Fig. 4. Determination of those metachronal types that are intermediate between two of the four basic types according to subdivisions of directions in the compass-rose (left). s (d, a, l) = symplectic (dexioplectic, antiplectic, laeoplectic) orientation of the beating with respect to the wave direction. For example: the power stroke occurs in a direction marked by dds; the type of metachrony will be "dexio-dexio-symplectic".

that metachrony in protozoan ciliary systems is occurring most frequently in the dexioplectic and symplectic types or in a range between these types. Adequate refinement in the determination of intermediate metachronal types is achieved by the method used in subdividing compass-rose directions (Fig. 4).

D. *Some problems of observing metachronism*

Correct assessment of metachronism requires that properties of wave propagation and ciliary activity are related to the plane of the metachronal system (i.e. parallel to the cell surface). It is not always feasible to observe the ciliated surface from above, so that metachronism is often studied in profile views with the line of sight nearly parallel to the focused portion of the cell surface. Single ciliary rows and membranellar bands are most conveniently seen in profile. In holotrichous protozoa or polymastigote flagellates, observation of waves at the cell-edges may also have certain advantages. It is important to realize that profile images of a metachronal system are only "sections" of three-dimensional events occurring in a zone parallel to the cell surface. As an example (Fig. 5a), waves travelling upwards and to the right may appear in profile as moving upwards; ciliary power strokes pointing downwards and to the right may appear to occur downwards in the plane of focus. Thus, a dexioplectic metachrony is counterfeited as an antiplectic type of metachronal coordination. With regard to possible misinterpretations of images of metachronal systems it has been proposed (Machemer, 1972a) to call apparent antiplectic patterns *antiplectoid* and apparent symplectic patterns *symplectoid*. The

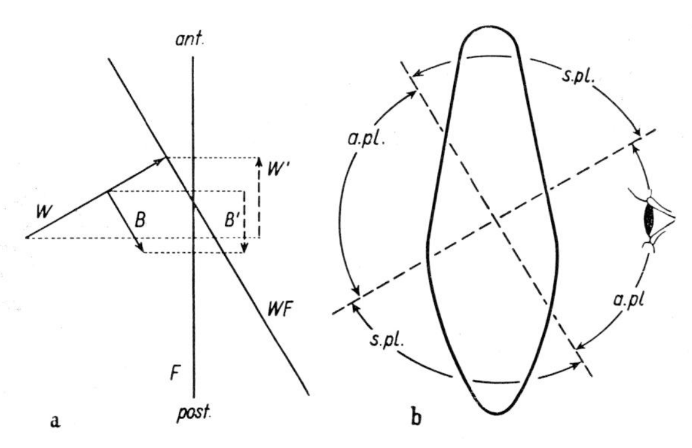

FIG. 5. Consequences of observation of ciliary metachronism in profile. a. A dexioplectic pattern, as seen in the surface view (complete arrows), may be antiplectoid in side view (dashed arrows). B′, projection of direction of power stroke (B). W′, projection of wave direction (W). WF, orientation of wave fronts. F, focal plane. b. Under conditions given in *a* profiles of metachronal waves are seen as antiplectoid (apl) or symplectoid (spl) in different sectors of observation. Compare with models in Fig. 6. From Machemer (1972a).

geometrical relations between surface and profile images of metachronism, as shown in Fig. 5a, are such that profiles of a certain type of metachrony will change from antiplectoid to symplectoid when viewed from different angles with respect to the wave fronts (Fig. 5b).

III. Methods of studying ciliary metachronism

Ciliary metachronism is observed in two principal planes: in profile at the cell edge or in surface view. In these planes metachronal parameters may be isolated from certain spatial configurations of cilia, as shown in the following. Some properties of different techniques in visualizing metachronal systems will also be briefly discussed.

A. How to understand surface views of waves

Dexioplectic metachronal patterns are very common in Protozoa. As an example, a model of the ciliated surface of *Paramecium* during forward

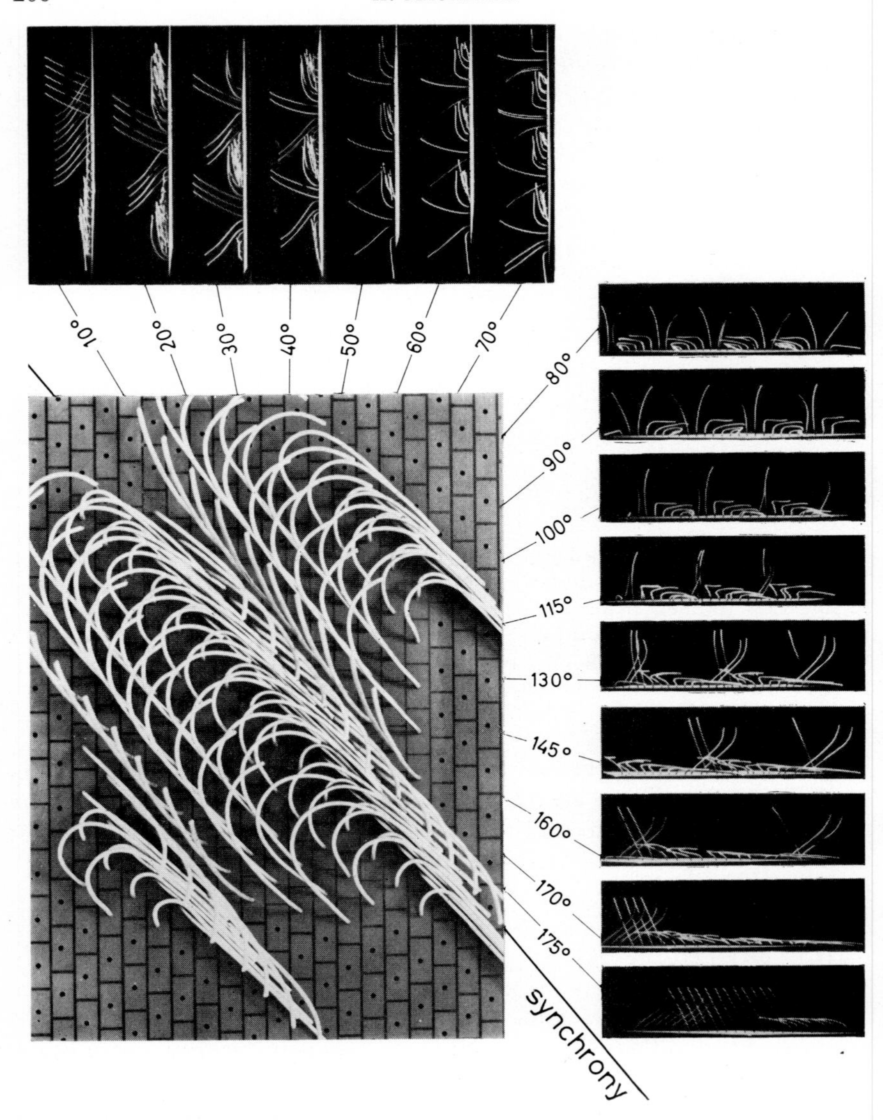

FIG. 6. Model of a dexioplectic ciliary pattern, as it is seen in a *Paramecium* during forward locomotion. Centre: Surface view, anterior direction up. The tips of individual cilia proceed in a counterclockwise cycle, with the power stroke towards the posterior-right. Waves move toward the anterior-right. Periphery: Profiles of this wave system, numbered according to the angular deviation of the observed plane from the line of ciliary synchronization. Patterns are antiplectoid at angles smaller than 90°, symplectoid at angles exceeding 90°. Profiles at angles between 20° and 40° are most commonly seen under the microscope. From Machemer (1972a).

swimming shows oblique lines of closely aggregated straight cilia alternating with broader bands of cilia which are more or less sickle-shaped (Fig. 6, centre). In the power stroke the cilia move towards the posterior and right, as seen immediately above the "dense" lines where cilia are most closely packed. If the observer looking at cilia proceeds in an anterior-right direction from those in the power stroke, he will see "earlier" stages of the cyclic movement: the band of sickle-shapes depicts recovery strokes of cilia which proceed in counterclockwise cycles. The numerical proportion of recovery stages to stages of the power stroke represents differences in angular velocity during one ciliary cycle. If the model cilia were set in motion, identical stages of bending would appear in progression across the surface in an anterior-right direction which is, therefore, the direction of wave transmission (normal to the lines of ciliary synchronization). Wavelength is identified as the perpendicular distance between the dense lines of ciliary aggregates. The power stroke occurs to the right with respect to the wave transmission, and the metachronal type is dexioplectic. In general, diaplectic systems may be easily identified from surface views by remembering that the openings of ciliary sickle-shapes approximately point in the direction of the power stroke and that waves travel towards the more crowded sides of the dense lines.

In protozoan symplectic systems the directions of "beating" and of the waves are identical with the overall orientation of the organelles. A typical power stroke is absent; locomotory effects occur along the main axis of helically wound organelles (Fig. 10). A surface view of a dexiosymplectic pattern has properties somewhat intermediate between the dexioplectic and symplectic patterns (see Fig. 29).

B. *How to understand profile views of waves*

Side views of the dexioplectic model (Fig. 6) show a great variety of ciliary configurations as a function of the angle (α) at which the plane of focus deviates from the line of ciliary synchronization ("wave angle"). Waves appear in the form of groups of "straight" cilia (power stroke) alternating with groups of bent-over cilia (recovery stroke). The apparent wavelength (λ') is shortest close to the mainline of metachrony ($\alpha = 90°$) and increases toward the line of synchrony. Convex sides of the straight cilia point in the apparent beat direction. The apparent wave direction is easily determined by the following method (Párducz, 1954). *Divergence* of the straight cilia performing the power stroke indicates that waves apparently move against the power stroke (antiplectoid metachrony; see profiles at angles up to 90°). *Convergence* of cilia in the straight groups shows coincidence of apparent directions of beat and waves (symplectoid metachrony; see wave profiles at angles between 90° and 180°). These properties are valid for either type of diaplectic metachronism. The discrimination between divergent and convergent groups of cilia performing the power stroke is more complicated in those metachronal systems that are more related to symplectic wave types, such as the dexiosymplectic pattern (Fig. 29).

C. *Can metachronal parameters be derived from profiles of waves?*

Systems of waves which are observed at defined angles (α) show unique patterns in profile, differing in their apparent wavelength (λ') and metachronal configuration. Observed wave profiles that fit to profiles of a

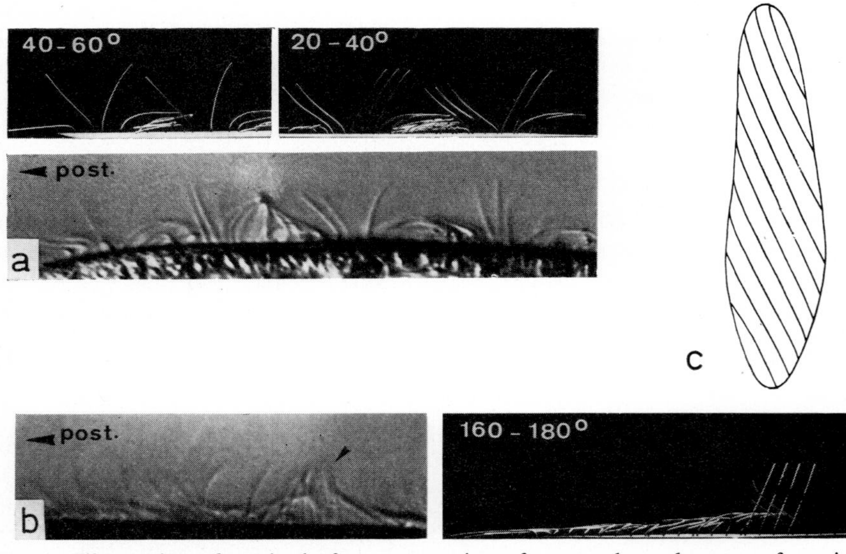

Fig. 7. Illustration of method of reconstruction of a metachronal pattern from its profile views by comparison with other profiles, which are "gauged" to certain wave angles. a. Observed and model profiles with antiplectoid pattern. b. Observed and model profiles with symplectoid pattern. c. Example of generalized reconstruction of wave system on the body surface of *Paramecium*, based on quantitative averaging of observed profiles. From Machemer (1972a).

known system can also be related to identical wave angles (Fig. 7). According to eqn 2 the true wavelength (λ) is calculated from λ' and α. Thus, the orientation and length of waves on the surface of a protozoan cell may be completely reconstructed (Fig. 7c), if, as is usual, the focal plane coincides with a certain body axis of the cell.

D. Direct observation of metachronism

In principle, metachronal waves can be microscopically observed, whenever (1) sufficient optical contrast for visualizing cilia is provided and when (2) the velocity of wave propagation does not exceed the resolving capacity of the human eye. This is realized by dark-field or interference-contrast illumination of cells such as opalinids or hypermastigote flagellates, where the beat frequency may be well below 10 Hz. In typical ciliates with normal ciliary frequencies of the order of 20–40 Hz, waves can be directly observed at reduced frequency in cells which are deteriorating, cooled down or treated with narcotics (lower alcohols, Bills, 1922; nickel ions, de Puytorac *et al.*, 1962). Especially suitable for observation are cilia within compound structures such as membranelles or cirri which are found in spirotrichs. Methylcellulose or other thickening, non-ionic substances reduce the frequency of cilia through increased viscosity of the medium. Metachronal properties, however, undergo considerable viscosity-induced changes (see section IV.C), or metachronism even breaks down if viscosity is increased beyond a critical level.

E. Stroboscopic observation

Short flashes of illumination synchronized with ciliary frequency create illusionary "frozen" images of single cilia as well as of metachronal waves. The frequency of cilia is determined during the course of continuous reduction of flashing rates at the *first* stopping of the cilia. At flashing rates 2 times (3 times, etc.) the correct frequency each cilium is seen 2 times (3 times, etc.) per cycle and the waves overlap each other. Profile images of stroboscopically stopped waves appear most clearly through optics providing interference contrast. Peristomal cilia and compound cilia within membranellar bands, which show little variation in frequency, are easily accessible for the stroboscopic method. Gray (1930) found that local gradients in frequency can be detected in the strobe light through apparent shifts of "frozen" waves toward a point where the waves come to rest (Fig. 8). This method is of great value for assessing transient or permanent gradients of frequency within metachronal waves along topographically defined lines.

F. Rapid fixation technique

Osmic acid may preserve metachronal systems by an almost instantaneous fixation of cilia and flagella. The history and details in procedure of this

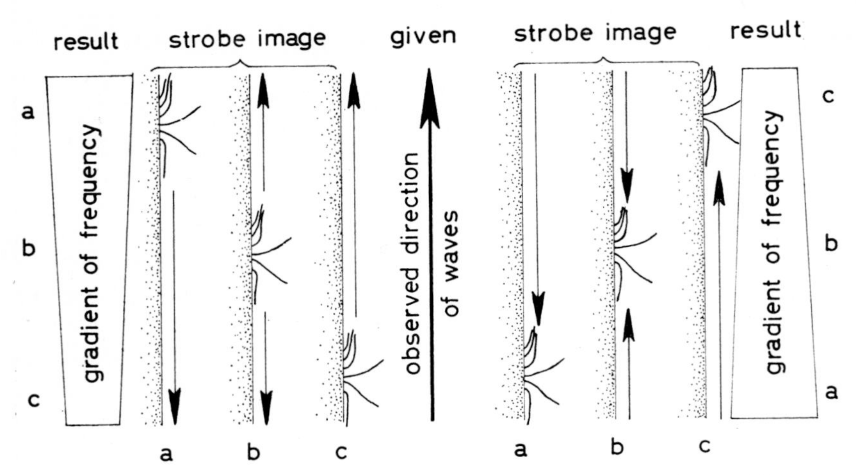

FIG. 8. Stroboscopic determination of gradients of frequency along ciliary rows seen in profile. Strobe-images of waves will move in a direction opposite to the observed transmission of the true waves, if the standing strobe-wave was fixed at a high level of the frequency-gradient (a). On the other hand, if the standing strobe-wave was located at the low end of the frequency gradient (c), strobe-images of waves will move in the direction of transmission of the true waves. Strobe-images fixed at the middle (b) of a frequency-gradient show diverging strobe-waves if the true waves move in a direction towards the frequency-maximum; strobe-waves converge toward a fixed image if true waves travel towards a minimum of frequency. Drawn profiles of waves: standing strobe-waves. Light arrows: directions of moving strobe-waves. Modified from Gray (1930).

important technique are described in the review by Párducz (1967). The first successful analyses of metachronal systems in ciliate Protozoa were based on rapid fixation and subsequent staining with haematoxylin (Párducz, 1952). A variation of the Párducz method is offered by Grębecki (1964), who employs nickel ions to improve haematoxylin lakes on fixed cilia. Since rapid fixation leads to preparations of complete ciliary systems of protozoan cells, certain forms of ciliary activity, such as beating in normal or reversed direction, and accompanying patterns of waves may be studied in detail from surface and profile images (Fig. 13). On the other hand, the assessment of transient events such as the "renormalization" from backward to forward swimming in ciliates is difficult, since the fixation cannot be aimed at specific cells at a specific time. Quantitative evaluations of entire samples of fixed cells help to identify transient metachronal events through the determination of relative frequencies of fixed stages (Machemer, 1969a). It should be remembered, however, that optimal concentrations of the fixative cannot be applied instantaneously and simultaneously over the whole cell surface. Therefore, the fixation may create a certain percentage of chemically induced responses of the cilia (Machemer, 1970a). The validity of the rapid fixation technique was recently questioned by Kuznicki *et al.* (1970), who found no metachronal waves in swimming paramecia, filmed with a high-speed camera. These findings were not confirmed by other authors, who used similar techniques (Allen, 1969; Tamm, 1970; Machemer, 1972a, 1974; compare Figs 15, 16, 23, 25). Successful use of rapidly fixed cilia in electron microscopic analysis of ciliary motion (Satir, 1963) was recently extended by combining rapid fixation, critical-point drying and scanning electron microscopy (Horridge and Tamm, 1969). Images of fixed metachronal waves in *Opalina* with excellent resolution of ciliary detail (Fig. 13) were confirmed by *in vivo* photographs (Tamm and Horridge, 1970).

G. *Photography of ciliary systems*

Appropriate technical installations allow the photographic documentation of rapid motions of such delicate structures as cilia and flagella. There are 6 basic requirements which, if combined together, may result in undistorted images of ciliary metachronism:

1. Optical devices have to transform phase differences at the protoplasmic boundaries of cilia into differences in light amplitude strong enough to be detected by photographic emulsions.
2. Numerical apertures of objectives and condensers should be such ($\geqq 0{\cdot}6$) as to resolve single cilia (approximate diameter: $0{\cdot}2\ \mu$m).
3. The duration of the photographic exposure has to be very short in order to obtain sharp images of the oscillating cilia.
4. The intensity of illumination should not change the normal functioning of cells, but must be sufficient for photographic exposure.
5. Appropriate sensitivity of the photographic emulsion should be combined with sufficient resolving properties.
6. Any mechanical impairment of the free motion of cilia has to be excluded.

Examples of ciliary metachronism photographed with a 35 mm still camera and on 16 mm ciné-film at 250 frames per second are shown in Figs 16 and 17.

IV. Survey of metachronal systems in Protozoa

At this time only a few metachronal systems in Protozoa have been studied in some detail. More often information is incomplete or restricted to qualitative description due to technical difficulties of observation.

Summarizing available data should help to understand ciliary function within the metachronal system and to test information which reinforces either the "mechanical" or the "neuroid" hypotheses of metachronal coordination.

A. Polymastigote flagellates

Cleveland and Cleveland (1966) described locomotory systems in three large forms of polymastigote flagellates, *Mixotricha*, *Koruga*, *Deltotrichonympha*, each of them living in the hind-gut of the termite, *Mastotermes*. Locomotion and metachronism in *Mixotricha* is interesting because of its particularly primitive features. In *Koruga* and *Deltotrichonympha* metachronal coordination of the flagella is accompanied by undulations of the cell surface. This raises the question of how this metachronism is generated.

1. Mixotricha

This large drop-shaped flagellate (400–500 μm) shows monotonous forward locomotion without helical rotation around the longitudinal body axis (Fig. 9a). Metachronal waves cover the whole cell surface except for a small ingestive area at the posterior end. While travelling rapidly from the anterior end posteriorly the wave fronts assume an orientation at right angles to the longitudinal axis. An apparent "slanting" of the waves in profile views is similarly seen in wave profiles of *Koruga* where waves also travel exactly along the longitudinal body axis (Cleveland and Cleveland, 1966). The short "ciliary" organelles (length 10 μm, diameter 0·15 μm) incorporated in the metachronal system are symbiotic* spirochaetes which are externally attached, singly or in groups of up to 4, to the posterior faces of small "brackets" arising at regular intervals on the cell surface (Cleveland and Grimstone, 1964). According to reconstructions of the waves by Cleveland and Cleveland, the helical form of spirochaete movement (Jahn and Landman, 1965) appears in profile to be sinusoidal, comprising one complete bending wave on each spirochaete (Fig. 10a). Since harmonious bending helices of $n\lambda$ produce resulting vectors of force parallel to the helical axis (Fig. 44), continuous locomotion should occur opposite to the direction of wave transmission; this corresponds to observations.

Metachronism in *Mixotricha* is *symplectic* since the directions of "beating" and of the transmission of waves are the same. "Beating" is understood

* The full extent of spirochaete symbiosis in *Mixotricha* has not yet been established. According to Cleveland and Cleveland, symbiosis may apply at least to the form of locomotion of the host as well as of the passenger.

as the net force resulting from helical undulations and causing locomotion in the opposite direction. The length of the metachronal waves is reported

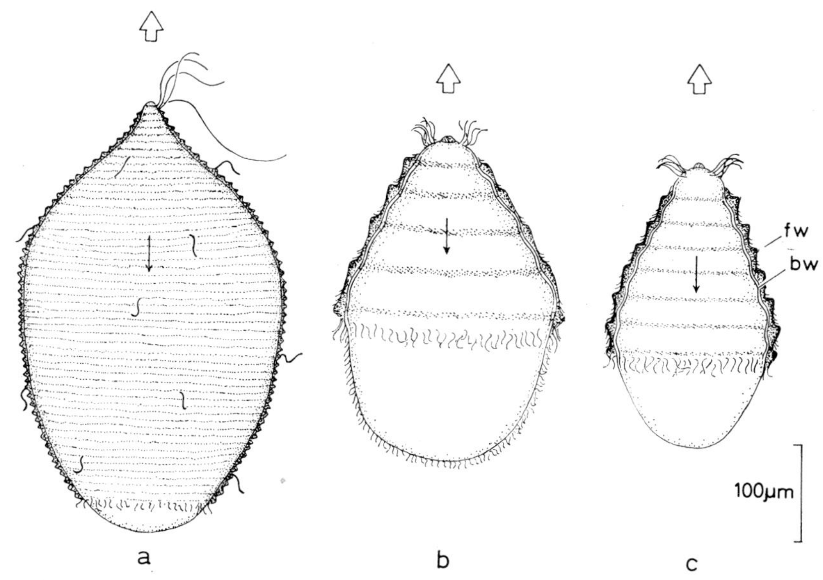

FIG. 9. Metachronal systems in the polymastigote flagellates, *Mixotricha* (a), *Deltotrichonympha* (b) and *Koruga* (c). Schematic representation.

a. Dense aggregates of symbiotic spirochaetes (10 μm long, 0·15 μm in diameter) cover the entire cell surface except a small ingestive area at the posterior end. Metachronal waves (λ = 7·5 μm) travel posteriad (thin arrow) in the direction of the forces resulting from helical oscillations of spirochaetes (symplectic metachronism). Consequently, *Mixotricha* swims forward in an unrotated path (big arrow). Four flagella below the anterior papilla, three of them pointing obliquely forwards, are assumed to influence swimming direction. A few larger spirochaetes (up to 25 μm long, 0·5 μm in diameter) are often associated with the common smaller type.

b, c. Anterior parts of *Deltotrichonympha* and *Koruga* are covered with flagella beating helically in a symplectic metachronal system similar to that of the spirochaetes of *Mixotricha*, and also resulting in unrotated forward locomotion. However, waves are longer (up to 40 μm) and flagellar metachronism (fw) is accompanied by phase-shifted undulations of the cell membrane ("body waves", bw). The two tufts of rostal flagella exhibit synchronized helical beating. The posterior part of *Deltotrichonympha* is coated with motile spirochaetes without metachronal coordination. Drawing constructed after figures and data from Cleveland and Grimstone (1964); Cleveland and Cleveland (1966).

to be 7·5 μm. The full range of beat frequency (and rate of passage of waves) was not determined, but frequency did not decrease below 5 Hz. According to observations by Cleveland and Cleveland, metachronism in the posterior direction may arise at any place on the cell surface where spirochaetes are found in aggregations close enough to allow the synchronization of sequences of individual helical waves (Gray, 1928; Machin, 1963). Optimal synchronization between the "peaks" and "valleys" of helical undulations does not only require similarity of direction, frequency and amplitude of individual waves but also an equal wavelength, i.e. similar properties of bending curvature throughout the length of the

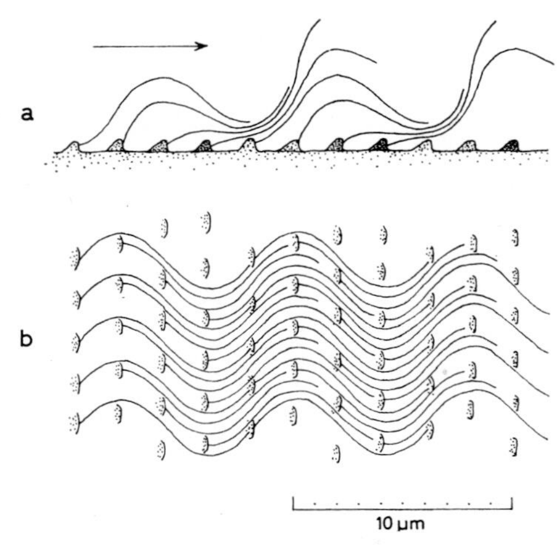

FIG. 10. Composition of metachronal waves in *Mixotricha*.

a. Profile view. Spirochaetes are attached to the posterior sides of "brackets" emerging from the cell surface. Helical bending waves travel along each spirochaete in a proximo-distal direction forming waves of metachronal coordination that travel in a posteriad direction (arrow). From Cleveland and Cleveland (1966).

b. Surface view (hypothetical). Transformation of wave pattern seen in profile (a) reveals that individual helical undulations harmoniously fit the metachronal pattern through (1) identity of individual and metachronal wavelength and (2) a specific pattern of brackets as revealed by electron microscopy (Cleveland and Grimstone, 1964). In theory, this would be the most simple form of metachronal coordination.

individual wave. Metachronism in *Mixotricha* is close to these conditions, since metachronal wavelength (λ) reflects the length of individual bending waves (Fig. 10) which is a rare phenomenon in metachronal systems. Under the circumstances of spirochaete movement provided by *Mixotricha*, the pattern of metachronal waves represents the continuation of individual sequences of helical bending supported by a distinct pattern of distribution of the "brackets" (Fig. 10b). The problem of metachronal coordination appears to be reduced, in this case, to the question of how spirochaetes are synchronized to each other (see section V.A).

2. *Koruga* and *Deltotrichonympha*

These flagellates are somewhat smaller (300–400 μm) than *Mixotricha* but have a similar shape (Fig. 9b, c). Metachronal waves covering the anterior one-half to two-thirds of the cell surface at right angles to the longitudinal axis are seen to move straight backward, while unrotated locomotion occurs at "several hundred μm per second" (Cleveland and Cleveland, 1966). While metachronism of the densely packed flagella (20–30 μm length) undoubtedly is related to locomotion, its mechanism is less clear. Cleveland and Cleveland (1966) found in *Koruga* that the "flagellar waves" (length: 22–40 μm) ride on top of "body waves", i.e. undulations of the cell surface, having the same wavelength and considerable amplitude ($\geqq$ 5 μm). Body waves precede flagella waves by less than 0·5 λ. Since flagellar waves (but not body waves) broke down* at surface areas where the cell met an obstacle, Cleveland and Cleveland concluded that "the flagella act purely passively and their movements are thrust upon them by the active body wave". This interpretation is not satisfactory in some respects. The obviously asymmetrical form of flagellar waves (Fig. 11d) would not result from passive flagella, tilted at their bases by some 40°. In addition, photographs of wave profiles in *Koruga* remind one of symplectic patterns of actively beating spirochaetes in *Mixotricha*. The synchronization of flagellar and body waves in *Koruga* and *Deltotrichonympha* may be a unique illustration of mechanical forces generated by actively beating flagella and exerted against the fluid medium as well as against the cell surface (Fig. 11). If it is assumed that harmonious helical bending in flagella exceeds the length of a complete wave (Fig. 11a), a small section of the flagellar shaft will produce, in the course of one cycle, vectors of force with different orientations, which cannot be counterbalanced by vectors from other sections of the shaft (compare Fig. 46b). A population of similar flagella which is tilted toward the cell surface will create "additional" forces which are directed vertically against the cell surface at definite stages of the beating cycle (Fig. 11b). Under conditions of metachronal coordination of these flagella, "waves" of additional vertical pressure will be swept along the cell membrane, synchronously with the metachronal flagellar waves. If the membrane lacks sufficient rigidity to resist the waves of pressure exerted by densely aggregated flagella, these waves of pressure will be seen as periodic undulations of the surface synchronized with the flagellar waves (Fig. 11c, d). Flagellar activity in *Koruga* and *Deltotrichonympha* appears to be controlled to

* The transient absence of flagellar waves close to mechanical obstacles does not imply that individual flagellar movement must be abolished. Accordingly, photographs presented by Cleveland and Cleveland suggest disturbance of the wave pattern rather than extinction in the zone of collision.

some extent by the cell, since frequency, wavelength and rate of locomotion are reported to vary. Even flagellar orientation seems not to be fixed since slanting of the wave fronts was observed during collision with an obstacle. These phenomena are not seen in *Mixotricha*. Another important difference from *Mixotricha* is the increased metachronal wavelength in comparison with that of an individual flagellum. Absolute and relative increase in wavelength may be caused by the flagellar length, at least twice that of *Mixotricha* spirochaetes. Long flagella apparently produce increased mech-

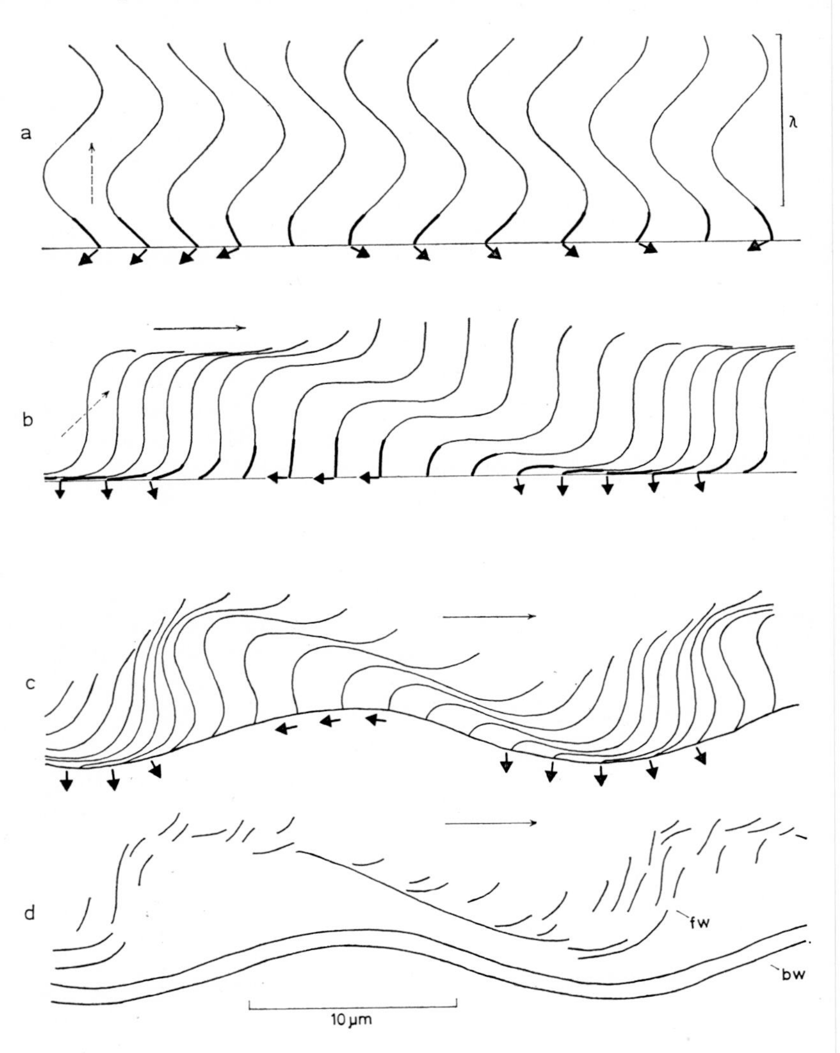

Fig. 11.—Caption on facing page.

anical coupling along the mainline of metachrony (section V.C.1). This increase in "synchronization" will be expressed by increased length of metachronal waves. Short wavelengths comparable with those in *Mixotricha* are found in metachronal systems of cilia having a length of about 10 μm. Since metachronal wavelength may exceed the length of individual undulations (as clearly seen in *Deltotrichonympha*; Fig. 9b), the form of metachronally coordinated flagellar beat cannot be that of a simple helix. Apparently, coupling forces in different directions of the metachronal field are such as to provide different degrees of "synchronization" and leading to a stable hydrodynamic balance which is described by the minimum interference principle (see section V.A.2). Machin's postulate (1963) that one flagellum propagates only one wave mode, does not apply to flagellar metachronism since proximal and distal curvatures of the flagella differ from each other at different stages of periodic bending (Fig. 11c; compare also Fig. 13).

B. *Opalina*

This large oval flagellate (200–500 μm), living in the rectum of amphibia, has a motile system which combines properties of typical flagellates with those of holotrichous ciliates. Its flagella (15–20 μm) are somewhat shorter than those found in *Koruga*, but they beat in patterns of metachronism which resemble metachronal waves in polymastigote flagellates. *Opalina* shares, on the other hand, the capability of ciliate protozoa to "reverse"

FIG. 11. Metachronism in *Koruga*. Three hypothetical stages of development of the metachronal waves of flagella and "body waves" as seen in photographic records.

a. Flagella beating in a helical bending wave produce "additional" forces of pressure against the cell membrane (heavy arrows) through sections of the flagellum (heavy line) that are not counterbalanced by other parts of the organelle. If a series of flagella shows shifting in phase, the additional forces will be re-oriented periodically.

b. The same series of flagella, when tilted by 45° toward the cell surface. Periodically, additional vectors of force assume a more vertical or a more horizontal orientation with respect to the cell surface, thus creating waves of additional pressure against the membrane.

c. In *Koruga* (and *Deltotrichonympha*) a similar mechanism of additional pressure causes undulation of the cell membrane which is synchronized with certain stages of flagellar bending and hence related to the metachronal waves by a certain phase-shift. Helical shape of flagella is somewhat modified due to discrepancy of individual and metachronal wavelengths.

d. Profile of flagellar waves (fw) and body waves (bw), drawn from photographs by Cleveland and Cleveland (1966).

Broken thin arrows: direction of bending waves; complete thin arrows: direction of metachronal and body waves.

the direction of beating spontaneously or in response to stimuli. This behaviour would justify the term of "protocilia" for opalinid motile organelles.

1. Patterns of wave transmission and swimming direction

An undisturbed *Opalina* swims in a more or less straight line with a small rate of rotation in a left-hand helix around its longitudinal axis

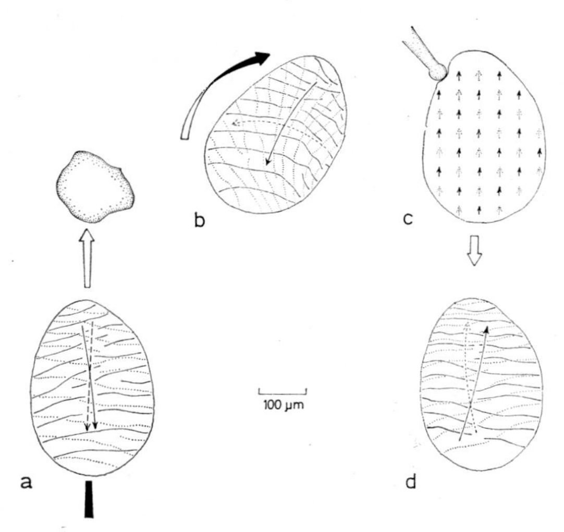

FIG. 12. Locomotion and patterns of metachronal waves in *Opalina*.

a. Forward swimming of unstimulated cell occurs in a straight line with a small rate of rotation in a left-hand helix. Metachronal waves starting at the anterior end are slanted from the transverse axis in a counter-clockwise direction. Propulsive forces, which are oriented approximately parallel to the mainline of metachrony in a posterior direction, deviate from each other on the two surfaces of the cell by a small angle thus creating the rotated path of swimming.

b. *Opalina* colliding with a solid object usually turns toward the right and intensifies its left-hand helical rotation. The amount of turning of the wave fronts in a counter-clockwise direction depends on the intensity of stimulation. Graded increase in stimulation may result in successive shifts and finally in a total inversion of the direction of beating and wave propagation, accompanied by swimming in backward direction (d).

c. "Reversal" of flagellar beating activity by a strong stimulus, for instance a microneedle touching the cell surface. Flagella perform one synchronous "stroke" toward the anterior end and then continue with helical beating in the reversed direction (d).

d. *Opalina* during backward swimming. Metachronal waves move anteriad. Complete lines: crests of waves on "dorsal" side of cell; dotted lines: waves on ventral side of cell. Note the increase in wavelength towards the posterior end. Arrows on the cell surface indicate wave direction and approximate orientation of propulsive forces of flagella. Flat arrows: swimming direction.

(Fig. 12a). Metachronal waves travel posteriorly on both surfaces of the flattened cell. When *Opalina* collides with a solid object it usually swerves toward the right, intensifying rotation. This behaviour is reflected in the orientation of the wave fronts on the dorsal and ventral side of the cell (Fig. 12b). The left-hand swimming helix of *Opalina* is understood if it is assumed that locomotory forces of the beating flagella are oriented approximately parallel to the direction of propagation of the metachronal waves. Graded shifts in wave orientation and swimming direction were observed in a range close to 180° (Okajima, 1953; Fig. 12d). Kinosita (1954) demonstrated similar shifts in the wave pattern of cells that were gradually depolarized by increasing amounts of external KCl concentration.

2. *Flagellar reversal*

Strong stimuli, such as tapping with a microneedle, may produce an immediate "reversal" of all flagella toward the anterior end (Fig. 12c), accompanied by an apparent loss of metachronal coordination (Okajima, 1953). Quasisynchronous reversal of flagella causes strong forces of mechanical coupling and hence increase in wavelength in the direction of reversal. Obviously, a metachronal field with a wavelength (λ) becoming very large will approach the case of total synchronization; this apparently occurs during "reversal". After the completion of one stroke in reverse a new metachronal pattern with waves moving toward the anterior end "is set up instantly all over the cell surface" (Okajima, 1953; Fig. 12d).

3. *Origin of metachronal waves*

As mentioned above, metachronism in *Opalina* starts simultaneously in different regions of the cell and waves may travel in different directions. This rules out that pacemakers control the origin and other properties of waves in *Opalina*. Okajima (1953) observed that metachronal coordination broke down in cells after application of anaesthetics such as ethyl alcohol in parallel with a reduction of beat frequency which is between 2 and 4 Hz under normal conditions. Cutting or pressing *Opalina* with a glass needle often resulted in different metachronal patterns in the separated regions. These observations were interpreted in terms of disturbed mechanical interaction between the flagella. Differences in time scale between instant "wave formation" and slow "wave conduction" (100–200 $\mu m\ s^{-1}$), as pointed out by Okajima, do not contradict but rather corroborate a hypothesis of the mechanical origin of waves, since similar flagellar activity should produce similar results of metachronal coordination at any point of the cell surface without delay. The rate of wave

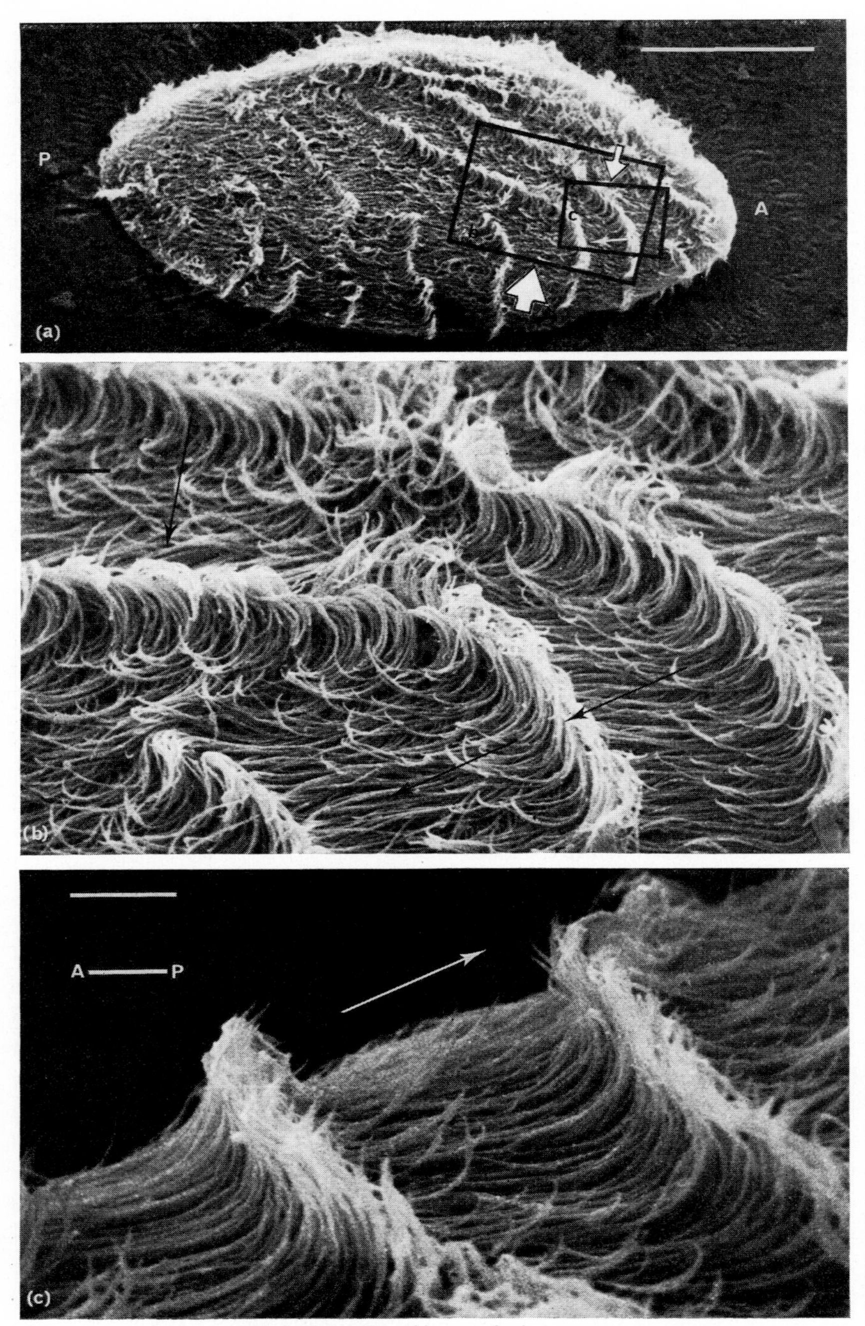

FIG. 13.—Caption on facing page.

propagation (product of frequency and wavelength), on the other hand, has no meaning beyond reflecting conditions of mechanical interaction between flagella (see section V.C.2).

4. Form of flagellar beat and properties of waves

Photographs taken from rapidly fixed animals as well as from living specimens show a complicated form of beat which combines features of helical undulation in the wave crests with "straight" flagellar stages in the troughs of the waves (Fig. 13a, b). Párducz (1967) and Tamm and Horridge (1970), referring to forms of ciliary beat, assume a counterclockwise cycle of the flagellum (as seen looking from tip to base), during which the "recovery stroke" covers most of the space between the wave crests while the straight power stroke is somehow "buried" under the waves. Tamm and Horridge are aware of the difficulties in explaining that "in the act of performing the effective stroke [the flagella] penetrate through the trailing side of the wave crest". Moreover, profiles of metachronal waves taken from photographs of *Opalina* do not show flagellar tips projecting into the waves (Sleigh, 1960; Tamm and Horridge, 1970). A reconstruction of the metachronal wave (Fig. 14) based on new photographic data reveals increasing helical coiling of the more or less straightened flagellum on the "lee side" of the crest, inducing shortening and raising of the helix. When the flagella tips have reached their highest position on the wave crest their bases on the "luff side" of the crest are already straightening out close to the cell surface in the wave direction with a slight declination toward the right. When the bending wave has passed the flagella in a distal direction, the flagella assume nearly straight positions in the posterior halves of the wave troughs. This is well seen in photographs showing longer waves (lower frequency) which are common in

---

FIG. 13. *Opalina*, system of metachronal waves as it appears in the scanning electron microscope. From Tamm and Horridge (1970).

a. Ventral side of specimen; wave system resembles state of locomotion shown in Fig. 12b. Horizontal bar: 100 μm.

b. Enlargement of region *b* in (a) showing waves of different length (30–50 μm) moving towards the lower left corner of the print. Note the orientation and form of flagella in the posterior part of wave-troughs in comparison to forms around the wave crests. Horizontal bar: 10 μm.

c. Enlargement of region *c* in (a) which is seen from behind. Waves move towards anterior-right corner of print. Steep uprising of anterior wave front and left-hand helical curvature of flagellum can easily be discerned. Horizontal bar: 10 μm.

A—P, orientation of antero-posterior axis; thin arrows, directions of wave transmission; broad arrows, directions of viewing of regions *b* and *c*.

the more posterior parts of the cell (Fig. 13b). This form of beat obviously features some properties of polarization of sequential bending in a counter-clockwise flagellar helix. During one period of the cycle, which may be termed "recovery stroke", the helix is wide in diameter and reduced in length; in the following period, partly overlapping in single flagella, the helical diameter is reduced but helical pitch is increased, reminiscent of

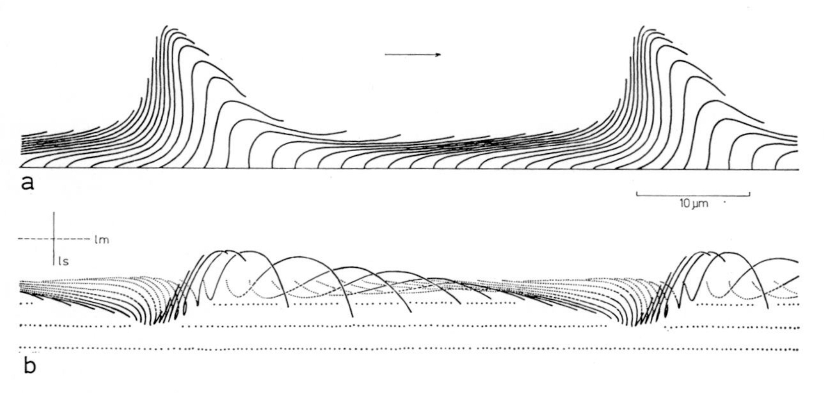

FIG. 14. Hypothetical reconstruction of metachronal wave in *Opalina*, based on photographs of living and fixed animals by Sleigh (1960) and Tamm and Horridge (1970).

Profile (a) and corresponding surface view (b) of waves as seen in the mainline of metachromy (lm) with waves travelling toward the right (arrow). Helical bending of flagella, appearing in one plane in form of sinusoid curvatures, increases in stages from the "posterior" part of the wave trough over its "anterior" part to the "lee side" of the wave crest. Overlapping of phases of curling (distally) and stretching (proximally, on "luff side" of wave crest). At increased frequency of beat (shorter wavelength) the proportion of stretching of flagella is reduced and eventually hidden under stages of new curling (compare Fig. 13c). Straight *and* winding flagella are slightly turned toward the right of the mainline of metachromy, indicating symplectism with tendency toward dexioplexy (dss-type). For reasons of clarity only every fourth or fifth flagellum is drawn. Orientation of the cortical pattern of flagella (bottom) has no influence on the form and direction of metachronal waves. ls, line of flagellar synchronization; lm, mainline of metachrony.

a "power stroke". The helical axis of the "power stroke" deviates from the wave direction by a small angle towards the right indicating a dss-type of metachronal coordination.* Assuming that vectors of force resulting from the flagellar beat also decline by a small angle from the mainline of metachrony towards the right, it is clear that helical rotation in swimming animals should be somewhat less than indicated by the orientation of waves on both sides of the cell.

* dexio-sym-symplectic; compare introductory definitions, section II.C.

C. *Paramecium*

This holotrichous ciliate with its well-known freshwater species *P. caudatum*, *P. multimicronucleatum* and *P. aurelia* is spindle-shaped, of moderate size (length 150–250 μm) and uniformly covered with cilia approximately 10 μm long. The ciliary system of *Paramecium* is a highly versatile machinery for locomotion whose forms of activity are shown by recent electrophysiological studies to be closely related to ionic events occurring across the cell membrane (see Naitoh and Eckert, this volume, Chapter 12).

1. *Patterns of metachronal waves and modes of swimming*

In cells that are swimming forward metachronal waves proceed from the posterior left to the anterior right side encircling the cell in left-hand helices of wave crests (Fig. 15a, b). The wave system is specifically interrupted at the right* margin of the oral groove (marked by the preoral suture of kineties) with no signs of transition between small peristomal waves (λ = 7–9 μm) and larger body waves (λ = 10–14 μm). Waves on the left* side of the groove are continuous with body waves. The swimming path of cells, characterized by metachronal patterns similar to those seen in Fig. 15a and b, is a left-hand helix, indicating that locomotory forces of the cilia should be directed more or less parallel to the wave fronts (Párducz, 1954). Swimming in *Paramecium* is not influenced by the right-wound shape of the anterior part of the cell body, since, at the dimensions of ciliary activity, forces of viscosity (and not of inertia) determine hydrodynamic behaviour (Seravin, 1970).† A photograph of metachronal waves on a forward swimming *Paramecium* is given in Fig. 16. Transmission of metachronal waves on the cell surface of *Paramecium* has recently been documented in high-speed films (Fig. 17).

Patterns of waves accompanying backward swimming were first detected in fixed specimens (Párducz, 1956a). Wave crests are more or less longitudinally oriented on the dorsal side of the cell, exhibiting reduced wavelength and moving from right to left. Waves on the ventral side may encircle the cell mouth while moving towards it (Fig. 15c). Correlations between the orientation of waves and the reversed swimming path were carried out during chemically induced continued backward swimming

* Anatomical properties of the cell are determined as if seen from the dorsal side.

† The Reynolds number, expressing the ratio of forces of inertia to forces of viscosity in a certain fluid system, is estimated at 10^{-3} for *Paramecium* and 10^{-4} for a single cilium (Jahn and Bovee, 1968).

(Párducz, 1959; Gręcki *et al.*, 1967), when paramecia rotate toward the right, if observed in the direction of swimming. Correspondingly, wave fronts are somewhat tilted, now travelling in the anterior left direction (Fig. 15d). This indicated that, similar to conditions during forward locomotion, forces resulting from the ciliary stroke are oriented parallel to the wave crests.

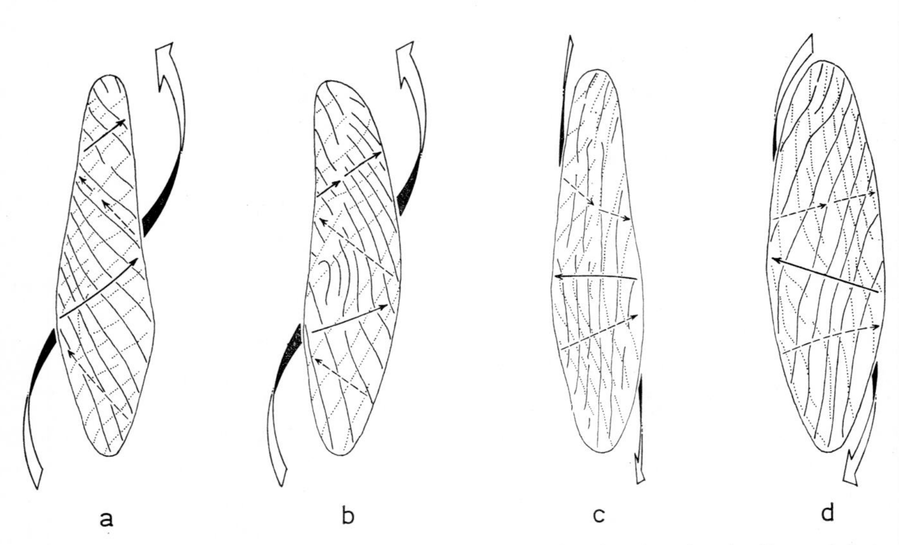

FIG. 15. Organization of the wave system and mode of swimming in *Paramecium*; all-round representation of waves fixed during forward (a, b) and backward swimming (c, d). a. Specimen in dorsal view, b. in ventral view; c, d. dorsal views. Wave fronts are represented by lines (dotted lines on the through-focused opposite side), and arrows indicate their helical course. The swimming path (helical flat arrows) is left-hand during forward swimming and right-hand during backward swimming. In all cases the propelling forces appear to be oriented parallel to the wave fronts. a, b, c, Redrawn from Machemer (1969a, 1969c); d, combined from photographs by Párducz (1959).

It is less well known that *Paramecium* performs a variety of manoeuvring movements such as swimming in curves (under rotation and without rotation) and "deflecting" from an obstacle under an obtuse angle (Alverdes, 1922). Apparently, these movements result from graded changes in orientation and frequency of the ciliary beat at different regions of the cell surface. Párducz (1956b) related patterns of waves with graded shifts in wave angles and wavelength to swimming of *Paramecium* in curves (Fig. 18). Transitions between waves of forward swimming (left sides of cells) and of backward swimming (right sides) in the preparations indicate graded responses of cilia to local stimulation.

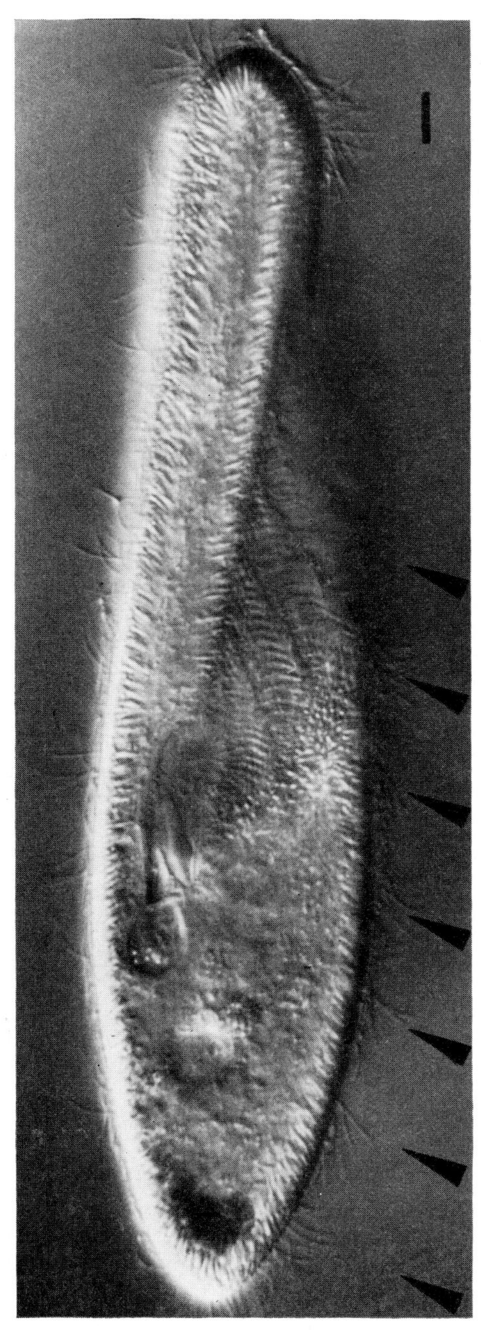

FIG. 16. Metachronism in live *Paramecium*, swimming forwards. Waves appear in profile and in surface view (arrowheads). Pattern corresponds to that of the fixed animal shown in Fig. 15b. Vertical bar: 10 μm. From Machemer (1972b).

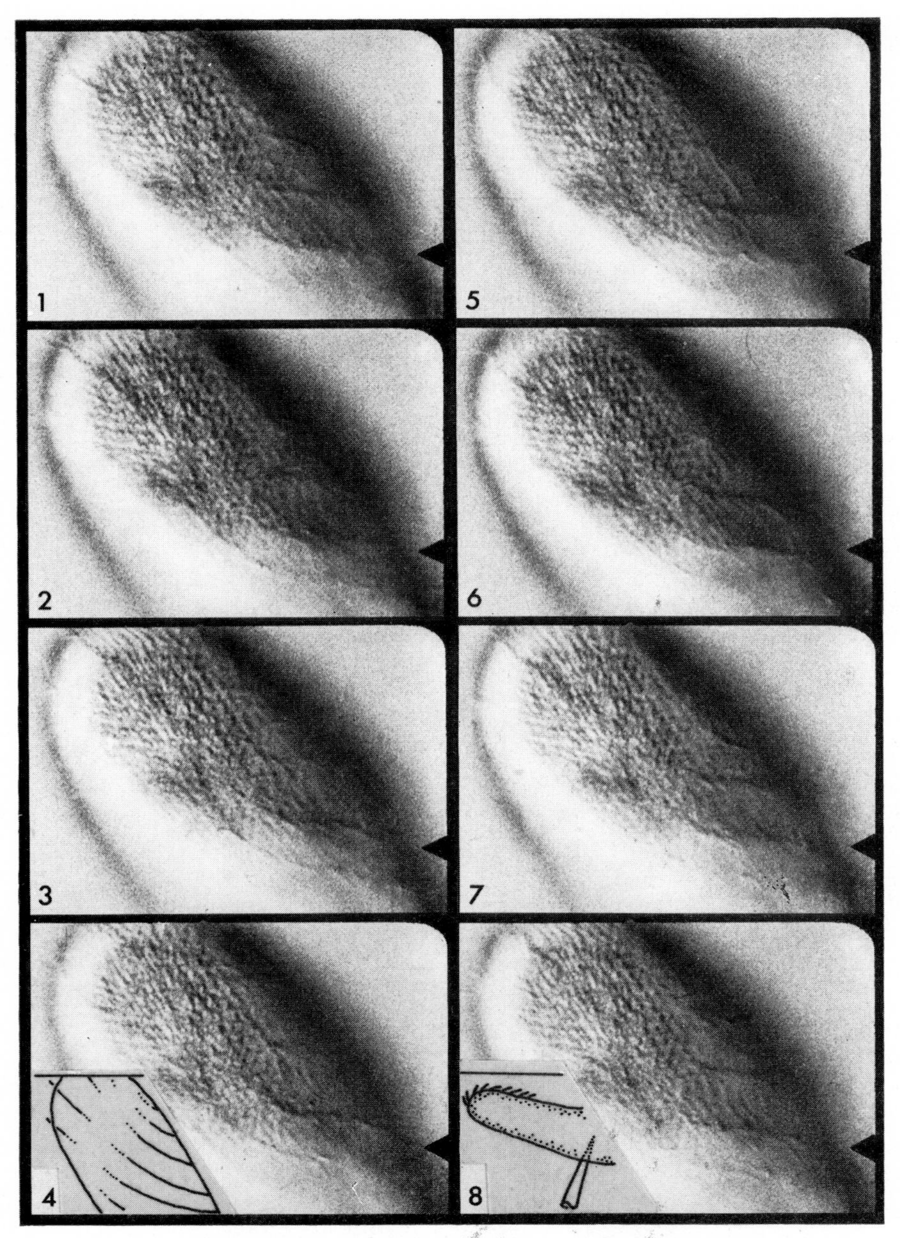

FIG. 17. A demonstration of wave transmission on the anterior surface of *Paramecium*. Sequence of frames (1–8) corresponds to every fourth frame of a high-frequency film taken at 250 frames per second. The animal was pinned by a glass capillary and removed from the coverglass to allow unrestricted ciliary motion (inset in frame no. 8). The wave pattern (inset in frame no. 4) corresponds to forward swimming. Note waves advancing in anterior right direction relative to the fixed triangular mark. From Machemer (1974).

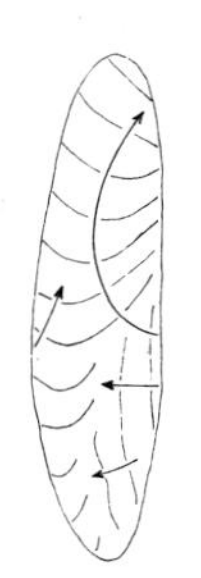
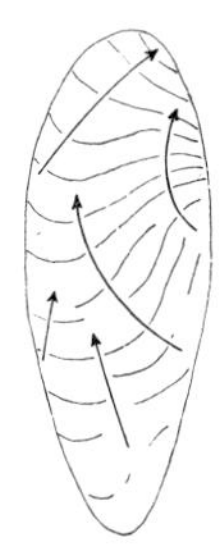

FIG. 18. Two examples of graded transition between normal waves, corresponding to forward swimming (left sides of specimens), and reversed waves, corresponding to backward swimming (posterior-right regions). These patterns of waves demonstrate that cilia on different parts of the cell may respond to stimuli with graded shifts in beat direction, thus enabling the cell to perform elaborate manoeuvring of swimming. Drawn from photographs by Párducz (1956b).

2. *Events of ciliary coordination during the avoiding reaction*

Ciliate Protozoa which make contact with a mechanical obstacle or a source of thermal or chemical stimulation may perform a series of specific movements which eventually result in the avoidance of the dangerous zone (Jennings, 1906). Phenomenologically, three stages of the "avoiding reaction" are found in *Paramecium*: (1) retreat from the source of stimulation through reversal of the swimming direction, (2) stationary swerving rotation around the posterior end, and (3) resumption of normal swimming in a new direction (Fig. 19). Párducz (1956a) offered a new explanation of this complicated sequence based on series of fixed preparations of stages of the movement in *Paramecium*. According to this author the cell reacts to the stimulus by a synchronous reversal of all cilia consisting of one vigorous stroke toward the anterior end of the cell; this stops forward locomotion and initiates darting backward. If the stimulus has sufficient strength, cilia continue to beat in the reverse direction in a metachronally coordinated way. Backward swimming is increasingly slowed down by a graded restoration of ciliary activity in the normal direction, which Párducz found to commence at the anterior end. Expansion of the zone of normally beating cilia and reduction of the area of reversed cilia leads to a transient equilibrium of anteriorly and posteriorly pointing forces with a resulting force towards the right; this accounts for the intermediary swerving phase. Forward swimming is resumed when forces of normal beating overcome those of beating in reverse (see inset in Fig. 19). This concept of the mechanism of the avoiding reaction has not been fully supported by Grębecki (1965) and Grębecki and Mikolajczyk (1968) who found renormalization of cilia starting at the posterior end of *Paramecium* but no metachronal waves in this region. Quantitative evaluations of samples of fixed specimens showed, on the other hand, that wave patterns of the transitory stage, if occurring, correspond to the Párducz scheme (Machemer, 1969a). Results of high-frequency filming have not unequivocally corroborated Párducz's findings: Kuznicki *et al.* (1970) could

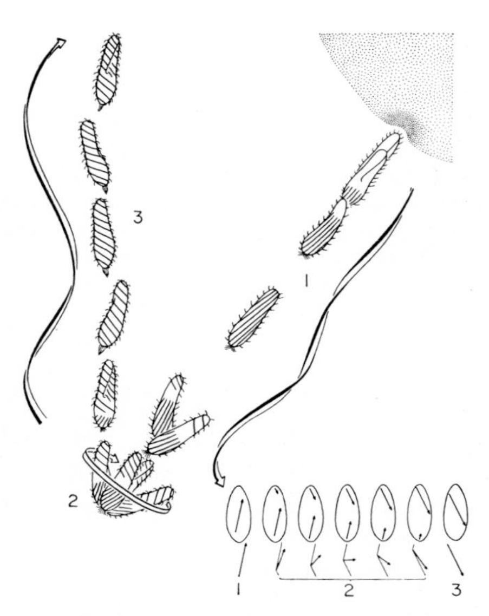

FIG. 19. Phenomena of metachronism and ciliary propulsion during the avoiding reaction of *Paramecium*, after Párducz. Phase 1: Retreat from a source of stimulation by reversal of beating accompanied by waves moving to the anatomical anterior and left. Phase 2: Transitory swerving movement by competing fields of anteriorly and posteriorly beating cilia. Phase 3: Resumption of forward swimming after complete restoration of normal beat of cilia with waves travelling anteriad and to the right. Inset: Distribution of forces and resulting vectors of propulsion during phases 1–3. Modified from Grell (1973) after Párducz (1956a).

not detect metachronal coordination of cilia before and after the synchronous reversal, a finding which might be due to a reduced activity of the filmed cilia. Following reversal induced by electric stimulation, the cilia pass through a period of inactivation (Fig. 22d, e) before normal beating and metachronal coordination are restored (Machemer, 1974). In summary, the Párducz scheme apparently only represents one possible variation of timing and local ciliary response in the course of the avoiding reaction; however, it seems to be correct in principle, assuming an instantaneous reversal response of the cilia and a graded restoration of the normal state of activity.

3. *Properties of ciliary beat and wave transmission*

As was noted above, the forms of helical swimming in *Paramecium* are such that propulsive forces of the cilia must be directed more or less parallel to the wave fronts (Fig. 15). The first conclusive description of this (dexioplectic) relation of waves to the ciliary power stroke, given by Párducz (1954)* on the basis of images of fixed cilia, was recently confirmed

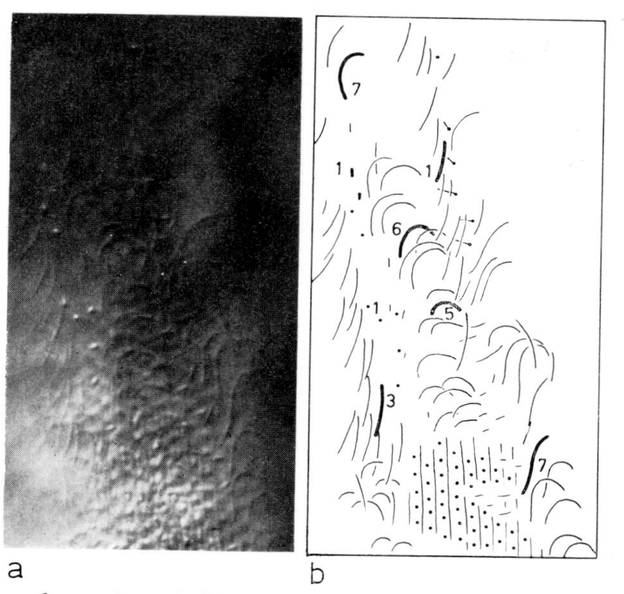

FIG. 20. Forms of metachronal ciliary beating in a forward swimming *Paramecium*; surface view. a. *In vivo* photograph, b. redrawing of ciliary outlines. Steep waves travel in an anterior-right direction; the power stroke (stages 1, 3, compare Fig. 48a) occurs in the direction of the wave fronts, and the recovery stroke (stages 3–7) with clear counterclockwise progression of the curved cilia. Some cilia are outlined heavily for identification of stages. a. From Machemer (1972a).

by *in vivo* analysis of patterns of ciliary beating (Machemer, 1972a; Figs 6, 20, 21). In surface view as well as in profile, stages of the "straight" power stroke are less frequent than stages of the curved recovery stroke, indicating higher angular velocity of the power stroke in comparison with the recovery stroke (polarization in time). The power stroke of normally swimming animals points in the posterior right direction and occurs parallel to the wave fronts. Cilia in their most upright position appear

* In his review paper Párducz (1967) describes this type of beating as dexio-antiplectic, based on definitions which differ somewhat from those given in this paper (section II.C).

optically cross-sectioned in surface view (Fig. 20, stage 1). Tips of cilia at the end of the power stroke (Fig. 20, stage 3) are slightly bent in the posterior left direction owing to ciliary deformation in a counterclockwise helix. The recovery period of the cycle occurs in a counterclockwise mode and, contrary to the power stroke, with the cilia bent down toward the cell surface and assuming a sickle-shaped curvature (polarization in space; stages 5–7 in Fig. 20). The polarization of dexioplectic ciliary beating

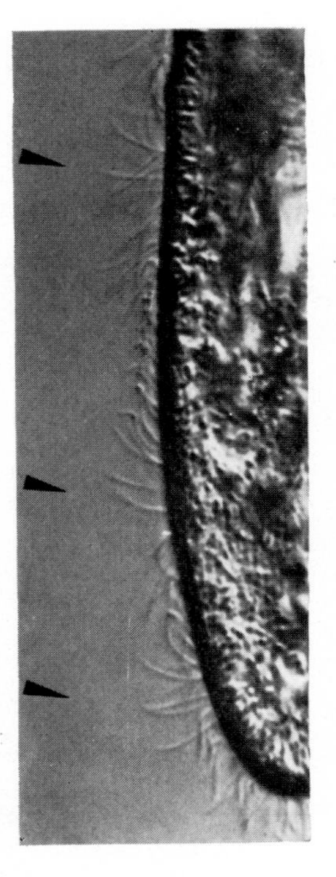

FIG. 21. A profile view of ciliary metachronism in a forward swimming *Paramecium*, posterior end. Waves pass the focal plane at an angle of about 20° in an anterior direction. The power stroke (arrowheads) occurs with a posteriad component towards the observer. Note the antiplectoid wave profile and the difference in form during the power stroke (straight, erect) and the recovery stroke (bent over).

presents its various aspects in different views of the cell surface (Figs 20, 21). A schematic representation of the dexioplectic metachrony in *Paramecium* is shown in Fig. 6.

Recent analysis of high-frequency films taken during electric stimulation of *Paramecium* (Machemer, 1974) confirmed that the dexioplectic organization of metachronism is valid at different states of ciliary activity and during forward and backward swimming (Figs 22, 23). Augmented ciliary frequency in the normal direction induced by stimuli hyperpolarizing the cell membrane (compare Chapter 12 by Naitoh and Eckert) is accompanied by a clockwise turning of the whole wave system (Fig. 23a, b), inducing an increase in the apparent wavelength in profile views (Fig. 22a, b; compare model profiles 20–40° in Fig. 6). Reversal of ciliary beating induced by depolarizing stimuli goes along with strong mechanical coupling between the cilia and hence a very long metachronal wave which moves toward the posterior left (Fig. 23c) and which can be detected only

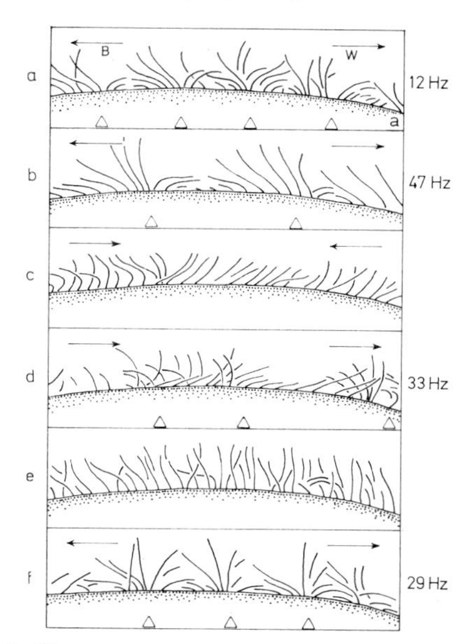

FIG. 22. Sequences of ciliary activity following electric stimulation of *Paramecium* (hyperpolarization: a, b, f; depolarization: a, c–f). Drawn from high-frequency films.

a. Low pre-stimulus activity (Frequency 12 Hz) with normal antiplectoid wave profiles and cilia beating away from anterior end (a).

b. Augmented frequency (47 Hz) as a result of hyperpolarizing stimulation. Note increase in apparent length of waves (hollow triangles).

c. Performance of a vigorous stroke towards the anterior end following a depolarizing stimulus. This frame shows a section of a long antiplectoid wave travelling posteriad.

d. Reversed beating of cilia towards the anterior end at increased frequency (33 Hz) after performance of the quasi-synchronous stroke depicted in (c). Note increased apparent wavelength of symplectoid waves.

e. Ciliary inactivation at the end of the period of reversed beating (d) preceding resumption of normal activity (f). The cilia show irregular and slow wavering movements with small amplitude.

f. Post-stimulatory renormalization of ciliary beating following hyperpolarization (b) or depolarization (c, d, e). Frequency is increased in comparison with the pre-stimulus condition.

Arrows left: direction of beat (B); arrows right: direction of waves (W). From Machemer (1974).

through the antiplectoid profile configuration (Fig. 22c). The frequency of the continued reversed beating which follows the vigorous "synchronous" stroke is higher than that in the prestimulus condition (Fig. 22a, d).

The power stroke is turned more clockwise in comparison with the first stroke in reverse (Fig. 23c, d), thus causing the symplectoid profile configuration of the cilia (Fig. 22d; compare with model profiles 145–170° in Fig. 6). The frequency of reversed beating is continuously reduced and may reach a point where metachronism breaks down due to partial or complete inactivation of the cilia (Fig. 22e). Beating and metachronism

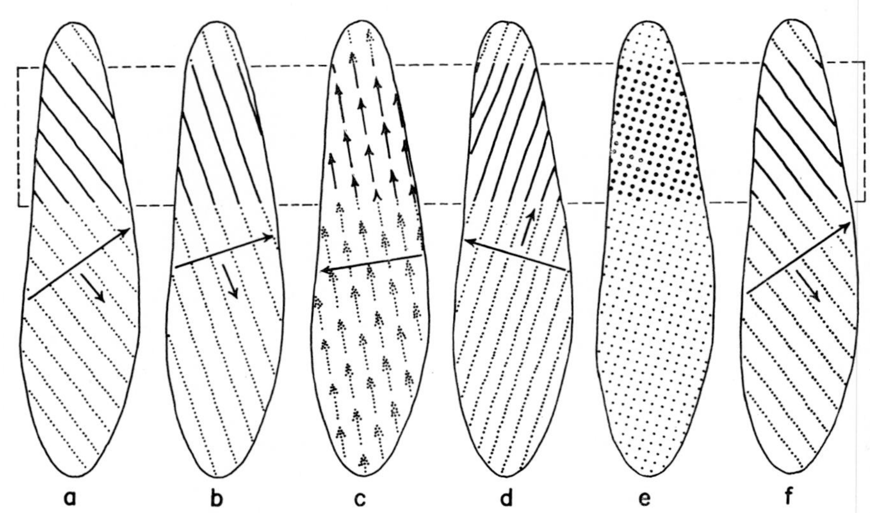

FIG. 23. Reconstruction of metachronal patterns following electric stimulation, after the sequence of ciliary profiles shown in Fig. 22. Metachronal waves (long arrows) and power strokes (short arrows) turn more clockwise after application of a hyperpolarizing stimulus (a→b), and then slowly return to normal orientation (f). The occurrence of a depolarizing stimulus leads to immediate reversal of ciliary orientation (a→c); cilia continue reversed beating while gradually shifting in a clockwise direction (d) and decreasing their frequency. The end of a period of reversal is marked by ciliary inactivation (e), after which normal beating is resumed (f). Note the persistence of dexioplectic organization of ciliary activity under different conditions. The dashed zone near the anterior end of *Paramecium* marks the location of frames shown in Fig. 22. From Machemer (1974).

in the normal direction is then gradually restored, often accompanied by a strong increase in frequency (post-stimulatory augmentation; Fig. 22f). These sequences of ciliary events have in common that the orientation of the power stroke shifts in the counterclockwise direction with increasing depolarizing stimulation, but shifts in the clockwise direction with increasing hyperpolarizing stimulation, a behaviour pattern reminiscent of correlations between membrane potential and flagellar orientation in *Opalina* (Kinosita, 1954). Regarding the avoiding reaction of *Paramecium*, sequence c–f in Figs 22 and 23 corresponds to the Párducz scheme including, however, an increased period of ciliary reorientation at the end of reversed beating.

4. Special conditions in the oral groove

Paired peristomal cilia in the oblique furrow leading to the cell mouth provide nutritive currents toward the cytostome of *Paramecium*. During

FIG. 24. Metachronal organization of paired peristomal cilia in the oral groove; fixed preparation, ventral view. a. Orientation of waves under normal conditions. The peristomal waves spread more radially in an anterior-right direction. Discontinuity of wave fronts is seen at the right margin of the groove (median in picture), where the preoral suture separates regions of single and paired cilia. b. Enlarged section of (a). The wave pattern corresponds to a dexioplectic arrangement of body cilia (Fig. 6); the power stroke is performed parallel to the wave fronts (arrowheads). Vertical bars: a, 50 μm; b, 10 μm. From Párducz (1955).

forward swimming wave patterns of the oral groove somewhat encircle the cytostome on the left side indicating a more radial spread of waves (Figs. 15a, b, 24). During backward swimming peristomal waves have a similar configuration (Fig. 15c) or they are radially oriented suggesting transmission in a more circular path (Fig. 15d). According to Párducz (1955) peristomal cilia, as opposed to body cilia, beat at right angles to the wave fronts which encircle the cell mouth on the left side thus exhibiting an antiplectic type of metachronal coordination, in the course of which water currents would be oriented *directly* toward the cell mouth. Patterns of fixed peristomal cilia, on the other hand, do not differ significantly from those of the body cilia (compare Fig. 24 with model wave Fig. 6). The pictures presented by Párducz seem to corroborate a dexioplectic rather than antiplectic type of metachrony, since cilia performing the power stroke are caught in positions which clearly indicate motion parallel to the wave crests (Fig. 24b). Accordingly, Grębecki (1965) expresses the view that during forward swimming and in the absence of

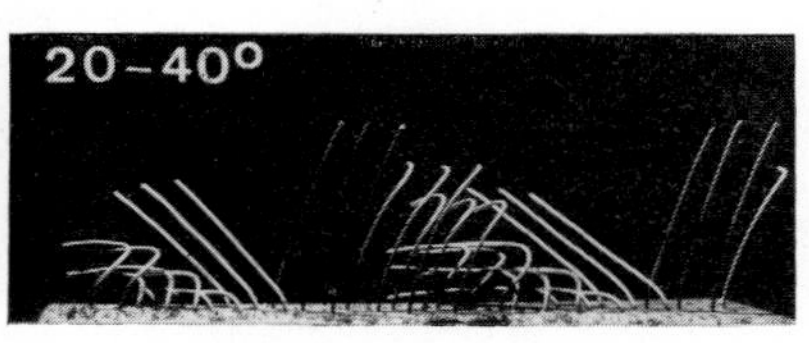

Fig. 25. Live photograph and model reconstruction of peristomal ciliary waves in longitudinal profile. Model cilia beat posteriad (post.) and towards the observer; so do photographed cilia. Note the pronounced straightening of the cilia during the recovery stroke and the reduced amplitude in comparison with body cilia (Fig. 6, profiles at 20–40°). From Machemer (1972a).

suspensions of particles peristomal water currents move along the lines of ciliary synchronization. If this interpretation is correct, the main nutritive current created by the peristomal cilia during forward swimming partly *bypasses* the cell mouth and is only partially diverted toward the mouth by buccal ciliature. This feeding mechanism would correspond to that observed in the hypotrich ciliate *Stylonychia* (Machemer, 1966; see section IV.E). Profiles of the peristomal waves show a reduced ciliary amplitude and less pronounced bending downward during the recovery stroke (Sleigh, 1962; Machemer, 1972a; Fig. 25). The latter property probably results from mechanical effects of close ciliary pairing. In summary, properties of peristomal ciliary beating have not yet been analysed sufficiently to justify the assumption that there are basic differences between metachronism of cilia in the oral groove and on the body surface.

5. *Effects of physical factors on metachronism*

Low *temperature* of the medium reduces the velocity of wave transmission without significantly changing the orientation of the metachronal pattern (Párducz, 1954). According to quantitative evaluations of temperature effects, low temperature not only reduces beat frequency (and hence rate of wave propagation) but also induces a slight counterclockwise turn

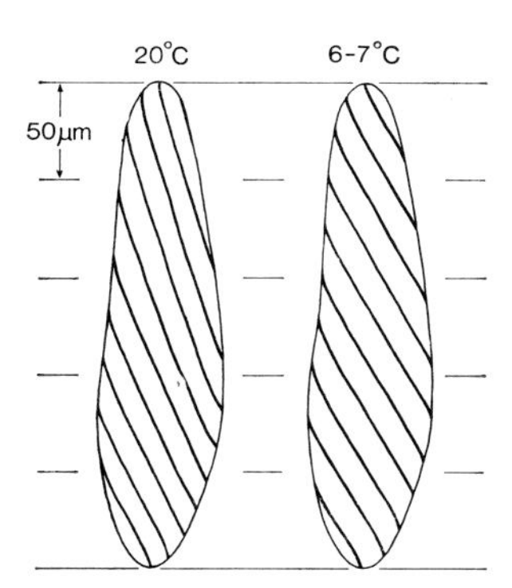

FIG. 26. Temperature-induced alteration of metachronism in *Parramecium*. Reduction of temperature from 20° to 6°C led to an overall *counterclockwise* turn of the directions of waves and beat by 9° and an increase in mean wavelength from 10·7 μm to 13·6 μm. Simultaneously the frequency of beat of the peristomal cilia dropped from 32 to 10 Hz. From Machemer (1972b).

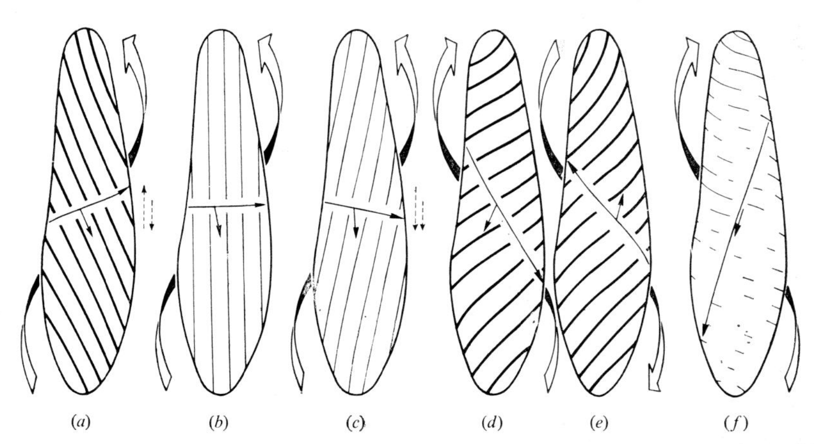

FIG. 27. Viscosity-induced alteration of metachronism in *Paramecium*. The direction of waves (long arrows) and beating of cilia (short arrows) were turned more *clockwise* when the viscosity of the medium was increased from normal (a) to 2·6 cP (b), 5·6 cP (c), 40 cP (d, e) and 135 cP (f). Swimming (flat arrows) in left-hand helices (a–c) changed to right-hand swimming paths (d–f). The angle between the directions of waves and the ciliary power stroke was increasingly reduced (transition from dexioplexy (a) to symplexy (f). Spontaneous changes in the direction of swimming (reversals) at 40 cP (d, e) were accompanied by 180°-inversion of the wave pattern. Increase in viscosity decreased frequency and increased wavelength (1 cP: 32 Hz, 10·7 μm; 40 cP: 14 Hz, 14·3 μm). Heavily outlined patterns are based on quantified data; waves in thin lines are taken from photographic data; waves in interrupted lines indicate partial breakdown of the metachronal system. From Machemer (1972c).

of the waves in combination with an increase in wavelength (Machemer, 1972b; Fig. 26). Modifications of beating properties with rising temperature correspond to phenomena accompanying electrically induced augumenta-

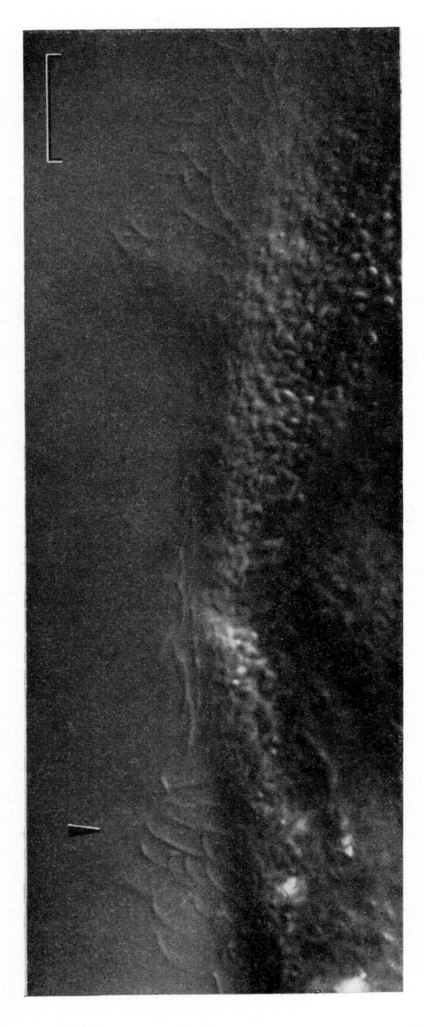

FIG. 28. The form of a ciliary beat in a symplectoid wave at moderate viscosity (5·6 cP); oblique view of cell edge of an animal swimming forward (anterior end up; compare with Fig. 27c). Polarization of the beat is still well preserved: note bent-over cilia performing the recovery stroke and straight cilia during the power stroke towards the rear end (arrowhead). Waves pass the plane of focus at 10–20° and move towards the observer. Compare with profiles 160–170° in Fig. 6. Scale: 10 μm. From Machemer (1972c).

tion of ciliary activity (compare Figs 22, 23a, b). Changes in the *viscosity* of the surrounding medium considerably influence properties of metachronal coordination (Machemer, 1972c). With increasing viscosity the normal wave pattern is turned in a clockwise direction towards a longitudinal orientation with waves travelling from left to right at viscosities between 2 and 3 centipoise (cP) (Fig. 27b). Beyond this viscosity waves travel from anterior

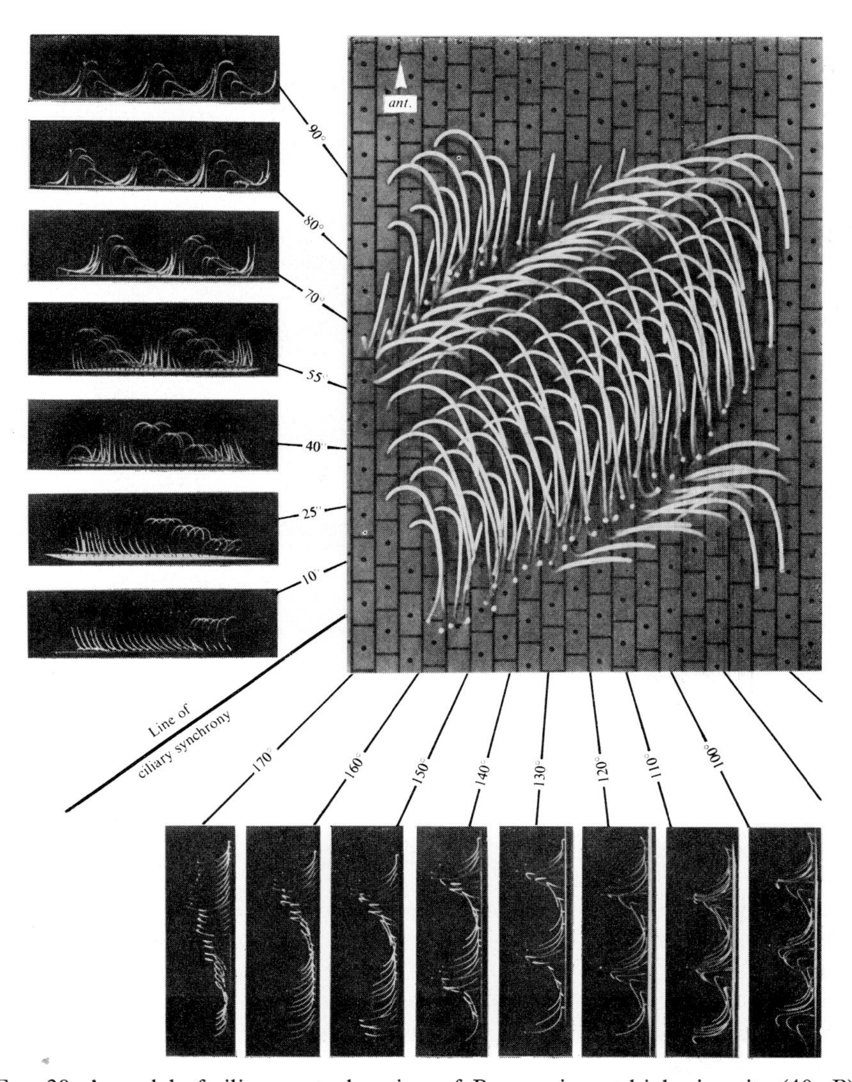

FIG. 29. A model of ciliary metachronism of *Paramecium* at high viscosity (40 cP) demonstrating transformation of the normal dexioplectic type of coordination (Fig. 6) to the dexiosymplectic type. Waves move in the posterior-right direction. The power stroke is directed towards the posterior and left, but is less pronounced within the counterclockwise helical cycle. Among possible ciliary profiles those at wave angles of 110–140° are most often observed. From Machemer (1972c).

left to posterior right (Fig. 27c, d). Between 40 and 100 cP wave direction has been shifted towards the posterior-left direction. At viscosities exceeding 100 cP the metachronal coordination deteriorates increasingly (Fig. 27f). In parallel with viscosity-induced turning of wave patterns the ciliary power stroke also shifts in a clockwise direction, thus transforming the left-hand swimming pattern of *Paramecium* to right-hand swimming Fig. 27a–f). Profiles of ciliary waves seen parallel to the longitudinal axis of

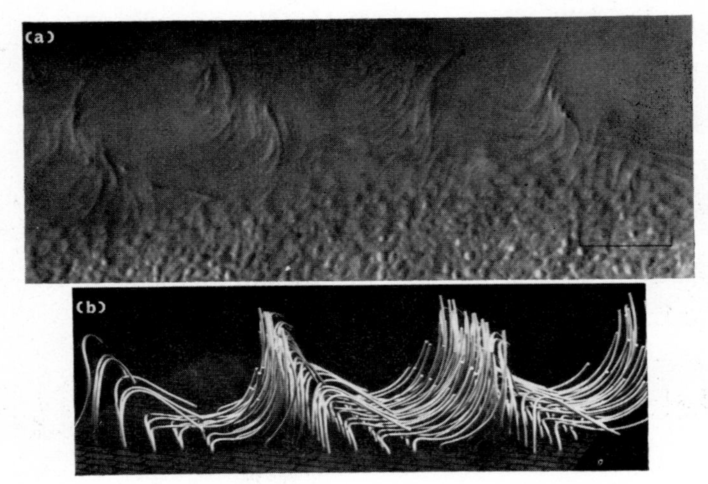

FIG. 30. An example of metachronism at 40 cP in a swimming *Paramecium* (a: oblique surface view) and corresponding model (b: identical to model in Fig. 29). The cilia exhibit a reduction of properties of polarization and an increase in helical winding. Note similarity to waves seen in *Opalina* (Fig. 13c). Scale: 10 μm. a. From Machemer (1972c).

the cell change from antiplectoid to symplectoid configurations even at small increases of viscosity (Sleigh, 1966; Fig. 28; compare dashed arrows in Fig. 27a and c). Since the rate of clockwise turning of the power stroke is smaller than that of the wave pattern, coordination changes from the dexioplectic (Fig. 27a) to the symplectic type (Fig. 27f).

Transformation of metachronal types induced by increase in viscosity is accompanied by a graded change of ciliary shapes during the beating cycle. The form of beat which is typically polarized in time and space at normal viscosity assumes increasingly the properties of continuous helical beating, i.e. with fading differences between the power stroke and the recovery stroke. Continuous helical beating of *Paramecium* was observed by Kuznicki *et al.* (1970) at high viscosity of the medium and was assumed to be existing also under normal conditions. This generalized conclusion is not supported by recent filming experiments (Machemer, 1974; compare Fig. 22a–f). Typical differences in the form of beat and consequences of metachronal coordination at normal and increased viscosity (40 cP) are illustrated by model reconstructions of cilia (Figs 6, 29) which were developed from photographic data (e.g. Fig. 30).

Ciliary beat in *Paramecium* at high viscosity has a striking similarity to the form of flagellar beat exhibited by *Opalina* (compare Figs 30a and 13c). This similarity may be explained in part, on a hydrodynamic basis, by assuming compensatory effects of mechanical coupling in the long and densely packed flagella of *Opalina** on the one hand and of increased medium viscosity of *Paramecium* on the other hand. The relationship of the form of ciliary beating to the origin of metachronism will be discussed in section V.B.

D. *Stentor*

This large, funnel-shaped ciliate (200–2000 μm) is usually fixed to solid objects by a tapering "foot" but may also be found free-swimming. As in other heterotrich ciliates, the locomotory system consists of ordinary cilia covering the largest part of the body and of longer compound cilia (30 μm) which are called membranelles, since they consist of 2–3 rows of about 20 closely packed cilia. Membranelles are lined up in transverse orientation in a band (the adoral membranellar band) which follows a spiral course towards the cytostome at the bottom of the bowl-like peristomal field. Membranellar functions in *Stentor* deserve particular attention, since metachronism in these organelles exhibits properties which led to the assumption of mechanisms of internal (neuroid) conduction (Sleigh, 1956, 1957).

1. *Locomotory behaviour*

During forward swimming *Stentor* describes a left-hand helical path (Fig. 31d). In rare cases swimming without revolution or in a right-hand helix may be observed (Jennings, 1899). Upon mechanical or chemical stimulation the animal rapidly contracts and starts swimming backward in a right-hand helix (if viewed in the direction of locomotion; Fig. 31a, b). At the end of this period *Stentor* swerves sideways (c), thus determining a new direction of forward swimming (d). This behaviour is comparable with the avoiding reaction in *Paramecium* and other ciliates (Jennings,

* 100 μm^2 of surface area of *Opalina* contain 100–200 flagella; the same surface unit on the body of *Paramecium* has 40–50 cilia.

1899). In attached animals water currents indicate that ciliary activity does not differ from that during free swimming (Fig. 31e); however, body cilia may also be seen in an inactive state.

2. *Metachronal coordination and ciliary activity*

Under conditions corresponding to forward swimming the *membranellar band* exhibits series of conspicuous metachronal waves which start in the

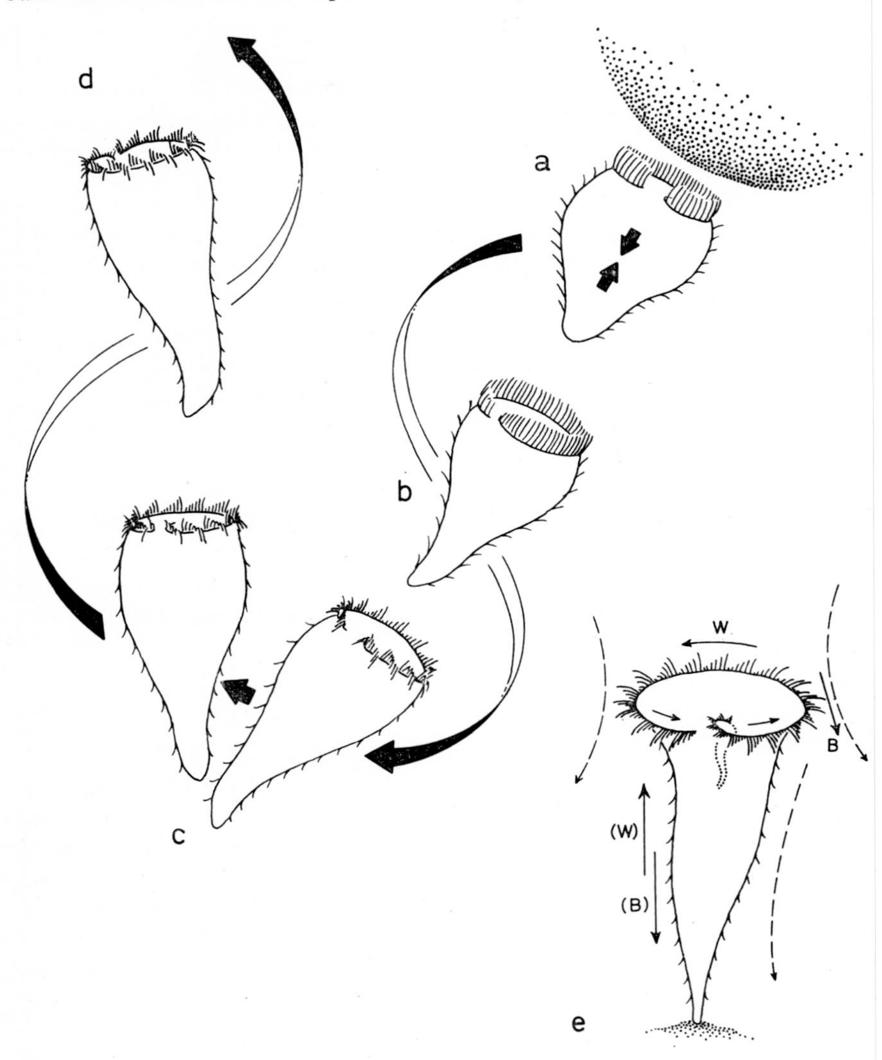

FIG. 31.—Caption on facing page.

gullet region and travel toward the peripheral end of the band (Figs 31e, 32). Examination of the compound cilia revealed a power stroke occurring centrifugally (with respect to the longitudinal axis) and toward the posterior (Sleigh, 1956) thus accounting for the whirl-like currents of water around the membranellar band. The course of the ciliary cycle is not restricted to one plane; during the recovery stroke a counterclockwise helical deformation of the cilia is observed (Fig. 32). It should be noticed that the membranelles beat toward their lateral edges, i.e. parallel to their larger diameter, which is a rather unexpected type of membranellar motion and, in fact, not a ubiquitous one (Fig. 34). Since the power stroke points toward the right, if the observer looks in the direction of wave propagation, metachronism of the membranelles was referred to as dexioplectic (Sleigh, 1962) on the assumption that the wave direction is identical with the course of the band. This appears to be correct since cilia in one membranelle are coupled to each other without apparent phase-shift, thus indicating that the line of synchronization is at right angles to the membranellar band. Uhlig (1960) saw membranelles beating *parallel* to the band in a proximal anlage of the band in *Stentor*. This seems to indicate that membranelles close to the gullet region gradually turn the direction of the power stroke more toward the proximal end of the row inducing a corresponding turn of the wave direction. Studies of the water currents close to the proximal membranelles would resolve this question but have not yet been reported. Chen (1972) observed that *body cilia*, while beating in the normal direction (posteriad) exhibit metachronal waves travelling anteriad in profile view. Though detailed observations are still lacking, the

---

FIG. 31. Ciliary activity and locomotion in *Stentor* (schematic). A free-swimming animal meeting a source of stimulation contracts and reverses the normal orientation of the body cilia and the membranelles (a). Continued reversed beating of cilia and membranelles (b) results in backward swimming in a right-hand helix. Following gradual restoration of normal beating of the cilia (c) the backward swimming is stopped, the *Stentor* swerves towards one side and resumes a left-winding forward swimming path in a new direction (d). Coordination of the body cilia and membranelles in a forward-swimming *Stentor* corresponds to that of an attached animal (e), which generates water currents for feeding purposes. On the body edge antiplectoid metachronal waves are observed moving toward the anterior end; the beat points posteriad. The membranellar power stroke has a centrifugal and posterior orientation, causing whirls of water around the membranellar band. Metachronal waves travel along the band from the cytostome towards the distal end; the coordination is dexioplectic. B, direction of power stroke; (B), apparent orientation of beat in profile; W, direction of waves; (W), apparent wave direction in profile. Dashed arrows indicate water currents; flat arrows show the orientation of the swimming helix. For properties of the membranellar beat compare with Figs 32 and 34a. Modified from Jennings (1899) and Chen (1972).

coincidence of available data with metachronism and propulsion in *Paramecium* (antiplectoid pattern, left-hand helical swimming) suggests that under normal conditions the cilia beat toward the posterior and right and metachronal waves move toward the anterior and right in a dexioplectic type of coordination.

During backward swimming the membranelles are pointing anteriad and vibrate with small amplitudes. Metachronism was observed (Chen,

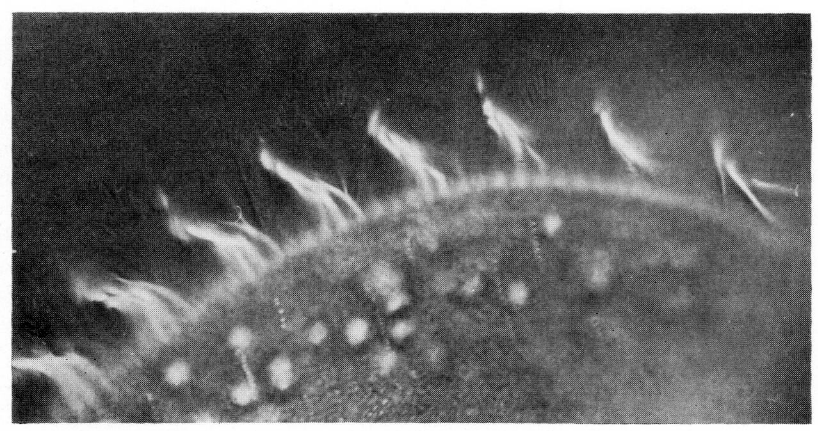

FIG. 32. Metachronal waves on *Stentor*-membranelles. A region of the membranellar band, corresponding to the upper part of the band in Fig. 31e, is observed from the outer side. Waves move from left to right; the power stroke occurs towards the observer. Note slight helical distortion of the cilia during the recovery stroke. From Sleigh (1973).

1972), but its direction (probably centripetal) was not reported. "Those body cilia which could be observed in profile . . . underwent a transient reversal in beat direction; . . . the metachronal wave appears disrupted" (Chen, 1972). These preliminary data on ciliary reversal in *Stentor* differ from the reversal behaviour in *Paramecium* only in that reversed membranellar beating appears to be reduced in comparison with normal beating.

3. *Effects of physical factors on membranellar metachronism*

A gradual reduction of *temperature* from 25 to 5°C led to a decrease in beat frequency (33→10 Hz; Sleigh, 1956). Simultaneously, the metachronal wavelength increased (22.0→27.5 μm). These results are corroborated by observations made on *Paramecium* (Machemer, 1972b). An increase in the *viscosity* of the medium reduced the membranellar frequency and increased the wavelength (1 cP: 28 Hz, 21·6 μm; 3·6 cP: 23 Hz,

26·1 μm). At these viscosities the product of frequency and wavelength (= the wave velocity) remained constant (600 μm s^{-1}; Sleigh, 1956). A viscosity-induced reciprocity between frequency and wavelength was also observed in *Paramecium* cilia (Machemer, 1972b). Since wave velocity tends to be constant at viscosities below 5 cP, Sleigh regards beat frequency and wave velocity as independent variables in the equation: frequency × wavelength = wave velocity (compare section V.C.2). However, wave velocity rose by 100% at increased viscosities up to 9 cP, mainly through a strong increase in wavelength (Sleigh, 1961).

4. *Effects of chemical agents*

Stability of the wave velocity also occurred after the introduction of frequency-modifying $MgCl_2$ and $AlCl_3$ (Sleigh, 1956). Mg^{2+} increased beat frequency by 20% at concentrations between 0·125 and 0·25 mM. Al^{3+} had similar effects, but was not effective at concentrations exceeding 0·05 mM. The increase in frequency was accompanied by a reduction in wavelength by an amount which left the wave velocity constant (590 μm s^{-1}).

The heart-accelerating glycoside *digitoxin*, on the other hand, had little effect on frequency (+6%) but considerably raised wavelength and wave velocity (+34%; +35%)* when applied at concentrations of 0·2 mg l^{-1}. From his experiments on the effects of physical and chemical factors on membranellar metachronism Sleigh (1956) concluded that most of these agents affect ciliary activity but not the process of wave conduction; inversely, the drug digitoxin specifically accelerates wave propagation without interfering with properties of the ciliary beat.

5. *Effects of cutting the membranellar band*

An incision across the band resulted in differences in frequency (mean: 8·4%) and of the wavelength on both sides of the operation (Verworn, 1889; Sleigh, 1957), but the wave velocity remained rather unaffected (variations from −3·7 to 4·3%). The frequency was usually reduced, but yet uniform in the section distal to the cut. The differences in frequency disappeared after healing of the incision. Discontinuities in frequency were also observed when the function of the membranellar row was impaired by a solid object. Uniformity of ciliary beating within isolated areas of the membranellar band may be the result of pacemaker cilia at the proximal end of the region dictating the rate of ciliary function. A hierarchy of pacemakers would guarantee the leadership of the most proximal

* According to Naitoh (1964) digitoxin (1–10 mg l^{-1}) increased swimming velocity and hence ciliary activity in *Paramecium*.

one, usually near the gullet region (Sleigh, 1957). However, there are also possible alternative explanations, based on membrane-controlled ciliary frequency.

6. *Excitation–transmission theory of metachronal coordination in* Stentor *membranelles*

Investigating the reduction of metachronal wavelength in the proximal end of the membranellar band, Sleigh (1957) found that, regardless of its absolute length, each wave comprised approximately 6 membranelles. The increase in wavelength from 12 to 27 μm in the distal direction thus has a parallel in the increase in inter-membranellar distance from 2·7 to 4·1 μm. These conditions led Sleigh to the conclusion that the rate of metachronal wave transmission is limited by the number of cilia to be excited. If the process of ciliary excitation, understood as starting ciliary motion and triggering a conducted impulse, is slow in comparison with rapid interciliary conduction, ciliary excitation will be the limiting factor. This "neuroid" theory of a step-by-step process of metachronism (Sleigh, 1957) explains all data about independence between ciliary activity (represented by frequency) and metachronal transmission (represented by wave velocity). Experimental evidence in favour of this theory, however, is indirect and does not exclude other explanations. It might be considered, for instance, that the number of 6 membranelles per wave is coincidental through changes in mechanical interciliary coupling of the smaller and closer packed membranelles, resulting in shorter wavelength. The neuroid theory is compared to a mechanical concept of ciliary metachronism in section V.C.3.

E. *Stylonychia*

This ciliate belongs to the group of hypotrichs which are generally regarded as having developed the most sophisticated ciliate systems for performing walking and swimming movements. The large freshwater species *S. mytilus* (300 μm) has been subjected to studies of its locomotory behaviour and ciliary functions (Machemer, 1965a, b, 1966, 1969b; Sleigh, 1968). An attempt will be made to screen the great complexity of reported data for some basic features of ciliary activity, which are found, in part, in other groups of the ciliate protozoa.

1. *Roles of ciliary organelles in locomotion*

The most common mode of locomotion is "*walking*" on solid substrates (Fig. 33). Walking is mainly performed by 18 powerful cirri (8–22 component cilia, up to 50 μm long) moving slowly during a "stiff" swing in the posterior direction (power stroke) and returning rapidly in a bent mode in the opposite direction (recovery stroke; Sleigh, 1968). The walking in any direction (forward, backward, straight or in arcs) and at variable speed (0·1–2·5 mm s^{-1}; Machemer, 1965b) is strikingly smooth, apparently because it is assisted (1) by two rows of marginal cirri (25 μm) and (2) by a band of membranelles which surrounds the frontal lip and then turns towards the cell mouth. The function of the 3 long terminal

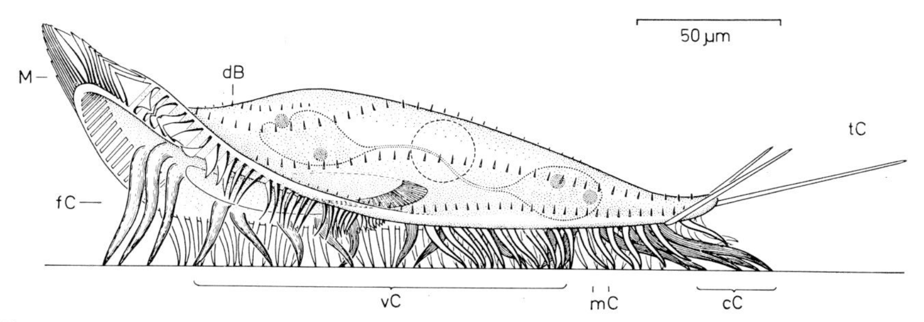

FIG. 33. Locomotory organelles in *Stylonychia mytilus*; side view of a resting animal. The frontal lip is lined by the membranellar band (M) which turns to the undersurface and toward the cell mouth on the anterior left side. Note the orientation of membranellar movement and the antiplectoid pattern on this section of the band. The cell is supported by 3 large frontal cirri (fC), 10 ventral cirri (vC) and 5 caudal cirri (cC); all of them are responsible for walking movements. Single rows of marginal cirri (mC) line the left and right body edges. 3 terminal cirri stick out stiffly from the posterior end. The dorsal membrane carries short bristles (dB) of unknown function. From Grell (1973) after Machemer.

cirri is probably to stabilize locomotion and to assist quick turning of the cell. *Swimming* in both forward and backward directions is elicited by stimulation but also appears to occur spontaneously. The common velocity of swimming (1 mm s^{-1}) is less than that of walking, since the cell proceeds in low-pitch left-hand helices with the ventral side pointing toward the axis of translation. During swimming all cirri are oriented anteriad or posteriad and perform oscillations at high frequency in a counterclockwise helical mode; alternatively, the 3 large frontal cirri may work in a lashing manner at reduced frequency, but also with a helical component. The membranelles support swimming through the continuous production of water currents. Movements of walking and swimming are regularly interrupted by a brief *synchronous reversal* and *synchronous reorientation* of the cirri, by which the animal is thrown backward by approximately one body length and then pushed forward in a direction to the right of the old one. In spite of similarities with the common avoiding reaction

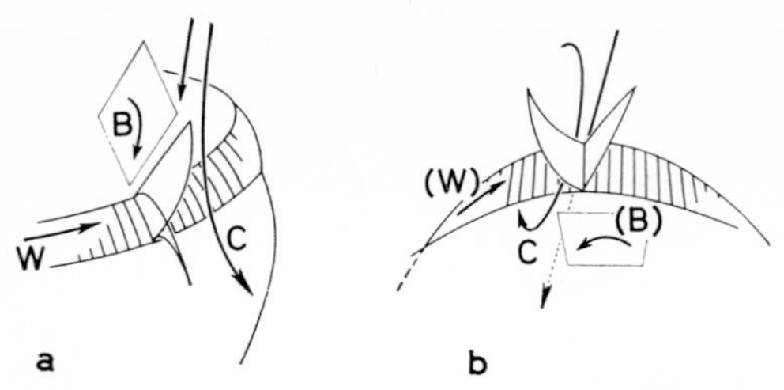

FIG. 34. Differences in membranellar function in *Stentor* (a) and *Stylonychia* (b). In both animals waves travel in proximo-distal direction along the band. In *Stentor* the power stroke (B) occurs vertically to the band and points downward; water currents take the same direction. In *Stylonychia* the beat appears to be oriented parallel to the band, but the water currents have a more transverse orientation. W, true; (W), apparent wave directions; (B), apparent beat direction; C, direction of water current. From Machemer (1966).

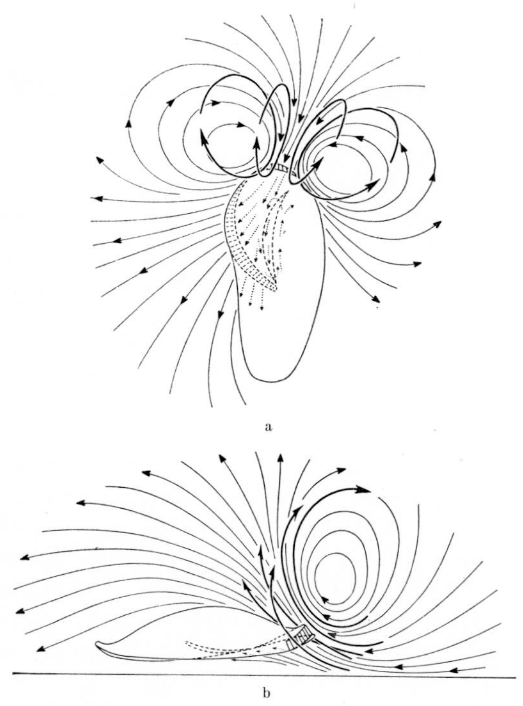

FIG. 35. Water currents produced by the membranelles of *Stylonychia* while stationary or during forward locomotion. a. Dorsal view; b. side view. A half-circular arc of whirls accompanies the frontal zone of membranelles indicating transverse or oblique vectors of force (with respect to the membranellar band) resulting from the beating of the membranelles. Oblique transport of water is also seen in the adoral zone of the membranelles, which provide currents by-passing the cell mouth. From Machemer (1966).

in ciliates this movement occurs commonly in the absence of stimulation*; after stimulation continued reversed beating and hence backward locomotion may follow the synchronous stroke of the cirri towards the anterior end, before a synchronous reorientation towards the rear end occurs.

2. *Ciliary function and metachronism in the membranellar band*

In resting and forward-moving animals membranelles are seen beating predominantly parallel to the course of the band. Metachronal waves travel in a proximo-distal direction, i.e. from the gullet toward the anterior right end of the band (Fig. 34b). Quantitative measurements from high-frequency films (Machemer, 1966) showed constant frequency but a wavelength decreasing toward the end of the band. In the frontal membranellar zone the angular velocity on the left side was maximal towards the left, but it was maximal towards the right on the right side. This explained the separation of water currents at the front end of *Stylonychia* (Fig. 35) and suggested antiplectoid beating of the left and symplectoid beating of the right membranelles (Fig. 36b). The orientation of the water currents somewhat transversely to the course of the band and to the main excursion of the membranelles (!) was hypothetically explained by the assumption of a "turbine-effect" of the asymmetrical distribution of forces of the membranelles, the broader (and hence stronger) edges of which serve to twist the membranelles during the power stroke and the recovery stroke, thus adding transverse vectors to the primary forces of beating (P′, R′ in Fig. 36b). At first sight, the metachronal properties of membranellar beating appear difficult to understand. How can metachronal wavelength so drastically decrease (from 40 μm to 28 μm), and metachronism apparently change in a 130 μm section of the band which carries similar membranelles spaced at equal distances (4·5 μm) and beating at identical frequency (range: 36–47 Hz)? The possible answer is that the observer sees only apparent wavelength (λ') which may be derived from a constant λ in a dexioplectic organization of metachronism (Fig. 36b). Changes in the apparent beat pattern may result from a gradual turning of the wave system with respect to the course of the band. Obviously, the waves follow changes in the orientation of *resulting* forces of beating (B) which, in turn, result from specific structural and functional ciliary properties.†

* Spontaneity was also observed in spike-like depolarizations of the cell membrane of *Stylonychia* (Machemer, 1970b) which resemble the regenerative calcium-response accompanying ciliary reversal in *Paramecium* (Machemer and Eckert, 1973). Okajima and Kinosita (1966) found spontaneity in the reversal of the cirri and membranelles in the related form *Euplotes*.

† It should be noticed that, as in other ciliates, the membranellar power stroke follows an antero-posterior axis, regardless of the asymmetric course of the band.

During backward locomotion the membranelles reverse the orientation of their power strokes (P′). Water currents, as far as observed in the dorsal view, are also reversed. The metachronal waves still travel proximo-distally and decrease in length in the same direction. Since the transverse component of beat appears to be morphologically determined (and thus non-reversible) reversal of the membranellar function should result in a 90°-reorientation rather than a 180°-inversion of water currents. This question cannot be settled with the data available at present. The example

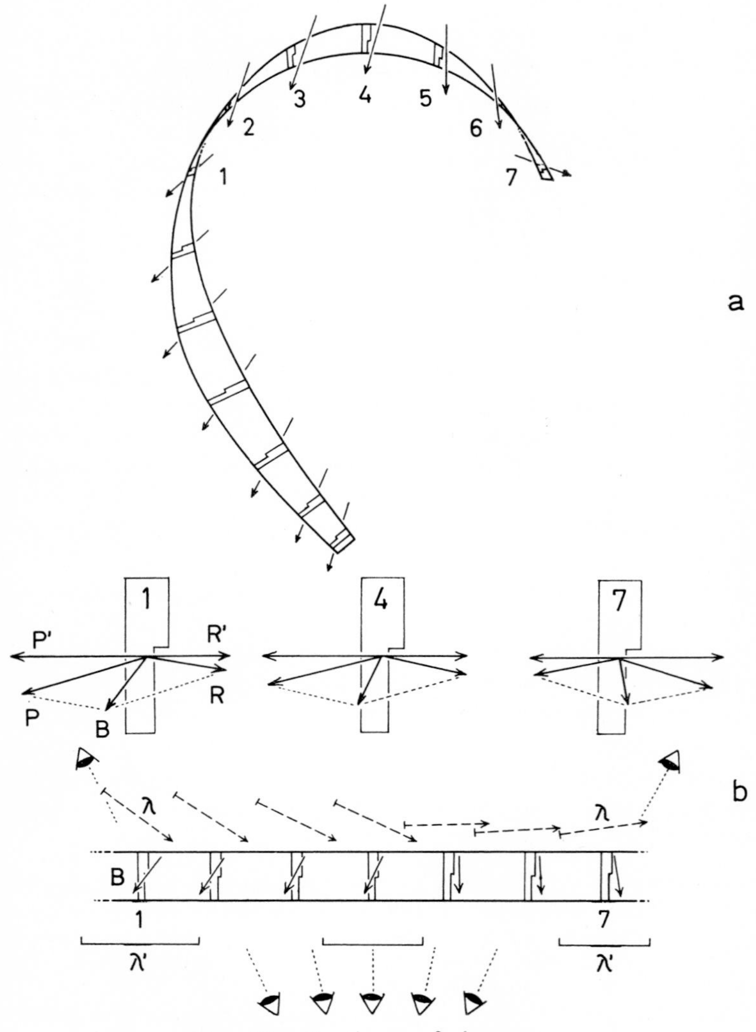

FIG. 36.—Caption on facing page.

of membranellar function in *Stylonychia* may demonstrate that, as in the ciliary bands in *Didinium*, metachronism has not to follow the course of a ciliary row, as coincidentally occurs in *Stentor* membranelles, but its basic properties (dexioplexy) nevertheless may remain identical.

3. *Ciliary function and metachronism in the rows of marginal cirri*

Contrary to the groups of ventrally inserted cirri, the marginal cirri do not directly participate in walking movements. Projecting from the sides of the cell and pointing somewhat downward, they assist walking as well as swimming by specific patterns of counterclockwise oscillatory movement, which may be different in the left and the right row or even in functional groups within one row (Machemer, 1965b). From the analysis of high-frequency films (Machemer, 1969b) 4 different patterns of beating were isolated in the marginal cirri:

1. Metachronic movement (Fig. 37a) occurs in resting animals at irregular frequencies up to 10 Hz without significant polarization of the strokes in the anterior or the posterior direction. Metachronal waves are short, usually covering 5–8 cirri and have no constant direction of propagation.

2. Synchronic movement (Fig. 37b) typically includes all marginal cirri and consists of one vigorous stroke of high amplitude and high angular velocity towards the anterior or posterior end. This type of movement is not strictly synchronous, since the first cirri (with respect to beat direction) are often advanced (sometimes retarded) in comparison with the cirri at the other end of the row, indicating an antiplectoid (or symplectoid) wave of extreme length. There are graded stages of transition between the synchronic and the metachronic movements (reduction in amplitude and

---

FIG. 36. Analysis of membranellar metachronism in *Stylonychia*. a. The course of the membranellar band in dorsal view with orientation of vectors of force (arrows) inferred from the water currents (Fig. 35). 1–7, the position of 7 membranelles of the frontal zone, the beating parameters of which were quantitatively determined. b. The portion of membranellar band between membranelles 1 and 7 spread in one plane with resulting forces of beat (B) and direction of waves (dashed arrows), assuming a dexioplectic type of metachronism. The apparent transmission of waves and the decrease of apparent wavelength (λ') in proximo-distal direction, as well as the change from antiplectoid (membr. 1–4) to symplectoid beating (membr. 5–7) are shown in correspondence with observations (Machemer, 1966). (Eye symbols in *b* indicate corresponding angles of observation in *a*.) The three enlarged cross-sections of membranelles indicate a hypothetical explanation of how ciliary excursions parallel to the membranellar band (P′, primary power stroke; R′, primary recovery stroke), may produce a transverse or oblique resultant force (B) through a "turbine-effect" of the asymmetrically beating membranelle by diverting forces of the power stroke and recovery stroke in one direction (P, R).

intensity of beating; decrease in apparent wavelength) which suggests that the metachronic pattern may consist of series of weak "synchronous" reorientations of the cirri.

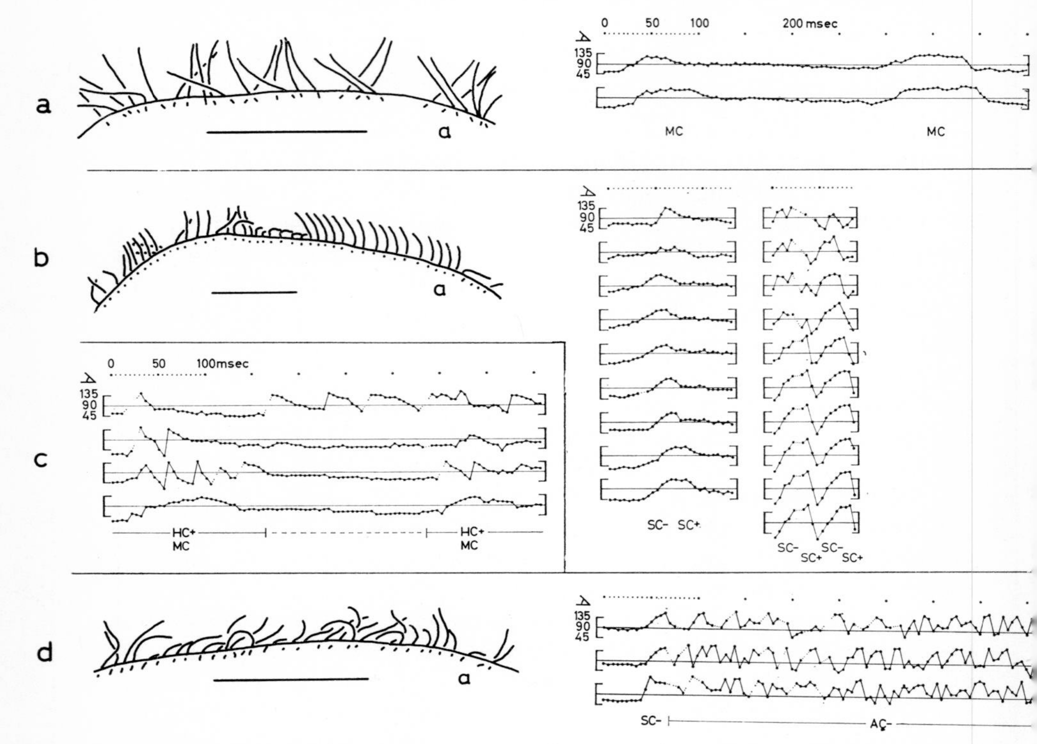

Fig. 37. Four patterns of beating in the marginal cirri of *Stylonychia* from high-frequency films. a. Metachronic type. Note the shift in phase between 2 neighbouring cirri in the accompanying diagram. b. Synchronic type. In the drawing a stroke occurs within one group toward the anterior end (a). Diagrams show common series of synchronous beating (SC) in opposite directions. c. Heterochronic type. The polarized nature and independence between neighbours is evident from the diagram. d. Achronic type. Cirri with pronounced anterior or posterior orientation and rapid helical oscillations; no metachronal coordination (see diagram). Drawings represent left marginal row (a, d) or right marginal row (b). a. Anterior end; horizontal bars: 50 μm. Diagrams illustrate time-related angular shifts of neighbouring cirri. MC, SC, HC, AC, occurrence of metachronic, synchronic, heterochronic and achronic functions; + (−): power stroke or orientation occurs toward the posterior end (anterior end). From Machemer (1969b).

3. Heterochronic movement of the cirri (Fig. 37c) occurs at moderate frequencies (5–17 Hz) and is characterized by distinct polarization of the beat towards the posterior or anterior end. A common feature is that the cirrus is rigid during the slow period of the cycle but exhibits bending during the period of high angular velocity (up to 5·5 × higher than in the slow period). The heterochronic movement which lacks metachronal coordination is reminiscent of the walking motion of the ventral groups of cirri. It is worth noting that the temporo-spatial polarization of the heterochronic activity is different from common ciliary polarization (where the power stroke combines high angular velocity with pronounced stiffness of the cilium).

4. *Achronic movement* (Fig. 37d) consists of high frequency helical oscillations of low amplitude (up to 70 Hz; 25–75°) which have a common orientation toward the posterior or the anterior end of the animal and promote fast locomotion. Metachronal coordination could not be detected. At reduced frequencies forms of transition between the achronic and heterochronic patterns were observed (Fig. 37c, third line).*

4. *Questions of interpretation of metachronism in the marginal cirri*

The understanding of metachronal properties in the marginal cirri of *Stylonychia* is obscured for 3 reasons: (1) the cirri lack uniformity in beating and (2) may be functionally individualized within one row; (3) beating of the cirri occurs in one line (and not in a field of regularly spaced organelles). Because of these complications the interpretation of metachronal phenomena is more tentative than substantial.

The synchronous reversal of the cirri and accompanying behavioural phenomena resemble ciliary reversal in *Paramecium* (Machemer, 1965b). Similarities are also found in the concomitant metachronism: profiles of cilia as well as the cirri usually show an antiplectoid configuration within a long wave. Assuming a dexioplectic organization of the underlying metachronal system the true wave direction is nearly perpendicular to the plane being observed (compare Figs 23c and 38a), while the beat deviates little from the longitudinal axis. The observed reduction in wavelength along the row of cirri might result, at least in part, from a graded increase in angular deviation of the power stroke from the longitudinal axis (Fig. 38b). There are alternative possibilities to explain phase-shift in the synchronized row of cirri, as for instance: mechanical triggering by the leading cirrus or gradients of sensitivity or membrane conductance to events following depolarization. Experimental evidence is not yet sufficient for a determination of the coordinating mechanism. The occurrence of short waves during the "metachronic movement" is difficult to understand in terms of a dexioplectic wave system. Since beating is often

* In the hypotrich *Euplotes* the frontal and anal cirri exhibit patterns of beating which are similar to the heterochronic and achronic movement of the marginal cirri in *Stylonychia* (Okajima and Kinosita, 1966).

restricted to groups of cirri, or "stationary" cirri are by-passed by the wave (Machemer, 1969b), a mechanical trigger-mechanism may take over the regulation of phase relations between neighbouring cirri. Grouping of the cirri is also observed in other patterns of beating, for instance during

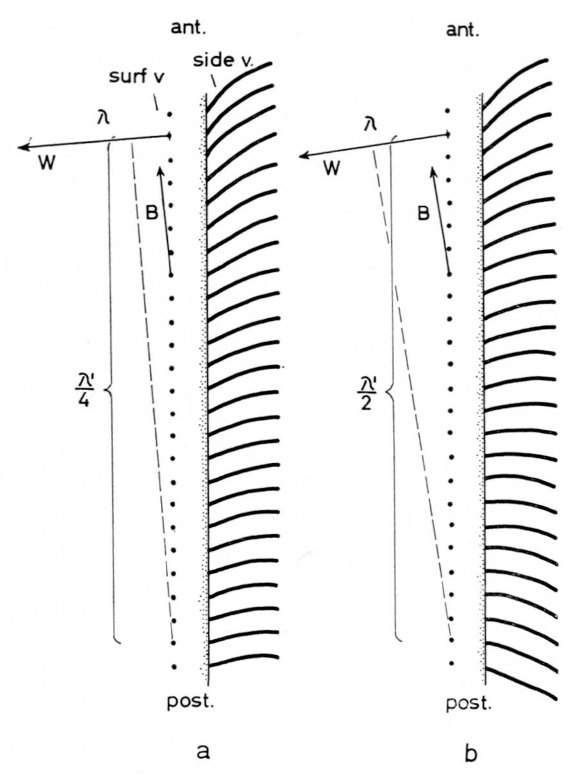

FIG. 38. Metachronal properties of the "synchronous" stroke of the marginal cirri of *Stylonychia*; hypothetical. a. A row of synchronized cirri shows a certain metachronal phase-shift, which may be understood as resulting from a wave occurring transversely to the row, while the cirri beat, in a dexioplectic manner, nearly parallel to the row. b. Apparent metachronal wavelength (λ') will decrease with increasing angular deviation of the beat (B) from the row. ant. (post.), anterior (posterior) end of the row; surf. v., surface view; side v. side view of the row. W, direction of wave; B, direction of beat.

a synchronous reversal at moderate strength (Fig. 37b). The functional individuality of the cirri may account for the fact that metachronism could not be observed during the heterochronic "stalking" and the achronic high-frequency gyrations of the cirri. Differences in beat frequency and amplitude and even switching between two types of motion

(Fig. 37c) prevent the establishment of common hydrodynamic parameters of beating and hence the formation of a regular metachronal system. In summary, metachronism in the cirri as well as in the membranelles of *Stylonychia* may be understood as dexioplectic when the compound ciliary organelles perform polarized beating in a common direction and with sufficient angular velocity. The possible occurrence of other mechanisms of coordination at lower frequencies of the cirri will be discussed in broader context in section V.B.5.

F. Other ciliates

The metachronal systems of some other ciliate protozoa within the orders Holotricha, Spirotricha and Peritricha have been studied in more or less detail. Although interpretations were not unanimous in all cases, it now appears that the dexioplectic pattern of metachronism is widely spread among the ciliates, including those forms, such as *Didinium*, *Metopus* or *Opisthonecta*, where ciliary rows have been organized into narrow bands.

1. Ophryoglena, Uronema, Tetrahymena, Colpidium

These hymenostome ciliates, though very different in size, show patterns of waves which are similar to those in the closely related *Paramecium* (Fig. 39). *Ophryoglena* (Párducz, 1964) has a large pear-shaped body (200–450 μm) with a circular cross-section, and the oral groove and cell mouth are found in the widened upper quarter of the cell (Fig. 39a). The animal swims in a left-hand helical path of varying steepness reaching velocities of up to 4 mm s^{-1}. Backward swimming is rarely seen, since a solid obstacle is usually avoided without reversing. Changes in the direction of locomotion may occur without obvious stimulation. The analysis of fixed preparations of *Ophryoglena* revealed densely packed cilia coordinated in a way which is similar to that seen in *Paramecium* (compare inset in Fig. 39a with Fig. 6). The metachronal wavelength is 10–13 μm. Local modifications of the regular wave pattern may account for observed alterations of the straight swimming path (deflections, swimming in arcs, etc.). Considering the rotational shape of *Ophryoglena*, helical swimming has to be related to asymmetries in ciliary activity which are probably located in the anterior oral groove. Metachronism in species of *Uronema*, *Tetrahymena* and *Colpidium* was also documented by the fixation technique (Párducz, 1954, 1967) showing ciliary coordination of the dexioplectic type (Fig. 39b–d, note insets). Transmission and length of waves and the direction of the power stroke correspond to those in *Ophryoglena*, underlining dependence of metachronism on ciliary parameters. The validity of polarized beating of cilia in *Tetrahymena* has recently been questioned by Preston *et al.* (1970), who saw continuous helical undulations of cilia in high-frequency films of *Tetrahymena*. Since ciliary activity incorporates properties of helical *and* polarized beating and is, in addition,

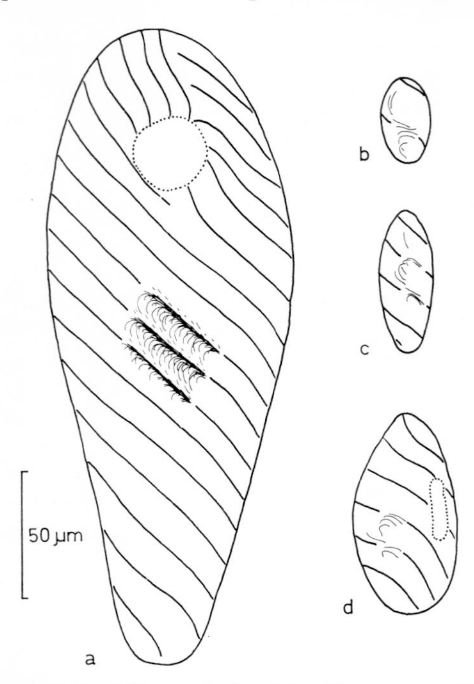

FIG. 39. Metachronism in different holotrich ciliates, corresponding to forward swimming. a. *Ophryoglena*, b. *Uronema*, c. *Tetrahymena*, d. *Colpidium*. Redrawn from figures of fixed preparations by Párducz (1954, 1964, 1967). In all examples cilia perform a counterclockwise cycle and beat in a polarized manner with the power stroke pointing toward the posterior and right (see insets). Waves move to the anterior and right in a dexioplectic pattern, like that described in detail in *Paramecium*. Note that metachronal wavelength is related to the size of the cilia rather than to the size of the cell.

heavily modified by mechanical factors such as viscosity, reported differences in the findings about *Tetrahymena* cilia may not have substantial importance.

2. *Didinium*

This ovoid ciliate (120 μm long), famous for the ferocious manner in which it attacks and devours paramecia, may hardly be regarded as a typical holotrich, since ciliation is reduced to two circular girdles, each consisting of a small number of ciliary rows (Fig. 40). Párducz (1961), who investigated locomotion and metachronism in *Didinium*, found right-hand helical swimming to be the prevailing form of locomotion in both forward and reverse directions, in spite of the rotational symmetry of the cell as well as of the ciliary apparatus. The velocity of locomotion may vary considerably, as do the steepness and diameter of the helical path; maximal

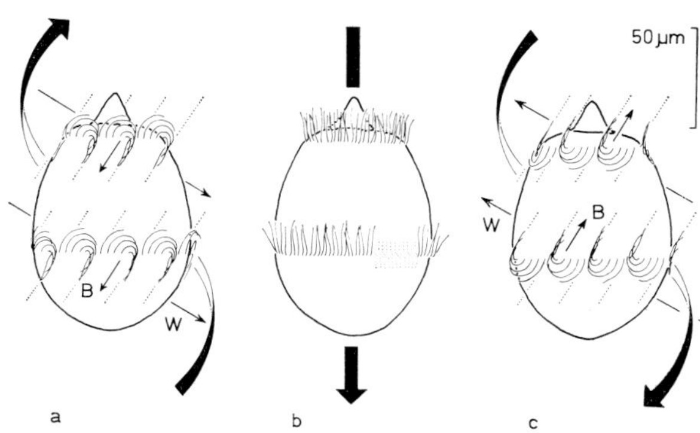

FIG. 40. The ciliary system and locomotion in *Didinium* (schematic). a. Forward swimming occurs in right-hand helices (flat arrow) opposite to the direction of beat (to the posterior and left; B) in two ciliary bands encircling the ovoid body. Metachronal waves appear to travel along the ciliary bands towards the right; the true wave direction (W), however, is at right angles to the beat direction (B). This dexioplectic pattern of waves can be inferred from the shapes of cilia corresponding largely to those seen in *Paramecium* and other holotrich ciliates. b. Unrotated synchronous ciliary reversal. All cilia perform one vigorous stroke towards the anterior end. The inset shows the arrangement of kinetosomes close to the ciliary bands; only kinetosomes in the oblique rows carry cilia. c. Reversed swimming of *Didinium*. Since the orientation of ciliary beating is reversed by 180°, the wave direction and the direction of helical swimming also undergo inversion. Note the clockwise shift of beat direction in comparison with (b) and compare with *Paramecium* (Fig. 23). Drawings combined and modified from Párducz (1961) and Sleigh (1966).

speed (up to 3 mm s^{-1}) was not found to be associated with linear swimming. Avoiding reactions, deflections and swimming in arcs were observed to be similar to those in other ciliates. Ciliary beating in forward-swimming *Didinium* is directed towards the posterior and left, thus accounting for the right-hand helical mode of locomotion (Fig. 40a). In partially narcotized animals Párducz observed a counterclockwise gyration of the cilia and metachronal waves travelling along the rows towards the right. This finding was confirmed by ciliary configurations of fixed specimens, where the counterclockwise shifts of neighbouring cilia proceed towards the left.

In the first stage of a spontaneous or induced reversal the cilia are synchronously "thrown" toward the anterior end, thus inducing unrotated rebounding of the cell (Fig. 40b). Continued backward swimming in a right-hand helix results from a complete reversal of ciliary beating and metachronal transmission in comparison with normal conditions (Fig. 40c).

Relationships between the direction of wave movement and the orientation of the ciliary power stroke have been differently interpreted because the true propagation of the metachronal system in *Didinium* is obscured (as is the movement of parallel stripes seen through a narrow slit: compare Fig. 1b). Sleigh (1966) proposed that the beat occurs at right angles to the ciliary bands and waves travel parallel to the rows (dexioplexy). Párducz (1967), on the other hand, by separating the wave direction from the course of the ciliary bands, assumed an antiplectic pattern of coordination with waves moving exactly against the direction of the oblique power stroke. Ciliary forms in the metachronal waves of *Didinium* are very similar to the dexioplectic organization of beating in *Paramecium* (compare Fig. 40a, c with Fig. 6). Hence the interpretation given by Sleigh may be modified by assuming that the invisible mainline of metachrony has an oblique orientation (from anterior left to posterior right; Fig. 40, W), which is in best agreement with all observations.

3. *Metopus*

The membranellar band of this heterotrich twists in a helical course around the drop-shaped cell. A densely ciliated band, comprising 5 rows of long cilia, follows the course of the membranelles at the margin of the umbrella-like anterior cap (Fig. 41a). Párducz (1954) observed "huge waves" travelling towards the anterior-right along the ciliary band and, in view of the sparse ciliation on other parts of the cell, concluded that the marginal cilia are primarily responsible for locomotion which occurs in a straight forward path. Párducz emphasized, quite correctly, the great similarity between the metachronal waves on the ciliary band of *Metopus* and of *Paramecium* (compare Fig. 41 with Fig. 6). The sequence of bending stages, as documented by Párducz, suggests that the true direction of metachronal waves is towards the right while the power stroke points in the posterior direction (Fig. 41b). This dexioplectic organization of beating in the ciliary band is reminiscent of similar conditions found in the 2 circular girdles of the holotrich *Didinium* (Fig. 40). Similarities to metachronism in *Metopus* were also observed in the posterior ciliary ring of free swimming stages of the peritrich *Vorticella* (Párducz, 1954).

4. *Opisthonecta*

This unstalked peritrich, resembling the free swimming stage of *Vorticella*, swims in right-hand helices with its aboral end forwards. Loco-

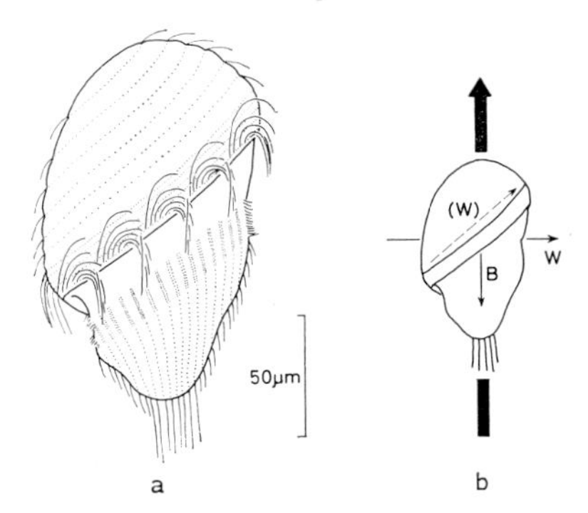

FIG. 41. Ciliary activity and metachronism in the helical ciliary band of *Metopus spiralis*. a. Schematic representation of the cell showing metachronal waves along the densely ciliated band (only cilia of the fifth, most marginal row are drawn). Short double rows of dots indicate the course of the membranellar band. b. Relations between locomotion (heavy arrow), direction of the power stroke ($\xrightarrow{B}$) and direction of wave ($\xrightarrow{W}$) passing the ciliary band at an acute angle; $\xrightarrow{(W)}$, apparent direction of wave transmission along the band. Modified from Párducz (1954).

motory forces are created by the aboral girdle of long cilia on which metachronal waves are seen (from the oral end) to encircle the cell in a clockwise direction, while the power stroke points toward the posterior and left (Jahn and Hendrix, 1969). The arrangement of cilia in the aboral girdle (short oblique rows), their metachronism and locomotory properties are very similar to those in the holotrich *Didinium*, so that a dexioplectic relation between waves and power stroke may also be expected for *Opisthonecta* (Fig. 42, compare with Fig. 40). Photographic and high-speed cinematographic recording of the activity of the aboral cilia by Jahn and Hendrix showed continuous vortices of oblique water currents encircling the girdle. Similar vortices have been observed in *Stylonychia* (Fig. 35) and may be a common concomitant of activity in ciliary bands. Analysis of films taken during the rapid turning movement of *Opisthonecta* revealed that while the animal performed a sharp turn of 135° towards the right within 200 ms, cilia seen in profile on the outside of the turning course beat 7 times in the normal direction (36 Hz), but at the same time the inside cilia completed only 2 cycles of reversed beating (10 Hz). "The action is comparable to that caused by reversing one oar of a simple row-boat without a keel" (Jahn and Hendrix, 1969). The local response of the membranelles exhibited in this example is no uncommon phenomenon in ciliates. Párducz has repeatedly (1956b, 1958, 1963) reported local responses of cilia in *Paramecium*. In high-frequency films of *Paramecium*, cilia were recorded with different activity and different latency to electric stimulation (Machemer, unpublished results). Okajima and Kinosita (1966) found spontaneous reversal occurring, coincidentally, only in 2 of the 5 frontal cirri of *Euplotes*. The nature of the local ciliary response is still unknown, though possible explanations may be offered. (See Chapter 12 by Naitoh and Eckert.)

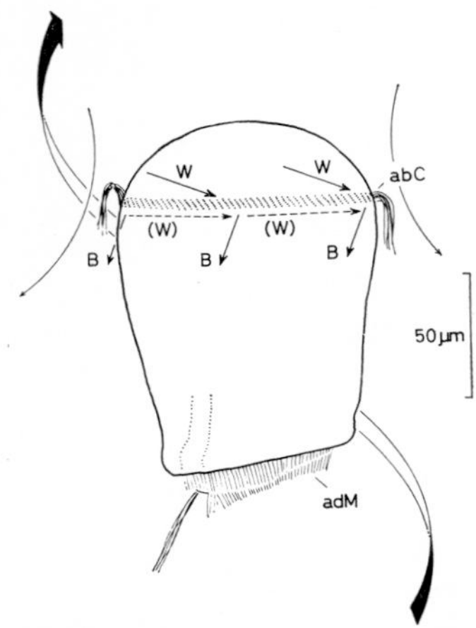

FIG. 42. Activity of cilia in the aboral girdle of *Opisthonecta*. During right-hand helical swimming in the "forward" direction, cilia of the aboral girdle (abG) beat toward the "posterior" and left. Metachronal waves appear to encircle the cell in a clockwise direction (dashed arrows), if seen from the oral region. Though the ciliary wave has not yet been analysed, the true wave direction will probably be toward the "posterior" and right, i.e. similar to the dexioplectic metachronism in *Didinium* and *Metopus* (Figs 40, 41). adM, adoral double row of membranelles. $\xrightarrow{B}$, direction of beat; $\xrightarrow{(W)}$, apparent direction of waves; $\xrightarrow{W}$, presumed direction of waves. Curved arrows, orientation of vortices created by the aboral cilia. Definition of "anterior" and "posterior" from normal function of the locomotory organelles. Modified from Jahn and Hendrix (1969).

5. *Balantidium* and *Nyctotherus*

Párducz (1967) presented fixed preparations of two ciliates from the rectum of the frog, *Balantidium* and *Nyctotherus*, which, according to his interpretation, show changes in the type of metachronal coordination. Specimens of the holotrich *Balantidium entozoon* swimming in right helices exhibited waves travelling posteriad and to the right (Fig. 43a). In another preparation *Balantidium* was found to swim in left-hand helices; the corresponding orientation of the metachronal system was such as to indicate wave transmission towards the anterior and right (Fig. 43b). Párducz described the metachronism of the first case as dexiosymplectic and called the metachronism shown in the second case dexioantiplectic.

However, both preparations appear to be basically similar in exhibiting dexioplexy (compare the sickle-shape of cilia during the recovery stroke with the dexioplectic model in Fig. 6) which may or may not have undergone slight modifications due to differences in ciliary activity or viscosity

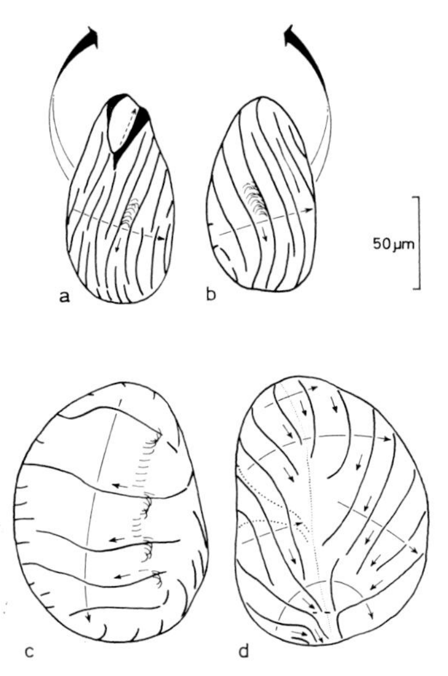

FIG. 43. Ciliary beat and metachronism in the entozoic ciliates *Balantidium* (a, b) and *Nyctotherus* (c, d). *Balantidium* swims in right-hand helices (a, flat arrow) with waves toward the posterior and right (long arrow), or the cell may swim in left-hand helices (b) with waves travelling towards the anterior and right. In both cases metachronism is dexioplectic (power stroke: short arrow) as indicated by the direction of swimming and the form of cilia during the recovery stroke (sickle-shapes in insets between wave fronts). Note the anteriad transmission of waves along the "membranelles" (dashed arrow). The laterally compressed *Nyctotherus* has a concave right side (c) with waves moving posteriad (long arrow) and a convex left side (d) showing a bilaterally symmetrical orientation of wave fronts with respect to the suture of kineties (fine dotted line). Dislocation of waves in (d) relative to the longitudinal axis results in apparent circulatory wave movement in a clockwise direction. The relation of the power stroke (short arrows) to the waves is dexioplectic. The ventral side of cell is marked by the cytostome (dotted lines). From photographs by Párducz (1967).

of the medium; we have seen in *Paramecium* that a clockwise turn of the power stroke accompanies increased frequency of normal beat (Fig. 23a, b) or is induced by increase in viscosity (Fig. 27).

A still more striking case of transformation of the metachronal pattern is reported from the heterotrich *Nyctotherus cordiformis*, which usually creeps with its flattened and somewhat concave right side oriented towards the solid substrate. This side of the cell shows long waves moving in the posterior direction (Fig. 43c). Párducz characterized these waves as symplectic, though the orientation of the ciliary pattern speaks in favour of dexioplexy. Two observations by Párducz may deserve attention: (1) The common mode of locomotion is described as a "slow swerving creep with a tendency to wind to the left"; (2) cells are attached to the substrate so that they "are hard to suck up in a pipette". Since the ciliary power stroke on the lower (right) side obviously has no posteriad orientation, the transmission of wave crests backwards may provide a means of exerting forces on the substrate and thereby assist locomotion. The ciliary beat occurring to the right (if observed from above) might contribute to the left-hand swerving of the cell and, eventually, help to evacuate water from the space between cell and substrate, thus supporting adhesion. Metachronism on the convex left side of *Nyctotherus* shows a bilaterally symmetrical orientation of the wave fronts with respect to the suture of kineties (fine dotted line) of the ventral and the dorsal region (Fig. 43d). Waves move towards the anterior and right on the ventral half, but turn more clockwise on the dorsal half of the surface, where they travel towards the posterior and right. This change in orientation of the waves is responsible for an optical effect of apparent circulatory wave transmission along the margin of the cell (Párducz, 1967). The orientation of the ciliary beat associated with this complex pattern of waves is rather uniform, if dexioplexy of the waves is assumed: all cilia have the power stroke directed towards the posterior with angular deviations between orientations on the two sides of the suture of generally less than 90°. Differences in orientation and intensity of beat (and thus metachronism) also occur on the two sides of the preoral suture of kineties in *Paramecium* (Figs 15a, b, 24a). Sutures may, in general, have a functional significance in that they separate gradients of ciliary activity (see section V.C.4). Párducz' interpretation of metachronal patterns changing between the ventral-left side (dexioantiplectic) and the dorsal-left side (dexiosymplectic) of *Nyctotherus* appears to be more difficult to understand than a graded modification in the direction of ciliary beating.

6. *Spirostomum*

The large heterotrich freshwater species *S. ambiguum* is in the form of a long cylinder (1–3 mm) and is seen swimming forward in steep left-hand helices. Locomotion is created by cilia occurring in longitudinal rows all over the body and by a band of membranelles which occupies the anterior half of the body length. Boggs *et al.* (1970) observed from high-frequency films that the body cilia beat in a polarized manner. The power stroke deviates slightly from the longitudinal axis toward the posterior and right, as would be expected from the left-hand swimming movement. At increased viscosity *Spirostomum* swims in steep right-hand helices; according to the films the power stroke has undergone a small turn toward the posterior and left. Similarities of the reported data with ciliary activity in *Paramecium* suggest a similar metachronal organization of ciliary beating. *Spirostomum* performs a typical avoiding reaction (Jennings, 1899) which may be accompanied, as in *Stentor*, by a rapid contraction after sufficient stimulation. Different properties of membranellar beating (frequency, amplitude) on both sides of a cut through the membranellar band (Verworn, 1889) have been an early argument in favour of a mechanical hypothesis of metachronal coordination. Pütter (1904) observed that stimulus-induced reversal in the beating of the membranelles occurred later than in the body cilia, but renormalization was earlier in the membranelles.

V. Relations between flagellar or ciliary activity and metachronism

All metachronal systems surveyed in the preceding paragraph have some properties in common, the most eye-catching of which are the synchronization and the specificity of the metachronal type. Parameters associated with the beating organelle itself (such as length of the shaft, frequency and amount of polarization of the beat) as well as physical properties of the medium contribute to the type of the waves, their direction, length and velocity. As will be shown below most of the available data on metachronism may be understood on the basis of a concept of mechanical interference between beating organelles. The examination of possible principles of metachronism will be guided by 3 questions: (1) How do flagella or cilia synchronize? (2) How is the metachronal wave generated? (3) How is metachronism regulated?

A. Synchronization between beating organelles

As shown in the introductory sections synchronization between neighbouring flagella or cilia is inseparable from any metachronal coordination in a field. Where synchronization between beating organelles does not appear in the plane of observation, as for instance in a paramecium viewed in profile, or the field is restricted to a narrow band, as occurs in the ciliary girdles in *Didinium*, synchronization is nonetheless a real phenomenon, as can be inferred from certain configurations of the cilia within the visible wave system.

The synchrony of beating between flagella or cilia is an invariable all-or-none phenomenon. Metachronism does not occur without synchronization of cilia, although in narrow bands the lines of synchrony may be very short or eventually reduced to a single cilium in single rows. It appears reasonable to assume that the physical laws governing the synchronization between two spirochaetes, flagella, sperm-tails or cilia are identical with those controlling metachronal fields. From this point of view, synchronization, occurring at right angles to bending waves travelling along individual spirochaetes or flagella, may be regarded as the initial step towards metachronal organization of beating in a field, a view which was already expressed by Gray (1928).

1. *Some properties of synchronized beating*

In an idealized case of synchronization between helically beating flagella (Fig. 44) forces generated by complete bending waves travelling distally will be such as to result in continuous locomotory forces parallel to the flagellar helix. The forces acting transversely to the helical axis

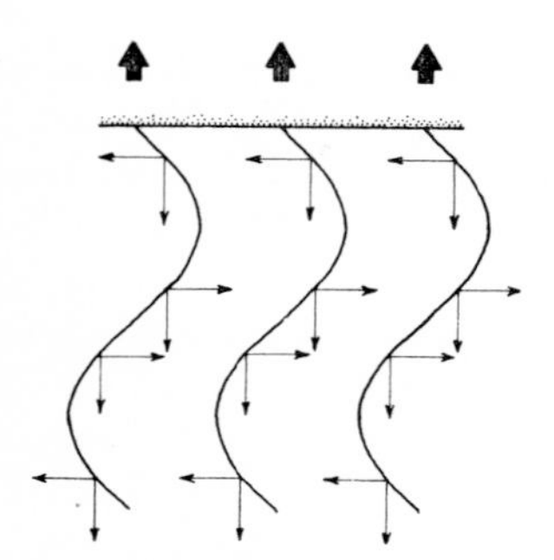

FIG. 44. Illustration of forces generated by 3 synchronized flagella; schematic. Vectors of force parallel to the helical axis of the flagella give rise to locomotory forces (heavy arrows). Vectors transverse to the helical axis, though balanced against each other in an individual flagellum, have a common orientation in different sections of the synchronized flagella which reduces mechanical interference between neighbouring flagella.

are counterbalanced, as a whole, in each individual organelle, but may nevertheless have considerable local effects (see below). During synchronization of the flagella,* all neighbouring transverse forces have the same orientation at any time of the cyclic movement or, in other words, the flagella display the minimum possible viscous interaction between one another. The concept of "minimum interference", originally postulated as a coordinating principle in a metachronal wave (Gray, 1930), appears to describe well, in qualitative terms, the result of synchronization of flagella or cilia.

Movement of the spirochaetes attached to *Mixotricha* may be an example

* For reasons of simplicity it is assumed that the neighbouring flagella beat at identical frequency, wavelength and amplitude.

of synchronization which is very similar to the schematic case illustrated in Fig. 44. According to reconstructions of the metachronal waves in profile view and in surface view (Fig. 10) the points of spirochaete insertion (brackets) are arranged in such a way as to provide nearly perfect synchronization between the spirochaetes at any place on the cell surface and at any time during the helical oscillation. Metachronal waves appear as direct continuations of the individual bending waves. Cleveland and Cleveland (1966), who observed synchronization in different free and attached spirochaetes, point out that "firm contact must be made over a good portion of their length in order for synchronization to occur", a postulate which appears to be realized in *Mixotricha*. The more sparsely distributed motile spirochaetes on the posterior end of *Deltotrichonympha*, on the other hand, exhibit no signs of synchronization.

2. *How synchronization may work*

The metachronal system in *Mixotricha* will be understood on a mechanical basis, if it is possible to explain spirochaete synchronization. This mechanism is not evident in itself, since the state of minimum mechanical interference during synchronization represents a high degree of self-organization of motion which might appear at first sight to violate the second law of thermodynamics. Physical laws are not broken, however, if a simple servo-mechanism inherent in closely spaced oscillating units is taken into consideration. Consider two flagella (1 and 2 in Fig. 45), each performing harmonious helical waves in the counterclockwise direction with identical frequency and a wavelength of bending of $\lambda = 1$,* but beating out of phase by $\frac{\lambda}{2}$ (Fig. 45, a, a′). In position (a) the flagella are stopped in their motion through mutual contact at (K). Flagella in position (a′) are free to move until they are blocked as in position (a) at point (K). Under practical conditions the positions (a) and (a′) have little chance to occur, since minor differences in frequency and force of beating will destroy the unstable balance of forces toward a phase relation of $<\frac{\lambda}{2}$. In position (b), for example, flagellum 1 is advanced by $\frac{\lambda}{6}$ in comparison to 2. Viscous coupling or even contact (K) between the flagella now causes acceleration of 2 by 1 and deceleration of 1 by 2 leading to the state of synchronization in position (c). If 2 accelerates too much it will soon be in a position similar to (b′) where new interference with 1

* This assumption neglects the complication of synchronization between two organelles beating with different wavelength. Machin (1963) developed a mathematical model to explain synchronization under these conditions.

causes deceleration of 2 and acceleration of 1 which again leads to synchronization of the flagella (positions c′→c″→c). Thus synchronization is stabilized through a negative feedback or servo-mechanism provided

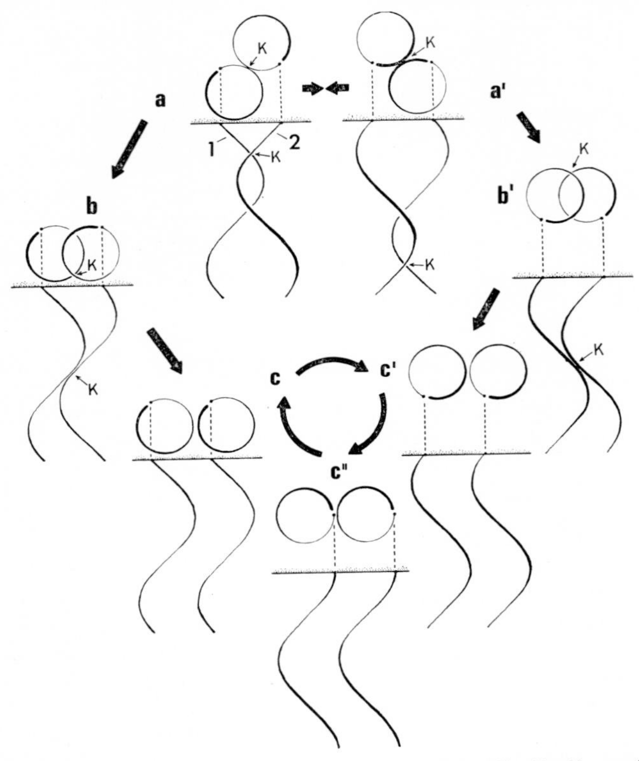

FIG. 45. How flagella synchronize; schematic. Two flagella (1, 2), projecting perpendicularly from a cell surface and beating helically in the counterclockwise direction, are depicted in various positions in side view and as viewed from the tips. A phase-shift between the flagella of $<\frac{\lambda}{2}$ increases viscous interaction or even leads to physical contact (K), by which the "later" flagellum is accelerated and/or the "advanced" flagellum is decelerated (positions b, b′). This leads automatically, by servo-regulation, to the synchronized state of beating (c→c′→c″→c). Phase-differences of exactly $\frac{\lambda}{2}$ (positions a, a′) would cause mutual mechanical blocking of the flagella, but these positions are unstable: through minor disturbances they will develop over stages of transition (b, b′) towards the stabilized state of synchronization.

by the unidirectional helical oscillation of the closely aggregated flagella. It should be noted that synchronization occurs as the *result* of an energy-consuming process of acceleration and deceleration of the flagella. Gray

(1930), in applying the minimum interference principle to phase differences between cilia in a metachronal wave, described a servo-regulation when writing that "the faster moving cilium will constantly tend to accelerate its slower neighbour and the slow cilium to slow up the more rapid cilium". Coordination between flagella or cilia in a metachronal field cannot be understood, however, without adding some new considerations to the idea of synchronization by minimum interference.

B. *Questions concerning the generation of the metachronal wave*

Contrary to the phenomenon of synchronization, the metachronal coordination of flagella and cilia varies with changing properties of beating and of the surrounding medium. There are nevertheless some general trends accompanying metachronism in the Protozoa. In the polymastigote flagellates waves are symplectic, coinciding with the direction of bending waves on the flagella. The metachronal wavelength tends to be progressively separated from the bending wavelength (*Mixotricha*→*Koruga*→*Deltotrichonympha*). In the Opalinids symplectic metachronism is somewhat modified toward the dexioplectic type. The flagella of *Opalina* assume, morphologically and functionally, an intermediate position between typical flagella and cilia; the first signs of abandoning harmonious helical undulations in favour of a polarized beating become apparent. In the ciliates dexioplexy is the general rule of metachronal coordination between short cilia which display polarization of the beat in time and space. The organization of the wave system is maintained independently of the orientation of the power stroke. The effect of increased viscosity on the cilia of *Paramecium* is comparable to that of increased length of the organelle and/or reduction in polarization of beating: dexioplexy is gradually transformed into symplexy. These observations suggest that the parameters of the metachronal wave are intimately related to the mechanical forces generated by the oscillating organelles. There are no data reported concerning the minimal mechanical requirements of metachronism to be established in a field of beating oscillators. Apparently a more or less common orientation in the direction of beating and a similar frequency above a critical level are preconditions for metachronal waves to occur. It should be considered, however, that observed beating properties of individual flagella or cilia within a field will be at no times primary ones but always already modified by the averaging effect of viscous interaction between the oscillators. At this time factors leading to certain properties of metachronism can be assessed only in a generalized and highly simplified way. Two factors associated with a flagellum or cilium seem to play an important role in the spatial orientation of mechanical forces: the length of the organelle and the amount of polarization of the cyclic movement.

1. *Influence of the length of the beating organelle on the generation of mechanical forces*

In Fig. 46 it is schematically assumed that 4 organelles differing in length are beating in harmonious helical waves with identical λ, amplitude and frequency and creating identical forces on comparable sections of the shaft. The flagellum (a) carries two complete wavelengths. As with the

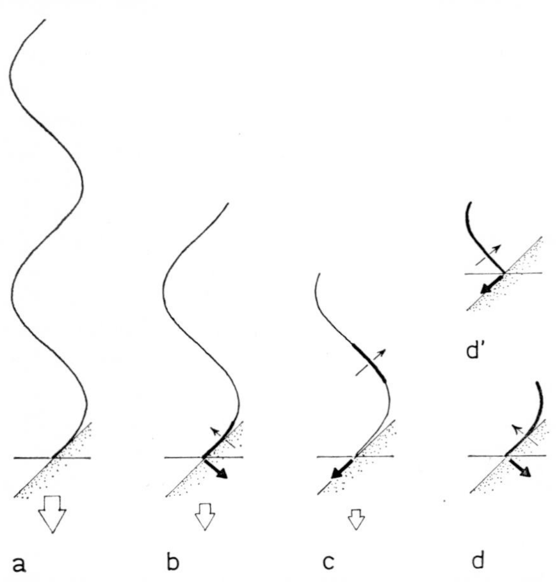

FIG. 46. Influence of the length of helically beating organelles on the generation of locomotory forces; schematic. a. Flagella generating complete waves of bending ($n\lambda$) are able to counterbalance all transverse forces thus producing only translational forces of dislocation (open arrows). b. Shorter flagella with a bending wave of e.g. $\frac{7}{6}\lambda$ generate "additional forces" through unbalanced sections of the shaft (heavy line), while forces parallel to the flagellar helix are reduced. The unbalanced vectors will oscillate with the flagellum. c. Further reduction of the flagellar length, e.g. to $\frac{5}{6}\lambda$, leads to an increasing proportion of the additional forces with respect to the (decreasing) translational forces. d. In very short organelles, e.g. $\frac{1}{3}\lambda$, assuming continuous helical motion and similar bending as in the flagella a-c, the whole shaft will be unbalanced and locomotory forces are determined by the actual position of the shaft (d, d′). Tilting of the organelles against the cell surface (dotted contour) may have the result that forces generated by short organelles become effective for locomotion only during a limited period of the cyclic movement.

flagella depicted in Fig. 44 all mechanical forces will be balanced in such a way that the resulting locomotory forces (hollow arrow) are oriented parallel to the flagellar helix. In a flagellum shown in (b) the shaft carries a wave of somewhat more than 1λ. As a consequence a certain section of the shaft (heavy line) is not counterbalanced on the flagellum thus adding a new vector of force (heavy black arrow) to the reduced translational forces; this additional vector changes orientation in the course of the cyclic oscillation. In principle, this case may apply to the flagella found in *Koruga* and *Deltotrichonympha*. Flagellum (c) is still shorter than (b) but is also unbalanced; the proportion of the "additional forces" with respect to the reduced translational force is increased. The flagella of *Opalina* may have some similarity with this type of beating organelle. In (d) the shaft of the short organelle is totally unbalanced on its helical course. Consequently, oblique mechanical forces changing orientation in harmony with the cyclic progress of bending (d′) determine the work of the organelle. If the helices shown in the diagram are, as usual, somewhat tilted toward the cell surface (dotted contour in Fig. 46), the locomotory work of an oscillating organelle will be more discontinuous with decreasing length. For instance, the ciliary organelle shown in (d) will be more effective for locomotion when moving from (d′) to (d) than from (d) to (d′).*

2. *Effects of unpolarized and polarized cyclic beating on the generation of mechanical forces*

Cilia functionally differ from flagella in that their movement tends to show periodic differences in angular velocity during the cycle ("polarization in time") and the curvature of bending undergoes periodic modifications ("polarization in space") leading to a depression of the circular envelope of beating, a pronounced straightening during the power stroke and bending during the recovery stroke. Protozoan flagella, on the other hand, tend to produce a continuous and undeformed (and hence "unpolarized") travelling helix. In a typical flagellum, as for instance shown in Fig. 44, dislocation of water along the helical axis will be rather continuous; the flagellum works like the screw of a ship. Transversely to the flagellar helix, on the other hand, each section of the shaft produces forces which change orientation by 180° each half-cycle and which increase with increasing frequency and amplitude of the oscillation. In qualitative hydrodynamic terms, the changes in viscomechanical forces of drag and pressure are high transversely to the flagellar helix, though locomotory

* This assumption neglects a number of complications which are associated with beating of organelles of different length, such as changing bending properties along the shaft and the occurrence of discontinuities in angular velocity.

forces, because of their common orientation, are primarily generated along the helix. This line of argument is underlined by the phenomenon of transverse synchronization between neighbouring spirochaetes and flagella which results from the interaction of the transverse forces on gyrating sections of the flagellar shaft (section V.A.2).

What are the consequences of ciliary polarization on the generation of mechanical forces? In a cilium water is no longer transported along the axis of the (virtually helical) cilium but in a direction determined by the power stroke. The power stroke is a period of increased angular velocity, depressed beating envelope and straightened form of the shaft occurring when the cilium moves toward the cell surface (compare path of idealized cilium in Fig. 46 from (d′) to (d)). These properties enhance the periodic directional output of the cilium which is already given without polarization. The recovery stroke, on the other hand, consists of bending stages of high curvature occurring at reduced angular velocity close to the cell surface. This low dynamic "profile" reduces forces of viscous resistance during the return to an upright position in the unpolarized cilium (Fig. 46, d→d′). Thus, polarized ciliary beating greatly amplifies effects which are already given through the reduction in length of the shaft: water is transported in the direction of discontinuous transverse forces and no longer parallel to the helical shaft. If the direction of generated water streams is assumed to be an indicator of "beat direction",* polarization of the cyclic oscillation, combined with a reduction of the length of the shaft, modifies "indirect" beating by means of a screw toward a "direct" type of beating, where propulsion occurs parallel to the leading forces of viscous interaction. This change in orientation of locomotory forces with respect to the orientation of forces relevant for viscomechanical interaction may be a decisive factor in the production of dexioplectic metachronism in the ciliates.

3. *Spatial distribution of generated forces and orientation of the metachronal field*

In a polymastigote flagellate the metachronal waves travel along a line which is parallel to the orientation of the helical flagellar axis and thus parallel to the beat direction. Synchronization between the flagella occurs at right angles to the direction of beating (Fig. 47a). In terms of coupling of the cyclic movement between neighbouring flagella, phase-shift is maximal in the mainline of metachrony and minimal (zero-shift) in the line of synchrony (compare Fig. 1). It becomes evident from Fig. 47 that

* This generalized term is useful to determine types of metachronism, irrespective of the different mechanism of "beating" in flagella and cilia.

synchronization occurs in a plane where one would expect maximal viscous interaction, and that the waves move in a plane where the forces of drag and pressure are expected to be minimized.

In a ciliate like *Paramecium*, on the other hand, synchronization occurs parallel to the power stroke and waves are propagated at right angles to it (Fig. 47b). Again, coupling between the cilia is maximal parallel to

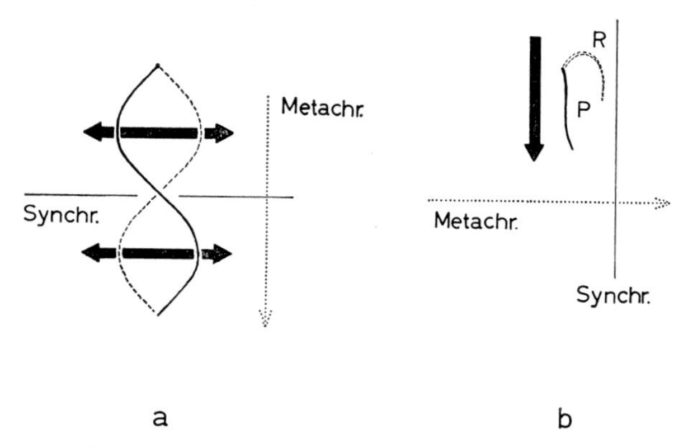

FIG. 47. Relation of the metachronal field to forces of viscous interaction generated by beating flagella (a) and cilia (b).

a. Flagellum in surface view beating downwards. Forces of viscous drag and pressure are greatest transversely to the helical axis (heavy arrows) though continuous forces and hence transport of water occur parallel to the helical axis. The accompanying metachronal coordination is such that strongest coupling between the flagella (synchr., synchronism) is found parallel to the leading viscomechanical forces. The metachronal wave (metachr.) travels at right angles to these forces.

b. Cilium in surface view beating downwards. The principal viscous forces are oriented parallel to the power stroke (P) which generates a discontinuous transport of water downwards. The recovery stroke (R), because of its properties in time and space, is less important for the production of mechanical forces. As in (a) the line of synchronization of the metachronal wave system is oriented parallel to the leading forces of viscous interaction; wave transmission occurs at right angles to these forces towards the side to which the cilium moves in the recovery stroke.

the dominant viscous forces generated during the power stroke, and coupling is minimal at right angles to this. This was tested on an empirical basis by reconstructing envelopes of ciliary beating in *Paramecium* (Machemer, 1972c). Seven stages of the model cilia of a normal dexioplectic pattern (Fig. 48a) were marked with points of reference along the shaft and tracks of these reference points were projected parallel and perpendicular to the cell surface (Fig. 49a–c). Ciliary excursions were maximal in a plane parallel to the line of synchronization and minimal parallel to the mainline of metachrony. In consideration of additional effects of increased angular velocity during the power stroke one must conclude that forces of viscous interaction are maximal parallel to synchronization. Increase in viscosity produces a more helical form of the ciliary beat (Fig. 48b). The tracks of reference points during the cycle assume the form of a somewhat elliptic paraboloid, the main transverse axis of which deviates from the line of synchronization by an acute angle (Fig. 49d–f). The helical ciliary axis is tilted against the cell surface in such a way as to point towards the direction of the metachronal wave.

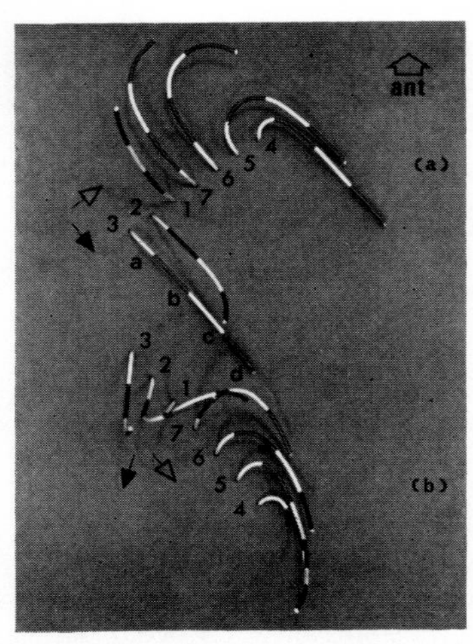

FIG. 48. Sequential stages of ciliary beating in *Paramecium* represented by model cilia. a. Dexioplectic coordination at normal viscosity (compare Fig. 6). b. Dexiosymplexy at high viscosity (compare Fig. 29). Numbers 1–3 show stages of the power stroke, numbers 3–7 stages of the recovery stroke. Black arrow, direction of power stroke; hollow arrow, wave direction. Note the increase in counterclockwise helical winding at high viscosity, associated with a transformation of metachronal coordination. From Machemer (1972c).

Thus, the observed orientation of the metachronal field is somewhat intermediate between the case of prevailing ciliary polarization (Fig. 47b) and prevailing helical beating (Fig. 47a). This result is expected, since mechanical vectors of force generated by the power stroke as well as by the transverse dislocations of the ciliary helix should contribute to the final orientation of ciliary synchronization in the metachronal field.

The coincidence of the orientations of leading vectors of viscous interaction with ciliary synchronization and of minimized interaction with the metachronal wave is seen in different types of metachronal coordination: symplexy, dexioplexy and dexiosymplexy. It may, therefore, be a causative link in the generation of metachronism on the basis of mechanical interference.

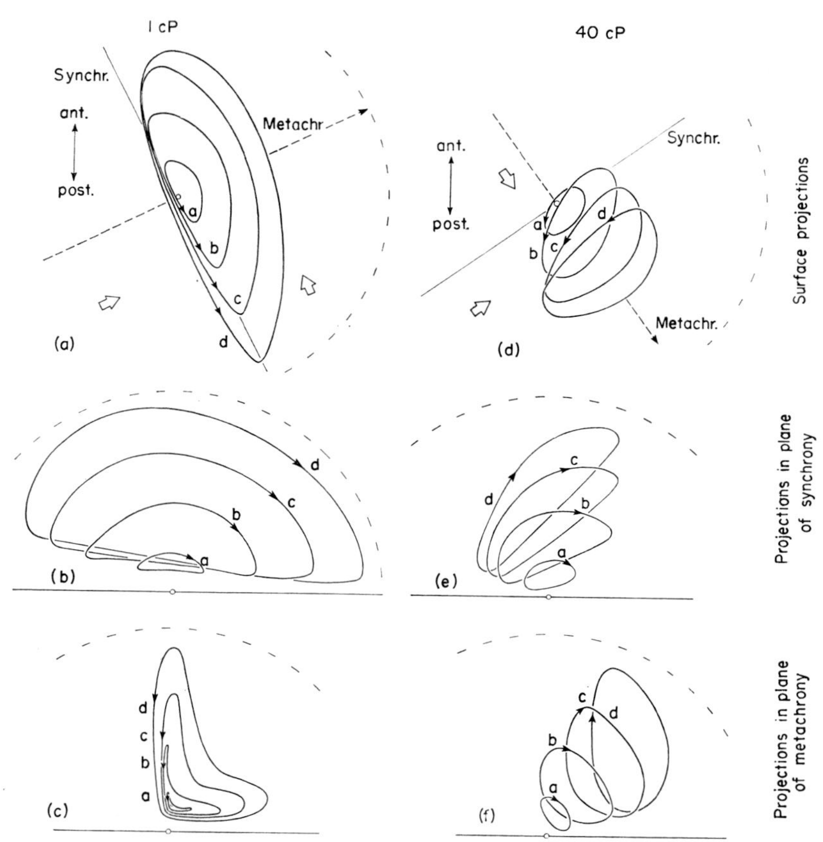

FIG. 49. Envelopes of beating cilia in *Paramecium* at normal viscosity (a–c) and at high viscosity (d–f) and effects on the orientation of the metachronal field. The envelopes are represented by the tracks of 4 reference points along the shaft of model cilia (Fig. 48) seen in surface projection (a, d), in side views parallel to ciliary synchronization (b, e) and to the mainline of metachrony (c, f). synchr., line of synchronization; metachr., mainline of metachrony. The arrowheads in the envelopes mark the position of the cilium during the power stroke (stage 2 in Fig. 48). The dashed arc around the point of ciliary insertion indicates the ciliary length (10 μm). From Machemer (1972c).

4. *Direction of ciliary gyration and determination of the wave type*

Considerations concerning the orientation of the metachronal field through the spatial distribution of viscous coupling forces do not directly explain the definite orientation of the wave. In *Paramecium*, for instance, synchronization along the direction of the power stroke might lead, as it appears, to laeoplectic as well as to dexioplectic metachronism. Which factors determine the dexioplectic orientation of the field, i.e. wave transmission to the left-hand side of the orientation of the power stroke? Answers to this question may be given in two ways: (1) Empirically, it

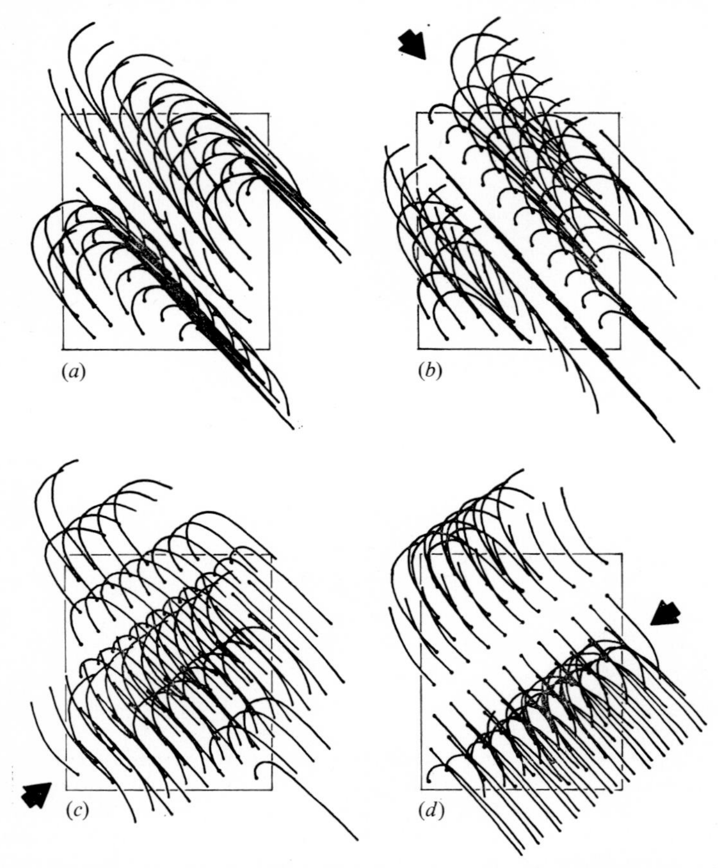

FIG. 50. Consequences of counterclockwise gyrations of the cilia on metachronism in *Paramecium*. Seven forms of model cilia representing stages of the beat cycle seen in surface view (Fig. 48), are arranged in such a way as to depict dexioplexy (a), laeoplexy (b), symplexy (c) and antiplexy (d). Arrows indicate zones of mechanical obstruction between cilia in the hypothetical types b–d. No obstruction occurs in case (a) which corresponds to the observed pattern. A laeoplectic mirror-image of (a), including a clockwise gyration of the cilia, would also provide unimpeded beating of the cilia. From Machemer (1972c).

was shown that, given the stages of sequential ciliary bending, only dexioplectic metachronism allows mechanically unimpeded beating of the cilia (Machemer, 1972c; Fig. 50a). All of the other three basic types of coordination including laeoplexy (Fig. 50b) require arrangements of bending stages which are mechanically impossible or lead to a definite obstruction between the movements of neighbouring cilia. Thus, dexioplexy appears to be a type of coordination offering the least mechanical interference in the ciliary field. (2) A second approach to an answer starts from the fact that cilia would not be obstructed in a type of coordination representing the mirror-image of dexioplexy. As a matter of fact, this type would be laeoplectic, but the sense of ciliary gyration would be *clockwise* instead of *counterclockwise*. Coupling between the 2 ciliary parameters, sense of gyration and type of diaplectic metachronism is well established in the Protozoa. All protozoan cilia that have been adequately studied gyrate in the counterclockwise direction and show dexioplexy. In support of this line of argument, it was recently shown in the lateral cilia of the lamellibranch *Mytilus* that the laeoplectic coordination of ciliary activity is associated with a clockwise gyration of the cilia (Aiello and Sleigh, 1972). Thus metachronal coordination in *Mytilus* as well as in the ciliate Protozoa has the property that waves travel towards the "recovery side" of the envelope of gyration (compare Fig. 49a). It appears reasonable to assume that the mechanism of this relation should be basically similar to the servo-regulation of minimum interference controlling synchronization. An important factor may be the reduced angular velocity of ciliary arcs in the "recovery sector" of the cyclic gyration. Transfer of identical stages of beating under maximal phase-shift (= metachronal wave) will produce a minimum of possible mechanical obstruction if occurring toward the recovery sector of the cycle where ciliary arcs continually overlap each other. The automatic selection of "possible" and "little interfering" phase relations between neighbouring cilia during metachronal organization of the field will, therefore, lead to wave transmission toward the recovery sector of individual cycles.

5. *Mechanical hypotheses on the origin of metachronism*

Experimental data on metachronism in *Paramecium* and considerations presented in the preceding paragraphs were recently summarized in a hypothesis (Machemer, 1972c), the postulates of which are as follows:

1. Metachronal coordination will arise in a population of oscillators if the viscomechanical forces generated by the oscillating units assume a predominant orientation across the field.

2. That type of metachronism will be realized by servo-mechanism which provides the least obstructed operation of the individual oscillators.

3. The strongest vectors of mechanical coupling across the field have the highest probability of synchronizing oscillatory movement.

4. The metachronal wavelength expresses the difference in the amounts of mechanical coupling parallel and perpendicular to the line of synchronization. Wavelength increases (i.e. tends toward synchronization) as forces of coupling along these coordinates become more similar; wavelength decreases with increase in the difference between these forces.

This hypothesis incorporates the concept by Gray (1930) that continuous mechanical interference is minimized in metachronal coordination of the cilia. Contrary to Gray's opinion, however, forces developed by the ciliary power stroke (or by any activity of a beating organelle) are assumed to affect metachronism specifically. Earlier hypotheses based on the interference principle considered possible events occurring along the mainline of metachrony. Jahn (1964) and Jahn and Bovee (1967) postulated that a delay in the transmission of mechanical forces through the medium generates the metachronal wave. Sleigh (1966) assumed that strong coupling should result from close aggregations of the cilia and lead to antiplexy or symplexy, depending on the occurrence of these aggregations during the recovery stroke or the power stroke. The concepts by Jahn, Bovee and Sleigh regard effects of the ciliary power stroke on the generation of the metachronal wave without sufficiently explaining, however, the simultaneous occurrence of synchronization and metachrony.

According to Murakami (1963) a metachronal system may be set up in a ciliary field by two possible mechanisms: either by continuous visco-mechanical control of certain phase-relations or by brief mechanical stimuli triggering the response of a neighbouring cilium. In the Protozoa cilia exhibit forms of graded activity. Autonomous beating of isolated protozoan cilia was repeatedly demonstrated (Alverdes, 1922; Párducz, 1954). In *Paramecium* the preoral suture of kineties marks a continuous break in orientation and length of waves (Figs 15, 24), a condition which can hardly be reconciled with a triggering concept. These observations suggest that mechanical triggering cannot be a major factor in establishing this metachronism. Complications also would arise to explain how simultaneous triggering occurs parallel to the wave fronts but not along the mainline of metachrony. Essentially, the whole organization of a metachronal field would have to be computed within each single cilium, if metachronism occurred exclusively on a mechanical triggering basis.

Ciliary mechanosensitivity, on the other hand, has been shown to occur in the Metazoa (Murakami, 1963; Thurm, 1968) and may also contribute to the phase relations between some ciliary organelles in the Protozoa. The marginal cirri of *Stylonychia* often show transient grouping of cirri with different activity (section IV.E.4), an observation that may be understood in terms of locally modified values of a threshold response to mechanical stimulation by neighbouring cirri. Modification or inhibition of the ciliary activity, however, must not necessarily occur in conjunction with a process of sensory transduction triggering the ciliary cycle.

Metachronism of spirochaetes in *Mixotricha* appears to be an excellent example in favour of a concept of continuous mechanical interference, since the spirochaetes are extracellularly attached to the host cell and thus have to "communicate" to each other through the fluid medium. In models of *Paramecium* treated with the detergent Triton X-100 Naitoh and Kaneko (1973) found metrachronal coordination of the reactivated cilia corresponding to that of normal cells. Waves were seen to accompany polarized ciliary beating even at very low frequency, but, after Triton exposure, waves were missing in cells showing high frequency with irregular bending of the cilia. These data may well be understood in terms of the concept of mechanical interference as outlined above.

6. *The "metachronal impulse"*

Since the days when metachronism was believed to be transmitted along intracellular fibre systems the rather nebulous concept of the "metachronal impulse" (Worley, 1934) has been a vehicle of ideas of a neuroid transmission of metachronal waves. Párducz (1954) discriminated between an autonomous "constant apolar rotation of the cilium" and a "wave of excitation of endogenous origin passing at regular intervals continuously along the periphery of the cell" (wording by Párducz, 1967). It is to the credit of Párducz that he has convincingly demonstrated that the direction of the metachronal transmission is independent of cortical structures and may actually proceed in any direction. The neuroid nature of the metachronal impulse was indirectly supported by observations such as the "unnatural" anteriad propagation of the waves (Párducz, 1954) and that "coordinating waves pass invisibly also through areas without cilia, as shown . . . in *Paramecium* partly denuded with chloroform" or "in *Didinium*, although their [the waves'] effects become visible only in the ciliary girdles" (Párducz, 1967). These points in support of the neuroid concept may equally be understood in terms of mechanical ciliary interference leading to a dexioplectic metachronal system. Partial deciliation of *Paramecium* should be a crucial test of the neuroid hypothesis, but waves fixed by osmium tetroxide are not appropriate evidence because some cilia may be lost in the procedure following fixation. Eckert and Naitoh (1970) gave electrophysiological evidence that electric pulses spread along *Paramecium* at a velocity (>100 mm s^{-1}) which is more than 100 times faster than metachronal waves (0·35 mm s^{-1} under normal conditions). "Thus, electrical activity conducted by the plasma membrane cannot account for the metachronism of ciliary beat."

C. *What factors determine metachronal characteristics?*

A metachronal system may undergo changes in the direction of the waves (with and without relation to beat direction), in wavelength and wave velocity. The direction of metachronal waves probably results from a spatial pattern of mechanical forces generated by the oscillating units and is thus a question of ciliary control. Other parameters such as length and velocity of the wave have a complicated relation to ciliary activity. Some investigators even tend to regard wave velocity as independent of mechanical circumstances of ciliary motion. Available information, however, supports a mechanical concept of metachronal regulation by focusing attention on a little understood parameter: ciliary polarization.

1. *Effects of ciliary polarization on the wavelength*

Ciliary polarization is a complex modification in time and space of harmonious helical beating (definition: section V.B.2) which is obviously aimed at the production of uni-directional transverse forces by short organelles. Polarization can be measured, so far, only incompletely by plotting planar aspects of the amplitude against time (Fig. 37). In fields of cilia the measurement of properties of polarization meets unsurmountable difficulties. Therefore, conclusions with respect to ciliary polarization are largely indirect in nature. Sleigh (1956, 1961) observed an increase in the wavelength of *Stentor* membranelles when the external viscosity was increased. In *Paramecium*, the viscosity-induced increase in wavelength is accompanied by specific changes in the form of ciliary beating (Machemer, 1972c). Reconstructions of bending stages at 1 cP and 40 cP (Fig. 48) showed a clear reduction in spatial aspects of ciliary polarization. Support for the assumption that an increase in ciliary polarization reduces wavelength (Machemer, 1972c) comes from the observation that the wavelength decreases in those cases where polarization is likely to increase:

1. Metachronal waves tend to be shorter at the anterior end of *Paramecium* in comparison with the middle part of the cell. At the same time frequency tends to increase anteriorly (Machemer, 1972a–c).

2. Metachronal waves are shorter in the oral groove (paired cilia) if compared to waves on the main body cortex (single cilia) (Párducz, 1955; Machemer, 1972a, b).

3. With increasing temperature the length of metachronal waves is reduced in *Stentor* (Sleigh, 1956) and in *Paramecium* (Machemer, 1972b). Cilia of cooled *Paramecium* show an increase in the proportion of the beating cycle spent in the power stroke.

4. Wavelength in backward swimming paramecia at normal and increased viscosity tends to be reduced in comparison to that in forward swimming specimens (Párducz, 1956a; Machemer, 1969a, 1972c). Backward and forward swimming at 40 cP occurred at the same ciliary frequency (14·1 Hz; Machemer, 1972c).

The increase in frequency itself, often accompanying a decrease in wavelength, does not provide an explanation for changing spatial relations of viscous coupling between the cilia. Moreover, there is no evidence that in *Paramecium* differences in frequency between body cilia and cilia of the

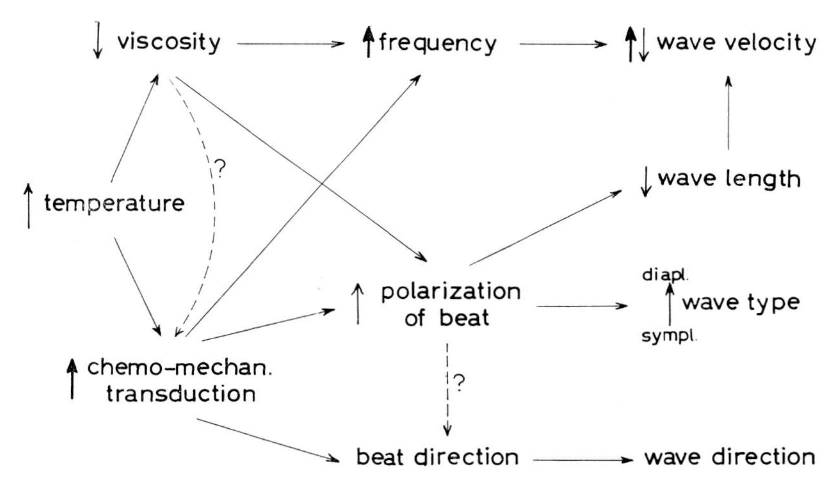

FIG. 51. A preliminary attempt to describe effects of physical factors (temperature, viscosity) and physiological factors (rate of chemo-mechanical transduction) on ciliary parameters (frequency, direction and polarization of beat) and properties of metachronism (direction, type, length and velocity of the wave). ↑ (↓), increase (decrease) of a factor; sympl.→diapl., tendency of transformation of the wave type from symplectic to diaplectic metachrony; heavy arrow indicates strong modification of a factor. Compare with the text.

peristome are such as to account for an observed 30% difference between wavelength on the body and the peristome at low temperature (Manchemer, 1972b). For these reasons ciliary polarization is the best candidate to assume responsibility for the size of coupling forces parallel to and normal to ciliary synchronization (point 4 of hypothesis in section V.B.5).

Possible lines of action of certain external and internal factors on the metachronal wavelength are illustrated in Fig. 51. The ciliary machinery converting chemical energy into mechanical work is assumed to influence simultaneously the frequency and polarization of beat which, in turn, are modified also by external viscosity. Pairing of the cilia would reduce the viscous interference, but not the gross energy output per cilium, and thus provides an alteration of ciliary polarization that is independent of the frequency. The relations outlined in Fig. 51 may to a certain degree also apply to metachronism in flagellates. Longer flagella tend to increase the wavelength, since longer shafts increase net forces along the helical axis with respect to the transverse forces (if other factors are kept constant; compare Fig. 46). The frequently observed antero-posteriad increase in wavelength in *Opalina* may be explained on a structural basis, but is probably also due to a decrease in flagellar polarization occurring along with decreasing frequency in the posterior direction. The tendency of the flagella of *Opalina* to show temporo-spatial polarized beating (section IV.B.4) is illustrated by comparison of waves in *Opalina* (Fig. 13c) with similar waves in *Paramecium* at 40 cP (Fig. 30).

2. *Determination of wave velocity*

Each metachronally organized cilium will return to the same stage of bending after the interval required to perform one cycle. This interval is identical with the time the metachronal wave takes to travel one wavelength. If, at constant frequency, the conditions of viscous coupling are such as to provide long waves, the rate of wave transmission or wave velocity (v) will be increased in comparison with conditions with smaller waves. Thus, wave velocity is determined by two variables: the frequency (f) and wavelength (λ) according to the equation

$$v = \lambda f.$$

Wave velocity decreases whenever the rate of decrease of one of the two variables exceeds the rate of increase of the other. At low temperature, for instance, wave velocity is reduced in *Stentor* and *Paramecium*, since the decrease in frequency is greater than the increase in wavelength (Sleigh, 1956; Machemer, 1972b). Increasing viscosity reduces frequency and increases wavelength at rates which may differ in different ciliary structures. In the membranelles of *Stentor* the rates of modification of the values of f and λ were similar between 1 cP and 3·6 cP, thus resulting in a constant v (Sleigh, 1956). At higher viscosities wave velocity increased through an increasing rate of growth of λ (Sleigh, 1961).* Wave velocity in *Paramecium* is reduced between 1 cP and 40 cP because of the sharp decrease in beat frequency (-57%) in comparison with changes in wavelength ($+25\%$; Machemer, 1972c).

The observation by Sleigh (1956) that wave velocity tends to be constant at low viscosities is corroborated by findings at normal viscosity in the

* A similar effect is reported by Gosselin (1958) who observed in cilia of mussel gills a viscosity-induced increase in length and velocity of the waves at constant frequency.

lateral cilia of *Mytilus* (Gray, 1930). The apparent independence of the rate of wave propagation from physical factors of ciliary beating is the result of a coincidental reciprocity of relations between wavelength and frequency. Figure 51 illustrates the ambiguous role of wave velocity resulting from combined influences of external and internal factors on the frequency and polarization of ciliary beating. Relative stability of wave velocity results from a 2-fold effect of changes in the rate of energy-transduction on ciliary frequency and polarization. A similar compensatory effect may be produced by changes in viscosity or temperature; while an increase in frequency tends to increase wave velocity, an increase in polarization tends to decrease it. As a consequence, in mechanically coordinated metachronism, wave velocity cannot be regarded as an appropriate indicator of ciliary activity, even if external physical conditions are kept constant. In addition wave velocity is no indicator of the propagation by mechanical triggering, since effects of viscous interference will be inseparably incorporated in an observed rate of dislocation of the metachronal wave. In summary, all properties of metachronism of the type discussed here appear to be under the direct control of parameters of ciliary beating which, in turn, are modified by internal *and* external factors. The most direct line of action occurs between chemo-mechanical transduction and wave direction, though external viscosity influences the direction of beat by a still undetermined pathway (Machemer, 1972c).

3. Stentor *membranelles*

Data on metachronism in the membranelles of *Stentor* led to the excitation-transmission theory (Sleigh, 1956, 1957), according to which beating of the compound cilia is coordinated by neuroid impulses which are relayed in each cilium by a process of ciliary excitation (compare section IV.D). Supported by indirect experimental evidence, this has been one of the first concepts rationalizing the origin and regulation of a specific case of protozoan metachronism. Because of the assumed stepwise nature of wave propagation this theory might be referred to as of the "neuroid trigger" type though "the function of the conducted 'impulse' is to maintain the phase relationship . . . rather than trigger the beat" and "neuroid interference" might therefore be a better description (Sleigh, 1969).

Metachronal coordination in the membranellar band of another spirotrich ciliate, *Stylonychia*, may be explained on the basis of principles of mechanical interference, but available evidence is also indirect (section IV.E.2). All data on membranellar metachronism in *Stentor* would, alternatively, fit a concept of mechanical interference, if wave velocity is understood as stabilized through reciprocal changes in frequency and wavelength (see preceding paragraph) and the constancy in the number of membranelles per wave is considered as a coincidental result of increased polarization in the smaller and closer spaced membranelles of the gullet: reduced wavelength through increased polarization of the shorter cilia (sections V.B.1 and V.C.1) would meet reduced inter-membranellar spacing. Restrictions in the assessment of metachronal parameters along membranellar bands and the complicated function of the membranellar organelles will be further obstacles in the way towards definite understanding of membranellar metachronism.

4. *Ciliary integration as revealed in wave patterns*

In *Opalina* and many ciliate Protozoa the patterns of metachronal waves undergo local modifications indicating graded alterations in the properties of beating of single organelles. Changes in the orientation of the waves signal corresponding changes in beat direction; modification of

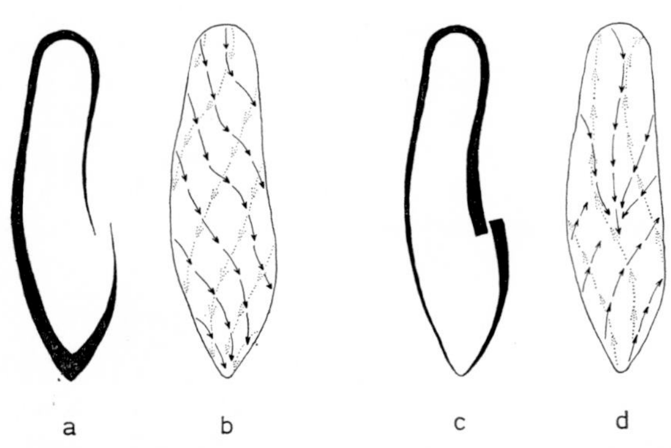

FIG. 52. Regulation of the orientation of metachronal waves and the ciliary power stroke in *Paramecium* as postulated by the hypothesis of a stomato-caudal gradient of excitability. a. Gradient in normal-swimming orientation decreasing in the caudo-stomatal direction. b. Consequences of the normal orientation of the gradient as seen in ventral view: waves and power stroke are oriented by the caudal pole of the gradient. c. Reversed orientation of gradient during backward swimming. d. Consequences of reversal of the gradient as seen in ventral view: determination of waves and power stroke by the stomatal pole of the gradient. Redrawn from Grębecki (1965). Compare with wave patterns in Fig. 15.

the wavelength may indicate changes in frequency and/or polarization of the beat. The problem of the overall integration of ciliary activity has two major aspects: (1) What is the principle of the topographic organization of specific wave patterns? (2) What is the nature of the organizing mechanism? In recent years attempts have been made to answer the first question largely to the neglect of the second. Earlier ideas of an antero-posterior gradient of sensitivity regulating ciliary activity in response to electric and other stimuli (Kinosita, 1936; Kitching, 1961; Seravin, 1962; Doroszewski, 1963) were modified by Grębecki (1965) with respect to regular asymmetries in the ventral pattern of metachronal waves during forward and backward swimming of *Paramecium*. According to the stomato-caudal hypothesis of Grębecki, the synchronized wave fronts follow the slope of a neuroid "gradient of excitability" which extends from the posterior towards the anterior end of the cell and then downward towards the cell mouth (Fig. 52a, c). The ciliary power stroke is assumed to point towards

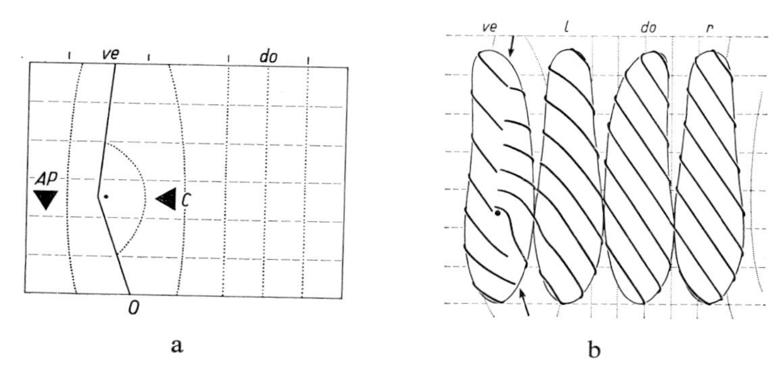

FIG. 53. Regulations of the orientation of metachronal waves in *Paramecium* as postulated by the 2-gradient hypothesis. a. Projection of the cell surface on a cylindrical envelope spread into one plane. Fieldlines (dashed) of the antero-posterior gradient (AP) extend between the anterior and posterior ends. Fieldlines (dotted) of the cytostomal gradient (C) spread clockwise following the orientation of the kineties; spreading is limited on one side by the oral suture (O). b. The same projection as (a) with resulting fieldlines connecting points of identical fieldstrength after addition of the AP-field and the C-field. Outlines of the ventral (ve), left (l), dorsal (do) and right view (r) of *Paramecium* are included to assist comparison with wave patterns of forward locomotion (compare Fig. 15a, b). Arrows show the course of the suture of kineties. From Machemer (1969c).

the caudal pole of the gradient along its slope (then generating forward locomotion) or towards the buccal pole of the gradient (then causing backward locomotion). This elegant concept of a bipolar and reversible gradient regulating metachronism and beat direction is not confirmed in its topographical aspects by patterns of metachronal waves in *Paramecium* (Fig. 52b, d; compare with Fig. 15). One consequence of the hypothesis, the inability of the peristomal cilia to beat in the anterior direction, does not correspond with observations of fixed and live paramecia (Machemer, 1969a; unpublished film data). As an alternative to the stomato-caudal hypothesis Machemer (1969c) postulated two gradients on the membrane of ciliates regulating metachronism: an antero-posterior gradient and a transverse cytostomal gradient which occurs in one direction with respect to the suture-line of kineties (Fig. 53a). Fieldlines of the gradients extending transversely between the anterior and posterior poles and spreading clockwise* from the peristome towards the somatic cortex were assumed to add to one another in such a way that an oblique pattern of fieldlines would result similar to the orientation of the metachronal waves (Fig. 53b). Alteration of the gradients relative to each other would produce the systems of waves observed during certain modes of stimulation. Though the 2-gradient hypothesis was founded on a neuroid basis, its concept may also be applied to the orientation of the ciliary power stroke. Francis (1969) observed that in *Opalina* the lines connecting points of equal sensitivity to electric stimulation (Okajima, 1953) or points of equal membrane resistance (Naitoh, 1958) correspond to the isopotential lines on a conducting paper model of *Opalina*, if the electric poles are located in the anterior-right and posterior sutures of the dorsal and ventral kineties. The nature of the relationship of the arrangement of kineties and their suture-lines to the activity of the cilia and flagella is still out of sight. Certainly gradients of ciliary activity are not related to local shifts in membrane potential; isopotentiality of ciliate cells was demonstrated experimentally (Eckert and Naitoh, 1970). It is a general postulate of different studies on gradients of ciliary activity that modifications in the function of the cell membrane and related ciliary structures have a field-like distribution. This postulate may eventually be re-interpreted on the basis of passive or active ionic transport across the cell boundary (see Naitoh and Eckert, Chapter 12, this volume).

VI. Metachronism and propulsion

Both flagella and cilia generate forces of propulsion, but the mechanism by which these forces are transmitted to the fluid medium is not well understood. In this early state of investigation of the hydrodynamics of metachronally organized oscillators, reviewing comments are premature. Three lines of research have been developed or may be anticipated:

1. Propulsion occurs as a direct consequence of directional forces generated by the individual oscillating units. Metachronism is not a factor in propulsion besides averaging mechanical parameters of single oscillators.

2. Mechanical forces resulting from the movement of flagella or cilia are mediated to the fluid environment through properties of the metachronal system. The undulating envelope of the metachronal waves provides a boundary which exerts forces on the water. The rate of propulsion is determined by parameters of the metachronal wave system.

* If seen from the posterior end.

3. Similar assumption as (1), but forces produced by individual organelles mechanically interfere with forces generated by the undulating envelope of metachronally organized oscillators.

Data summarized above support a view of ciliary propulsion similar to assumption (1). For instance, the sense of winding of the swimming cell appears to result directly, in all known cases, from the orientation of the ciliary power stroke, irrespective of the properties of the accompanying wave system (compare Fig. 27). Observations on the relations between ciliary activity and metachronism, summarized in Fig. 51, led to the conclusion that metachronal waves provide only limited information on the work done by the cilia.

Seravin (1970) has drawn attention to the fact that the ciliary layer in Protozoa is surrounded by systems of small whirl-streams which wind in a direction opposite to the winding of the cell body. It remains to be determined whether perturbations of water streams close to the cilia are causative (Seravin's assumption) or circumstantial for propulsion, and if whirl-streams reduce or enhance the locomotory effect of the cilia. In studying the movement of water by protozoan and metazoan cilia Sleigh and Aiello (1972) suggest that an important effect of metachronism is to convert discontinuous motion of the cilia into steady propulsion of water. It should be considered, however, that in the fluid space close to the ciliary tips small particles perform progressive cycloid motions (Seravin, 1970), and similar motion is observed also during metachronally coordinated beating (Machemer, unpublished film data). A function of metachronism in terms of locomotion is assumed by Blake (1971a, b, c) in a mathematical approach to the mechanics of ciliary propulsion. This author treats the ciliary layer as a continuous and waving surface, which does not allow slippage of water across the boundary marked by the ciliary tips. The limitations of this "envelope model" in comparison with other concepts of propulsion dealing with effects of individual cilia on the fluid are discussed by Blake (1972). It appears that the understanding of metachronism in conjunction with ciliary activity will exert an important influence on further studies of ciliary propulsion.

VII. Conclusion

All flagellate and ciliate Protozoa exhibiting metachronal coordination show, as far as is known, specific relations between the physical parameters of the oscillating units and the wave system. Differences between symplectic metachrony seen in flagellates and dexioplexy prevailing in ciliates are not fundamental in nature, since forms of transition are produced in one cell at different viscosities. The relationship between forces generated by a flagellum or cilium and the resulting metachronism is so close that neuroid mechanisms of coordination are rendered superfluous. Among mechanical hypotheses on the generation of metachronism the concept of continuous mechanical interference is the most promising one, allowing detailed understanding of properties of an observed metachronal system. Possible involvement of mechanical triggering in establishing phase relations cannot be excluded. Temporal and local modification of metachronism may be understood as determined by membrane-controlled ciliary activity. From an evolutionary point of view the dexioplectic metachronism may have been developed from symplectic forms of coordination by progressive polarization of the counterclockwise flagellar beating accompanying reduction in length of the shaft.

References

AIELLO, E. and SLEIGH, M. A. (1972). *J. Cell Biol.* **54**, 493–506.
ALLEN, R. D. (1969). Communication at the Third Int. Congr. Protozool., Leningrad 1969.
ALVERDES, F. (1922). *Arb. Gebiet exp. Biol.* **3**, 1–127.
BILLS, C. E. (1922). *Biol. Bull.* **42**, 7–13.
BLAKE, J. R. (1971a). *J. Fluid Mech.* **46**, 199–208.
BLAKE, J. R. (1971b). *J. Fluid Mech.* **49**, 209–222.
BLAKE, J. R. (1971c). *Bull. Austr. math. Soc.* **5**, 255–264.
BLAKE, J. R. (1972). *J. Fluid Mech.* **55**, 1–23.
BOGGS, N., JAHN, T. L. and FONSECA, J. R. (1970). *J. Protozool.* **17** (Suppl.), Abstr. 16.
CHEN, V. K. (1972). Dissertation, Univ. New York, Buffalo.
CLEVELAND, L. R. and CLEVELAND, B. T. (1966). *Arch. Protistenk.* **109**, 39–63.
CLEVELAND, L. R. and GRIMSTONE, A. V. (1964). *Proc. R. Soc.* B **159**, 668–685.
DE PUYTORAC, P., ANDRIVON, C. and SERRE, F. (1962). *J. Protozool.* **9** (Suppl.), Abstr. 23.
DOROSZEWSKI, M. (1963). *Acta Protozool.* **1**, 313–319.
ECKERT, R. and NAITOH, Y. (1970). *J. gen. Physiol.* **55**, 467–483.
ENGELMANN, T. W. (1868). *Jena Z. Naturw.* **4**, 321–478.
FRANCIS, D. (1969). *J. Protozool.* **16**, 498–500.
GOSSELIN, R. E. (1958). *Fed. Proc.* **17**, 372.
GRAY, J. (1928). "Ciliary Movement." Cambridge University Press.
GRAY, J. (1930). *Proc. R. Soc.* B **107**, 313–332.
GRĘBECKI, A. (1964). *Acta Protozool.* **2**, 375–377.
GRĘBECKI, A. (1965). *Acta Protozool.* **3**, 79–100.
GRĘBECKI, A., KUZNICKI, L. and MIKOLAJCZYK, E. (1967). *Acta Protozool.* **4**, 398–408.

Grębecki, A. and Mikolajczyk, E. (1968). *Acta Protozool.* **5**, 297–303.
Grell, K. G. (1973). Protozoology. Springer Verlag, Berlin.
Horridge, G. A. and Tamm, S. L. (1969). *Science, Wash.* **163**, 817–818.
Jahn, T. L. (1964). 8th Meeting Biophys. Soc., Abstr. FE 13.
Jahn, T. L. and Bovee, E. C. (1967). *In* "Research in Protozoology" (T.-T. Chen, ed.), Vol. 1, pp. 39–198. Pergamon Press, Oxford.
Jahn, T. L. and Bovee, E. C. (1968). *In* "Infectious Blood Diseases of Man and Animals" (D. Weinman and M. Ristic, eds), Vol. 1, pp. 393–436. Academic Press, New York.
Jahn, T. L. and Hendrix, E. M. (1969). *Rev. Soc. Mex. Hist. nat.* **30**, 103–131.
Jahn T. L. and Landman, M. D. (1965). *Trans. Am. microsc. Soc.* **84**, 395–406.
Jahn, T. L. and Votta, J. J. (1972). *A. Rev. Fluid Mech.* **4**, 93–116.
Jennings, H. S. (1899). *Am. Nat.* **33**, 373–389.
Jennings, H. S. (1906). "Behaviour of the Lower Organisms." Columbia University Press, New York.
Kinosita, H. (1936). *J. Fac. Sci. Tokyo Univ.*, Sect. IV. **4**, 189–194.
Kinosita, H. (1954). *J. Fac. Sci. Tokyo Univ.*, Sect. IV. **7**, 1–14.
Kinosita, H. and Murakami, A. (1967). *Physiol. Rev.* **47**, 53–82.
Kitching, J. A. (1961). *In* "The Cell and the Organism" (J. Ramsay and V. Wigglesworth, eds), pp. 60–78. Cambridge University Press.
Knight-Jones, E. W. (1954). *Q. Jl microsc. Sci.* **95**, 503–521.
Kuznicki, L., Jahn, T. L. and Fonseca, J. R. (1970). *J. Protozool.* **17**, 16–24.
Machemer, H. (1965a). *Arch. Protistenk.* **108**, 91-107.
Machemer, H. (1965b). *Arch. Protistenk.* **108**, 153–190.
Machemer, H. (1966). *Arch. Protistenk.* **109**, 257–277.
Machemer, H. (1969a). *J. Protozool.* **16**, 764–771.
Machemer, H. (1969b). *Z. vergl. Physiol.* **62**, 183–196.
Machemer, H. (1969c). *Arch. Protistenk.* **111**, 100–128.
Machemer, H. (1970a). *Acta Protozool.* **7**, 547–551.
Machemer, H. (1970b). *Naturwissenschaften* **57**, 398–399.
Machemer, H. (1972a). *Acta Protozool.* **11**, 295–300.
Machemer, H. (1972b). *J. Mechanochem. Cell Motility* **1**, 57–66.
Machemer, H. (1972c). *J. exp. Biol.* **57**, 239–259.
Machemer, H. (1974). (In preparation.)
Machemer, H. and Eckert, R. (1973). *J. gen. Physiol.* **61**, 572–587.
Machin, K. E. (1963). *Proc. R. Soc.* B **158**, 88–104.
Murakami, A. (1963). *J. Fac. Sci. Tokyo Univ.*, Sect. IV. **10**, 23–35.
Naitoh, Y. (1958). *Annotnes zool. jap.* **31**, 59–73.
Naitoh, Y. (1964). *Zool. Mag.* **73**, 207–212.
Naitoh, Y. and Kaneko, H. (1973). *J. exp. Biol.* **58**, 657–676.
Okajima, A. (1953). *Jap. J. Zool.* **11**, 87–100.
Okajima, A. (1954). *Annotnes zool. jap.* **27**, 40–45.
Okajima, A. and Kinosita, H. (1966). *Comp. Biochem. Physiol.* **19**, 115–131.
Párducz, B. (1952). *Ann. Hist. nat. Mus. natn. Hung.* **2**, 5–12.
Párducz, B. (1954). *Acta biol. Acad. Sci. Hung.* **5**, 169–212.
Párducz, B. (1955). *Ann. Hist. nat. Mus. natn. Hung.* **6**, 189–195.
Párducz, B. (1956a). *Acta biol. Acad. Sci. Hung.* **7**, 73–99.

Párducz, B. (1956b). *Ann. Hist. nat. Mus. natn. Hung.* **7**, 363–369.
Párducz, B. (1958). *Acta biol. Acad. Sci. Hung.* **8**, 219–251.
Párducz, B. (1959). *Ann. Hist. nat. Mus. natn. Hung.* **51**, 227–246.
Párducz, B. (1961). *Ann. Hist. nat. Mus. natn. Hung.* **53**, 267–280.
Párducz, B. (1963). *Acta biol. Acad. Sci. Hung.* **13**, 421–429.
Párducz, B. (1964). *Acta Protozool.* **2**, 367–374.
Párducz, B. (1967). *Int. Rev. Cytol.* **21**, 91–128.
Preston, J. T., Jahn, T. L. and Fonseca, J. R. (1970). *Am. Zool.* **10**, 505.
Pütter, A. (1904). *Z. allg. Physiol.* **3**, 406–455.
Satir, P. (1963). *J. Cell Biol.* **18**, 345–365.
Seravin, L. N. (1962). *Tsitologiya* **4**, 545–554.
Seravin, L. N. (1970). *Acta Protozool.* **7**, 313–323.
Seravin, L. N. (1971). *Adv. comp. Physiol. Biochem.* **4**, 37–111.
Sleigh, M. A. (1956). *J. exp. Biol.* **33**, 15–28.
Sleigh, M. A. (1957). *J. exp. Biol.* **34**, 106–115.
Sleigh, M. A. (1960). *J. exp. Biol.* **37**, 1–10.
Sleigh, M. A. (1961). *Nature, Lond.* **191**, 931–932.
Sleigh, M. A. (1962). "The Biology of Cilia and Flagella." Pergamon Press, New York.
Sleigh, M. A. (1966). *Symp. Soc. exp. Biol.* **20**, 11–31.
Sleigh, M. A. (1968). *Symp. Soc. exp. Biol.* **22**, 131–150.
Sleigh, M. A. (1969). *Int. Rev. Cytol.* **25**, 31–54.
Sleigh, M. A. and Aiello, E. (1972). *Acta protozool.* **11**, 265–277.
Sleigh, M. A. (1973). *Abstr. 4th Intern. Congr. Protozool., Univ. Clermont*, p. 387.
Tamm, S. L. (1970). Personal communication.
Tamm, S. L. and Horridge, G. A. (1970). *Proc. R. Soc.* B **175**, 219–233.
Thurm, U. (1968). *Zool. Anz.* Suppl. **31**, 96–105.
Uhlig, G. (1960). *Arch. Protistenk.* **105**, 1–109.
Valentin, A. (1842). *In* "Handwörterbuch der Physiologie" (R. Wagner, ed.), Vol. 1, pp. 484–516.
Verworn, M. (1889). "Psycho-physiologische Protistenstudien." Jena, 1–219.
Verworn, M. (1891). *Pflügers Arch. ges. Physiol.* **48**, 149–180.
Worley, L. G. (1934). *J. cell. comp. Physiol.* **5**, 53–72.

Chapter 11

Metachronism of cilia of Metazoa

MICHAEL A. SLEIGH

Department of Zoology, University of Bristol, England

I. Introduction

The phenomenon of metachronism and methods of studying it have been described fully by Machemer in the preceding chapter. The purpose of this chapter is to supplement that report by consideration of features of metachronism peculiar to Metazoa. These include the effect on the character of metachronal coordination of the presence of cell boundaries, and types of metachronism described in metazoans that are not known in protozoans, such as laeoplectic metachrony and metachrony by excitation of individual beats, as seen in ctenophores. The use of cilia to carry mucus for feeding or the removal of foreign materials is better developed in Metazoa than Protozoa, and the propulsion of mucus by cilia will be compared with the propulsion of water. Reviews concerning metachronism in Metazoa occur in recent articles by Kinosita and Murakami (1967) and Sleigh (1969).

The relationship between the direction of ciliary beating and the direction of movement of metachronal waves in ciliary systems of many Metazoa was described by Knight-Jones (1954), who also named the four principal forms of metachronism, as Machemer has reported (p. 202). Of these four patterns, symplectic metachronism is certainly the least common and there are probably no authentic reports of its occurrence in Metazoa, while the other orthoplectic pattern, antiplectic metachronism, has been widely reported in Metazoa, where the cilia concerned usually function in the transport of particles, normally trapped in mucus. Exceptionally, as in the comb plates of ctenophores, cilia that show antiplectic metachronism are used in the propulsion of water. The use of cilia for swimming or for the propulsion of large amounts of water in feeding or respiratory currents is usually associated with diaplectic metachronism, some organisms using dexioplectic metachronism and others laeoplectic metachronism.

Studies of the metachronism on the ciliary pads of *Mytilus* gill led Murakami (1963) to conclude that these cilia, which retained a constant direction of beating, displayed any of the four metachronal patterns, although metachronism usually appeared symplectic. The recording system used for these observations was a slit camera, with which, as Murakami points out, the records only portray the components of movement of cilia parallel to the slit. In his study of the metachronism of *Paramecium*, Machemer (1972; see p. 223, this volume) found it necessary to make models of the three-dimensional form of ciliary waves in order to understand the planar profile views of the waves—he was thereby able to explain how a small change of orientation could make the profile views of dexioplectic waves change from an apparent antiplectic type to an apparent symplectic type. In the absence of precise knowledge of the three-dimensional form and constancy of beat of cilia on the ciliary pads, the records of a slit camera are probably at least as difficult to interpret as the profile views of *Paramecium* were previously. Murakami clearly found some promising material, but we now know that it needs studying in three dimensions for the results to be meaningful.

II. Metachronism and cell boundaries

Ciliary systems in Metazoa commonly involve many cells and the question arises of whether the presence of cell boundaries influence in any way the activity of the component cilia of a functional group. In some examples of ciliary systems described by Sleigh (1969) it was noted that all of the cilia borne on a cell beat together, e.g. latero-frontal cilia of gills of bivalve molluscs and cilia of ctenophore comb plates, while in other cases the cilia borne on a single cell beat at different times in metachronal sequence. It is currently believed that in Protozoa the metachronal activities of cilia are not associated with any propagated membrane potential change, but depend on hydrodynamic interaction between the moving cilia, while there are additional mechanisms for control of rate and direction of beating, which are whole-cell responses that are associated with propagated changes of membrane potential (see the chapters by Machemer, p. 199 and by Naitoh and Eckert, p. 305). In metazoan systems, with the possible exception of those cases where all cilia of a cell beat together, it is also now assumed that metachronal coordination is hydrodynamic and not dependent upon electrical activity of the cells. Evidence for this assumption includes the observations that the lateral cells of *Mytilus* gill are about the same length as a metachronal wave, so that all stages of beat occur simultaneously on a single cell (Lucas, 1932), that metachronism was retained in lateral cilia of glycerol-extracted models of *Modiolus* gill filaments (Child, 1965) and that in some systems where a large number of cilia move together as a single compound structure, a compound organelle may contain some cilia from one cell and some from a second cell, while other cilia from those cells form parts of other compound cilia (Sleigh, unpublished). Metachronal waves normally pass smoothly over a ciliated epithelium, no discontinuity or irregularity being observed at cell boundaries, so that the number and orientation of cell boundaries within the epithelium is without influence on the metachronism; the presence of non-ciliated cells does, of course, disturb the metachronal pattern and, for example, the increase in proportion of non-ciliated cells in some respiratory ailments must limit metachronal integration over the epithelium.

The regulation of rate or direction of beating in such multicellular systems, under the direction of nerves and probably also hormones, presumably acts at the cell level and is assumed to affect all of the cilia on one cell simultaneously; such control is discussed by Aiello (p. 353, this volume). Each compound latero-frontal cilium of *Mytilus* contains all of the 30 or more cilia on a latero-frontal cell and all of these cilia normally move together as a unit; such synchronous beating could be controlled at the whole cell level, which is interesting because the manner of coordination of alternate latero-frontal cilia reported by Dral (1967), in which cilia of the first, third, fifth, etc., latero-frontal cells beat together and 180° out of phase with those of the second, fourth, sixth, etc., cells, would appear to demand a mechanism more complex than hydrodynamic metachronism. However, the activity of these latero-frontal cilia needs further investigation because photographs such as Fig. 1 clearly show a long sequence of these cilia forming part of a single metachronal wave. In ctenophore comb plates all of the cilia of many cells move together as a single unit; these giant compound cilia, which provide a special case for a number of reasons, will be considered in a separate section (p. 292).

III. Diaplectic metachronism in Metazoa

In his review on metachronal patterns, Knight-Jones (1954) described the distribution of dexioplectic and laeoplectic metachronism in the majority of metazoan groups. He concluded that it was normal for cilia

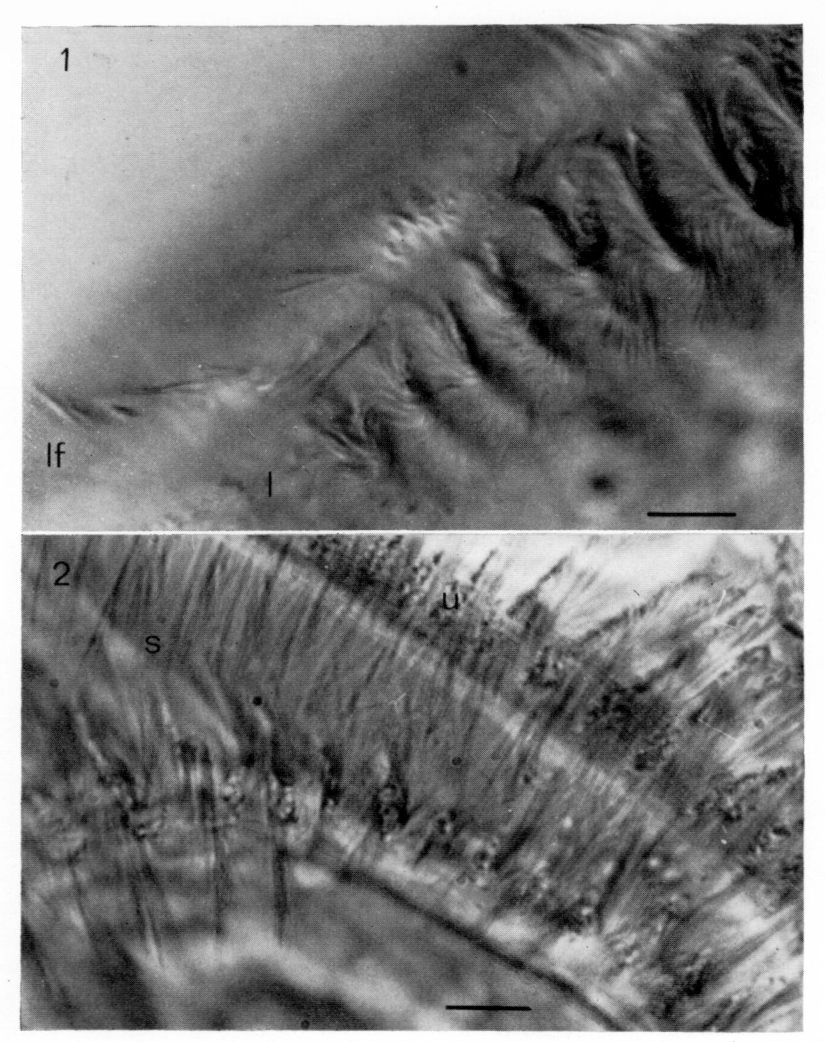

FIG. 1. Photomicrograph of part of the lateral surface of a gill filament of *Mytilus* showing long waves of coordination of latero-frontal cilia (lf), and a series of metachronal waves of lateral cilia (l) in which the plane of the photograph cuts cilia nearer to the cell surface at the right. Lateral cilia beat towards the observer and are turned to the observer's left during the recovery stroke; metachronal waves travel towards the left. (cf. Fig. 7. Scale = 10 μm. Photograph by Aiello and Sleigh (unpublished).

FIG. 2. Photomicrograph of two bands of gill cilia of *Corella*. Cilia of the upper band (u), are beating towards the observer and downwards towards the left, propelling water through the gill slit (s); waves are propagated towards the observer's right and in the recovery stroke the cilia are also turned towards the right. Cilia of the lower band are beating away from the observer at the opposite edge of the same gill slit. Scale = 10 μm.

with diaplectic metachronism to show the same pattern—either dexioplectic or laeoplectic—in all species of a taxonomic group. Thus the lateral gill cilia of molluscs always show laeoplexy, while the cilia around the stigmata (gill slits) of ascidians always show dexioplexy. Occasionally it is found that one group within a larger taxonomic category shows the opposite pattern, e.g. the velar cilia of the veliger larvae of eulamellibranch molluscs show dexioplexy, while the velar cilia of other molluscs show laeoplexy (metachronism of the lateral gill cilia of eulamellibranchs is laeoplectic as in other molluscs). One interesting case which Knight-Jones reports (and which has also been shown to me by Dr. R. Strathman) is the metachronism of the ciliated epaulettes of echinopluteus larvae, in which the cilia on one side of the body show laeoplexy and those of the other side show dexioplexy.

The functions of cilia with diaplectic metachronism seem to be served equally well by dexioplectic and laeoplectic patterns, but the patterns are consistent; it is, therefore, pertinent to ask what it is that determines which pattern occurs in a particular case. Knight-Jones concluded that no feature of gross anatomy was involved, but that the micro-anatomy of the ciliated cells or "some stereochemical peculiarity of the ciliary basal apparatus" could be the source of determination of metachronal pattern. It now seems likely that the form of metachronism is determined within the cilia themselves by internal characteristics, as yet unidentified, which determine the three-dimensional configuration of the ciliary beat. Machemer (p. 205, this volume) has carefully described the movement in three dimensions of cilia of *Paramecium* and has related the occurrence of dexioplexy to the fact that in the recovery stroke the cilium moves back to the left of the line it followed in the effective stroke; indeed in his Fig. 50 Machemer demonstrates that a cilium performing this type of beat could not participate in any of the other three metachronal patterns without obstruction of some cilia by others. The lateral cilia of the gill of *Mytilus* show a three-dimensional beat in which the cilia move back in the recovery stroke to the right of the line followed in the effective stroke (Fig. 1), and this pattern of beating has been related to the occurrence of laeoplectic metachronism by Aiello and Sleigh (1972). Cilia in such diaplectic systems apparently do not show a planar beat, and it is concluded that where the tip of the cilium (when seen by an observer looking down on the epithelium) traces out an anticlockwise loop the metachronism is dexioplectic, while in those cases where the ciliary tip traces out a clockwise loop the metachronism is laeoplectic. Preliminary observations on the dexioplectic metachronism of the ascidian *Corella* (Fig. 2) and on the laeoplectic metachronism of velar cilia of *Archidoris* veliger larvae support this conclusion.

The characteristics of laeoplectic and dexioplectic metachronism, as described for metazoan examples, are entirely consistent with the hypothesis that the coordinated activity of the cilia depends upon hydrodynamic interaction between the moving cilia; indeed the viscous-mechanical coupling between the close-set cilia of such a system as the band of lateral cilia of *Mytilus* must be so tight that there is no freedom for independence of movement of any cilia except those at the margins. Diaplectic metachronism occurs over extensive ciliated areas, such as the body surface of the flatworm *Convoluta*, where it is laeoplectic (Dorey, 1965), as well as over bands of cilia on gills of *Mytilus* and *Corella* and single rows of compound cilia on the velum of veliger larvae of *Archidoris*.

IV. Antiplectic metachronism of ctenophores

The metachronism of such symmetrical ctenophores as *Pleurobrachia* and *Beroe* will be described first, since they are known the best (Sleigh, 1968, 1972), and variations shown by other forms will be mentioned later. The ctenophores named have eight meridional rows (combs) of rectangular paddles (comb plates), with the plates arranged parallel to one another within the rows. Each plate is a giant compound cilium based upon hundreds of ciliated cells that form a ridge upon which the plate is borne. The larger plates of a *Pleurobrachia* 6·5 mm long were found to have a length of 0·5 mm, a width of 0·7 mm and a thickness of about 20 μm; on the basis of data given by Afzelius (1961) for *Mnemiopsis*, it is assumed that such a plate contains in the order of 100,000 cilia. Rows of cilia oriented along the longer axis of the base of the plate are connected throughout much of the length of their shafts by interciliary material associated with intraciliary "compartmenting lamellae" (Afzelius); these connections are supposed to help maintain the integrity of the plate. The separation of comb plates within the comb row is a little less than half the length of a comb plate. A band of cells carrying simple cilia extends from the aboral end of each comb row towards the apical organ; these ciliary tracts (ciliated grooves) join in pairs before entering the apical organ, where each of the four tracts terminates at one of the four balancer cilia that support the statolith of the sense organ (Horridge, 1965). The activity of a comb row is normally initiated by the movement of the balancer cilium that stands at its head, and a single wave of beating passes down the associated ciliated grooves and down the two comb rows to which they lead—for one beat of a balancer cilium, each cilium and comb plate of that quadrant of the body gives a single beat. The waves of activity therefore pass down the row from the apical organ towards the mouth and the effective strokes of the ciliary beats are directed aborally. Some of these features are illustrated in Fig. 3.

Comb plates are unusual among cilia in that they remain stationary unless stimulated, while most other types of cilia are spontaneously active; each beat of a comb plate is triggered by a stimulus originating outside the cilium, the normal stimulus resulting from a beat of the next plate above (aborally). The stimulus exciting the comb plate (in these species of ctenophore at least) is believed to be mechanical and to result from the hydrodynamic drag effect exerted on the lower plate by the upper plate as it performs an effective stroke. This triggering stimulation is

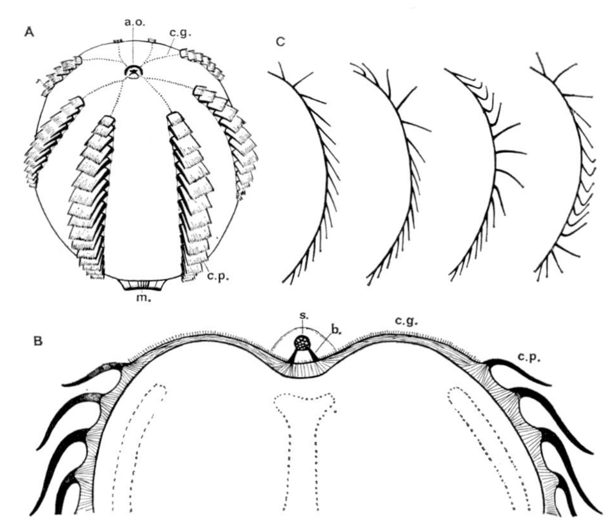

FIG. 3. Features of the structure of *Pleurobrachia* and the movement of its comb plates. (A) an entire ctenophore showing the arrangement of the rows of comb plates (c.p.) and ciliated grooves (c.g.) in relation to the apical organ (a.o.) and the mouth (m). (B) The ciliated grooves extend from the top comb plate of the row to the balancer cilia (b) that support the statolith (s) in the apical organ. (C) Waves of beating pass down the comb rows, and each cilium performs an upward effective stroke. The comb plates seen in each figure are about 0·5 mm long. (A) and (B) from Sleigh (1972) and (C) from Sleigh (1968), with permission.

believed to occur in the following way: comb plates at rest lie close on top of those lower in the row and when one plate moves upwards in its effective stroke water is drawn in behind it—this inflow of water will draw up the next plate below and the initial passive movement of the lower plate leads into an active effective stroke. Evidence in favour of this explanation includes the following:

1. Tracings from high-speed ciné films show that the lower part of a plate lifts as the plate above it performs an effective stroke, and a few milliseconds later the lower plate commences its effective stroke (Fig. 4; Sleigh, 1972).

2. The velocity of propagation of metachronal waves increases linearly with increase in frequency of beat, i.e. the interval between the beat of

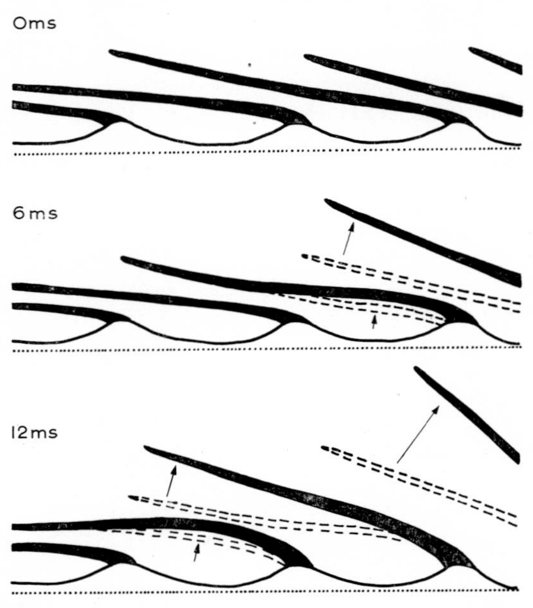

FIG. 4. Profiles of ctenophore comb plates traced from a high-speed ciné film and showing the way in which the basal part of the plate is drawn upwards as the plate above it moves away in its effective stroke; only later does the tip of the lower plate move upward as that plate commences its active swing. From Sleigh (1972), with permission.

one plate and the beat of the plate below decreases progressively as frequency increases (Fig. 5; Sleigh, 1968); it is known that the angular velocity of the effective stroke increases with increase in frequency (Sleigh and Jarman, 1973) and therefore the excitation of the lower plate would be expected to follow more quickly after the movement of the upper one if the frequency rises.

3. A comb plate may be excited mechanically by (a) being moved when touched by a microneedle, (b) being moved when a microneedle lying close above the plate is moved away from the plate without touching it, (c) being moved by a jet of water from a micropipette; rhythmic movement of a microneedle (or rhythmic jets of water) stimulate rhythmic beating over a wide frequency range (Sleigh and Jarman, 1973).

4. The stimulation of one plate by that above can be prevented if the movement of either the upper plate or the lower plate is restricted by a microneedle. A needle at positions *a* or *b* in Fig. 6 prevents plate 0 coming

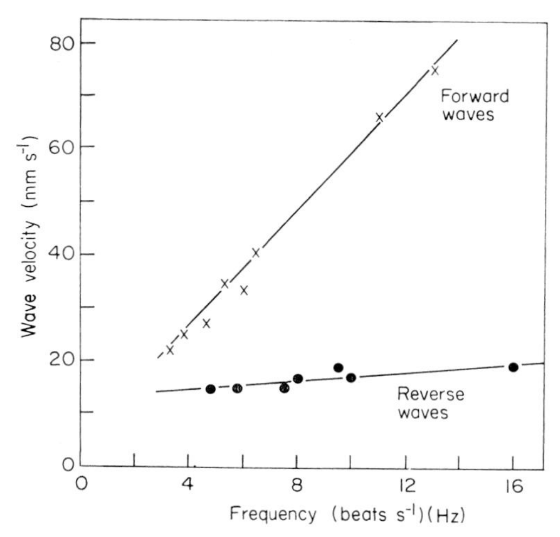

FIG. 5. The relationship between the frequency of beating of *Pleurobrachia* comb plates and the velocity of propagation of metachronal waves for forward waves passing towards the mouth and for reverse waves passing away from the mouth. From Sleigh (1968), with permission.

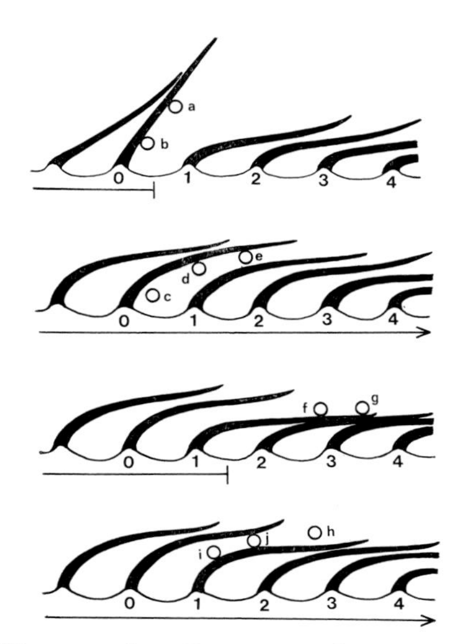

FIG. 6. Diagrams to illustrate the effects upon the propagation of metachronal waves of the obstruction of movement of *Pleurobrachia* comb plates by microneedles held in various positions; the lines and arrows below the comb plates indicate whether the metachronal waves are blocked or whether they pass the point of obstruction. From Sleigh (1972), with permission.

sufficiently close to plate 1 to cause appropriate water flow when plate 0 moves, while a needle at *c*, *d* or *e* permits the necessary water flow and the beating of plate 1 is stimulated; needles at *f* or *g* hold down plate 1 preventing its response to movement of plate 0, but a needle at *h* allows plate 1 to slip beneath the needle and needles at *i* or *j* allow movement of the upper part of the plate so that in these cases plate 1 responds and the metachronal wave is transmitted (Sleigh, 1972).

5. When a sheet of celluloid is interposed between two comb plates (0 and 1 in Fig. 7) in such a position that it prevents the flow of water behind plate 0 affecting plate 1 (e.g. at *m*, *n*, or *o*), the excitation of plate 1 is prevented, provided the celluloid does not touch plate 1. With the celluloid at positions *p* or *q* plate 1 beats after plate 0, presumably because the movement of the celluloid in contact with plate 1 has stimulated that plate, while a flow of water beneath the celluloid in position *r* could lead to the observed stimulation of plate 1 (Sleigh, 1972).

6. Numerous examples of failure of transmission of metachronal waves have been seen in high-speed ciné films; they occur because of lack of overlap between two adjacent plates as the upper plate begins to beat, usually resulting from this plate beating a second time before its previous recovery stroke has been completed so that the distal region of the plate is bent forward and not lying close above the lower plate.

7. Occasionally a plate within a quiescent row may beat without stimulation from the ciliated groove—such spontaneous beating results in a normal metachronal wave being conducted down the row and a reverse (symplectic) wave being conducted up the row towards the apical organ. In the reverse wave a plate is stimulated when it is pushed upwards by the effective stroke of the plate lying below it. The conduction speed of reverse waves is slower than normal waves because the stimulating plate must move further before it produces sufficient movement in the responding plate to trigger an active beat (Figs 5, 8).

The conclusion that comb plates of *Pleurobrachia* and *Beroe* are coordinated mechanically is in agreement with the findings of Verworn (1890). However, Child (1933) studied specimens of the same genera and concluded that the coordination must depend upon internal conduction, since the prevention of the movement of a plate did not always block metachronal wave transmission. The observations of Child were consistent with metachronal coordination by the passage of an internal impulse from plate to plate along the row by a path which is not continuous along the comb row because it involves some activity in the bases of the plates as well as conduction between the bases. The results reported above (Fig. 6) indicate that under some conditions a restrained plate may be able to stimulate the next plate mechanically, and it is believed that something like this may have occurred in Child's experiments, since it is

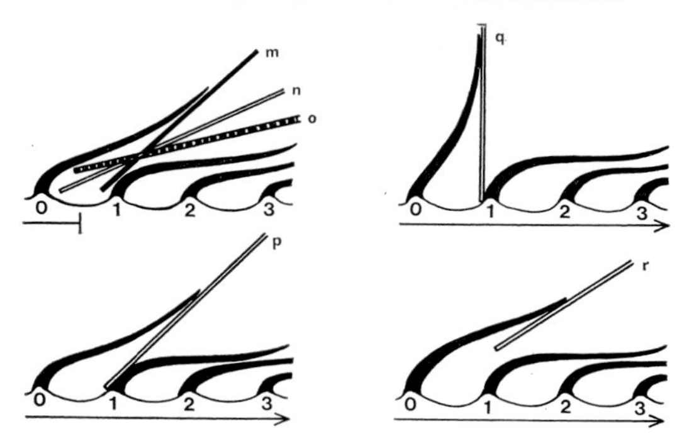

FIG. 7. The effects upon the propagation of metachronal waves of *Pleurobrachia* of the insertion of a piece of celluloid film in various positions between adjacent comb plates. An obstruction at m, n or o blocks the wave, but the wave passes if the celluloid is in positions p, q or r. From Sleigh (1972), with permission.

difficult to be certain whether or not plates move without ciné film records. If this belief is correct, the coordination does take place from plate to plate, but involves mechanical triggering of one plate by the preceding

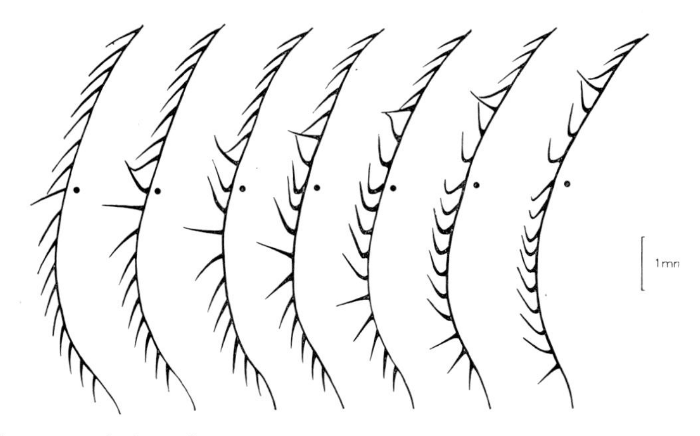

FIG. 8. The transmission of a reverse metachronal wave of *Pleurobrachia* seen when a comb plate within the row (indicated by the dot) shows a spontaneous beat and pushes up the overlying comb plate above it; this then beats and moves the next plate in an aboral direction, which passes the stimulation up the row. A forward wave travels down the row from the spontaneously active plate at a faster rate than the reverse wave passes up the row. From Sleigh (1968), with permission.

one. The experiments on the effects of ions and drugs on *Pleurobrachia* reported by Sleigh (1968), and interpreted as supporting Child's conclusions, could have the alternative explanation that the metachronal wave velocity is increased in solutions with increased magnesium ions, ouabain and curare because these substances increase the angular velocity of the effective stroke of the cilia, although this explanation is as yet unconfirmed. The effects of increased viscosity on the activity of the plates has not yet been adequately investigated; it would be expected that the angular velocity of the plates would be reduced by an increase in viscosity, but the viscous coupling between plates would be increased, so that the effect of viscosity on the relation between beating and wave velocity is likely to be complex, and the finding by Sleigh (1966) that increased viscosity reduces the angular velocity without affecting the wave velocity cannot be regarded as proof of metachronism by internal conduction. Synaptic connections found between nerve fibres and the inner ends of the ciliated cells by Horridge and Mackay (1964) are assumed to be concerned in the inhibition response of the ctenophores, in which stimulation of the body results in an immediate stoppage of comb plate activity all over the body, this inhibition response does not occur in solutions of $MgCl_2$ (Horridge, 1965).

The rows of comb plates in *Cestus* all occur at the upper edges of the strap-shaped ctenophore, the bases of the plates having been turned in development by almost 90° so that all plates beat towards the dorsal mid-line and the metachronal waves are effectively dexioplectic in some rows and laeoplectic in others (Fig. 9). The plates are brought closer together by the developmental movements and as a result the coupling between adjacent plates could be much increased; this correlates with the observation that the speed of metachronal waves is substantially faster than in other ctenophores studied (up to 600 mm s^{-1}; Sleigh, 1972). It was from the ciliated cells of the comb plates of this ctenophore that Horridge (1965) obtained intracellular recordings with a microelectrode. These records showed a resting potential of up to 40 mV and negative-going potentials of up to 28 mV which coincided with the beat of the plate. It was not possible to determine whether these potentials precede or follow the beat, and there is no clear evidence about the origin of the potential changes. Further studies are clearly needed to establish whether the ciliary beats are accompanied by a true depolarization; since the power stroke of the beat is strong enough to break off the electrode tip (Horridge), and the base of the plate moves noticeably at each beat, potentials could arise by mechanical distortion of the electrode tip. Attempts to repeat this experiment have not so far been successful (Tamm, personal communication).

The metachronal waves of *Mnemiopsis* were found by Parker (1905) to be conducted past a place whose beating was prevented by mechanical restraint or local cooling, or past the position of a missing plate. Horridge (1966) and Tamm (personal communication) have pointed out that the difference between the comb rows of such ctenophores as *Pleurobrachia* and *Beroe* and those of the lobate forms like *Mnemiopsis*, *Bolinopsis* and *Eucharis* is that in the former the ciliated groove ends at the first plate, while in the latter a tract of short cilia occurs between adjacent plates within the comb row, following the line of the ciliated groove. In these lobate forms Tamm (personal communication) has found that the prevention of movement of several plates together does not stop the passage of the wave to plates beyond the blockage and in this the wave transmission

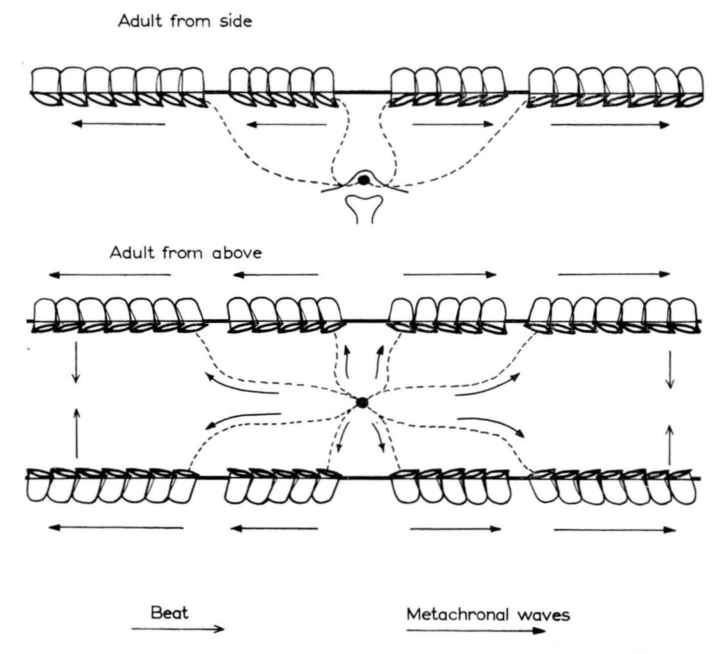

FIG. 9. Diagrams showing the arrangement of comb plates of Venus' Girdle, *Cestus*. The direction of propagation of metachronal waves and the direction of the effective stroke of the beat of the comb plates are indicated by arrows. The dotted lines represent the ciliated grooves that pass outwards from the apical organ of this ctenophore to the first comb plate of each row. From Sleigh (1972), with permission.

differs from that in *Beroe* and *Pleurobrachia*. The observation that the short cilia between plates participate in the waves of beating is probably important, but their role in the transmission of the wave from plate to plate is not clear. It is relevant in this context to note that it is also not clear how the beating of the short cilia of the ciliated groove of any ctenophore triggers the activity of the first plate at the head of the comb row.

Although there is confidence that the coordination between comb plates of some ctenophores can be explained best by mechanical triggering of every beat, this group of organisms still provides unanswered questions about ciliary coordination.

V. Ciliary metachronism and the propulsion of water and mucus

Different metachronal patterns are characteristically associated with different functions of the ciliary system involved. The symplectic metachronal pattern of the protozoan *Opalina* (see Machemer, p. 217) is associated with slow-moving waves with almost unbroken outlines, providing a wave surface which may act effectively as an undulating envelope that exerts a propulsive force on the water; from calculations based upon this assumption Blake (1971a, b) was able to predict swimming velocities within the range observed for *Opalina*. In the diaplectic and antiplectic patterns the individual cilia move separately through the water in the effective stroke, although in the recovery stroke they tend to be bunched. These patterns allow each cilium to make use of a substantial part of the work output of the effective stroke in productive water propulsion, both because little of the energy of that stroke is dissipated in contacts with the other cilia and because the tips of the cilia in the effective stroke extend well above the bent-over cilia of the recovery stroke; the latter point is emphasized in diaplectic systems where the recovering cilia are bent down towards the cell surface at one side or the other so that the profiles of the bunched recovery waves are kept very low and provide reduced resistance to the forward flow of water maintained by the effective strokes. In the fluid overlying a ciliated epithelium we can therefore recognize a lower region in which both effective and recovery strokes move the fluid, while an upper region is influenced only by the effective strokes. In developing a more sophisticated model for the propulsion of fluids by cilia, Blake (1972) took account of the movements of individual cilia and pointed out not only that such a two-layer situation will exist but also that the movement in the upper layer will be substantially larger than that in the lower layer because the influence of the presence of the cell surface (which introduces viscous drag) decreases as one moves further from the surface.

Measurements of the flow of fluid produced by cilia of several types (Sleigh and Aiello, 1972) indicate that the velocity of water propelled by cilia increases with distance from the cell surface to reach a maximum towards the ciliary tip, where the average water velocity was found to be between one-fifth and one-third of the tip velocity of the cilium during its effective stroke; beyond the ciliary tip the flow rate decreases although forward movement of water was still detectable 2 or 3 cilium lengths from the cell surface (Fig. 10). In the case of the lateral gill cilia of *Mytilus*, 15 μm long and beating at 20 Hz (20°C), the ciliary tips move at about 4 mm s^{-1} in the effective stroke and the estimated average speed of water flow near the ciliary tips was 1·2 mm s^{-1}. Sleigh and Aiello concluded that the metachronal relationships of the cilia are important in maintaining a continuous rapid flow, because no part of the layer of water immediately overlying the ciliated surface was ever very far from effective-stroke cilia. In the case of *Mytilus* lateral cilia, not only are two adjacent filaments separated by only about twice the length of the lateral cilia, so that in beating the tips of the lateral cilia may touch those of the opposite filament,

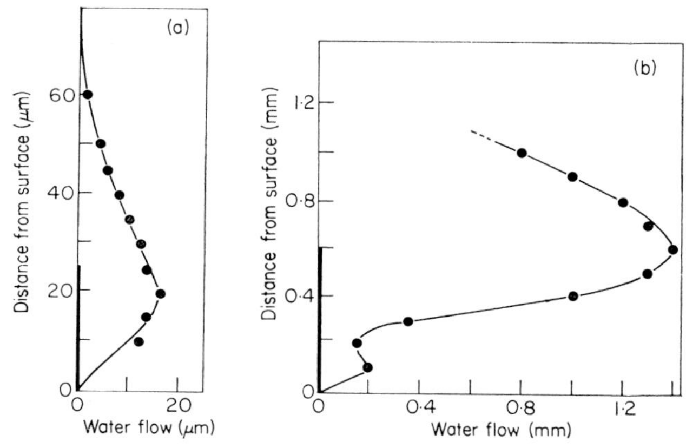

Fig. 10. The movement of water caused by the activity of (a) membranelles (compound cilia) 25 μm long of the ciliate *Stentor* and (b) comb plates 0·6 mm long, of the ctenophore *Pleurobrachia*. In each case the distance moved by particles in the water flow in one cycle of beating is plotted against the distance of the particles from the cell surface. Data obtained from the analysis of ciné films. From Sleigh and Aiello (1972), with permission.

but the fanned-out cilia of one effective stroke are separated from those of the next effective stroke on the same filament by less than the length of a cilium. In addition, the metachronal waves of opposed filaments travel in opposite directions, thereby doubling the frequency with which a body of water between the filaments is propelled by ciliary effective strokes in comparison with the action of a single band of cilia; the same effect could probably be achieved by waves of the two filaments travelling in the same direction, and strictly alternately, but this would require a complex regulating mechanism. Every aspect of the configuration and motion of the metachronal waves seems to be geared towards transmitting as much as possible of the power of the ciliary beating to the propulsion of fluid.

Some different problems are encountered in the propulsion of mucus by cilia. In many organisms cilia move mucus in water, e.g. the frontal cilia of *Mytilus* gill or the locomotor cilia of larger flatworms, while in terrestrial animals cilia may move mucus over surfaces that are exposed to humid air, e.g. in the human respiratory tract. As far as the cilia are concerned, these two situations are probably more similar than they appear because it is reported that a sheet of mucus is moved over the epithelium at about the level of the ciliary tips, and there is a more fluid water layer beneath the mucus in which the ciliary recovery strokes take place (Fig. 11). The existence of two layers in these mucous systems was first described

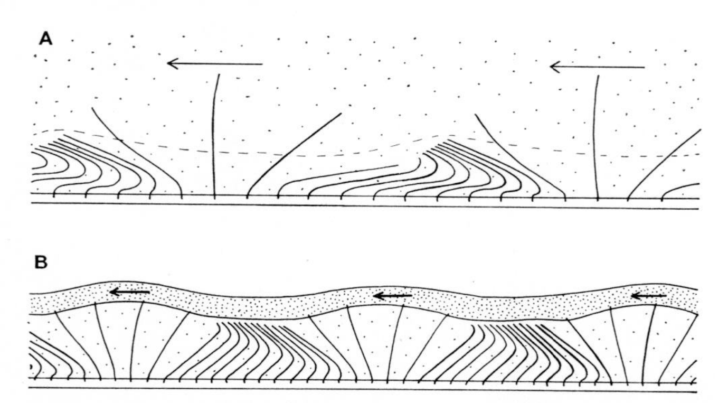

FIG. 11. A comparison between the propulsion of water (A) and the propulsion of mucus (B) by cilia that are coordinated into antiplectic metachronal waves that move towards the right. There is little resultant flow of fluid near the cell surface in either case, but in A the propulsion of the water around and beyond the ciliary tips is indicated by the arrows, while in B the carpet of mucus that is penetrated by the tips of the active-stroke cilia is propelled in the direction shown. (B, Based on a figure by Lucas and Douglas, 1934).

for ciliated epithelia of the monkey by Lucas and Douglas (1934), who observed that a layer of mucus might remain stationary while the cilia were beating vigorously beneath it.

Mucus has considerable elasticity and Litt (1970) has pointed out that elasticity rather than viscosity governs the transfer of energy from cilia to mucus; in Newtonian fluids like water, viscosity is all-important. It is assumed that when cilia propel mucus, only the tips of the effective stroke cilia penetrate the mucus and "claw" it forward. Because of the elasticity of the mucus this clawing action should be continuous so that the elastic action pulls more mucus forwards rather than drawing back that which has just been moved forward by the cilia.

Reports on the metachronism of cilia that propel mucus suggest that it is always antiplectic, e.g. on mammalian nasal mucosa (Proetz, 1933), frog palate epithelium (Seo, 1931) and the frontal ciliated epithelium of *Cardium* gill (Knight-Jones, 1954). This metachronism appears to be rather weakly developed in comparison with that in systems propelling water; the cilia are very short but are packed close together. Rhodin and Dalhamn (1956) found that rat tracheal cilia were 5 μm long and occurred at a density of 8·4 μm^{-2}; cilia of *Opalina* are 15 μm long and spaced at about 1 μm^{-2} and *Paramecium* cilia are 12 μm long and spaced at about 0·5 μm^{-2}. Measurements by Dalhamn (1956) on the activity of rat tracheal cilia show that the propulsion of mucus by the cilia is quite effective; mucus was transported at between 200 and 250 $\mu m\ s^{-1}$ (at 37°C) when the cilia beat at 21–22 Hz with a cycle in which the duration of the recovery stroke was about 3 times that of the effective stroke, so that the tip velocity of the cilium in the effective stroke was 500–800 $\mu m\ s^{-1}$. It would be very valuable to have more data on the propulsion of mucus by cilia and especially on the beating and metachronism of the cilia involved.

References

Afzelius, B. (1961). *J. biophys. biochem. Cytol.* **9**, 383–394.
Aiello, E. and Sleigh, M. A. (1972). *J. Cell Biol.* **54**, 493–506.
Blake, J. (1971a). *J. Fluid Mech.* **46**, 199–208.
Blake, J. (1971b). *J. Fluid Mech.* **49**, 209–222.
Blake, J. (1972). *J. Fluid Mech.* **55**, 1–23.
Child, C. M. (1933). *J. comp. Neurol.* **57**, 199–252.
Child, F. M. (1965). *Proc. 2nd Intern. Conf. Protozool., London* p. 110. Excerpta Medica, Amsterdam.
Dalhamn, T. (1956). *Acta Physiol. scand.* **36**, Suppl. 123, 1–161.
Dorey, A. E. (1965). *Q. Jl microsc. Sci.* **106**, 147–172.
Dral, A. D. G. (1967). *Neth. J. Sea Res.* **3**, 391–422.
Horridge, G. A. (1965). *Am. Zool.* **5**, 357–375.
Horridge, G. A. (1966). *Symp. Zool. Soc. Lond.* **16**, 247–266.
Horridge, G. A. and Mackay, B. (1964). *Q. Jl microsc. Sci.* **105**, 163–174.
Kinosita, H. and Murakami, A. (1967). *Physiol. Rev.* **47**, 53–82.
Knight-Jones, E. W. (1954). *Q. Jl microsc. Sci.* **95**, 503–521.
Litt, M. (1970). *Archs intern. Med.* **126**, 417–423.
Lucas, A. M. (1932). *J. Morph.* **53**, 265–276.
Lucas, A. M. and Douglas, L. C. (1934). *Arch. Otolaryngol.* **20**, 518–541.
Machemer, H. (1972). *Acta Protozool.* **11**, 295–300.
Murakami, A. (1963). *J. Fac. Sci. Tokyo Univ.* Sec. IV, **10**, 23–35.
Parker, G. H. (1905). *J. exp. Zool.* **2**, 407–423.
Proetz, A. W. (1933). *Ann. Otol. Rhinol. Laryngol.* **42**, 778–817.

Rhodin, J. and Dalhamn, T. (1956). *Z. Zellforsch.* **44**, 345–412.
Seo, A. (1931). *Jap. J. med. Sci., Biophys.* **2**, 47–75.
Sleigh, M. A. (1966). *Am. Rev. resp. Dis.* **93**, Suppl., 16–31.
Sleigh, M. A. (1968). *J. exp. Biol.* **48**, 111–125.
Sleigh, M. A. (1969). *Int. Rev. Cytol.* **25**, 31–54.
Sleigh, M. A. (1972). *In* "Essays in Hydrobiology" (R. B. Clark and R. Wootton, eds), pp. 119–136. University of Exeter.
Sleigh, M. A. and Aiello, E. (1972). *Acta protozool.* **11**, 265–277.
Sleigh, M. A. and Jarman, M. (1973). *J. Mechanochem. Cell Motility* **2**, 61–68.
Verworn, M. (1890). *Pflügers Arch. ges. Physiol.* **48**, 149–180.

Note added in proof

Tamm (1973) has compared the coordination of comb plates in two groups of ctenophores, those lacking a ciliated groove between adjacent plates (e.g. *Beroe*) and the lobate forms which possess a ciliated groove between adjacent plates throughout the comb rows. He found that in lobate ctenophores the conduction of a wave along the comb row could be prevented by interrupting the propagation of the metachronal wave along the ciliated groove between two plates, but that wave condition could not be prevented by eliminating mechanical interaction between adjacent plates. It is still not clear how beating of cilia of the ciliated groove leads to excitation of a comb plate.

Tamm, S. L. (1973). *J. exp. Biol.* **59**, 231–245.

Chapter 12

The control of ciliary activity in Protozoa

YUTAKA NAITOH and ROGER ECKERT
Department of Biology, University of California, Los Angeles, California, U.S.A.

I. Introduction

The locomotor behavior of many ciliates and flagellates depends on the movements performed by cilia or flagella. Jennings (1906) found that all kinds of tactic behavior depend on transient changes (i.e. reversal) in the beating direction of cilia or flagella in response to a stimulus. The transient reversal in direction of beating causes the organism to swim backwards for a distance. Before the organism resumes its forward locomotion it rotates to a variable extent about its rear end due to an asymmetrical restoration of normal beating of the cilia, and perhaps also due to asymmetries in the morphological arrangement of cilia over the cell surface. Thus, the organism resumes forward swimming in a new direction, and thereby avoids further contact with the source of the stimulus (Fig. 1). Jennings (1906) called this response the "avoiding reaction". The change in beating direction has not been reported to occur in cilia of vertebrates and most invertebrates, but it does occur in response to electric stimulation in echinoderm larvae and pelagic tunicates (Mackie *et al.*, 1969; Galt and Mackie, 1971). The ability of protozoan cilia to change their beat direction appears to be an adaptive differentiation of ciliary function essential for behavioral responses in Protozoa.

Toward the end of the last century Verworn (1889a) noted that the reversal in beat direction of cilia ("ciliary reversal") occurs on the cell

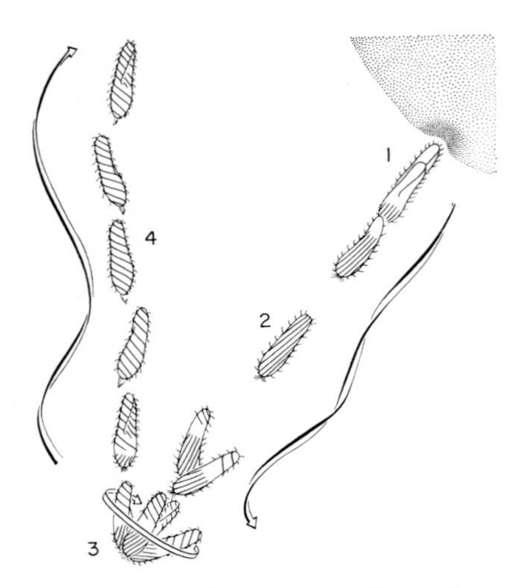

FIG. 1. The avoiding reaction of *Paramecium*. Stage 1, sensory transduction of the mechanical stimulus produced by collision with an obstacle and subsequent excitation of the membrane to produce the calcium response and reversed beating of cilia; Stage 2–3, backward swimming and renormalization beginning at cell anterior; Stage 4, forward locomotion. After Grell (1968) modified from Párducz (1967).

surface facing the cathode when *Paramecium* is placed in the path of a galvanic current. Bancroft (1905) argued later that ciliary reversal at the cathodal side corresponds to an excitation of that part of the cell by analogy with Pflüger's polar (cathodal) excitation in nerve and muscle cells. Ciliary reversal, then, became a subject for studying cell excitation as well as the cellular control of ciliary movement.

Electrophysiological approaches to the problems of ciliary movement in protozoans were first carried out at Tokyo University by Kamada, Kinosita and their students (Kinosita and Murakami, 1967). They found that ciliary activity in the Protozoa is correlated with the electric activity of the surface membrane. In recent years we have extended these studies with more detailed electrophysiological examinations of the ciliates, and have demonstrated that the direction of ciliary beating is regulated by Ca^{2+}-dependent bioelectric events in the membrane. We have also examined ionic mechanisms by which the ciliates detect external stimuli and signal the regulatory mechanisms controlling the direction and frequency of ciliary beating. This chapter will deal primarily with the electrophysiology of the surface membrane as it applies to the stimulus sensitivity of the cell and to the response of the locomotor apparatus to the electric behavior of the cell membrane.

II. Control of ciliary orientation

A. *Responses to electric stimuli*

1. *Stimulation by external electrodes*

(*a*) *Polar effect* (*cathodal ciliary reversal*). Verworn (1889a) found in his studies on galvanotaxis of *Paramecium* that the beating direction of cilia

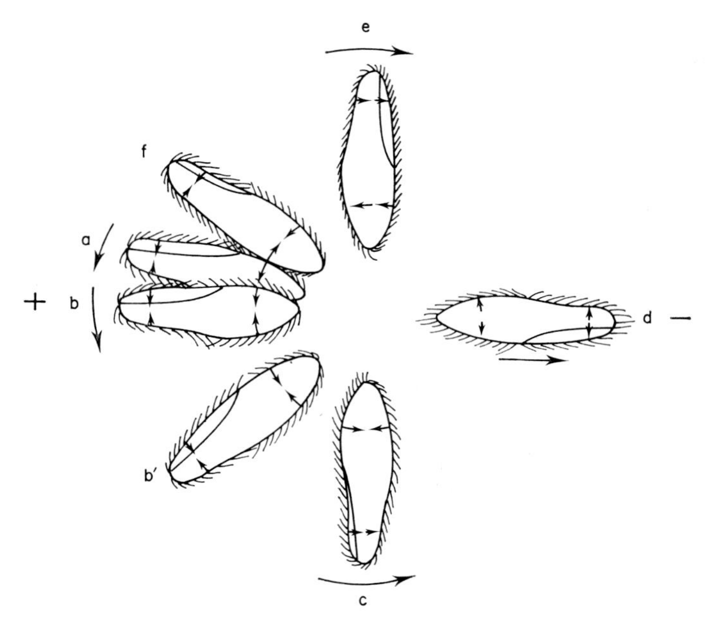

FIG. 2. Effects of the electric current on the cilia of *Paramecium*. The large arrows show the direction towards which the animal turns. The small internal arrows indicate the direction in which the cilia of the corresponding quarter of the body tend to turn the animal. In all positions save *c* and *e* the cilia of different regions oppose each other. From *a* to *d* the turning is toward the aboral side; from *d* to *f*, toward the oral side. Jennings (1906).

on the cell surface facing the cathode reverses when the current is turned on. Other investigators (Ludloff, 1895; Statkewitch, 1904; Jennings, 1906) confirmed this and demonstrated that orientation of the specimen toward the cathode depends on the reversed beating of cilia (Fig. 2).

Bancroft (1905) proposed that cathodal ciliary reversal is analogous to cathodal excitation in nerve and muscle, and argued that reversed beating corresponds to an excitation of the cilia. Kinosita (1936a, b, c) demonstrated a strength–duration relationship, summation of subthreshold

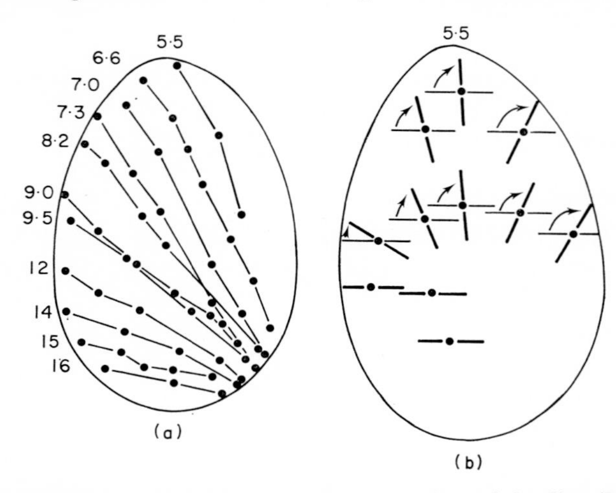

FIG. 3. (a) The "equi-threshold lines" on the cell surface of *Opalina*. The numerals indicate relative values of applied current.

(b) The local change in orientation of the metachronal wave when points on the ventral surface of *Opalina* are stimulated by a current of intensity represented in relative value by 5·5 in (a). Thin and thick straight lines represent the orientation of the metachronal wave before and after stimulation respectively. Arrows indicate the direction of changes in orientation of the wave. Okajima (1953).

currents and a hyper-sensitive phase after stimulation for cathodal ciliary reversal in *Paramecium*. These findings led him to stress the similarity of ciliary reversal to excitation of excitable cells in the Metazoa.

(*b*) *Regional differences in electric sensitivity*. In contrast to *Paramecium*, *Opalina* orients and migrates toward the anode. Ciliary reversal, however, takes place at the surface of the specimen facing the cathode, as in *Paramecium*. Wallengren (1903) found that cilia on the right anterior part of *Opalina* reverse their beat direction at lower cathodal current intensities and more strongly at a given intensity than the cilia over the rest of the cell. This, together with shape of this cell causes it to orient and migrate towards the anode (Jennings, 1906).

Regional differences in electric sensitivity of cilia was demonstrated in

Opalina by Okajima (1953), who applied localized current to small areas of the cell surface with an extracellular micropipette electrode. The electric threshold for ciliary reversal was found to be lowest at the anterior right portion and highest at the posterior left (Fig. 3a). The degree of ciliary reversal in response to a cathodal current was also greatest at the anterior right portion (Fig. 3b).

Topographic differences in electric sensitivity of cilia can be seen also in *Paramecium*, and may occur generally among the ciliated Protozoa. Cilia on the anterior half of *Paramecium* are more sensitive and respond to stimulation more strongly than those on the posterior half (Eckert and Naitoh, 1970).

2. *Stimulation by intracellular electrodes*

In an electric field current enters the cell through the membrane nearest the anode and leaves through the membrane nearest the cathode. The distribution of current densities and direction across the membrane depends primarily on the electric resistances of the membrane, the cytoplasm and of the external solution as well as the shape and orientation of the cell. The membrane resistance can change as a function of time in response to polarizing currents and is non-linear with membrane voltage (see section II.A.3). These factors result in very complex spatio-temporal distributions of current, contributing to the complexity of galvanotaxis found in many protozoans.

Application of current through a microelectrode inserted into the cytoplasm of the cell makes the problem far simpler, for the total amount and the direction of current is carefully controlled. The electrotonic properties of a ciliate such as *Paramecium* cause the applied current to produce nearly uniform changes in electric potential across the entire cell surface (II.A.3). Thus, ciliary responses of a given type can be expected over the entire cell surface if the polarizing current is injected with a micro-electrode. This is confirmed by the finding that an outward current through the membrane of *Opalina* induces ciliary reversal on the whole surface of the specimen, while an inward current induces an increase in beat frequency of cilia in the normal direction (Naitoh, 1958). In the resting state the membrane of *Opalina* is electrically negative on the side of the cell interior (II.A.3). Thus, we can say that a depolarization (decrease in intracellular negative potential) of the membrane by outward current induces ciliary reversal and a hyperpolarization (increase in intracellular negative potential) by inward current leads to an increase in beating frequency in the normal (forward swimming) direction.

Regional differences in electric sensitivity of cilia are also seen in the case of intracellular application of current (Naitoh, 1958; Eckert and

Naitoh, 1970). This means that cilia on different parts of the cell have a different sensitivity to a uniform depolarization of the membrane.

3. *Electric characteristics of the membrane*

(*a*) ***Resting membrane potential.*** Kamada (1934) measured the intracellular potential of *Paramecium* in various ionic solutions by employing a quartz capillary microelectrode filled with saline solution, and found that the degree and polarity of the membrane potential depend on the extracellular

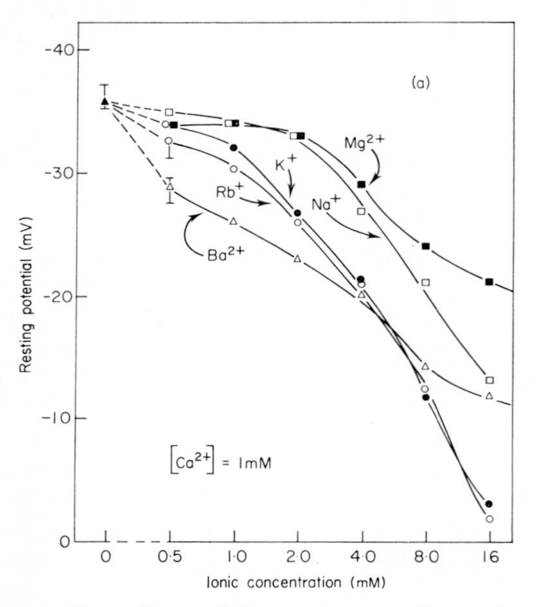

FIG. 4. (a) Concentration effects of five cations on the resting potential of *P. caudatum*. The calcium concentration was 1 mM throughout. Each point is the mean of five measurements on different specimens. Vertical lines give the range of measured values.

cation concentration. In culture or dilute saline solution the cytoplasm is electro-negative with reference to the outside solution. The cation dependency of the membrane potential was confirmed by Yamaguchi (1960), Kinosita *et al.* (1964a) and Naitoh and Eckert (1968a). Figures 4a and b show the concentration effects of various cations on the membrane potential in *Paramecium*. An increase in the concentration of any of the cations tested depolarizes the membrane, although the degree of the depolarization differs among the cation species. K^+ and Rb^+ are the most effective. These data suggest that the resting membrane is permeable to most of the cations, though it is most permeable to K^+ ions.

The most common cations in the culture or pond water are Ca^{2+}, K^+, and Na^+; their concentrations are generally several millimoles each or less. The cytoplasm of *Paramecium* is high in K^+ (about 20 mM KCl equivalent; Dryl and Grębecki, 1966; Dunham, 1973; Naitoh and Eckert, 1973), while very low in Ca^{2+}, which appears to be below 10^{-7} M (Naitoh

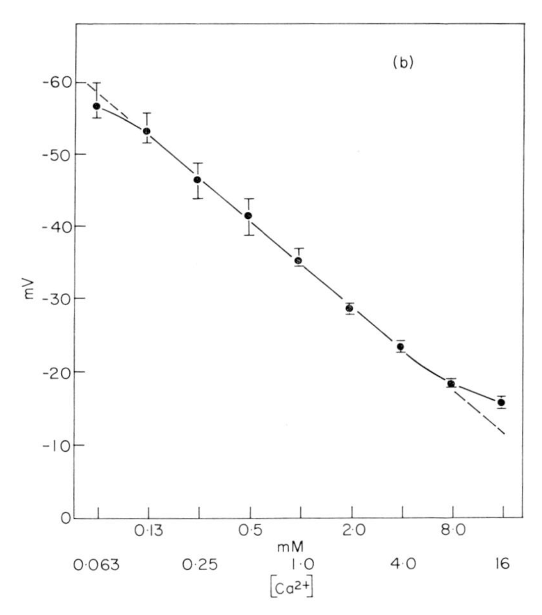

FIG. 4. (b) Resting potential as a function of the calcium concentration. Each point is the mean of five measurements on separate specimens. Vertical lines give the ranges of fluctuation in measurements. Naitoh and Eckert (1968a).

and Kaneko, 1972). The membrane potential, therefore, is thought to be a function of all these ions according to the equation (Hodgkin and Horowicz, 1959):

$$Vm = E_K(g_K/G)+E_{Ca}(g_{Ca}/G)+E_{Na}(g_{Na}/G) \ldots$$

where Vm is the membrane potential; E_K, E_{Ca}, E_{Na} are the respective Nernst equilibrium potentials due to concentration differences of these ions between the inside and outside of the cell; g_K, g_{Na}, g_{Ca} are the respective partial conductances of the membrane to these ions; and G is total conductance of the resting membrane. In most of our experiments we simplify the medium to eliminate all cations except Ca^{2+} and K^+. The actual value of the potential appears to be closer to the K^+ potential, presumably because of relatively higher permeability of the resting membrane to K^+ ions. The Na^+ concentration does not strongly affect the membrane potential, so it appears that sodium permeability is relatively low.

The resting potential of *Opalina* is very sensitive to external K^+ concentration but even less sensitive to Na^+ than that of *Paramecium*. Ca^{2+} increases the resting potential (Kinosita, 1954; Naitoh, 1958, 1964; Ueda, 1961). Selective permeability to K^+ ions is as well developed in *Opalina* as in metazoan nerve and muscle cells.

The resting membrane potentials of some protozoans, such as *Noctiluca* (Eckert and Sibaoka, 1968) and *Stentor* (Chen, 1972) are difficult to study quantitatively owing to the presence of large and extensive cytoplasmic vacuoles. The potential record from inside a vacuole is, of course, the algebraic sum of potentials across all the membranes intervening between the cell exterior and the vacuole.

(*b*) *Active membrane potentials*

(*i*) *Regenerative calcium response*. The membrane of *Paramecium* behaves as a simple ohmic resistance with a parallel capacitance in response to an injection of current pulse of sufficient intensity to produce an electrotonic potential shift of several millivolts (Fig. 5A).* Depolarizations of increasing

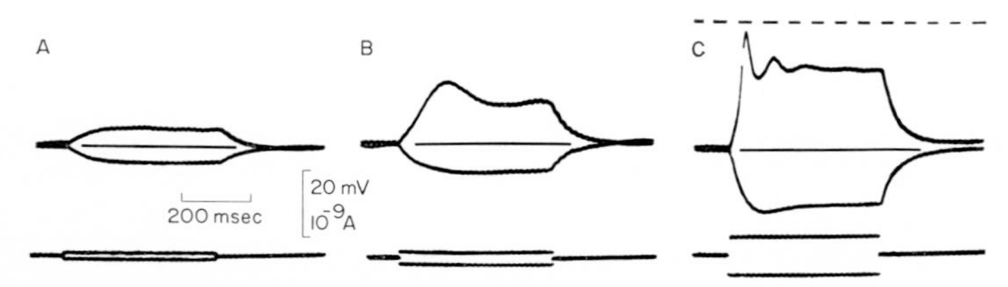

FIG. 5. Membrane responses to long current pulses in *P. caudatum*. A–C, increasing current intensities. Record A shows a nearly pure electrotonic depolarization with almost no regenerative component. Regenerative behavior increases in a graded manner as passive depolarization increases in rate of rise and amplitude in B and C. Hyperpolarization shows delayed anomalous rectification. The dashed line in C gives the level of reference (zero) potential. Naitoh *et al.* (1972).

intensity show an inflection on the upstroke, which indicates the onset of an active response arising from the electrotonic depolarization. The potential reaches a peak, then drops to a lower level with a damped oscillation (Fig. 5B, C). The maximum rate of rise and the peak value of the regenerative component gradually increase to quasi-saturated levels with increasing stimulus intensities.

The ionic mechanism of the regenerative component was investigated

* This is consistent with the ultrastructure, which indicates a single surface membrane separating the cytoplasmic compartment from the extracellular world. The alveoli do not form a continuum (Allen, 1971).

by testing various concentrations of cations for their effect on the saturated peak value of the potential (Naitoh *et al.*, 1972). The peak value increases linearly with a logarithmic increase in the external Ca^{2+} concentration with a slope of about 25 mV for each 10-fold increase in concentration (Fig. 6a). This approaches the ideal value of 29 mV predicted from the

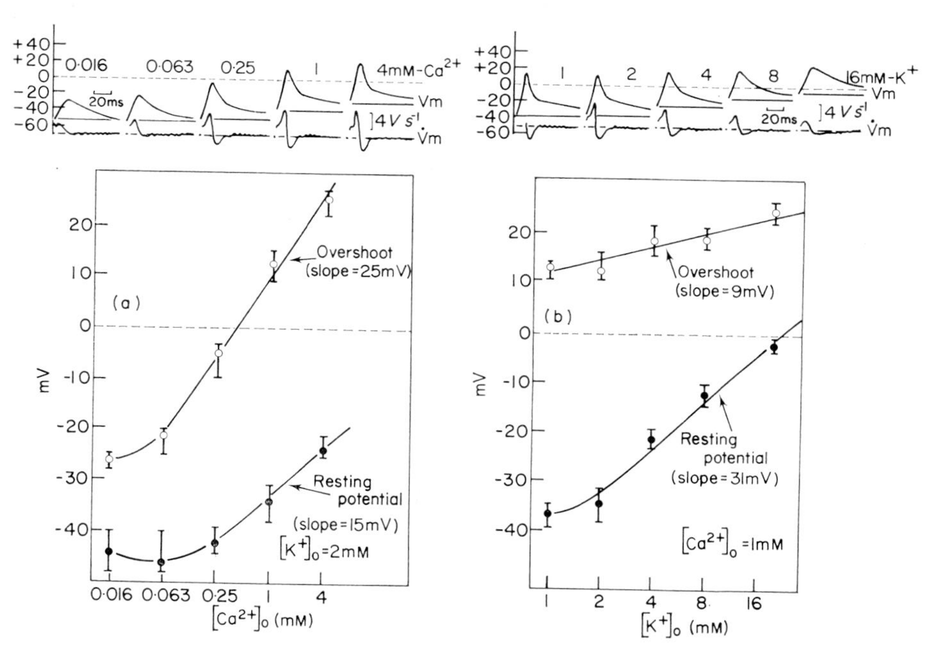

FIG. 6. The electrical responses of *P. caudatum* in media of different ionic constitution to 2 ms current pulses of supramaximal intensity.

a. The electrical response as a function of extracellular calcium. Potassium was held constant at 2 mM while $CaCl_2$ concentration was varied as shown.

b. The electrical response as a function of extracellular potassium. Calcium held constant at 1 mM throughout, while $[K]_0$ was varied. Top, potential (Vm) and derivative ($\dot{V}m$) recordings at $CaCl_2$ or KCl concentrations indicated. Each point on the graphs is the mean of five measurements, with deviations indicated by vertical lines. Naitoh *et al.* (1972).

Nernst equation for diffusion potentials of divalent cations. The effect of the K^+ concentration on the peak value is much smaller (Fig. 6b). Na^+ shows little effect on any parameter of the regenerative component.

These findings indicate that depolarization of the membrane produces an increase in permeability of the membrane to Ca^{2+} ions, which leads to an inward current carried by Ca^{2+} ions in accordance with the ionic hypothesis of Hodgkin, Huxley and Katz (Hodgkin, 1957) for nerve action potentials. Since it is not an all-or-none action potential we will term the regenerative depolarization a "calcium response".

The intracellular concentration of free (unbound) Ca^{2+} has not been measured directly, but is assumed to be lower than 10^{-7} M, which is four or more orders of magnitude lower than the external Ca^{2+} concentration. This would predict a calcium equilibrium potential of 120 mV or more, internally positive, for an extracellular calcium concentration of 1 mM; the peak of the calcium response never approaches this value. This suggests that the increased calcium permeability produced by a depolarizing

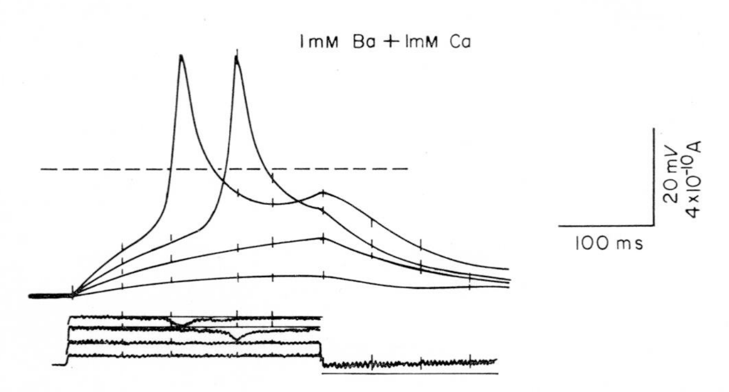

FIG. 7. Electrical responses of *P. caudatum* to outward current pulses (*ca.* 250 ms in duration) in a mixture of 1 mM Ba + 1 mM Ca. The dotted line shows the potential level of a reference electrode in the bath. Downward deflections on the current traces (below) are due to current leakage through the 10^9 ohm current-passing circuit during the action potential. Naitoh and Eckert (1968b).

stimulus is not more than several times greater than the potassium conductance. This would also explain why the calcium response is graded rather than all-or-none.

In the presence of external Ba^{2+} or Sr^{2+} the membrane of *Paramecium* generates all-or-none action potentials in response to a depolarizing current (Fig. 7) (Naitoh and Eckert, 1968b). Concentration effects of these ions on the action potential suggest that these ions carry most of the regenerative inward current. Conversion of the membrane to produce all-or-none action potentials might result in part because these ions carry inward current more readily than Ca^{2+}, and in part because they produce a decrease in membrane permeability to potassium ions. These electrical characteristics of the membrane of *Paramecium* resemble those of crustacean muscle cells (Werman and Grundfest, 1961).

Inward depolarizing current which is responsible for the upstroke of the action potential in *Paramecium* can be demonstrated with the voltage-clamp technique (Fig. 8), first used on squid axon by Cole (1949) and

Hodgkin, Huxley and Katz (1949) to measure inward Na^+ current. In *Paramecium* a sudden shift of the membrane potential level in the depolarizing direction evokes a change in membrane conductance which results in a transient inward membrane current carried by Ca^{2+} (Naitoh and Eckert, unpublished). This is followed by a steady outward current. The peak

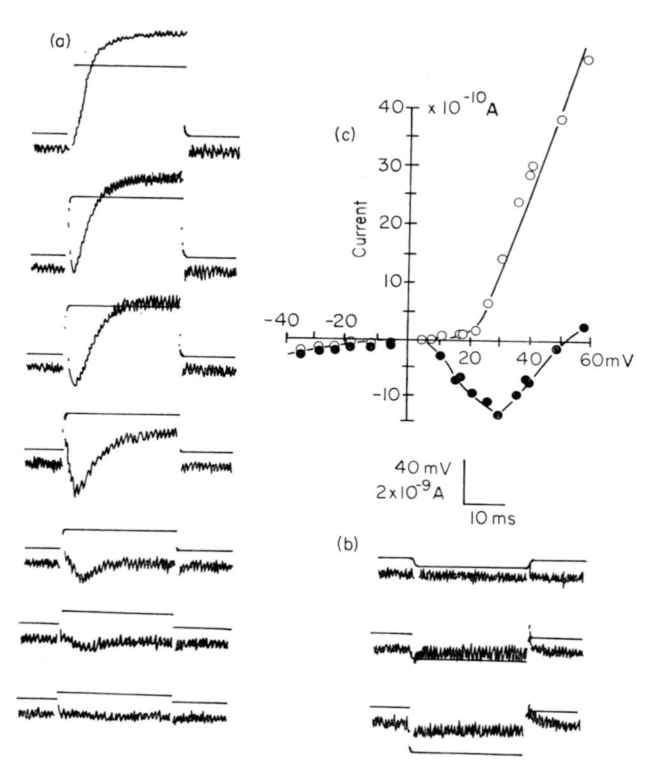

FIG. 8. Current–voltage relations in *P. caudatum.* (a) Positive-going voltage clamped pulses (smooth traces) of 20 ms duration increasing in strength from bottom to top of series. Membrane current ("noisy" traces) shows initial inward transient phase (downward deflection) followed by steady state outward phase. (b) Negative-going (hyperpolarizing) voltage clamp pulses. (c) Transient (solid circles) and steady state (open circles) current intensities plotted against voltage displacements. Naitoh and Eckert (unpublished).

value of the initial inward Ca^{2+} current increases with increasing depolarization to a maximum value, then decreases with further increase in the depolarization, producing a classic "N"-shaped current–voltage curve with a region of negative resistance. It is this portion of the current–voltage relation which is responsible for the regenerative nature of the calcium response (Fig. 8c). This resistance to steady outward current

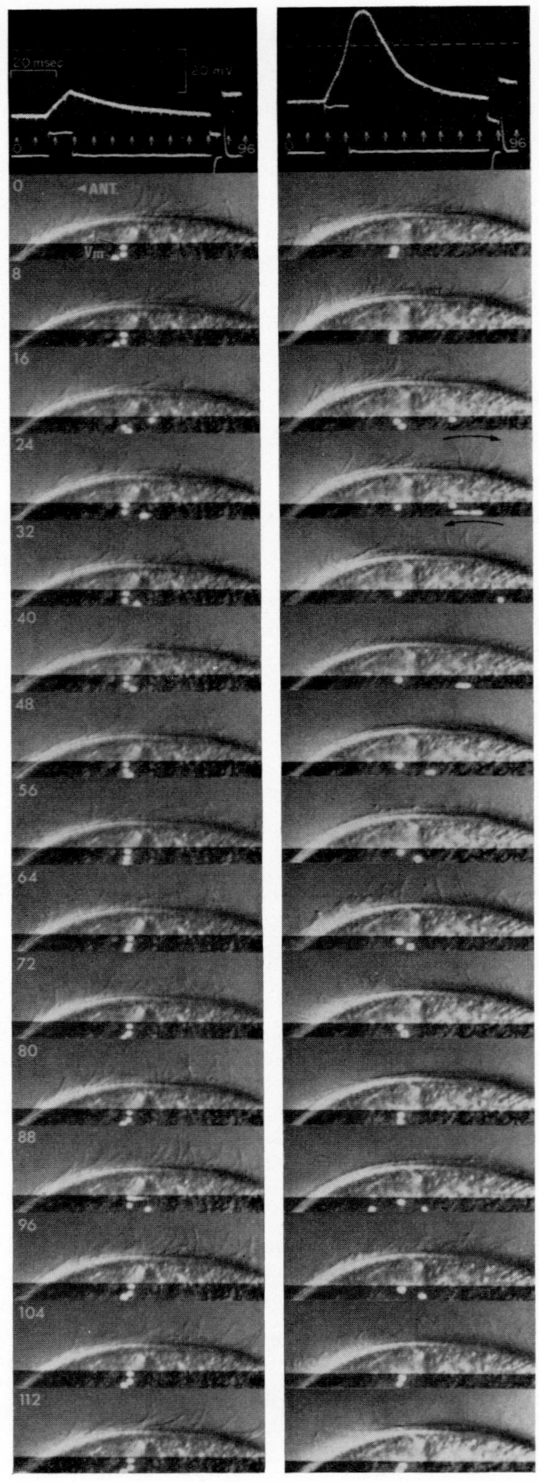

FIG. 9.—Caption on facing page.

decreases very much with depolarizations beyond about 20 mV. By analogy with nerve membrane this is presumed to be due to a delayed increase in K^+ conductance of the membrane. Current–voltage relations for both the initial Ca^{2+} current and the delayed K^+ current are illustrated in Fig. 8c.

(*ii*) *Ciliary reversal evoked by membrane calcium response* (*Ca-current hypothesis*). Simultaneous recording of ciliary reorientation in non-beating cilia and of the regenerative calcium response clearly showed that ciliary reversal of *Paramecium* occurs in association with the regenerative response of the membrane with a latency of about 20 ms (Eckert and Naitoh, 1970). Machemer and Eckert (1973) recently employed high-speed cinematography for recording ciliary movement in *Paramecium* simultaneously with electric responses of the membrane to outward currents (Fig. 9). Ciliary reversal occurs only when the Ca^{2+}-dependent regenerative response takes place. The duration of the period of reversed beating becomes longer with an increase in the amplitude of the regenerative component (Fig. 10). Such data, together with numerous observations that extracellular calcium is required for ciliary reversal in response to all forms of stimulation (Bancroft, 1906; Kamada, 1940; Okajima, 1954b; Kanno, 1958), formed the basis of the calcium current hypothesis (Eckert, 1972) which states that all stimuli which evoke ciliary reversal do so by causing a net influx of calcium from the external medium to the cell interior. This influx is regulated primarily by membrane permeability to Ca^{2+}, and secondarily by membrane permeability to K^+ and other ions which can carry charges out of the cell to compensate for charges carried in by Ca^{2+}. According to this scheme (Fig. 11) any stimulus (e.g. outward current, high K^+ concentration, etc.) which depolarizes the membrane produces an increase in the calcium permeability (or more precisely, the calcium conductance, g_{Ca}). If all else remains equal this will cause calcium

---

FIG. 9. Ciliary activity in *P. caudatum* in response to a purely electrotonic potential (column A) and in response to a calcium response (column B). Specimen bathed in 1 mM $KCl + 1$ mM $CaCl_2 + 1$ mM Tris-HCl, pH 7·2. Oscilloscope traces of electric responses of cell membrane (upper trace) to depolarizing current pulse (lower trace). Stimulus in A, 10^{-9} A for 10 ms; in B, 2×10^{-9} A for 10 ms. Dashed line gives reference potential. Calibration pulses for potential and electrode resistance occur at far right of voltage trace. Arrows indicate artifacts which correspond to ciné frames reproduced below. The numbers on the ciné frames indicate time in milliseconds starting with the first frame taken during the CRO sweep. The dark band at the bottom of each ciné frame resulted from additional exposure during printing to enhance the contrast of the superimposed CRO beams (white spots) displaying stimulating current (I, top) and membrane potential (Vm, bottom). Between these is a fixed beam for positional reference. The anterior end of the paramecium is toward the left. Machemer and Eckert (1973).

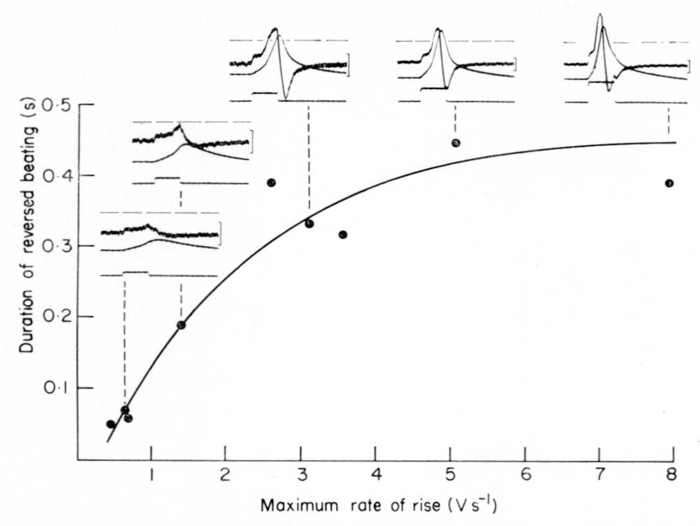

FIG. 10. Duration of the period of reversed beating in *P. caudatum* as a function of the maximum rate of rise of the calcium response in one specimen. Representative recordings are shown corresponding to five points along the plot. In these the upper trace is dVm/dt, the middle trace is Vm and the lower trace indicates a 20 ms current pulse. The vertical calibration marks equal 2 V s^{-1}; note the change in scale for the fourth and fifth recordings. Machemer and Eckert (1973).

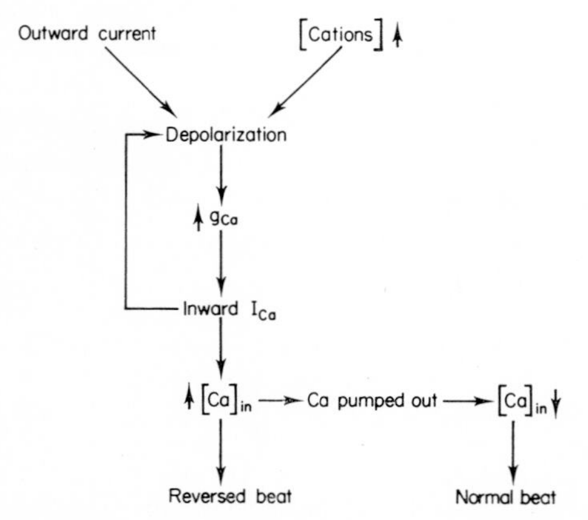

FIG. 11. The sequence of steps leading from stimulation to reversal of ciliary beating. Any depolarizing stimulus, such as an electrotonic outward current, results in a transient increased conductance to Ca^{2+}. This results in an influx of Ca^{2+} (I_{Ca}). As a consequence the concentration of Ca^{2+} in cilia and cortex ($[Ca]_{in}$) rises, activating the mechanism for reversal. Subsequent removal of Ca^{2+} by active processes restores the normal beating orientation of the cilia. The feedback arrow from "Inward I_{Ca}" to "Depolarization" indicates the regenerative nature of the calcium response. After Eckert (1972).

ions, which have a strong electrochemical gradient tending to drive them into the cell, to show a net influx which results in an increase in the intracellular and presumably in the intraciliary calcium concentration.

Evidence is presented elsewhere from the work of Naitoh and Kaneko (1972, 1973) that ciliary reversal occurs when the intraciliary Ca^{2+} concentration rises above 10^{-6} M (see section II.F). By making the assumption that the membrane calcium response (i.e. regenerative influx of Ca^{2+} in response to depolarization) occurs in the portions of the surface membrane covering the cilia as well as in the interciliary portions of the surface, it can be calculated that a calcium response only 1 mV in amplitude will result in a transient increment of 10^{-6} M in the intraciliary calcium concentration (Eckert, 1972). Thus, the influx of calcium ions which accompanies even small calcium responses of the surface membrane is sufficient in theory to produce reversed ciliary beating.

(*iii*) *Spontaneous depolarization.* Many ciliates and flagellates frequently show changes in direction of locomotion (avoiding reactions) without any evident external stimuli. Kinosita (1954) found in *Opalina* that changes in beat direction, which occur fairly regularly in frog Ringer solution, are closely associated with depolarization of the membrane (Fig. 12).

Spontaneous ciliary reversal of *Paramecium* was also found to occur in association with depolarization. In solutions containing both Ba^{2+} and Ca^{2+}, *Paramecium* repeatedly shows strong periodic ciliary reversal (Dryl, 1961). Repetitive all-or-none type action potentials which are similar to those induced by outward current in Ba^{2+}, Ca^{2+} mixtures (Fig. 7) were found to coincide with the reversals (Kinosita *et al.*, 1964a). The resting potential fluctuates slowly in both depolarizing and hyperpolarizing directions (Kinosita *et al.*, 1964a; Naitoh, 1966). When the slow spontaneous depolarization reaches a certain critical level the membrane induces a regenerative response which is in turn followed by ciliary reversal.

It is interesting to note here that microsurgically dissected fragments of *Opalina* can show spontaneous ciliary reversal in normal frog Ringer providing the fragment contains at least a part of the electrically sensitive right anterior portion. If that part is missing, the fragment only swims forward (Okajima, 1954a). This suggests that spontaneous depolarizations originate in a specialized portion of the cell membrane.

Although the mechanism of endogenous perturbations of the resting level is not known, it seems to have an important role in controlling locomotor activity.

(*iv*) *Genetic modification of electric mechanisms and locomotor behavior.* Ciliates are especially attractive for the study of membrane mutants because modifications of their locomotor behavior can be used to screen for genetically altered membrane function. Furthermore, they can be genetically manipulated by established techniques involving mating types, conjugation, autogamy, etc. (Sonneborn, 1970).

The genetic approach has provided further data concerning the role of the cell membrane in the regulation of ciliary activity. Kung (1971a, b) generated and selected several behavioral mutants of *P. aurelia*. Of these,

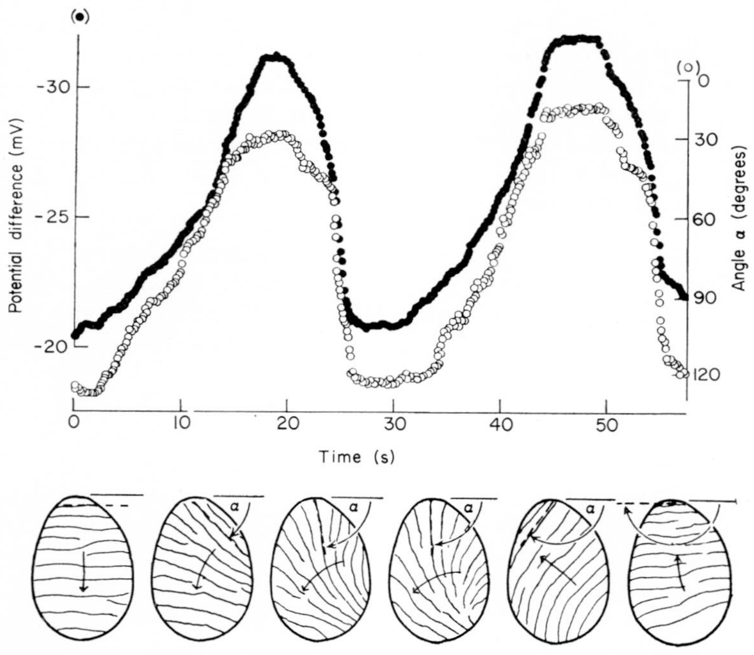

FIG. 12. The time change of membrane potential (solid circles) and beating direction of cilia (angle α) (hollow circles) during the spontaneous activity of *Opalina* in Ringer's solution obtained from a microcinematographic record. The sign of potential difference is that of the cell interior with reference to the external surface. Kinosita (1954).

a single-gene mutant termed "pawn" was characterized by the failure to show the avoiding reaction or to swim in reverse when stimulated mechanically or by transfer to solutions of high K^+ concentration. Electrophysiological studies revealed that "pawn" lacked the membrane mechanism for producing regenerative increases in calcium permeability (Kung and Eckert, 1972). This is manifested as the failure to generate both graded calcium responses and all-or-none barium action potentials (Fig. 13). On the other hand, Triton-extracted models (see section II.F) of the mutant "pawn" swim backwards in the presence of the same Ca^{2+} concentrations which produce reversed swimming in models of the wild type (Kung and

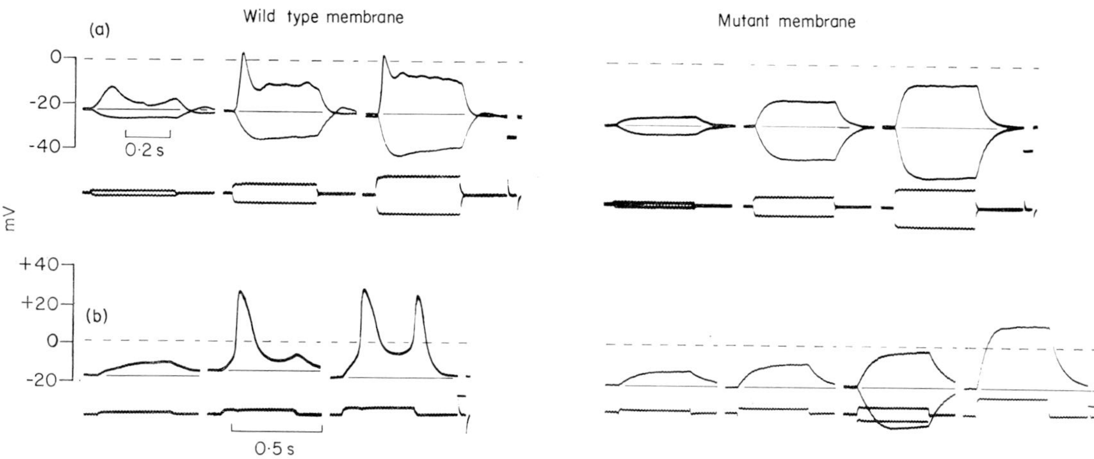

FIG. 13. Electric excitability of a wild type *P. aurelia* (left) compared with inexcitability of the membrane mutant "pawn" (right). (a) Bath solution consisted of 1 mM $CaCl_2$, 4 mM KCl and 1 mM Tris-HCl (pH 7·2). The wild type shows graded calcium responses to depolarizing current, while the "pawn" mutant shows only electrotonic responses. (b) Bath solution consisted of 1 mM $CaCl_2$, 4 mM $BaCl_2$, 1 mM Tris-HCl. The wild type exhibits all-or-none barium spikes (Naitoh and Eckert, 1968b), while the "pawn" shows only electrotonic responses. The lower set of traces in each record indicates the stimulus current. After Kung and Eckert (1972).

Naitoh, 1973). These findings taken together suggest that (1) the phenotypic lesion expressed in "pawn" occurs in the surface membrane, (2) this lesion is a failure of the calcium gating mechanism, and (3) the calcium gating mechanism is necessary for both the regenerative depolarization and for reversal of ciliary beat in the live ciliate. Ciliary reversal is uncoupled from membrane depolarization in "pawn" because of the lack of or the malfunction of a voltage-sensitive calcium gating mechanism. The correlation between the loss of the calcium response by the membrane of "pawn" and the loss of ability to swim in reverse is further evidence for the calcium current hypothesis of excitation–response coupling outlined in Fig. 11.

(*c*) *Passive electric properties*

(*i*) *Membrane constants*. As shown in Fig. 5, membrane of *Paramecium* can be approximated by an ohmic resistance in parallel with a capacitance, providing the current is small enough not to evoke a regenerative response or a delayed anomalous rectification. The overall resistance in ohms of a specimen (i.e. input resistance) is calculated according to Ohm's Law by dividing the steady state potential shift (V) by the injected current (A). The parallel capacitance (μF) can be calculated by dividing the time constant (i.e. the time in seconds until the potential level reaches 63% of the final steady level) by the input resistance (MΩ).

Membrane areas are calculated on the assumption that the specimens are covered by a smooth membrane. The actual surfaces, however, are more or less sculptured (see Allen, 1971). Moreover, ciliates have many cilia, and the surface membrane is continuous over these. Thus, in *Paramecium caudatum* two-thirds of the surface membrane covers the cilia. Since the cable properties of the cilia are not known, the true values of the membrane resistance and membrane capacitance are uncertain.

(*ii*) *Topographic differences in membrane resistance*. Naitoh (1958) measured topographic differences in the membrane resistance of *Opalina* by measuring electric current (inward across the membrane) through a suction pipette electrode placed against various portions of the cell surface. A voltage was applied between inside (positive) and outside of the pipette. Variations in the current passed out of the pipette were due in large part to variations in the membrane resistance.

As shown in Fig. 14, the right anterior part of *Opalina* has a lower electric resistance than the left posterior part. This indicates that the electric characteristics of protozoan cells need not be uniform over the cell surface. Furthermore, it is interesting that cilia in the area where the resistance is the lowest show the highest sensitivity to electric stimulation. Some regional differences in ciliary activity may therefore depend on regional differences in membrane properties.

(*iii*) *Cable properties*. Jahn (1961) called attention to the cable properties of *Paramecium* to explain polar effects of externally applied electric current (see section II.A.1). Eckert and Naitoh (1970) calculated the space constant of *Paramecium* according to the standard cable equation (Hodgkin and Rushton, 1946) by assuming *Paramecium* to be an infinite cable with diameter of 40 μm, and by introducing the values of membrane resistance ($2{\cdot}2 \times 10^9\ \Omega\ \mu$m), internal resistance (1200 $\Omega\ \mu$m; Gelfan, 1926–27) and external resistance (negligible). The value calculated was 1350 μm. This

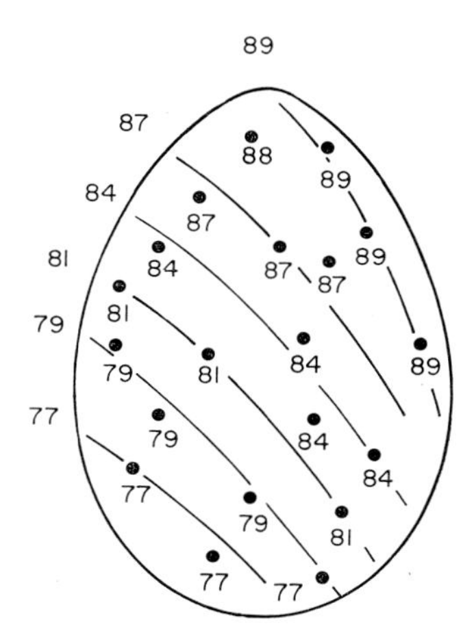

FIG. 14. Topographic differences in electric conductivity of the membrane of *Opalina*. Numerals represent the current intensities expressed in percentage of that measured without the organism. The applied voltage and the negative hydrostatic pressure at the electrode tip were kept constant throughout the measurement. Room temp.: 26·4°C. Naitoh (1958).

is more than 5 times the overall length of *Paramecium* (250 μm). A 1% decay of an electrotonic potential over the full length of *Paramecium* was calculated from the space constant according to Weidmann's (1952) equation for a short cable.

The actual decay of the electrotonic potential was determined by injecting an inward current near one end of a paramecium while monitoring the resulting hyperpolarization simultaneously from both ends (Fig. 15a). The measured decay ranged from 2 to 5% (Eckert and Naitoh, 1970), which is somewhat larger than the calculated value. Since the resistance encountered by extracellular current was ignored in the calculation, the discrepancy between calculated and measured decrement may have been due to the relatively high resistance of the dilute extracellular medium. In either case, the decrement is so small that we may consider the cell interior essentially isopotential.

The potential decay along the long axis of *Paramecium* was also determined in the case of the regenerative response. As shown in Fig. 15b, almost no distortion and time delay of action potential was observed between the two ends of the specimen.

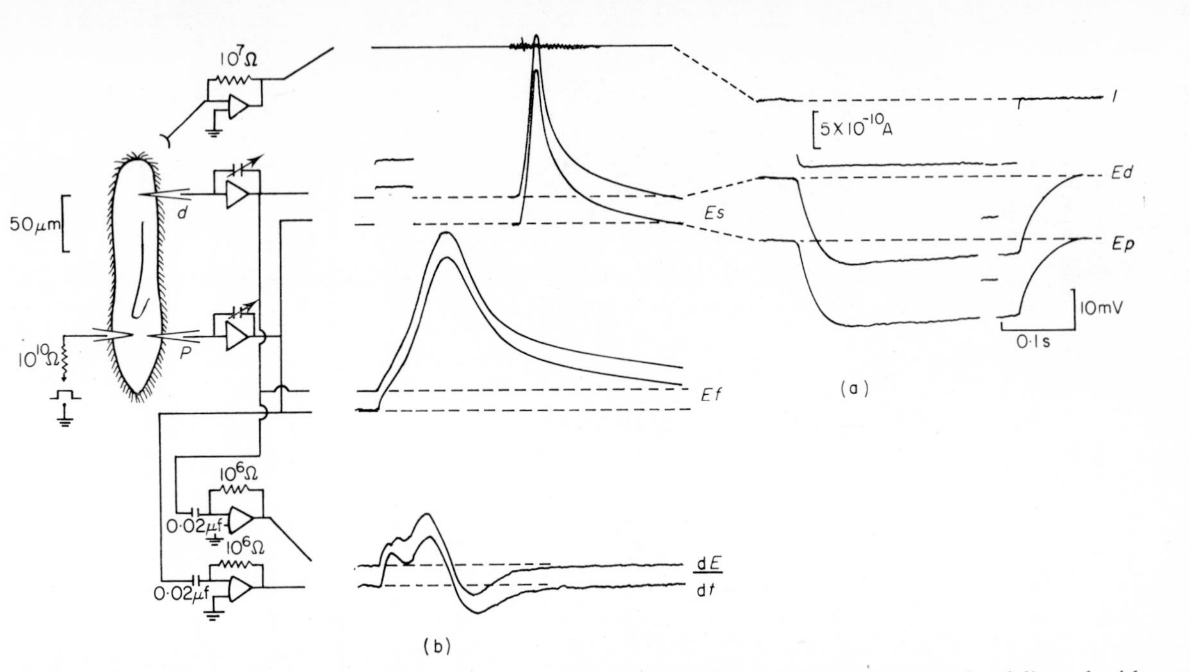

FIG. 15. (a) Electrotonic potentials in *P. caudatum* in response to a 300-ms hyperpolarizing current pulse delivered with a microelectrode near one end of the cell. Trace *I*, current monitor; trace *Ep*, signal from recording electrode (proximal) adjacent to the current-passing electrode; trace *Ed*, signal from recording electrode (distal) inserted near the other end of the cell. The two recording electrodes were 125 μm apart in a specimen measuring 210 μm overall. The corrected deflection in trace *Ed* was 100% of that in trace *Ep*.

(b) Depolarizing potential evoked by a 1 ms current pulse. The diagram shows the position on the specimen of insertion of proximal (*p*) and distal (*d*) recording electrodes and current electrode. The upper two sets of traces are shown at slow (*s*) sweep velocity. Trace *I* shows stimulating current; trace *Es* shows proximal and distal voltage records. The lower two sets of sweeps are at four times faster (*f*) velocity. The signal originating from the proximal electrode is displayed on the lower trace of each pair. The discontinuity in the rising phase of the depolarizing potential represents the transition from a passive RC (electrotonic) response to the regenerative depolarization of the graded calcium response. The square calibration pulses on the slow potential (*Es*) traces are 10 mV, 20 ms. Eckert and Naitoh (1970).

(*iv*) *Relation to control of ciliary activity.* The isopotential nature of ciliates permits electrotonic propagation of an electric signal from any portion of the cell membrane to the rest of the surface membrane with virtually no distortion and time delay, and therefore without need of all-or-none impulses such as are required for conduction over large distances in nerve and muscle cells. As a result, the entire ciliary population of the cell can be regulated by electrotonic spread of signals which can be graded rather than all-or-none.

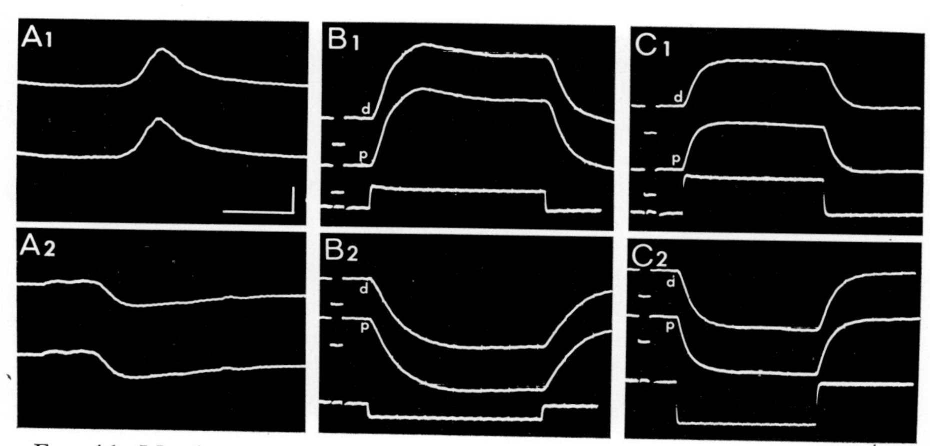

FIG. 16. Membrane potentials recorded from *Euplotes* with two intracellular electrodes inserted near opposite ends of a specimen. The separation of the electrodes was 60–80% of cell length. (A_1) Spontaneous depolarizing wave; (A_2) spontaneous hyperpolarizing wave in the same specimen; (B_1) outward current (bottom trace) injected with polarizing electrode inserted near the proximal (*p*) recording electrode; (B_2) the same specimen showing response to hyperpolarizing current; (C_1) similar to B_1 but with "neuromotor" fibrils visibly transected by a massive incision halfway between the proximal (*p*) and distal (*d*) recording electrodes; (C_2) the same specimen as C_1, showing membrane potential recorded from both sides of incision in response to hyperpolarizing current injected near the proximal electrode. Calibrations are 10 mV, 200 ms in A; 10 mV, 20 ms in B and C. Naitoh and Eckert (1969b).

Taylor (1920) reported that coordinated changes in the direction of beating of widely separated cirri and membranelles of *Euplotes* was lost as a result of microsurgical transection of "neuromotor" fibrils which run through the cytoplasm from a "motorium" to the bases of the cirri and membranelles. He concluded that the fibrillar system performs a nerve-like function to coordinate ciliary activity. These experiments were repeated with the aid of cinematography by Okajima and Kinosita (1966), but these workers were not able to confirm Taylor's report. Sectioning the fibrils did not destroy the coordination between the organelles at opposite ends of the cell. Naitoh and Eckert (1969b) demonstrated the isopotential nature of *Euplotes*, and showed that the isopotential condition is maintained after transection of the fibril system (Fig. 16). The coordination of ciliary orientation was closely correlated with the electrotonic spread of spontaneous and evoked changes in membrane potential. Coordinated (synchronous) reversal of the cilia on the two sides of an incision through the cortex was reported in *Paramecium* by Worley (1934) and in *Opalina* by Okajima (1954a).

The cable properties of ciliates such as *Paramecium* are also relevant to the problem of metachronal coordination of the ciliary beat. Grębecki (1965) and Párducz (1967) have suggested that impulses progress over the cell surface to coordinate metachronal waves. It should be noted, however, that metachronal waves in ciliates have velocities of less than 1 μm ms^{-1} (Sleigh, 1962), which is at least 100 times slower than the demonstrated rate of propagation of signals across the plasma membrane of *Paramecium* (Eckert and Naitoh, 1970). Signals travelling at velocities above 100 μm ms^{-1} could hardly coordinate the phasing of a wave of activity which proceeds at 1 μm ms^{-1} or slower. The nearly isopotential condition of the membrane also makes it difficult to conceive of a way in which the direction of membrane-borne impulses might be altered to coordinate changes which occur in the direction of the metachronal wave. It therefore seems most unlikely that the propagation of metachronal waves is coordinated by electric signals borne by the surface membrane.

B. Responses to mechanical stimuli

1. Collision with an obstacle

(*a*) *Avoiding reaction.* When a forward-swimming (or forward-creeping) protozoan collides with an obstacle (e.g. solid object, surface of the medium, another specimen) with its anterior end, a typical avoiding reaction occurs (Fig. 1). The problems posed by the reaction are (1) how the anterior end transduces the mechanical stimulus, and (2) how the signal is transmitted to the cilia over the entire cell surface causing them all to reverse their direction of beating.

(*b*) *Mechanoreceptor region.* Jennings (1906) touched the tip of a fine glass rod to various portions of the cell body of some protozoans (*Paramecium*, *Oxytricha*), and found that mechanical stimulation leads to the avoiding reaction only if the anterior surface is touched. A light touch evokes a weak ciliary reversal, so that the specimen merely turns aborally because of intrinsically stronger beating of the cilia on the oral groove. The turning of the specimen is always in the aboral direction irrespective of which side of the anterior end the mechanical stimulus is applied. A stronger touch produces stronger ciliary reversal, so that the specimen swims backwards for a distance before turning aborally and finally resuming forward locomotion in the new direction.

A mechanical touch to the posterior portion accelerates the forward movement because of increased frequency of ciliary beating. The midregion of *Paramecium* is insensitive to mechanical stimulation. Similar differentiation of mechanoreceptor properties has been demonstrated in *Dileptus* by Doroszewski (1961, 1963, 1965, 1968, 1970). Touching the anterior half of the specimen with a microneedle always induces ciliary reversal, causing the specimen to move backwards. The same stimulus applied to the posterior half induces normal beating, so as to move the specimen forward (Fig. 17). A microsurgically separated anterior half always responds with ciliary reversal to mechanical stimulation, while

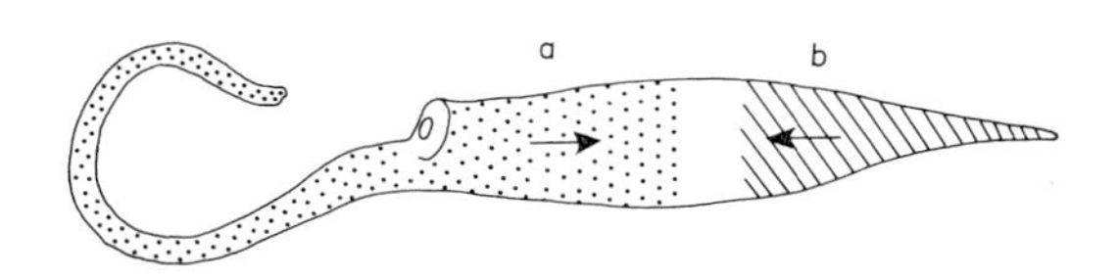

FIG. 17. *Dileptus cygnus*. a. Backward response area; b, forward response area. Doroszewski (1970).

the posterior half shows only normal ciliary beating. The posterior half regains the ability to reverse ciliary orientation in response to mechanical stimulation upon regeneration of a new anterior region. Thus, it appears that ciliates have two mechanoreceptor regions which respond to the similar stimuli in opposite ways. Stimulation of the anterior leads to ciliary reversal identical to that produced in response to a depolarizing current. Stimulation of the posterior leads to an increase in the frequency of beating which resembles the ciliary response to a hyperpolarizing current.

(*c*) *Mechanoreceptor potential*

(*i*) Anterior and posterior responses. Naitoh and Eckert (1969a) recorded potential changes in response to mechanical stimulation of *Paramecium*. A small glass stylus (10–20 μm in tip diameter) was driven against the cell surface by an electrically activated piezoelectric crystal. Mechanical stimulation of the anterior portion induces a transient depolarization of the membrane (Fig. 18a) which is closely followed by ciliary reversal. The degree of depolarization increases with increasing mechanical stimulus and reaches a saturation level. Mechanical stimulation of the posterior portion induces a transient hyperpolarization. The hyperpolarization increases in amplitude with increasing mechanical stimulation until it

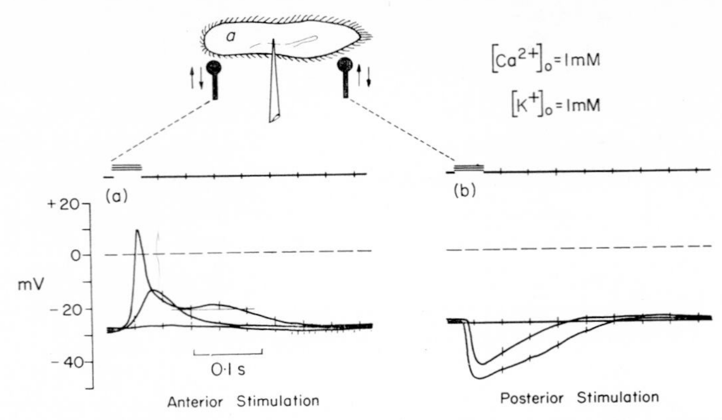

FIG. 18. Mechanoreceptor potentials in *P. caudatum*. The diagram shows an immobilized specimen impaled by an intracellular recording electrode. Mechanical stimuli of varying displacements were applied to either the anterior (*a*) or the posterior regions of the cell. (a) Anterior receptor potentials elicited by three intensities of mechanical stimulation of the anterior end. (b) Posterior receptor potentials elicited by three intensities of mechanical stimulation of the posterior end. Deflections in the upper trace show the duration and relative intensity of pulses activating the piezoelectric crystal. Naitoh and Eckert (1969a).

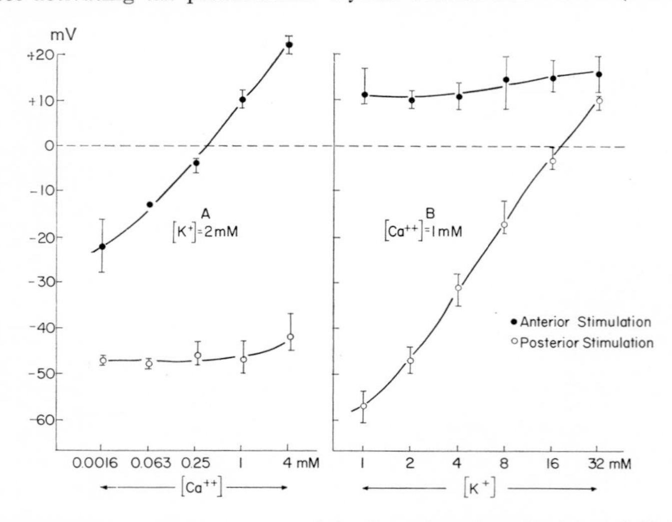

FIG. 19. Peak values of receptor potentials plotted against Ca^{++} and K^{+} concentrations. (Ordinate) Intracellular potential recorded at the peak of the anterior receptor potential (●) and the posterior receptor potential (○). In graph A [K^{+}] was held at 2 mM, and in graph B [Ca^{++}] was held at 1 mM. Vertical bars show the range of data from 5 to 12 measurements on as many different specimens. Only maximum responses were used. A maximum response is defined as one which shows no increase with further increase in stimulus intensity. Naitoh and Eckert (1969a).

becomes saturated (Fig. 18b). Similar anterior and posterior mechanoreceptor potentials also occur in *Euplotes* (Naitoh and Eckert, 1969b).

(*ii*) *Ionic mechanism for the anterior response.* Closer analysis of the anterior response revealed that it consists of an initial slow depolarization followed by a fast spike-like potential (Eckert *et al.*, 1972). General features of the fast component are identical with the regenerative calcium response evoked by a depolarizing current pulse. Effects of external cations on the fast component are similar to those on the regenerative response; that is, the peak value of the saturated fast component increases linearly with logarithmic increase in the external Ca^{2+} concentration, while the external K^{+} affects the value very little (Fig. 19). Thus, the fast component

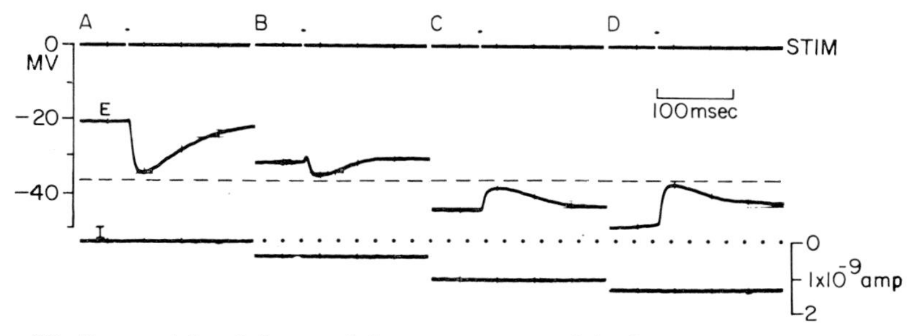

FIG. 20. Reversal level for caudal receptor potential of *P. caudatum* bathed in 1 mM $CaCl_2$ and 4 mM KCl_2. *Upper trace*, voltage pulses applied to the piezoelectric transducer. *Middle trace*, potential records. *Lowest trace*, hyperpolarizing currents applied with the intracellular electrode. No current applied in A. The caudal receptor potential reverses sign at about −37 mV. Naitoh and Eckert (1973).

appears to be a Ca^{2+}-dependent regenerative depolarization (i.e. the calcium response—see section II.A.3) in response to the initial slow depolarization induced by mechanical stimulation. The initial slow component is presumed to be due to a local inward ionic current through an increased permeability of the stimulated membrane of the cell anterior. While final evidence is still lacking for the identity of the ionic species carrying this receptor current, unpublished evidence (Friedman and Eckert) indicates that it is Ca^{2+}.

(*iii*) *Ionic mechanism for the posterior response.* The effects of external cations on the saturated peak value of the hyperpolarization (Fig. 19) indicate that a mechanical stimulus to the posterior surface causes the membrane of that region to undergo a selective increase in sensitivity to the potassium ion concentration (Naitoh and Eckert, 1973). If the membrane is hyperpolarized beyond a certain level by the injection of a constant inward current the posterior response inverts its polarity, producing a depolarization (Fig. 20). Such a reversal potential is characteristic of diffusion potentials produced by a transient increase in membrane permeability to one or more ions (Katz, 1966). These two lines of evidence indicate that mechanical stimulation of the posterior membrane increases its permeability to K^+, with the result that potassium ions carry current across the membrane to shift the membrane potential toward the K^+ equilibrium potential. Since the latter is more negative than the resting potential the potassium current flows out across the membrane, producing a hyperpolarization.

(*iv*) *Relation to ciliary activity*. The receptor depolarization produced as a direct consequence of mechanical stimulation of the cell anterior (i.e. anterior receptor potential) spreads uniformly over the cell, owing to the cable properties of the cell (see section II.A.3). This depolarization evokes, in turn, the voltage-sensitive regenerative increase in Ca^{2+} permeability throughout the entire cell membrane (i.e. the calcium response). As a consequence of the distributed influx of calcium ions, ciliary reversal takes place over the entire cell. When the anterior end of a free-swimming organism collides with a solid object the entire series of bioelectric steps described above ensue, and the resulting ciliary reversal produces the avoiding reaction.

The K^+-dependent hyperpolarizing posterior receptor potential also spreads electrotonically to produce a distributed hyperpolarization of the membrane. As in the case of a hyperpolarization produced by artificially injected inward current, this causes an increase in frequency of beating and accelerated forward locomotion. Thus, both anterior and posterior receptor potentials lead to adaptive locomotor behavior.

2. Geotaxis

Paramecium and certain other free-swimming ciliates, when introduced into a long vertical tube, show a tendency to swim upward against the direction of gravity (negative geotaxis). Any stimulus which increases swimming velocity of the specimen (such as gentle mechanical agitation, which evokes a hyperpolarization, or external application of Ca^{2+} ions) accelerates the accumulation of specimens at the upper end of a vertical tube. While the geotactic response of the specimens is very distinct when it occurs it is also quite labile. Thus, specimens from one culture may show strong geotaxis while those from another do not. It also varies in any given culture from day to day. This variability makes the quantitative study of geotaxis difficult. Various theories for the mechanism of geotaxis have been proposed (see Kuznicki, 1968). Close observation of individuals by Jennings (1906) revealed that avoiding reactions are absent or rare while specimens are swimming upward. If an individual begins to swim downward after reaching the top of the tube, avoiding reactions are repeated until it swims upward again. *Paramecium*'s center of gravity is reported to be located behind the center of the specimen (Verworn, 1889b; Dembowski, 1929a, b; Jahn and Votta, 1972). If this is true, upward orientation should be more stable than downward.

Why is it that downward oriented specimens show frequent avoiding reactions? Perhaps distortion of the membrane by gravitational force in the downward oriented specimen due to asymmetrical distribution of the body weight stimulates the mechanosensitive membrane and produces a depolarization.

C. *Responses to chemical stimuli*

1. *Chemosensitivity*

Forward-swimming ciliates show the avoiding reaction when coming into contact with certain kinds of chemicals in solution. Jennings (1906) examined the responses to various chemicals and found that (1) cations produce the avoiding reaction more effectively than anions, and (2) relative increases in the cation concentration induce the reaction.

If a drop of dilute salt solution is put in a thin layer of relatively concentrated solution of the same salt on a glass slide containing many of the ciliates, those that encounter the boundary of the diluted area continue to swim forward and enter the diluted solution. However, when these reach the opposite boundary they show avoiding reactions in response to the more concentrated salt of the surrounding solution. Consequently, specimens once entered into the region of dilute solution cannot leave it. The result is a collection of specimens in that region. In contrast, when a drop of a more concentrated salt solution is placed in the layer, the ciliates cannot enter the drop. The intensity of the avoiding reaction is a function of the difference in salt concentrations. The steepest concentration gradients evoke the strongest reactions.

Careful observation reveals that *Paramecium* increases its forward swimming velocity when it passes into a more dilute solution. The effect of pH is more complex. A change of pH from an optimum value (near neutral) in either direction at a border always induces an avoiding reaction in *Paramecium*. Therefore, when a drop of acid is placed in an alkaline layer (or vice versa) the ciliates collect in a ring-like region outside the acid (or alkaline) drop where the pH is optimum.

2. *Chemoreceptor region*

By analogy with the electric mechanism of mechanoreception in ciliates it seems likely that electric events in the membrane also are involved in behavioral responses to chemical stimulation. That is, a depolarization of the anterior membrane occurs when a forward-swimming specimen reaches the boundary of a more concentrated salt solution, spreading electrotonically over the cell to produce ciliary reversal (i.e. the avoiding reaction). In contrast, a hyperpolarization and resulting increase in beat frequency of cilia (and thus swimming velocity) occurs when an organism enters a more dilute salt solution.

It is noteworthy that the anterior of the ciliate is reported to be most sensitive to chemical stimuli by earlier workers (cf. Jennings, 1906). Naitoh (1961) applied KCl solution (15 mM) to various parts of *Opalina* with a micropipette, and found that in frog Ringer solution ciliary reversal

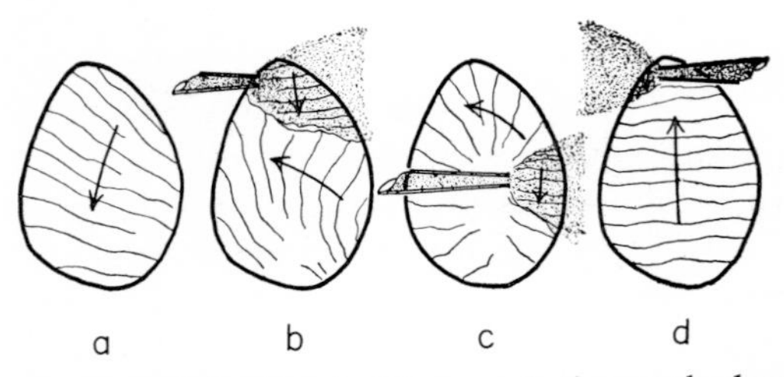

FIG. 21. Ciliary responses of *Opalina* in Ringer solution to the localized application of 15 mM KCl to various regions (b, anterior right; c, posterior right; d, anterior left) on the cell surface. a, no stimulus. Arrows indicate the direction of propagation of metachronal waves. Naitoh (1961).

takes place on those parts of the cell surface where the KCl solution is not applied (Fig. 21). The degree of ciliary reversal in the portions of the cell bathed in Ringer is greatest when the KCl depolarizing solution is applied to the left anterior part (Fig. 21d). In contrast, application to the right posterior portion evoked only a weak ciliary reversal (Fig. 21c). If a small part of the cell surface bathed in Ringer is electrically insulated from the external solution by a suction pipette, the cilia in the suction pipette do not show ciliary reversal upon application of the KCl solution to the cell anterior unless a conducting wire is connected between the inside of the pipette and the external solution. Measurement of the electric current through the wire during local application of KCl showed that the current is outward (depolarizing) through the membrane under the suction pipette. This current is stronger when the KCl is applied to the anterior left portion than to the posterior portion. On the other hand, if 15 mM KCl is applied to the anterior part of an *Opalina* immersed in 99 mM KCl (in which the membrane potential is almost zero) an increase of beating frequency occurs, and the current direction is inward through the cell under the pipette. These facts indicate that: (1) there is a high K^+-sensitive area on the cell membrane (the anterior region) where the membrane potential is most strongly depolarized by potassium ions; (2)

the depolarization spreads electrotonically to the rest of the cell and controls the ciliary activity; (3) electric and chemical (i.e. cationic) sensitivities are anatomically segregated in the membrane of *Opalina*, for the K^+-sensitive area (left anterior) does not correspond to the electrically sensitive area (right anterior) (Fig. 3a).

3. *Overall application of cations*

(*a*) *Difference between localized and overall application. Paramecium* shows a strong avoiding reaction when it encounters an area of relatively concentrated Ca^{2+} (or Mg^{2+}). However, when specimens are transferred suddenly into a high Ca^{2+} (or Mg^{2+}) solution they do not show any ciliary reversal and they continue to swim forward. On the other hand, ciliary reversal occurs in response to transfer of specimens into solution of high K^+ or Rb^+. It continues for scores of seconds in certain KCl solutions, causing the specimen to swim backward for long distances. It is well known that Ca^{2+} antagonizes this effect of K^+; if a high concentration of Ca^{2+} is present in the solution together with K^+, the transfer of the specimen into the solution will not result in ciliary reversal. K^+ as well as Ca^{2+} are known to depolarize the membrane of *Paramecium* (and other protozoans) (Fig. 4). Thus, one may ask why membrane depolarization upon the transfer of a specimen into a concentrated $CaCl_2$ solution does not produce ciliary reversal. Furthermore, ciliary reversal can be induced upon a transfer of specimens from concentrated mixture of Ca^{2+} and K^+ to a diluted one if concentration ratios between these two ions are adequately controlled (Naitoh, 1968). In this case the membrane is hyperpolarized upon the transfer.

These observations have led Naitoh (1968) to propose a mechanism for ciliary reversal by overall application of cations different from that by electric current depolarization of the cell membrane (see next section).

(*b*) *Gibbs–Donnan ratio and ciliary reversal.* Many investigators (Mast and Nadler, 1926; Oliphant, 1938, 1942) examined the effect of overall applications of various chemicals on ciliary movement in *Paramecium* and other ciliated protozoans. Their general conclusions are (1) monovalent cations in increased concentration are all more or less effective in evoking ciliary reversal, (2) anions only modify slightly the duration of ciliary reversal by cations, (3) divalent cations, especially Ca^{2+}, antagonize the stimulating effect of monovalent cations. Kamada and Kinosita (1940) examined the relation between the duration of ciliary reversal and the concentrations and ratios of K^+ ions and Ca^{2+} ions in the external media. The duration of ciliary reversal was dependent on both the ionic composition of the stimulation medium ($[K^+]/[Ca^{2+}]$ and concentrations of each cation) to which the specimens are transferred and that of the equilibration medium in which the specimens are kept prior to transfer. Jahn (1962) analysed their data with the assumption that ciliary reversal depends on the amount of cations bound to the membrane in accordance with the Gibbs–Donnan principle, in which the ratio of the amount of K^+ to that of Ca^{2+} bound by anionic sites of the membrane is proportional to the ratio of $[K^+]$ to the square root of $[Ca^{2+}]$ in the external solution. He found that ciliary reversal occurred in the published data when the ratio $\frac{[K^+]}{[Ca^{2+}]^{\frac{1}{2}}}$ in the external solution is raised, independent of the absolute concentrations of

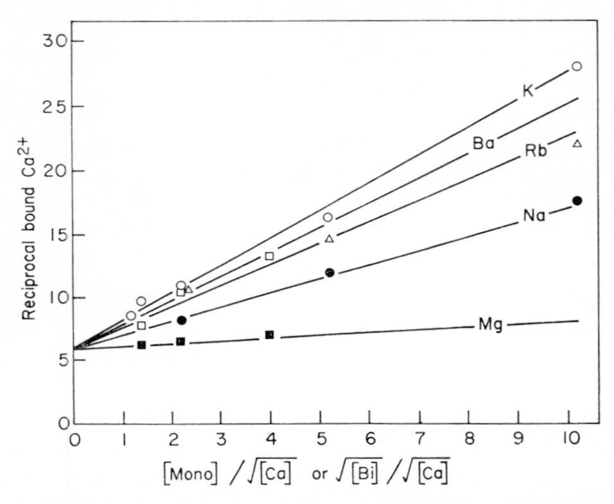

FIG. 22. The reciprocal value of the number of calcium ions bound to *Paramecium* as a function of the ratio of [monovalent cation] or [divalent cation]$^{\frac{1}{2}}$ to $[Ca^{2+}]^{\frac{1}{2}}$ in the external medium. The straight line relation for each cation species shows that cation binding obeys the Gibbs-Donnan principle. The slope of each line indicates the binding affinity of each cation species to the anionic binding sites on *Paramecium*. After Naitoh and Yasumasu (1967).

K^+ and Ca^{2+}. He proposed that the release of bound Ca^{2+} in response to an increase in the ratio $\frac{[K^+]}{[Ca^{2+}]^{\frac{1}{2}}}$ is responsible for ciliary reversal. Naitoh and Yasumasu (1966) supported Jahn's prediction that cationic bindings by *Paramecium* obeys the Gibbs-Donnan principle. They used Ca^{45} to measure binding affinities of several cations to *Paramecium* (Fig. 22). It should be noted that cations such as K^+, Rb^+ and Ba^{2+}, with high affinities for binding to the anionic binding sites of *Paramecium*, induce more prolonged ciliary reversal when they are applied externally than do those ions which bind less well. Cations with a very low affinity, such as Mg^{2+}, do not induce ciliary reversal.

(*c*) *A calcium hypothesis for ciliary reversal by cationic stimulation.* Naitoh (1968) compared the duration of ciliary reversal to the amount of Ca^{2+} bound by *Paramecium*, and found that the duration of ciliary reversal is correlated both with the amount of Ca^{2+} previously bound by the specimen and with the amount liberated upon transfer. On the basis of the findings an hypothesis was proposed for the mechanism controlling reversed ciliary beating in response to stimulation by externally applied cations. It proposes that externally applied cations bind to anionic sites on *Paramecium* in exchange for bound calcium in a manner consistent with the law of mass action (thermodynamically equivalent to the Gibbs–Donnan principle). The calcium ions liberated increase the cytoplasmic calcium concentration around the ciliary apparatus, activating its calcium-sensitive reversal mechanism. Activation of the reversal mechanism, in turn, results in the reversal of the beating direction of the cilia. The presence of a calcium-sensitive reversal mechanism in the ciliary apparatus has been demonstrated in extracted models of *Paramecium* (Naitoh, 1969; Naitoh and Kaneko, 1972, 1973; see section II.F). The duration of reversed beating is thought to be the time between the beginning of calcium release (when the $\frac{[K^+]}{[Ca^{2+}]^{\frac{1}{2}}}$ ratio is experimentally raised) and the time when the rate of release of Ca^{2+} drops below a critical level as the binding system approaches an equilibrium with the stimulation medium. The initial amount of Ca^{2+} is thought to affect the time course of liberation of the Ca^{2+}.

(*d*) *Electrical properties of the membrane and bound cations.* It has been proposed that coulombic interactions between ions and charged sites in membranes is the basis for membrane selectivity in ion permeation Hagiwara and Takahashi, 1967; (Diamond and Wright, 1969). Thus, coulombic ion binding might be expected to influence the electric properties of a membrane. In *Paramecium* membrane conductance was found to increase with increasing $\frac{[K^+]}{[Ca^{2+}]^{\frac{1}{2}}}$ in the external solution, which decreases the amount of bound Ca^{2+} (Naitoh and Eckert, 1968a). The time course of the action potential is also a function of external $\frac{[K^+]}{[Ca^{2+}]^{\frac{1}{2}}}$ (Naitoh *et al.*, 1972; Friedman and Eckert, 1973). The duration of the barium action potential appears to be proportional to the amount of bound Ba^{2+} (Naitoh and Eckert, 1968b).

In order to explain both electrically stimulated and ionically evoked ciliary reversal by a common mechanism, Eckert (1972) interpreted ciliary responses to cationic stimulation (Naitoh, 1968) in terms of membrane potentials and conductances. In this view the common mechanism coupling ciliary reversal to various forms of stimulation is an increase in the rate at which calcium enters the cilia through the surface membrane, driven by its electrochemical gradient. Thus any factor, such as increased calcium permeability, which causes Ca^{2+} to enter at a rate faster than it is removed should lead to an increase in the intraciliary concentration of Ca^{2+}, and, if this is sufficient, to a reversal of the ciliary beat. The simplest case would be a cationic stimulus (e.g. increased K^{+} concentration) which produces membrane depolarization, and thereby an increase in the calcium conductance of the membrane. However, the effects of cationic stimuli are complicated by the sensitivity of membrane conductance to the ratio between the ambient calcium concentration and the concentrations of other ions in the medium (Naitoh and Eckert, 1968a; Friedman and Eckert, 1973). The bioelectric explanation of ion antagonism in ciliary reversal is examined in more detail elsewhere (Eckert, 1972).

We have summarized above two alternative explanations for Ca–K antagonism in the ionic stimulation of ciliary reversal; first the cation-exchange hypothesis of Jahn (1962) and Naitoh (1968) and second, the calcium current hypothesis of Eckert (1972). Both of these pose certain problems. The ion-exchange hypothesis was proposed for the special case where changes in the external solution evoke reversed beating. Thus, it requires that there be two different mechanisms leading to an increase in intracellular free calcium, one in which depolarization of the membrane by a variety of stimuli produces an increase in calcium permeability, and another in which calcium bound to the surface is liberated by ion exchange. It also requires that Ca^{2+} liberated from the membrane move into the cilium rather than diffuse into the medium. The calcium current hypothesis is the more parsimonious of the two since it seeks to explain all cases of ciliary reversal (i.e. those produced by cationic stimulation as well as those in response to outward electric current) according to familiar bioelectric principles: that is, by the influx of calcium through the membrane down its electrochemical gradient. However, further experiments must be done before certain cases of cationic stimulation can be rigorously explained by the calcium current hypothesis.

Regardless of our views on the role of ion antagonism in ciliary reversal, the validity of the calcium current hypothesis for physiologically normal stimuli (i.e. mechanoreception or chemoreception by the anterior receptor membrane) in ciliates is accepted by us both (Eckert and Naitoh, 1972).

D. Responses to photostimulation

1. Phobic responses

Some species of Protozoa are sensitive to light. Engelmann (1882) found that when the green flagellate *Euglena* passes from a lighted area into a shaded area the beating form of its flagella undergoes a sudden change, driving the organism backward for a distance (Fig. 23). The flagellate then resumes forward locomotion in a new direction. As a result, the flagellate tends to remain in the lighted area. Engelmann called this response the "shock movement". The shock movement does not occur when specimens pass from a dark to a lighted area, thus they collect in the lighted area. The same kind of photophobic response is observed in some species

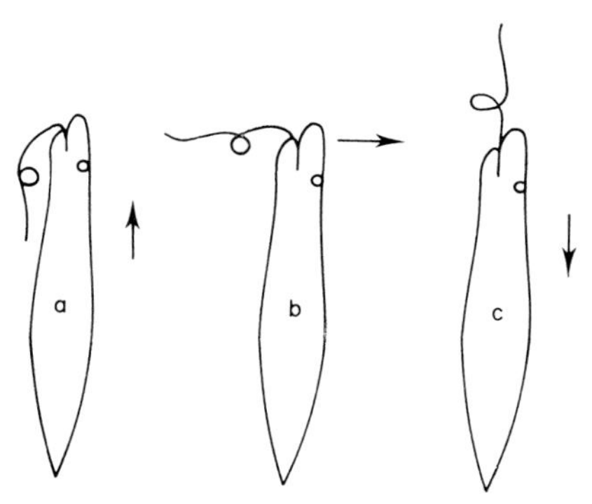

FIG. 23. Diagram showing the position of the flagellum as seen in a viscous medium; *a*, when *Euglena* is swimming forward in a narrow spiral; *b*, when swerving sharply towards the dorsal side; *c*, when moving backwards. Bancroft (1913).

of colored *Stentor*. Free-swimming specimens of *S. niger* show reversal of the body cilia when they come to a shaded region, propelling the specimens away from the shadow, and causing them to remain in the lighted area (Tuffrau, 1957). In contrast, free-swimming specimens of *S. coeruleus* show ciliary reversal when they come to a lighted area from a shaded area. Thus, the lighted area remains empty (Jennings, 1906; Mast, 1911).

2. *Photoreceptor organelle*

Engelmann (1882) and others (Jennings, 1906; Mast, 1911) found that the anterior region of *Euglena* is very sensitive to changes in light intensity (as manifested in the phobic reaction in its flagellum) but that the stigma, commonly called "eye spot", is not photosensitive. More recently the stigmaless mutants of *Euglena* have been found to be photosensitive (Luntz, 1931; Gossel, 1957). The photosensitive region is thought to be at the base of flagellum near the stigma, where the flagellum is somewhat swollen.

The anterior region of some species of *Stentor* (*S. coeruleus*; *S. niger*) is more sensitive to light than the rest of the cell body (Jennings, 1906; Mast, 1911).

3. *Phototaxis*

Photosensitive protozoans frequently swim towards a light source (positive phototaxis) or away from a light source (negative phototaxis) (Buder, 1919). As early as 1882, Engelmann explained the mechanism for phototaxis in *Euglena* as follows: *Euglena* swims forward in a narrow spiral, keeping its dorsal side (the side bearing the stigma) on the outside of the spiral. When the light is applied perpendicularly to the direction

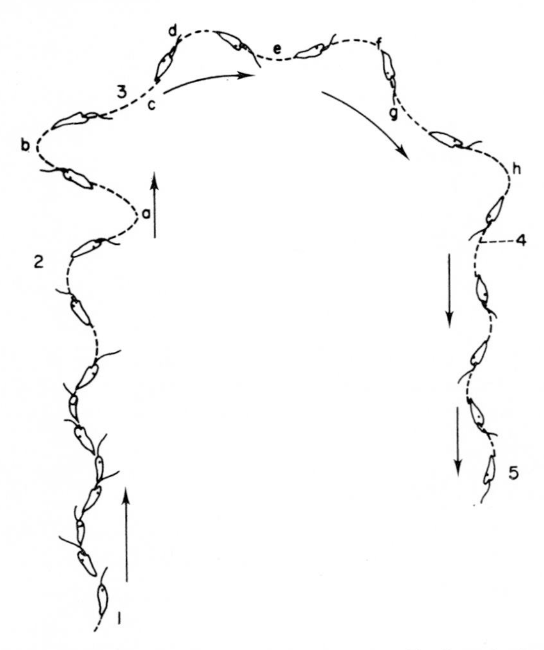

FIG. 24. Illustration of the devious path followed by *Euglena* in becoming oriented when the direction of the light is reversed. From 1 to 2 the light comes from the top of the figure; after 2 it comes from the bottom of the figure. The amount of wandering (a–h) varies in different cases. Jennings (1906).

of locomotion, shading of the photoreceptor region by the stigma during the course of spiral movement leads to a modification of the flagellar movement which rotates the specimen toward its dorsal side. By repeating the phobic response the axis of locomotion becomes aligned parallel with the rays of light, and the cell swims toward the source of illumination. If it deviates from a path parallel to the beam of light the stigma will cast a shadow on the sensitive portion and a correctional reorientation is executed (Fig. 24).

When the light intensity is stronger than 10^5 erg cm^{-2} s^{-1} (Diehn, 1969a;

see also Stahl, 1880) phobic responses occur upon further increases in the light intensity (negative phobic response), causing *Euglena* to swim out of the illuminated region. The phobic response ceases when the specimen is positioned so that the photoreceptor is then shaded by the cell body.

The shaded receptor hypothesis of Engelmann is also applicable to the *phototaxis* of *Stentor* (Jennings, 1906), and is generally accepted by modern investigators. In this case it is assumed that the light-sensitive anterior portion is shaded by the rest of the body. Diehn (1969a) compared the action spectra for the phobic response and for the phototaxis in *Euglena* by making use of a "phototaxigram" (Lindes *et al.*, 1965), in which a population of *Euglena* is photometrically monitored. Diehn (1969b) demonstrated that phobic responses are significantly accelerated in intermittent light with a frequency close to that of the rotation of *Euglena* during forward swimming. This may be due to an entrainment of the rotation of the specimens. These findings support Engelmann's hypothesis that phototaxis is based on the phobic response.

4. *Mechanism of locomotor responses to photostimulation*

There are no reported electrophysiological studies of the mechanism of photoreception in Protozoa, nor have there been extensive observations on ciliary or flagellar responses to photostimulation. However, it is reasonable to suspect that photostimulation evokes locomotor responses through bioelectric events originating in a photosensitive membrane as in visual receptor cells of higher organisms. The swimming direction of *Euglena* changes in an electric field due to changes in the direction and form of the flagellar movement similar to the phobic response of the flagellum in response to photostimulation (Bancroft, 1913). This suggests that electric events are involved in the photo responses of the flagellum.

The photoreceptor pigment in *Euglena* appears to be a carotenoid, for portions of the action spectrum of the photophobic response match the absorption spectrum of carotene (Diehn, 1969a). Differences in the spectra at the shorter wavelengths can tentatively be attributed to the absorption spectrum of screening pigments.

E. *Thermal sensitivity*

Mendelssohn (1902) demonstrated that when the ends of a glass tube containing paramecia are held at different temperatures the specimens collect near the end of the tube with the temperature closer to that under which they were cultured. The response of the specimens is so sensitive that a temperature difference of 3°C along a tube 10 cm long leads to an unequal distribution of the specimens toward the optimum temperature.

Closer observation of the movement of individuals indicates that the specimens moving from an optimal temperature toward a hotter or colder region show more frequent avoiding reactions than in an optimum temperature (Jennings, 1906). This results in an unequal distribution.

Oliphant (1938) found that *Paramecium* does not show any ciliary reversal upon transfer of a specimen equilibrated at one temperature to a test solution of the same ionic composition but with a higher or lower temperature. Thus it appears that temperature changes do not produce large potential changes. This is supported by the finding of Yamaguchi (1960) that a 10°C change in temperature produces a potential change of only 2 mV.

F. Control of ciliary orientation in extracted models

It has proved useful for our understanding of the coupling mechanism between the membrane events and ciliary activity to examine the effect of cations involved in membrane events on extracted cell models in which the surface membrane is functionally disrupted. In this approach externally applied chemicals have direct access to the cell interior without regulatory interference by the membrane. Hoffmann-Berling (1955) introduced the technique of glycerination (first used on muscle by Szent-Györgyi, 1949) to make extracted models of cilia and flagella, and found that the extracted organelles are reactivated by application of ATP and Mg^{2+}. Many investigators (see Arronet, 1971, for detailed references) confirmed the findings of Hoffmann-Berling in various extracted models from many kinds of ciliates, flagellates, and metazoan ciliated tissues. Recently Gibbons and Gibbons (1972) used a neutral detergent, octylphenoxymethoxyethanol (Triton X-100), to obtain excellent extracted models of sea urchin spermatozoa. Naitoh and Kaneko (1972, 1973) used Triton X-100 to make extracted models of whole paramecia. Cilia of models extracted for 30 min in 0·01% (by volume) Triton X-100 are reactivated in solutions which contain ATP and Mg^{2+}. If the Ca^{2+} concentration in the ATP-Mg^{2+} reactivation medium is buffered with EGTA below 10^{-7} M, the direction of swimming is forward as in the unextracted live specimens in the absence of stimuli. When the Ca^{2+} concentration is raised above 10^{-6} M the models swim backward due to reversed beating of the reactivated cilia; this corresponds to the backward swimming of live specimens in response to depolarizing stimuli (Figs 25, 26). In solutions containing ATP (without Mg^{2+}) the cilia do not beat. If the calcium concentration is below 10^{-6} M in such a solution the static cilia point toward the cell posterior. If the calcium ion concentration is raised above 10^{-6} M the cilia shift their

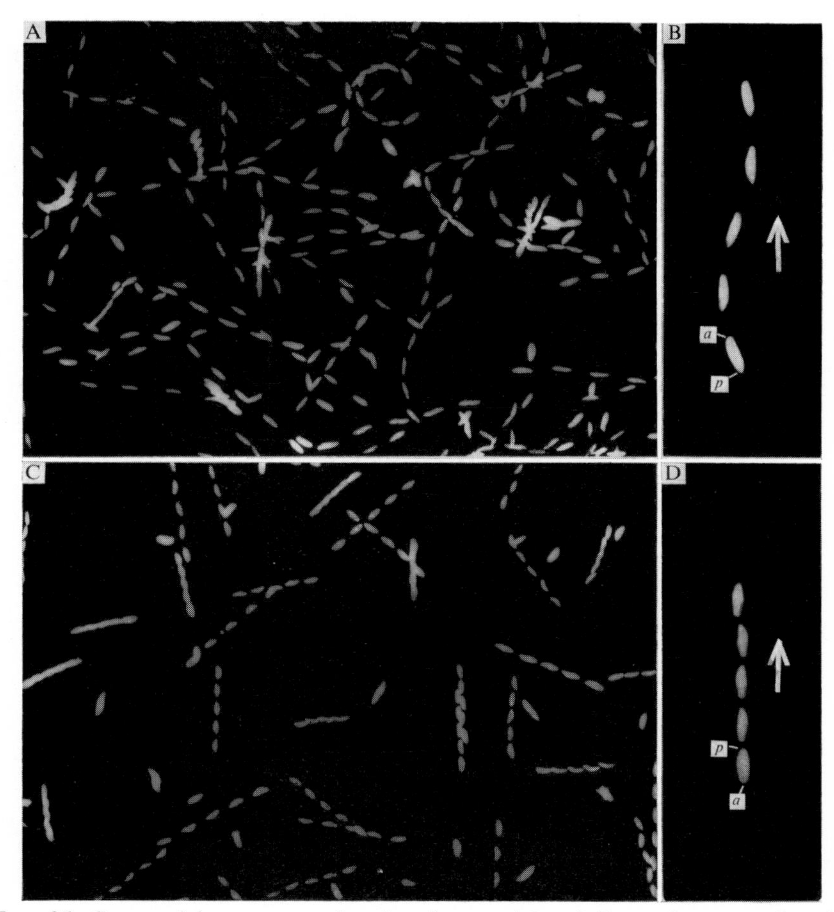

FIG. 25. Sequential exposure of swimming models of *P. caudatum* made with five successive xenon flashes, 1 s apart, in a thin (70 μm) layer of reactivation medium. Since the first flash was the strongest, the first image in each sequence is brightest. A. Forward-swimming models in a medium of 4 mM ATP, 4 mM $MgCl_2$ and 3 mM EGTA (thus, free Ca^{2+} below 10^{-8} M) (about $\times 9.2$). B. A forward-swimming model at higher magnification (about $\times 22$). C. Backward-swimming models in 4 mM ATP, 4 mM $MgCl_2$ and 5×10^{-5} M $CaCl_2$ (about $\times 9.2$). D. Backward-swimming model at higher magnification (about $\times 22$). After Naitoh and Kaneko (1973).

positions so as to point toward the cell anterior without beating; this corresponds to the orientation of cilia beating in reverse. A similar effect of Ca^{2+} is found in models of other protozoans, such as *Opalina* (Naitoh, unpublished) and *Euplotes* (Epstein, 1972). It appears, then, that there are at least two kinds of motile components in the protozoan cilium, one of which is concerned with cyclic bending of the cilium and another which governs the orientation of the effective power stroke. Both components

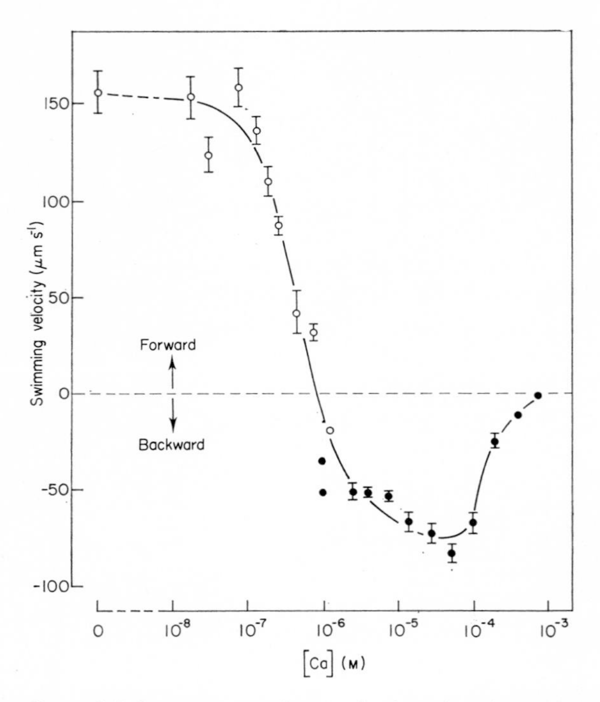

FIG. 26. The effect of Ca^{2+} concentration on both swimming velocity and swimming direction of models of *P. caudatum* reactivated by ATP and Mg^{2+}. The graph gives means and standard errors for measurements on 100–500 specimens. Concentrations of Ca^{2+} were adjusted by Ca^{2+} buffers in the range between 10^{-8} and 10^{-6} M (open circles) and by simple addition of $CaCl_2$ in the range above 10^{-6} M (solid circles). Naitoh and Kaneko (1972).

use ATP as an energy source, while the former requires Mg^{2+} and the latter Ca^{2+} as cofactors for their ATPase activities (Naitoh, 1969). Naitoh (1966) demonstrated that Ni-inhibited non-beating cilia of live *Paramecium* reverse their orientation in response to K^+ or electric stimulation.

The demonstration that the calcium concentration of the ciliary environment regulates ciliary reversal in extracted models supports the hypothesis that it is Ca^{2+} which couples the activity of cilia to the activity of the surface membrane in living cells (see sections II.A.3, II.C.3).

III. Control of beating frequency

A. *Electrophysiological correlates*

1. *Hyperpolarization*

Kamada (1931) noticed in galvanotaxis experiments that an increase in the beating frequency of cilia in the normal direction takes place on the surface of *Paramecium* facing the anode. Naitoh (1958) reported that an injection of hyperpolarizing (inward through the membrane) current into *Opalina* induces an increase in the beating frequency. This has been reported also in other protozoans (*Paramecium*: Eckert and Naitoh, 1970; *Euplotes*: Naitoh and Eckert, 1969b; Epstein, 1972). Machemer and Eckert (unpublished) recorded an increase in the beating frequency of cilia in *Paramecium* in response to a brief intracellular injection of a hyperpolarizing current. The increase is greater and lasts longer in response to stronger hyperpolarizations.

As already described (section II.B.1), mechanical stimulation of the posterior membrane of ciliates produces a hyperpolarizing mechano-receptor potential, which leads to an increase in the beating frequency of cilia in the normal direction, and in the free-swimming ciliate accelerates forward locomotion.

Decrease in the external cationic concentration results in an increase in the steady state membrane potential (hyperpolarization) (Kamada, 1934). Kinosita *et al.* (1964b) determined the beating frequency of cilia in *Paramecium* equilibrated in various salt solutions with different cationic species and concentrations. The beating frequency is higher in specimens with higher membrane potential equilibrated in more dilute solutions, irrespective of the cation species. Thus, it appears that hyperpolarization by any means stimulates ciliary activity. However, the beating frequency also increases in high calcium concentrations which produce some depolarization. The mechanism of frequency control is not known.

2. *Depolarization*

Sudden depolarization of the membrane produces a transient increase in the beating frequency associated with the reversal. Kinosita *et al.* (1965) reported that more than 300% increase in the frequency occurs in association with ciliary reversal in response to spontaneous depolarizing action potentials in a Ba^{2+}–Ca^{2+} mixture. Machemer and Eckert (unpublished) recorded an increase in the beating frequency in association with ciliary reversal induced by a depolarizing current pulse. In contrast to the increase produced by hyperpolarizing pulses, the latency to peak frequency is very short (several milliseconds), and similar to that for ciliary reversal. Reversal and rapid beating are terminated by a quasi-inactive period followed by a final period of rapid beating in the normal direction (see Machemer, p. 231, this volume).

Kinosita and Okajima (1968) and Epstein (1972) reported that in *Euplotes* the membrane potential modulates the frequency of normal

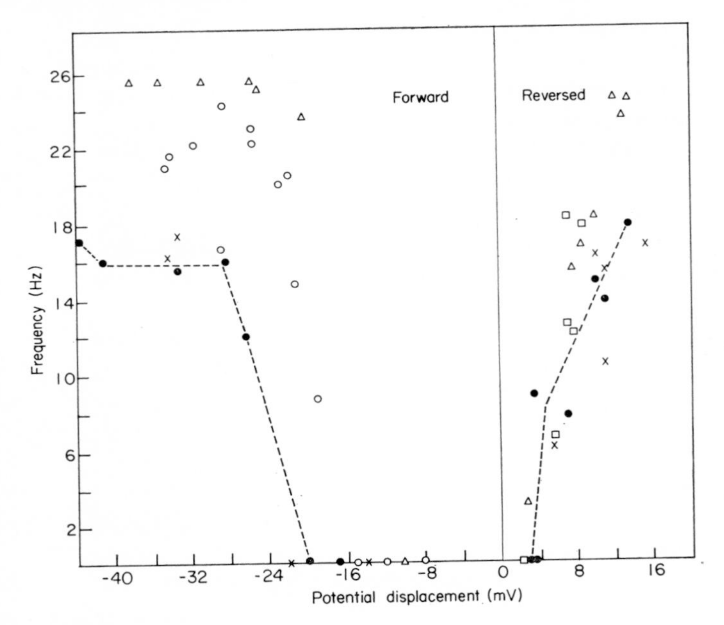

FIG. 27. The frequency of beating of the anal cirri of *Euplotes* sp. in response to electric stimulation. Evoked frequencies are plotted against the shift in membrane potential produced by a 300-ms pulse of injected current. Reversed beating, in response to depolarization, is shown on the right side of the graph, and normal beating, in response to hyperpolarization, is shown on the left. Different symbols represent different specimens; the dashed line connects the values recorded for one specimen. Epstein (1972).

beating of the cirri from zero to maximum frequencies of up to 30 beats per second. The beating stops between membrane potentials of about −20 and + several millivolts (Fig. 27). At higher depolarizations the cirri beat in reverse. The voltage–frequency relations are quite steep for both normal and reversed beating, as seen in Fig. 27. In *Paramecium* the frequency is lowered by reducing the membrane potential with raised extracellular cationic concentrations other than Ca^{2+} or Mg^{2+} (Kinosita *et al.*, 1964b). However, beating in *Paramecium* (and also in *Opalina*:

Naitoh, 1958) cannot be stopped by this method even though the ionic strength is raised sufficiently to depolarize the membrane completely. Under natural conditions, however, paramecia sometimes stop swimming when they gently come into contact with the surface of a solid object. In this state all the cilia on the cell body stop beating ("contact reaction" of Jennings, 1906). It is not known how this is regulated by the membrane.

B. *Control of beating frequency in extracted models*

Glycerol- or detergent-extracted models of ciliates and flagellates are reactivated in the presence of ATP and Mg^{2+} (Brokaw, 1961, 1963; Seravin, 1961; Gibbons, 1965; Naitoh and Kaneko, 1972, 1973). Beating frequency in such reactivated models is a function of Mg^{2+} and/or ATP

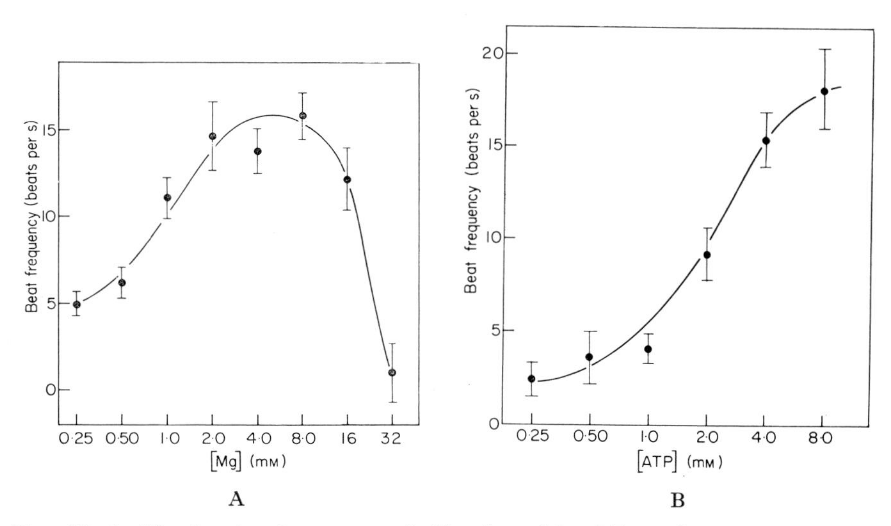

FIG. 28. A. The beating frequency of cilia of models of *P. caudatum* as a function of Mg^{2+} concentration in the reactivation medium. Concentrations of ATP (4 mM), H^+ (pH 7·0) and EGTA (3 mM) were kept constant throughout.

B. The beating frequency of cilia of the models as a function of ATP concentration in the reactivation medium. Concentrations of Mg^{2+} (4 mM), H^+ (pH 7·0) and EGTA (3 mM) were kept constant throughout. Naitoh and Kaneko (1973).

concentrations (Fig. 28). The model cilia can be reactivated in an ATP-Mg^{2+} medium with EGTA (Ca^{2+} concentration below 10^{-9} M) (Gibbons, 1965; Naitoh and Kaneko, 1973). This indicates that calcium ions are not required for the reactivation of the model cilia. The intraciliary concentration of Mg^{2+} regulated bioelectrically by the surface membrane may be involved in the control of beating frequency in live specimens.

Naitoh and Kaneko (1972) further examined the effect of Ca^{2+} concentration on the beating frequency of ATP-Mg^{2+} reactivated cilia of Triton-extracted paramecia. As shown in Fig. 29, the beating frequency shows a minimum at a Ca^{2+} concentration of about 10^{-7} M. An increase in concentration significantly raises the frequency, and a maximum occurs at about 10^{-6} M. At this concentration the model cilia reversed their beat direction (Fig. 26). Although the effect of Ca^{2+} on the beating frequency

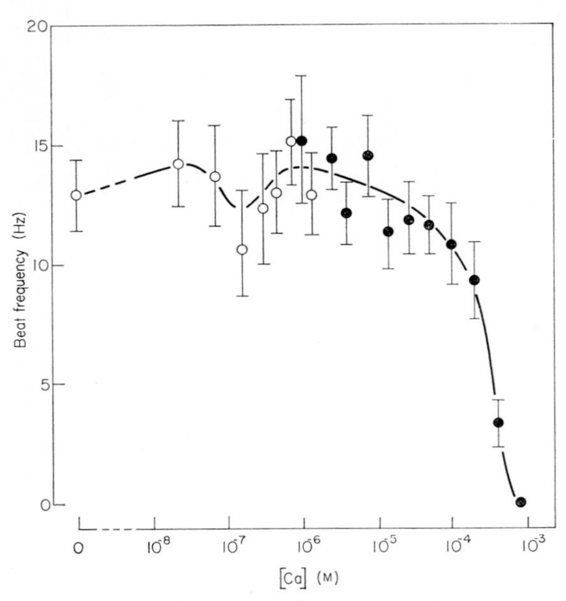

FIG. 29. Effect of Ca^{2+} concentration on beat frequency of cilia in models of *P. caudatum* reactivated by ATP and Mg^{2+}. Symbols as Fig. 26; means and standard errors for measurements on 50–100 specimens are given. After Naitoh and Kaneko (1972).

of the reactivated model cilia is not as striking as that associated with ciliary reversal evoked by membrane depolarization in the living cell, this intrinsic behavior of the ciliary apparatus may underlie the frequency increase which occurs in association with Ca^{2+}-dependent depolarization.

There is no topographic differentiation in the sensitivity of the extracted cilia to reactivation by ATP and Mg^{2+}. The frequency of well reactivated extracted cilia in an optimum ATP-Mg^{2+} medium is similar on all portions of the cell body (Fig. 30) (Naitoh and Kaneko, 1973). In contrast, the beating frequency differs slightly from region to region on the cell surface of live specimens (Okajima, 1953). The rate of change in beating frequency in response to electric stimuli also differs at different locations on the cell surface (Machemer and Eckert, unpublished). These observations indicate that localized differences in the nature of ciliary beating, which are characteristic of the live ciliate, are dependent on localized differentiation of membrane functions controlling beating frequency (Naitoh and Kaneko, 1973).

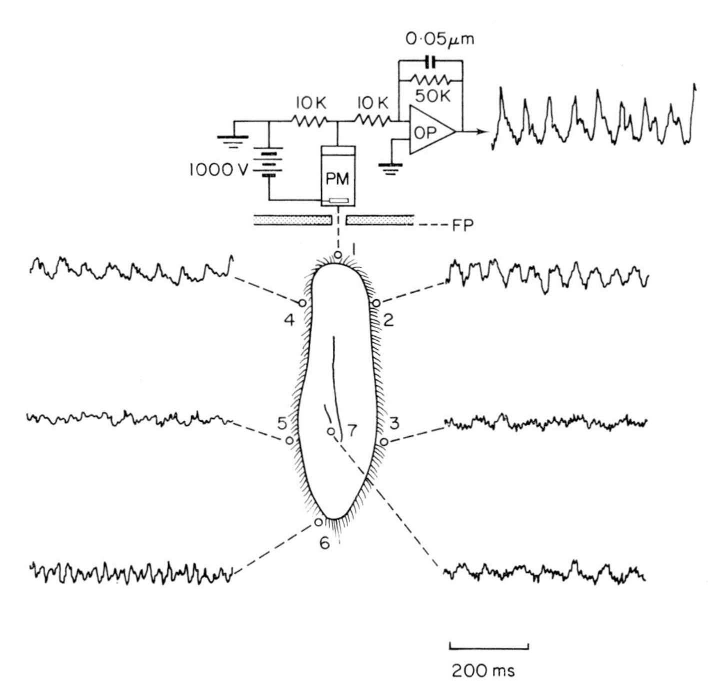

FIG. 30. Photometric monitoring of the beating frequency of reactivated cilia of a *Paramecium* model. The photometer sampled small areas of the image (small circles with numbers beside or on the model show the locations). This portion of the image was projected on to a photomultiplier (*PM*) through a small hole (0·5 mm in diameter) in a screen (*FP*). Changes in light intensity due to metachronal waves were amplified, displayed on an oscilloscope and photographed. *OP*, operational amplifier. Frequency of the electrical signal from location 6 was doubled due to lashing type beat (without metachronal waves) of the cilia at that location. Naitoh and Kaneko (1973).

IV. Concluding remarks

Bioelectric events in the cell membranes of Protozoa control both beating frequency and the spatial orientation of the movements of cilia and flagella. The coupling of locomotory activity to bioelectric behavior of the surface membrane endows the ciliated protozoans with the ability to react adaptively to various stimuli. These stimuli are detected by means of the transducer characteristics of differentiated portions of the surface membrane. An adequate stimulus applied to a receptor region changes the membrane permeability of that region and an ionic current flows across the membrane which produces a receptor potential (i.e. a transient depolarization or hyperpolarization). The ionic mechanisms of these receptor events are similar to those in some of the receptor cells of metazoans.

Electric currents spread within protozoan cells according to the same cable principles that apply in nerve and muscle cells. The membrane resistance, the conductances of the intra- and extra-cellular fluids and the size and shape of protozoan cells leads to a nearly isopotential condition for many species. *Paramecium*, for example, shows an electrotonic decay of only a few per cent over its length. Thus, receptor potentials spread electrotonically over the entire cell membrane with little time delay or decay, without the need for all-or-none action potentials or specialized conducting structures analogous to the nerve fibers of metazoans.

In the best known example, *Paramecium*, the cell membrane produces a regenerative increase in the membrane conductance to certain cations (primarily Ca^{2+}) in response to depolarization. Thus, depolarizations due to receptor current are amplified by the graded regenerative calcium response of the cell membrane. The distributed influx of Ca^{2+} causes a concomitant increase in the cytoplasmic calcium concentration which activates the reversal mechanism in the ciliary apparatus. This chain of events produces the avoiding reaction so fundamental in the locomotor behavior of ciliates.

Hyperpolarizing receptor potentials also spread electrotonically and induce an increase in the beating frequency of cilia. This results in rapid forward swimming of the specimen. The coupling of ciliary frequency to membrane potential remains unknown. The membrane potential of Protozoa can also fluctuate spontaneously, producing spontaneous changes in ciliary activity. It is suggested that certain kinds of behavior of protozoans (e.g. geotaxis and thermosensitive reactions) result from stimulus-dependent modulation of the rate of "spontaneous" potential fluctuations and thereby the frequency of occurrence of avoiding reactions in one direction of locomotion (e.g. with increasing or decreasing temperature) as opposed to the opposite direction.

Ion exchange phenomena in protozoan membranes may also contribute to the regulation of ciliary activity. It has been conjectured that Ca^{2+} liberated from membrane anionic sites by the competitive binding of monovalent cations increases the cytoplasmic Ca^{2+} concentration, and thereby induces ciliary reversal. Such a mechanism may supplement the influx of Ca^{2+} from the extracellular fluid when stimuli produce a depolarization of the membrane and a regenerative calcium response.

The evidence suggests that the protozoan ciliary system has at least two kinds of motile components, one of which corresponds to cyclic bending (Mg^{2+}-dependent) while the other governs ciliary orientation (Ca^{2+}-dependent). Ciliary bending appears to be mediated by mechanochemical interaction between pairs of outer tubules in cilia (Summers and Gibbons, 1971). Tamm and Horridge (1970) proposed a hypothesis that the orientation of the central fibrils governs the direction of effective power stroke (see also Fawcett and Porter, 1954; Gibbons, 1961). Dynein which is extracted from the arms of the outer tubules, shows a high Mg^{2+}-dependent ATPase activity (Gibbons, 1966), and is activated by Ca^{2+} to a lesser degree. At present, there are no biochemical data at the level of ciliary fine structure to which we can turn for insight concerning the mechanism of Ca^{2+}-activated ciliary reversal.

Cilia on different portions of the cell surface show different degrees of ciliary reversal in response to an overall uniform membrane depolarization. This results, at least in part, from intrinsic differences in the sensitivity of the ciliary apparatus in different locations to calcium ions. In contrast, differences in the beating frequency of the cilia in different locations on the surface appear to be due to factors extrinsic to the ciliary apparatus. These may include regionally differentiated regulatory functions of the membrane. The topographic differences in the sensory and regulatory functions of the membrane, topographic differences in Ca^{2+} sensitivity of the ciliary apparatus itself, and morphological specialization of cilia in different locations on the cell surface all contribute to the complexity and adaptiveness of the locomotor apparatus and the movements of the organism in its fluid environment.

References

ALLEN, R. D. (1971). *J. Cell Biol.* **49**, 1–20.
ARRONET, N. I. (1971). "Muscle and cell contractile models." Academy of Sciences of the U.S.S.R.
BANCROFT, F. W. (1905). *Arch. ges. Physiol.* **107**, 535–556.
BANCROFT, F. W. (1906). *J. Physiol.* **34**, 444–463.
BANCROFT, F. W. (1913). *J. exp. Zool.* **15**, 383–428.
BROKAW, C. J. (1961). *Expl Cell Res.* **22**, 151–162.
BROKAW, C. J. (1963). *J. exp. Biol.* **40**, 149–156.
BUDER, J. (1919). *Jb. wiss. Bot.* **58**, 105–220.
CHEN, V. K. (1972). Ph.D. dissertation. University of New York, Buffalo.
COLE, K. S. (1949). *Arch. Sci. Physiol.* **3**, 253–258.

DEMBOWSKI, J. (1929a). *Arch. Protistenk.* **66**, 104–132.
DEMBOWSKI, J. (1929b). *Arch. Protistenk.* **68**, 215–260.
DIAMOND, J. M. and WRIGHT, E. M. (1969). *A. Rev. Physiol.* **31**, 581–646.
DIEHN, B. (1969a). *Biochim. biophys. Acta* **177**, 136–143
DIEHN, B. (1969b). *Expl Cell Res.* **56**, 375–381.
DOROSZEWSKI, M. (1961). *Acta Biol. exp.* **21**, 15–34.
DOROSZEWSKI, M. (1963). *Acta Biol. exp.* **23**, 3–10.
DOROSZEWSKI, M. (1965). *Acta Protozool.* **3**, 175–182.
DOROSZEWSKI, M. (1968). *Acta Protozool.* **5**, 291–296.
DOROSZEWSKI, M. (1970). *Acta Protozool.* **7**, 353–362.
DRYL, S. (1961). *J. Protozool.* **8**, (Suppl.), 16.
DRYL, S. and GRĘBECKI, A. (1966). *Protoplasma* **62**, 255–284.
DUNHAM, P. (1973). *In* "*Biology of* Tetrahymena" (A. M. Elliott, ed.). Dowden, Hutchinson and Ross.
ECKERT, R. (1972). *Science, Wash.* **176**, 473–481.
ECKERT, R. and NAITOH, Y. (1970). *J. gen. Physiol.* **55**, 467–483.
ECKERT, R. and NAITOH, Y. (1972). *J. Protozool.* **19**, 237–243.
ECKERT, R. and SIBAOKA, T. (1968). *J. gen. Physiol.* **52**, 258–282.
ECKERT, R., NAITOH, Y. and FRIEDMAN, K. (1972). *J. exp. Biol.* **56**, 683–694.
ENGELMANN, T. W. (1882). *Arch. ges Physiol.* **29**, 387–400.
EPSTEIN, M. (1972). Ph.D. dissertation. University of California, Los Angeles.
FAWCETT, D. W. and PORTER, K. R. (1954). *J. Morph.* **94**, 221–281.
FRIEDMAN, K. and ECKERT, R. (1972). *Comp. Biochem. Physiol.* **45A**, 101–114.
GALT, C. P. and MACKIE, G. O. (1971). *J. exp. Biol.* **55**, 205–212.
GELFAN, S. (1926–27). *Univ. Calif. Publ. Zool.* **29**, 453–465.
GIBBONS, B. H. and GIBBONS, I. R. (1972). *J. Cell Biol.* **54**, 75–97.
GIBBONS, I. R. (1961). *J. biophys. biochem. Cytol.* **11**, 179–205.
GIBBONS, I. R. (1965). *J. Cell Biol.* **25**, 400–402.
GIBBONS, I. R. (1966). *J. biol. Chem.* **241**, 5590–5596.
GÖSSEL, I. (1957). *Arch. Mikrobiol.* **27**, 288–305.
GRAY, J. (1928). "Ciliary Movement", p. 162. Cambridge University Press.
GRĘBECKI, A. (1965). *Acta Protozool.* **3**, 79–101.
GRELL, K. G. (1968). "Protozoologie." Springer Verlag, Berlin.
HAGIWARA, S. and TAKAHASHI, K. (1967). *J. gen. Physiol.* **50**, 583–601.
HODGKIN, A. L. (1957). *Proc. R. Soc.* B **148**, 1–37.
HODGKIN, A. L. and HOROWICZ, P. (1959). *J. Physiol.* **148**, 127–160.
HODGKIN, A. L. and RUSHTON, W. A. H. (1946). *Proc. R. Soc.* B **133**, 444–479.
HODGKIN, A. L., HUXLEY, A. F. and KATZ, B. (1949). *Arch. Sci. Physiol.* **3**, 129–150.
HOFFMANN-BERLING, H. (1955). *Biochim. biophys. Acta* **16**, 146–154.
JAHN, T. L. (1961). *J. Protozool.* **8**, 369–380.
JAHN, T. L. (1962). *J. cell. comp. Physiol.* **60**, 217–228.
JAHN, T. L. and VOTTA, J. J. (1972). *A. Rev. Fluid Mech.* **4**, 93–116.
JENNINGS, H. S. (1906). "Behavior of the Lower Organisms." Columbia University Press, New York.
KAMADA, T. (1931). *J. Fac. Sci. Tokyo Univ.* IV, **2**, 285–298.
KAMADA, T. (1934). *J. exp. Biol.* **11**, 94–102.

KAMADA, T. (1940). *Proc. Imp. Acad. Japan* **16**, 241–247.
KAMADA, T. and KINOSITA, H. (1940). *Proc. Imp. Acad. Japan* **16**, 125–130.
KANNO, F. (1958). *Dobutsugaku Zasshi* (Zool. Mag., Tokyo), **67**, 165–168.
KATZ, B. (1966). "Nerve, Muscle, and Synapse." McGraw-Hill, New York.
KINOSITA, H. (1936a), *J. Fac. Sci. Tokyo Univ.* IV, **4**, 163–170.
KINOSITA, H. (1936b). *J. Fac. Sci. Tokyo Univ.* IV, **4**, 171–184.
KINOSITA, H. (1936c). *J. Fac. Sci. Tokyo Univ.* IV, **4**, 185–188.
KINOSITA, H. (1954). *J. Fac. Sci. Tokyo Univ.* IV, **7**, 1–14.
KINOSITA, H. and MURAKAMI, A. (1967), *Physiol. Rev.* **47**, 53–82.
KINOSITA, H. and OKAJIMA, A. (1968). *Symp. Cell Chem.* **19**, 191–197.
KINOSITA, H., DRYL, S. and NAITOH, Y. (1964a). *J. Fac. Sci Tokyo Univ.* IV, **10**, 291–301.
KINOSITA, H., DRYL, S. and NAITOH, Y. (1964b). *J. Fac. Sci. Tokyo Univ.* IV, **10**, 303–309.
KINOSITA, H., MURAKAMI, A. and YASUDA, M. (1965). *J. Fac. Sci. Tokyo Univ.* IV, **10**, 421–425.
KUNG, C. (1971a). *Z. vergl. Physiol.* **71**, 142–162.
KUNG, C. (1971b). *Genetics* **69**, 29–45.
KUNG, C. and ECKERT, R. (1972). *Proc. natn. Acad. Sci. U.S.A.* **69**, 93–97.
KUNG, C. and NAITOH, Y. (1973). *Science, Wash.* **179**, 195–196.
KUZNICKI, L. (1968). *Acta Protozool.* **6**, 109–117.
LINDES, D., DIEHN, B. and TOLLIN, G. (1965). *Rev. Sci. Instr.* **36**, 1721–1725.
LUDLOFF, K. (1895). *Arch. ges. Physiol.* **59**, 525–554.
LUNTZ, A. (1931). *Z. vergl. Physiol.* **14**, 68–92.
MACHEMER, H. and ECKERT, R. (1973). *J. gen. Physiol.* **61**, 572–587.
MACKIE, G. O., SPENCER, A. N. and STRATHMANN, R. (1969). *Nature, Lond.* **223**, 1384–1385.
MAST, S. O. (1911). "Light and the Behavior of Organisms", p. 109. John Wiley, New York.
MAST, S. O. and NADLER, J. E. (1926). *J. Morph.* **43**, 105–117.
MENDELSSOHN, M. (1902). *J. Physiol. Path. gén.* **4**, 393–410.
NAITOH, Y. (1958), *Annotnes zool. jap.* **31**, 59–73.
NAITOH, Y. (1961). *Dobutsugaku Zasshi* (Zool. Mag., Tokyo) **70**, 435–446.
NAITOH, Y. (1964). *Dobutsugaku Zasshi* (Zool. Mag., Tokyo) **73**, 233–238.
NAITOH, Y. (1966). *Science, Wash.* **154**, 660–662.
NAITOH, Y. (1968). *J. gen. Physiol.* **51**, 85–103.
NAITOH, Y. (1969). *J. gen. Physiol.* **53**, 517–529.
NAITOH, Y. and ECKERT, R. (1968a). *Z. vergl. Physiol.* **61**, 427–452.
NAITOH, Y. and ECKERT, R. (1968b). *Z. vergl. Physiol.* **61**, 453–472.
NAITOH, Y. and ECKERT, R. (1969a). *Science, Wash.* **164**, 963–965.
NAITOH, Y. and ECKERT, R. (1969b). *Science, Wash.* **166**, 1633–1635.
NAITOH, Y. and ECKERT, R. (1973). *J. exp. Biol.* **59**, 53–65.
NAITOH, Y. and KANEKO, H. (1972). *Science, Wash.* **176**, 523–524.
NAITOH, Y. and KANEKO, H. (1973). *J. exp. Biol.* **58**, 657–676.
NAITOH, Y. and YASUMASU, I. (1967). *J. gen. Physiol.* **50**, 1303–1310.
NAITOH, Y., ECKERT, R. and FRIEDMAN, K. (1972). *J. exp. Biol.* **56**, 667–681.
OKAJIMA, A. (1953). *Jap. J. Zool.* **11**, 87–100.

OKAJIMA, A. (1954a). *Annotnes zool. jap.* **27**, 40–45.
OKAJIMA, A. (1954b). *Annotnes zool. jap.* **27**, 46–51.
OKAJIMA, A. and KINOSITA, H. (1966). *Comp. Biochem. Physiol.* **19**, 115–131.
OLIPHANT, J. F. (1938). *Physiol. Zool.* **12**, 19–30.
OLIPHANT, J. F. (1942). *Physiol. Zool.* **15**, 443–452.
PÁRDUCZ, B. (1967). *Int. Rev. Cytol.* **21**, 91–128.
SERAVIN, L. N. (1961). *Biokhimiya* **26**, 160–164.
SLEIGH, M. A. (1962). "The Biology of Cilia and Flagella." MacMillan, New York.
SONNEBORN, T. M. (1970). *In* "Methods in Cell Physiology" (D. M. Prescott, ed.), Vol. IV. Academic Press, New York.
STAHL, E. (1880). *Bot. Ztg* **38**, 298–443.
STATKEWITSCH, P. (1904). *Z. allg. Physiol.* **4**, 296–332.
SUMMERS, K. E. and GIBBONS, I. R. (1971). *Proc. natn. Acad. Sci. U.S.A.* **68**, 3092–3096.
SZENT-GYÖRGYI, A. (1949). *Biol. Bull.* **96**, 140–161.
TAMM, S. L. and HORRIDGE, G. A. (1970). *Proc. R. Soc. Lond.* B **175**, 219–233.
TAYLOR, C. V. (1920). *Univ. Calif. Publs Zool.* **19**, 403–471.
TUFFRAU, M. (1957). *Bull. Soc. zool. Fr.* **82**, 354–356.
UEDA, K. (1961). *Annotnes zool. jap.* **34**, 99–110.
VERWORN, M. (1889a). *Arch. ges. Physiol.* **46**, 267–303.
VERWORN, M. (1889b). *Psychophysiologische Protistenstudien.* G. Fischer Verl., Jena.
WALLENGREN, H. (1903). *Z. allg. Physiol.* **3**, 22–32.
WEIDMANN, S. (1952). *J. Physiol., Lond.* **118**, 348–360.
WERMAN, R. and GRUNDFEST, H. (1961). *J. gen. Physiol.* **44**, 997–1027.
WORLEY, L. G. (1934). *J. cell. comp. Physiol.* **5**, 53–72.
YAMAGUCHI, T. (1960). *J. Fac. Sci. Tokyo Univ.* IV, **8**, 573–591.

Chapter 13

Control of ciliary activity in Metazoa

EDWARD AIELLO

Department of Biological Sciences, Fordham University, Bronx, New York, U.S.A.

I. Introduction

In this article I will examine the question of control by metazoan organisms of ciliary activity in their various tissues. It will not be a comprehensive review and references to the early literature will usually not be given if cited in the more recent works discussed here. Several examples of ciliary control have received considerable attention and these will be described and critically examined. The nature of control at the molecular level will be covered specifically by other authors in this volume so that except for alluding to the general nature of the ciliary mechanism and the factors that might influence beating at the level of the contractile apparatus itself, I will concentrate on mechanisms operating at the level of the organism, tissue and whole cell. The eventual goal will be to explain how such regulatory mechanisms influence ciliary beating.

II. The evidence

A. Frog oro-pharyngeal cavity and esophagus

The most extensively studied system in vertebrates in terms of control mechanisms is the ciliated epithelium of the frog oral cavity and esophagus. Alexandrov and Arronet (1956) and Satir and Child (1963) reported that glycerol-extracted cilia of the frog palate beat rhythmically when ATP was added, even when detached from the basal protoplasm. Vorhaus and Deyrup (1953) added ATP to normal pharyngeal tissue and observed a doubling of the rate of beating. The effect was dose dependent for ATP concentrations from $1{\cdot}5 \times 10^{-5}$ to $1{\cdot}5 \times 10^{-3}$ M. There was no attempt made to ascertain the locus of action of ATP; the authors simply noted that it accelerated particle transport. What normally determines the metabolic rate in frog ciliated cells has not been determined, but there is good evidence that the ciliary beating is under nervous control. There is some discrepancy in the literature regarding this issue. McDonald *et al.* (1927) reported that stimulation of a sympathetic nerve to the pharynx increased ciliary activity, whereas stimulation of a parasympathetic nerve decreased it. Lucas (1935), however, reported that pharyngeal cilia were normally quiescent but could be activated by stimulation of the palatine nerve, which he said was parasympathetic. The discrepancy regarding the nerve does not seem to have been clarified.

The role of excitatory and inhibitory nerves to the frog palate, (i.e. the region of the dorsal surface of the oro-pharyngeal cavity anterior to the pharynx), was studied in detail by Seo (1931, 1937) using *Rana nigromaculata*. In the 1931 paper he referred to previous work in which he found reflex acceleration of cilia following tactile stimulation of the tongue or electrical stimulation of the glossopharyngeal nerve. Local application of particles directly on the palate also causes an increase in ciliary beating. Although the response waned during repeated application of particles, stimulation of the tongue gave cilioexcitation again. He then reported an extensive series of nerve stimulation and transection experiments which elucidate the path for the reflex, namely: the glossopharyngeal nerve carries afferent fibers from the tongue and floor of the mouth: the glossopharyngeal, facial and trigeminal nerves carry both afferent and cilioexcitatory efferent fibers to the palate; the vagus nerve is not involved:

the medulla, but not the forebrain or spinal cord, is required for the reflex. Histological examination showed that the palate contained a nerve net and was innervated mainly by relatively few fibers from the palatine branch of the facial. Details of innervation and a description of two distinct nets is given by Gaupp (1904). An isolated palate-nerve preparation responded to local stimulation by a local excitatory response which became more extensive as the stimulus intensity (tetanic, induced current) increased. This included an increase in amplitude and rate of beating, a faster recovery stroke and a faster metachronal wave, which Seo noticed was composed of the recovery strokes and travelled opposite to the beat (antiplectic). The response to stimulation began after 1 s, rose rapidly to a peak, lasted 10 s or more and then faded. It could also be elicited by touching the isolated palate. It was noticed that although increased mucus secretion accompanied the reflex it was not an important factor in the ciliary response as determined by the rate of transport of blood cells over the surface. Seo also gave extensive data to support his claim that in the relatively intact animal, the excitatory response to mild stimulation of the tongue occurred only on that part of the palate which corresponded to the part of the tongue that was stimulated. It is not really clear how the nerve net, which appears to mediate local responses to local stimulation of the palate, is functionally related to the parasympathetic innervation. This latter seems capable of mediating a local response on the palate following local stimulation of the tongue as well as palate-wide response to intense stimulation of the tongue or glossopharyngeal nerve. The work awaits confirmation by other workers, although some of the essential features were confirmed and reiterated by Seo himself in 1937. In his review of this topic Gosselin (1966) presented some original data describing the activation of frog (*Rana pipiens*) cilia by D.C. stimulation. Single square-wave pulses of various voltage and duration were applied and the strength–duration curve for threshold activation was drawn. From this he estimated a chronaxie of about 2·5–8 ms which he stated is consistent with that for small unmyelinated fibers such as those usually found in the autonomic nervous system. Ciliary activity lasts from 30 to 60 s after each shock and can be kept continuous by appropriately spaced repetitive stimulation.

Information on the pharmacology of frog ciliated tissue is extensive, but somewhat controversial and poorly related to the nervous mechanisms described above. Satoh (1959a, b) found acetylcholinesterase (AChE) activity in the ciliated epithelium and in the whole mucosa of the palate of the toad *Bufo vulgaris formosus* using the Hestrin method. Ciliated epithelium alone split 0·0125 M acetylcholine (ACh) solution at the rate of 0·141 ($\pm$0·024) mg AChCl per 30 min per 5 mg tissue at 38°C. Acetyl-beta-methylcholine was split less rapidly and benzoylcholine not at all.

He located the activity histochemically using myristoylcholine chloride as substrate and DFP 10^{-8}–10^{-5} M as inhibitor. On the basis of both inhibitor and substrate relative specificity he concluded that the submucosal muscle fibers and blood vessels contained mainly true (specific) ChE whereas the ciliated mucus epithelial cells contained pseudo-ChE. In the absence of electron microscopy it is difficult to determine exactly where the observed ChE activity actually resides and Seo (1931) had specifically mentioned the absence of muscle fibers from the palate mucosa itself, but the possibility of a cholinergic innervation of the submucosa and possibly the mucosa is certainly present.

Hill (1957), Burn and Day (1958), and Milton (1959) tried to clarify the action of cholinergic and cholinergic-blocking agents on frog esophagus. As in the experiments by Seo on the palate, Hill found that mechanical irritation of the esophagus increased ciliary activity, as evidenced by the transport of carborundum particles and poppy seeds, both being moved at the rate of about 2·1 cm min^{-1}. However, she obtained no effect with *d*-tubocurarine up to 2 mg ml^{-1} or with ACh from 1 to 100 μg ml^{-1} but did get a 40% increase in rate with the unusually high concentration of 200 μg ACh ml^{-1}. This stimulatory effect of ACh was removed by washing. She reviewed the literature on drug effects and considered that ciliary activity, unlike muscle contraction, is not initiated by excitation of the cell surface and for this reason is relatively refractory to several agents that are generally very potent on nerve or muscle. Burn and Day (1958) disagreed with these findings and reported that ACh stimulated particle transport in frog esophagus about 50% from 12–21 mm per 100 s to 19–32 mm per 100 s at the much lower concentration of 10 μg ml^{-1}; greater concentrations were inhibitory. They also found that *d*-tubocurarine (*d*-TC) slowed the rate from 21–31 mm per 100 s to 10–19 mm per 100 s at 1–10 μg ml^{-1}. 100 μg ml^{-1} *d*-TC reduced the rate 50%. Recovery in all experiments was accomplished by washing. Burn and Day attributed the difference between their results and those of Hill to experimental techniques: they used only poppy seeds because carborundum was less reliable; they rinsed the esophageal membrane but did not keep it immersed in Ringer's: they got their best results with June frogs, autumn ones being less reliable. Milton (1959) used poppy seeds or moist garnet chips on immersed esophagus and found that 1 μg ml^{-1} *d*-TC reduced the rate 22% and 1 μg ml^{-1} ACh increased it 25–100%. A retesting of the various media used by previous authors indicated that the ionic balance was important, especially the level of calcium ions, which when present gave better stability, and phosphate ions, which seemed to diminish the response, perhaps by removing the calcium. In the absence of actual measurements of the activity of any of the ions, their effect on the ciliated cell, and the penetration of the drugs into the cells, the work may point up the importance of considering the ionic composition of the bathing medium, but does not adequately explain the different results.

Taken all together, past work suggests the presence of a cholinergic cilioexcitatory mechanism probably related to a nerve net and parasympathetic innervation, some part of which is sensitive to *d*-TC. The extent to which information on the palate, pharynx and esophagus applies to the other tissues, possible differences between species of frogs, and the possible functioning of ACh as a local hormone not directly related to the innervation remain unsolved. It also seems that it would be helpful to re-examine the tissue using histological techniques capable of identifying various kinds of endings at the electron microscope level.

B. *Trachea*

The mucosa of the respiratory tract of several vertebrate species has been studied extensively, usually more in terms of ciliary-mucoid clearing mechanisms than actual control. Using the anaesthetized rat with exposed trachea, Dalhamn (1956) studied various factors that influenced rates of ciliary beating, which ranged approximately from 1100 to 1500 beats min^{-1}, and mucus flow, from 10 to 15 mm min^{-1}. The latter was influenced by the nature and quantity of mucus as well as the rate of ciliary beating. He seems to have attributed most of the effects brought about by changes in temperature, humidity, irritants and so on, to direct action on the epithelium, with no particular reference to a possible involvement of the nervous system, nor does he mention the presence of nerves in describing his light and electron micrographs. Rhodin (1966) observed fine, unmyelinated nerve endings around the basal ends of ciliated and mucus-producing cells of the human tracheal epithelium and classified them as sensory nerve endings. He also described, without comment on function, nerves which travel through the lamina propria, mostly in the company of blood vessels. Some of these enter the thick basement membrane but none was seen passing through the thin basement membrane to the epithelium. In the cat trachea unmyelinated fibers appear to end as knobs on ciliated cells (Messerklinger, 1958). As to actual control, Laschkow (1955), quoted by Gosselin (1966), refers to Russian literature which describes vagal cilioacceleration and sympathetic cilioinhibition in warm blooded animals. Lommel (1908) found that dog tracheal cilia did not respond to vagal cutting or stimulation and Wardell *et al.* (1970) reported that in the dog trachea parasympathetic fibers go only to smooth muscle. Wardell proposed that goblet cells respond only to direct irritation whereas mucus glands respond to vagal reflex stimulation. Lucas and Douglas (1935) found that cilia of the turtle trachea were also unresponsive to vagal stimulation.

Pharmacological experimentation involving ACh and *d*-TC on the trachea again presents conflicting data as reviewed by Gosselin (1966). This is probably due, at least in part, to the difficulty of measuring ciliary beating and mucus secretion independently, and to the fact that the secretion and properties of mucus can be influenced both by direct effects and by those mediated through the vagus nerve. Corssen and Allen (1959) reported inhibitory effects of ACh and *d*-TC on isolated human tracheal ciliated cells, an experimental procedure that avoids this complexity, but the required concentrations seem too high to be meaningful (2·5% ACh and 0·5% eserine salicylate). Their results with 0·1% ACh, which gave a steady increase in the rate of rotation of cell clumps from about 18 to 30 rotations per min over a 40 min period are interesting but difficult to evaluate. In the absence of complete data and any statistical analysis, their data on individual explants of human tissue indicate that the cell is sensitive to ACh but not that ACh is normally present or acts as a mediator. However, Corssen's work on the effects of local anaesthetics using this system (see his discussion following the paper by Gosselin, 1966) suggested that membrane activity may be important in regulating ciliary beating on these cells.

Krueger and Smith (1960) in their study of the action of air ions found that positive ions inhibit ciliary activity in the trachea of several mammals (mouse, rat, guinea pig, rabbit and monkey) and presented indirect evidence that this was due to the release of endogenous 5-hydroxytryptamine (5-HT), which also constricted submucosal blood vessels. Negative ions antagonize or reverse this effect, perhaps by accelerating the oxidation of 5-HT (Krueger, 1962). Neither the permeability of the basement membrane nor the role of the blood supply in supporting ciliary activity has been determined and, since the same results were obtained *in situ* as with trachea extirpated after treatment, it would be difficult to attribute the effects solely to a reduction of adequate blood flow due to vasoconstriction. It seems that while there is no compelling evidence for direct nervous control of ciliary activity in the trachea, there is the possibility that neuro-transmitters or local mediators might diffuse from structures below the basement membrane to the ciliated cell, upon which they could have some effect. Considering the doses used to get effects and the relatively weak response obtained, it is doubtful that these would be normal regulatory mechanisms. In the absence of proof that positive air ions do release 5-HT and that 5-HT can influence the ciliated cell directly, it might be less misleading to consider the effects coincidental and consider the possibility that air ions act directly on the ciliated cell. It is known for example, that cigarette smoke inhibits ciliary activity and

ATPase of ciliated cells of the rabbit trachea, as shown histochemically by Cress *et al.* (1965). Most of the available information is compatible with the general notion that cilia of the vertebrate oral cavity, respiratory tract and esophagus are not under nervous control except in the frog, in which there is some degree of activation by a cholinergic mechanism involving a nerve net and some parasympathetic fibers.

C. Oviduct

Cilia of the oviduct of vertebrates have been studied less extensively. Eckert and Murakami (1972) penetrated the ciliated cells of the oviduct of the urodele *Necturus maculosus* with a microelectrode. Dimpling of the membrane by the electrode caused a transient acceleration, lasting less than 30 s. Penetration gave several minutes of acceleration. Washing in calcium-free medium slowly reduced the beating rate from about 7 s^{-1} to zero: beating was rapidly restored by the iontophoretic introduction of calcium through the electrode. Magnesium antagonized the calcium effect. The membrane resting potential was about 10 mV and no membrane potential changes above 1 mV were recorded during the experiment, indicating that changes in beating rate were not dependent on the membrane potential. The authors concluded that beating rate was determined by the intracellular concentration of calcium ions. It is unlikely that this cellular control mechanism is directly influenced by the innervation of the oviduct which is primarily noradrenergic to the smooth muscle and stimulates muscle movement (Brundin, 1969). In a review of the histochemical literature Friedricsson (1969) noted that alkaline phosphatase, but not esterase, is located in the ciliated cells, which do not appear to be innervated. The esterase that was detected using naphthyl acetate as substrate was concentrated in non-ciliated cells.

There are still some unanswered questions regarding the role of cilia in the oviduct. Blandau (1969) agrees with early observations by Parker (1931) that tracer particles may move in both directions on different parts of the folded mucosa of the rabbit oviduct but concludes that gamete transport must be primarily controlled by muscular movement of the oviduct. Although the presence and perhaps vigour of the ciliated cells is under long-term hormonal control (Brenner, 1969), there is no evidence for short-term control and little likelihood that a nervous mechanism for the control of oviduct cilia will be found.

D. Lower chordates

In certain of the lower chordates cilia of specific tracts of epithelium seem to be under nervous control. Bone (1960) described the organization of the atrial nervous system which supplies fibers to the gill bars in amphioxus (a cephalochordate). The lateral cilia, which by their beating expel water through the gill slits, and the frontal cilia, which carry mucus and entrapped particles to the pharyngeal groove, appear to be innervated by branches of the pharyngeal nerves. There is also a branchial plexus which is apparently sensory. Bone states that this innervation is not homologous with any division of the vertebrate autonomic nervous system. Unfortunately, there seem to be no published physiological experiments on how this system functions, but from their histological relationship Bone concluded that the lateral cilia are probably under nervous control. According to Jørgensen (1966), Bone has observed that unlike most other filter feeders that have been studied, amphioxus pumps water through the pharynx at slower rates when more food particles are present. Graphite does not have this effect, suggesting that the animal can sense the food value of suspended particles. Regulation of the feeding current implies nervous control of the lateral cilia.

Observations by Knight-Jones (1952) on the trunk cilia and the behavior of *Saccoglossus cambrensis* and *S. horsti* (hemichordates) reveal that these cilia can reverse and, in conjunction with peristaltic movements of the body, enable the animal to move backward in response to various stimuli. The author found that heavily ciliated areas were innervated by fibers from the ventral cord whereas non-ciliated patches of epithelium received few fibers. No direct proof for the apparent cilio-motor innervation was obtained, but the author seemed fairly certain that the cilia actually reversed their beat. The larva of *S. horsti* shows periodic, sudden, complete cessation of ciliary beating of the telotroch, a circumferential ciliated tract, but reversal did not occur and the author did not mention the presence or absence of nervous tissue (Burdon-Jones, 1952).

More complete studies have been done on urochordates (tunicates) and it is in this group that nervous control of ciliary activity seems to have been most convincingly demonstrated. MacGinitie (1939) described the feeding mechanism in several tunicates and pointed out that in *Ascidia californica* the stomatal cilia, which drive water through the ciliary-mucoid particle-trapping device in the pharynx, stop in response to adverse stimuli and lie flat, out of the way of the stomatal opening. In this position feeding currents stop and contraction of the body muscle expels the pharyngeal contents. Galt and Mackie (1971) observed a similar phenomenon in another ascidian, *Diplosoma macdonaldi*. Using external suction electrodes they recorded electrical activity associated with an apparent reversal of the feeding current, but interpretation of their results was complicated by the contraction of body muscle. However, in the appendicularian tunicates *Oikopleura dioica* and *O. labradoriensis* the absence of muscle contraction allowed them to associate the occurrence of 2 mV negative-going spikes, called ciliary reversal potentials, with reversal of stomatal cilia occurring either spontaneously or in response to stimuli. They were unable to ascertain the exact changes in ciliary beat that produced the reverse flow of water. With internal microelectrodes they also measured 70–80 mV positive-going spikes in, or perhaps adjacent to, the ciliated cells. These cells receive 0·5 μm diameter nerve terminals from the brain, which in turn receives sensory nerves from the lip, the body area most sensitive to stimuli that elicit the reversal response. Small fluctuations correlated with the actual beat were also recorded, but the authors considered them to be most probably microphonic artifacts due to movement of the cell membrane or the microelectrode.

Recently, Sleigh, Mackie, Singla and Williams (unpublished) investigated the ascidian *Corella* in which such stimulation as touching a siphon causes a single reverse beat followed by cessation of beating of the stomatal cilia and closing of the siphons. They were also able to obtain the ciliary response in isolated pieces of the branchial basket, which contains an autonomous pacemaker for generating repetitive inhibitory (reversal) responses. Cutting experiments demonstrated that this pacemaker is under reflex control in the whole animal. The electrical activity in the branchial basket correlates with the occurrence and duration of cilioinhibition. Experiments with drugs suggest that cholinergic synpases are involved. Nerve fibers run in the branchial bars close to the ciliated cells of the stomata and sometimes appear to make contact with those cells. Fedele (1926) observed ciliary reversal in *Doliolum*, a representative of the Thaliacea class of tunicates. It seems likely that nervous control of ciliary beating, to cause either reversal or short-term inhibition of feeding currents or movement, is relatively widespread throughout the non-vertebrate chordates.

E. Ctenophores

It has been known for many years that the various swimming and righting maneuvers of ctenophores are due to changes in the rate of beating of the cilia which comprise the comb rows. Each beat, which is directed aborally, travels orally along the comb rows in metachronal progression, and is initiated in *Pleurobrachia* (Sleigh, 1968) and in *Bolinopsis* (Horridge, 1955a) by a beat of the balancer cilium of the aboral sense organ that is associated with that row. Reverse waves sometimes occur but except for an unconfirmed observation by Child (1933) reverse beating has not been reported. Sleigh (personal communication) has observed backward swimming by *Callianira bialata* which indicates a reversal of the ciliary beat. There is in all ctenophores a general, rapidly spreading inhibition of all comb row cilia in response to various stimuli. In *Beroe* this is associated with muscular closing over of the comb rows, but cilio-inhibition also occurs in other species such as *Pleurobrachia* which have no such musculature (Mackie, 1970) and is therefore a true cilioinhibition. Horridge (1965b) found that electrical stimulation of the animal's surface caused general inhibition that could be prevented by the presence of excess magnesium ions in the water. After such treatment, electrical stimulation of a row could induce a beat. In 1966 Horridge proposed four different pathways for coordination of ctenophore behavior concerned with control of the following: the spread of bioluminescence; discrete muscular movement; cilioinhibition of the comb rows and general muscle contraction; cilioactivation of the comb rows. Cilioinhibition was ascribed to an ectodermal nerve net and cilioactivation to a pathway leading from the sense organ to the first comb plate of each row. In *Pleurobrachia* this path is along the meridional groove, at the base of which lie relatively empty, axon-like cylindrical structures up to 10 μm diameter making a chain of elongated cells with close, overlapping contacts but no obvious synapses (Horridge and Mackay, 1964). They make similar close contact (10–12 nm space) with the base of the polster cells (ciliated cells of the comb plate). Also shown by electron microscopy in the same paper are fine nerve endings 0·3–1 μm in diameter, lying close to the polster cells. They are filled with empty vesicles and are believed to be the inhibitory endings of the nerve net.

Intracellular recording electrodes in the polster cells of *Cestus* reveal a resting potential of about 40 mV negative, which reverses and goes about 25 mV positive during each beat of the comb (Horridge, 1965b). Small potentials were correlated with weak beats; no pacemaker or generator potential was seen and it was not possible to determine whether the beat or the change in membrane potential began first. Horridge and Mackay (1964) discussed the nerve-like morphological characteristics of the polster cells and the general difficulty of categorizing cell types of different phyla from light and electron microscope information. The polster cell is up to 100 μm long and could conceivably conduct an action potential like the one observed. The empty axons of the meridional groove arise from cell bodies that are ciliated. Although the beat ordinarily originates at the apical end, it can originate anywhere and the wave may travel in both directions. The ciliated cells seem to have combined nerve and ciliary function, with the ciliary beat being initiated by membrane depolarization. This could occur either spontaneously, or by deformation of cilia at the apical sense organ, or by local irritation, but could be prevented from occurring by the stabilizing influence of an inhibitory transmitter released by the nerve net. Despite considerable work, further electrophysiological and pharmacological analysis of this interesting system seems to be required so that the actual mechanisms of excitation and inhibition can be elucidated. The present data strongly indicate the presence of nervous control but do not include information on possible transmitters.

F. Echinoderm larvae

Reversal of beating of cilia on the arms of the pluteus larva of the sea urchin *Strongylocentrotus droebachiensis* was observed by Mackie *et al.* (1969) to occur in response to touch, electrical stimulation, or a superabundance of algal cells present in the medium. During reversal, which lasted for 4–18 s, 20 mV monophasic potentials occurring at the rate of about 3 s^{-1} could be recorded with external suction electrodes. This electrical activity and the reversal itself were not seen in the presence of 50% replacement of sea water with isotonic magnesium chloride. No nerve fibers could be found in the area of the cilia, and removal of apical and oral nerve cells did not alter the response which was concluded to be neuroid, in the sense that electrical potential changes are conducted through epithelial cells.

G. Anthozoa

Parker (1928) observed that cilia on the oral disc of *Metridium* usually beat outward as indicated by the movement of small pieces of paper placed there. Paper soaked in glycogen is carried toward the mouth, indicating ciliary reversal. Orally directed transport was also observed when mussel extract was added (Parker and Marks, 1928). They do not report having directly observed the cilia and it seems to me that the possibility should not be overlooked that muscular movement might bring oppositely directed, adjacent tracts into position in different situations. The same reservation can be held when interpreting the data of Baba (1968) concerning ciliary reversal on ridges of excised strips of the pharynx of *Actinia equina*. After 1–3 h the normal inwardly directed beating reversed to beat outward. The original direction was restored locally by weak direct current. Ciliary reversal and peristaltic movement spread inward from the current source during prolonged stimulation. No nerves were detected histologically, the ciliated epithelium and mucus cells resting on mesogloea. Baba suggested that the effect was propagated by ciliated cells.

H. Bivalve gill

There are some differences between species and many unconfirmed observations but *Mytilus edulis*, the edible blue marine mussel, has received enough attention to make it a useful starting-point in this field. Aiello (1960) reported the following findings: 5-hydroxytryptamine (5-HT, serotonin) stimulated lateral ciliary beating and oxygen consumption of excised gills; gill extract contained 5-HT-like activity when tested on the clam heart; cutting the branchial nerve to one gill in de-shelled but otherwise intact animals caused a more rapid cessation of beating on the cut side than on the uncut side. He postulated that lateral cilia might be under

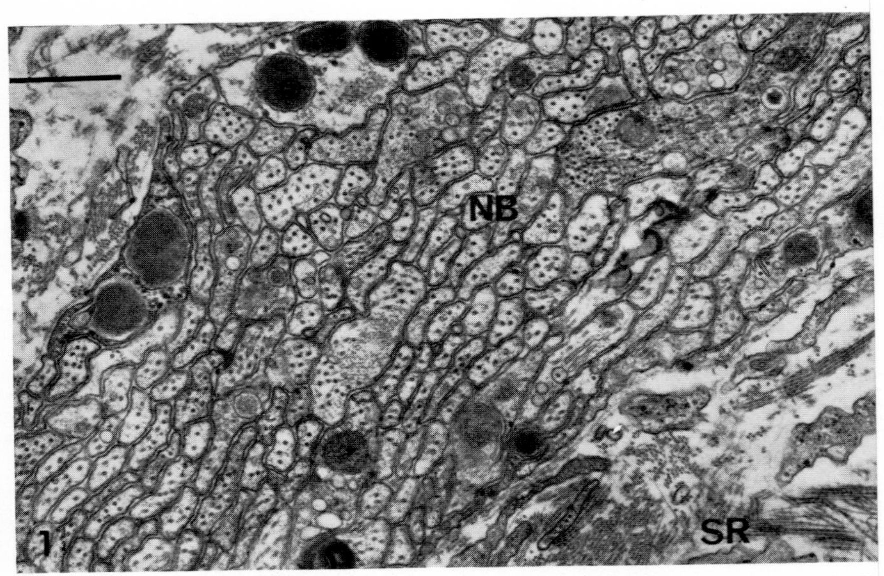

FIG. 1. Electron micrograph of a section through the branchial nerve bundle (NB) of the gill axis showing individual axons entering the gill filament (lower right). SR indicates the supporting rod of the filament. Neurotubules, several mitochondria and various vesicles are present. ×81,400. Scale marker = 0·2 μm. (Paparo, 1972.)

some kind of serotonergic nervous control. Gosselin (1961) presented extensive data on the dose/response relationship of 5-HT cilioacceleration, drawing typical S-shaped curves over two or three log units of concentration, with marked effects manifested at micromolar concentrations. The presence of 5-HT in gill was confirmed by paper chromatography and spectrofluorometry (Aiello, 1962; Gosselin *et al.*, 1962). The cilioexcitatory effect of electrically stimulating the branchial nerve was described by Aiello and Guideri (1964) and nerve fibers were observed to leave the branchial nerve in the gill axis and penetrate the filaments, with fine branches appearing beneath the ciliated cells (Aiello and Guideri, 1965). The injection of reserpine into whole mussels was found to decrease the content of 5-HT in the gills from 0·9 to 0·5 μg g^{-1} wet weight and significantly reduce the effect of nerve stimulation without interfering with the response to added 5-HT (Aiello and Guideri, 1966). This last paper also noted that bromolysergic acid diethylamide (BOL) reduces the response to both nerve stimulation and 5-HT.

A detailed description of the exact distribution of the branchial nerve has not been published but electron micrographs of *M. edulis* confirm the earlier light microscope findings discussed above and prove the presence

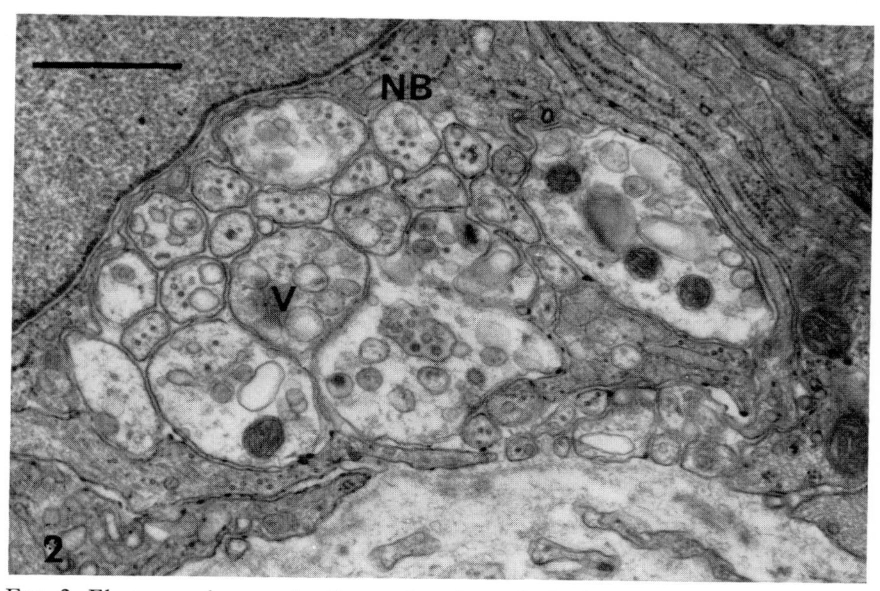

FIG. 2. Electron micrograph of a section through the base of a lateral cell of *M. edulis* (upper left), showing a group of axons from the branchial nerve. Mitochondria, and several kinds of vesicles (V) are seen in several axons which may be terminating. ×111,000. Scale marker = 0·2 μm. (Paparo, 1972)

of nerve axons entering the filaments and running under the lateral cells (Figs 1, 2; Paparo, 1972). Similar electron micrographic studies have been done on *E. complanatus* (Paparo, in press; Satir, personal communication; Satir and Gilula, 1970).

The metabolic effects of 5-HT on the ciliated lateral cell are difficult to distinguish from those on gill tissue as a whole. Preliminary histochemical studies using neotetrazolium (Bouffard and Aiello, 1969) indicate that most of the succinic dehydrogenase activity of the gill is localized in the lateral cells. Moore *et al.* (1961) reported that 10^{-5} M 5-HT increased oxygen consumption by about 80% (from 192 to 370 and from 226 to

391 μl per 100 mg per 3 h in two experiments) and increased the glycogen utilization (from 112 to 170 and from 58 to 195 μg) in isolated gills. Respiratory stimulation was blocked by BOL. Moore and Gosselin (1962) showed that 5-HT stimulated anaerobic glycolysis but were unable to demonstrate any increase in phosphorylase activity, using the release of phosphate from glucose-1-phosphate in the presence of glycogen as the basis of the assay. Malanga and Aiello (1971) reported that when gill pieces were incubated under nitrogen in Warburg vessels, in the same procedure that was used for measuring anaerobic glycolysis, the lateral cilia stopped beating. The addition of 10^{-4} M 5-HT caused sustained ciliary activity. This was observed by placing the Warburg vessel over an inverted microscope at various times during the incubation. The major acid end product was identified by thin-layer chromatography as succinic acid, and the addition of succinic acid to aerobically respiring whole gill or gill homogenate markedly increased the rate of oxygen consumption (Malanga and Aiello, 1972).

In working out the relationship between the frequency of branchial nerve stimulation (square wave, 0·1 V, biphasic pulses of 2 ms duration) and ciliary response, Paparo and Aiello (1970) noticed that low frequencies (2–10 Hz) activated quiescent cilia and accelerated slowly beating cilia, whereas higher frequencies (25 and 50 Hz) depressed and eventually stopped ciliary beating. Dopamine had a similar slow-acting cilioinhibitory effect that was dose dependent from 10^{-3} to 1 μg ml^{-1}. The inhibitory effects of both dopamine and 25 or 50 Hz stimulation were blocked partially or wholly by phenoxybenzamine in microgram per ml concentrations. Phenoxybenzamine had no effect on the cilioexcitatory action of 5-HT or low frequency stimulation. The presence of dopamine in gill tissue was demonstrated by Malanga *et al.* (1972) using specific isolation procedures and the spectrofluorescence analysis of oxidation products. Strangely enough, Malanga (1971) found that dopamine increased anaerobic glycolysis at concentrations that inhibited ciliary activity; its effect on aerobic respiration has not been reported.

The question of why different frequencies of stimulation should apparently activate different mechanisms has not been elucidated. An added complication is the presence of a cholinergic mechanism in the gill. Bulbring *et al.* (1953) found acetylcholine, cholinesterase, and choline acetylase in gill tissue and reported biphasic effects of acetylcholine and cholinesterase inhibition on the activity of frontal cilia. Gosselin (1966) confirmed the findings of Bulbring *et al.* and reported that lateral cilia had a similar but weaker response. Paparo (1969) and Paparo and Aiello (unpublished data) have found that 10^{-10}–10^{-6} M ACh can activate quiescent cilia and bring the beating rate to about 10 s^{-1}. This effect is blocked by 1 μg ml^{-1} BOL. After high concentrations of ACh, such as

0·01 M, washing with sea water produces a decline below the control level, an effect that is prevented by phenoxybenzamine at 5 μg ml^{-1}. Similar effects can be demonstrated with physostigmine, suggesting that the serotonergic and dopaminergic neural mechanisms are somehow related to the cholinergic mechanism in the gill.

There is some histochemical information bearing on this proposal. Fluorescence studies of freeze-dried gill treated with formaldehyde gas

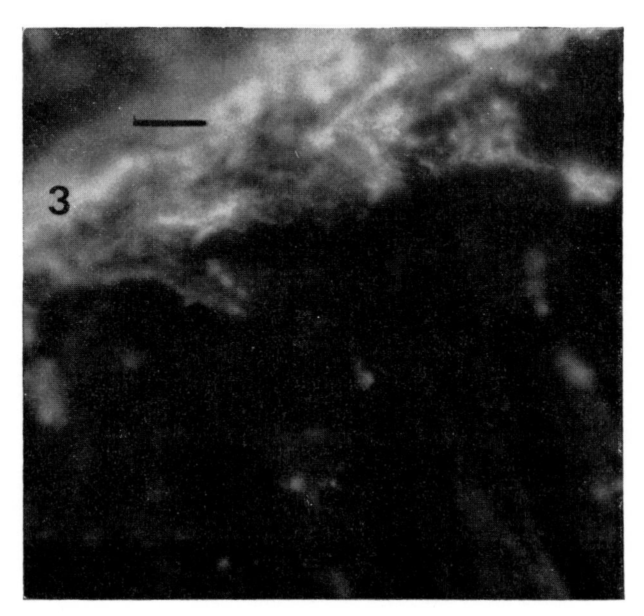

FIG. 3. Photomicrograph section through the gill axis of *M. edulis*. Normal tissue freeze-dried, treated with formaldehyde vapour and viewed with u.v. light. The branchial nerve at upper left fluoresces mainly green, indicating a predominance of dopamine. At regular intervals, fibers leave the branchial nerve and enter the gill filaments which also contain some non-specific autofluorescent orange, yellow and green material, especially in the blood sinus contained within each filament. ×960. Scale marker = 20 μm. (Stefano, unpublished.)

reveal a great mass of green fibers constituting the branchial nerve, with numerous branches travelling down the filaments, apparently between the chitinous rod and the overlying single layer of epithelium (Paparo and Finch, 1972). The specific fluorescence was practically eliminated in 48 h by injection of intact mussels with 50 mg kg^{-1} reserpine. Stefano and Aiello (1972) reported yellow fibers in the predominantly green fluorescing visceral ganglion, and unpublished data from experiments with 6-hydroxydopamine include the finding that this agent selectively reduces green fluorescence, revealing yellow fluorescing fibers in the gill filaments (Figs 3, 4).

All fibers in the neuropile of the visceral ganglion, the branchial nerve and the gill filaments give a marked reaction for cholinesterase using acetylthiocholine as substrate (Bouffard, 1969) suggesting that serotonergic

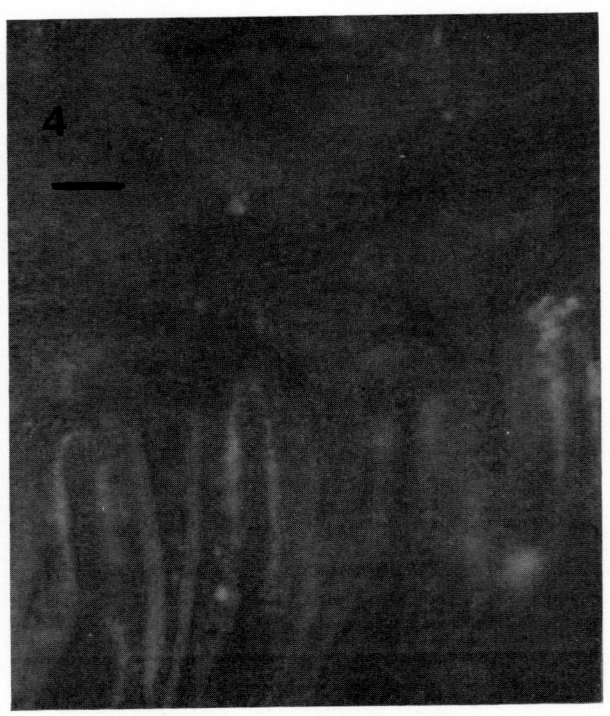

FIG. 4. Photomicrograph of a section of the gill axis of *M. edulis*. Animal pretreated by immersion in sea water containing 6-OH-dopamine (17 μg/ml) for 5 days before preparing tissues as in Fig. 3. The branchial nerve in upper half of picture shows little fluorescence due to selective destruction of dopamine containing axons. The green autofluorescence of the supporting rod on each side of the blood sinus in each gill is unchanged. $\times$960. Scale marker = 20 μm. (Stefano, unpublished.)

and dopaminergic fibers are also cholinergic. We have been unsuccessful in attempts to prove that mediators or transmitters are released by nerve stimulation but nerve histo-fluorescence is decreased by denervation (Paparo and Finch, 1972; Stefano and Aiello, 1972) and enhanced by nialamide 500 mg kg^{-1}, a monoamine oxidase inhibitor (Paparo and Finch).

A third effect of nerve stimulation of any kind is an immediate stoppage of lateral ciliary beating for less than a second (Aiello and Guideri, 1965). It was studied in detail by Takahashi and Murakami (1968) who concluded that the effect was truly neuronal, not simply due to current flow through the tissue, and involved fibers distinct from those that cause cilioexcitation. Murakami (1968) obtained the same effect by local electrical stimulation of the filament, the effect of a single 5 ms pulse causing abrupt stoppage after a delay of about 0·09 s, with recovery beginning about 0·1 s later. The effect occurs in the presence of stimulatory concentrations of 5-HT but not in the presence of extra KCl. Since KCl is generally found to depolarize the cell membrane, it may be that stoppage of ciliary beating is associated with hyperpolarization. Unfortunately, there are no published reports of anyone having measured intracellular potentials in these cells. There is, however, good reason to believe that the phenomenon itself occurs physiologically, for sudden cessation followed by a resumption of normal beating has been observed to occur spontaneously (Aiello and Guideri, 1965; Dral, 1967; Takahashi and Murakami, 1968).

The effect of ions on lateral cilia was studied extensively by Gray. He found that beating of lateral cilia was uniquely sensitive to acceleration by added K ions (Gray, 1922) and that Mg ions were inhibitory but antagonized by K ions (Gray, 1926). More recently Felton and Aiello (1971) showed that with the innervation intact no external ion was required for sustained beating, and that although K, Mg and Ca ions all had complex effects on the response to 5-HT, that agent could exert a marked cilioexcitatory effect in an ion-free medium such as iso-osmotic sucrose. In fact 5-HT caused a release of calcium from gill tissue, followed within 5 min by a rapid re-uptake, and studies with ^{45}Ca indicated that the initial effect of added 5-HT was to increase membrane permeability to calcium in both directions (Felton, 1972). There is good reason to think that 5-HT causes a change in the intracellular distribution of calcium which leads to the production or activation of some metabolic stimulant. Two sets of data go against the alternative hypothesis that 5-HT acts directly inside the cell: Aiello (1962) showed that cilioexcitation occurred before detectable amounts of 5-HT were taken up; 5-HT stimulation of anaerobic glycolysis and aerobic respiration is easily observed in whole gill but can not be demonstrated in homogenates. The influence of calcium levels on these various systems has not been fully explored.

Many of the studies described above have been performed with other species of bivalves. Aiello (1970) reported that 11 species, including *Crassostrea virginica*, *Modiolus demissus*, *Mytilus californianus*, *Mya arenaria* and *Elliptio complanatus*, responded to electrical stimulation of the branchial nerve by increasing the rate of beating of lateral cilia; two species also showed the slowly developing inhibition. Unfortunately, the work was done before the difference in effect of high and low frequency of stimulation was fully appreciated. All species were stimulated by 5-HT; dopamine was not tested. Work by Gosselin (1961), Moore *et al.* (1961), Moore and Gosselin (1962), Malanga and Aiello (1972), Malanga *et al.*

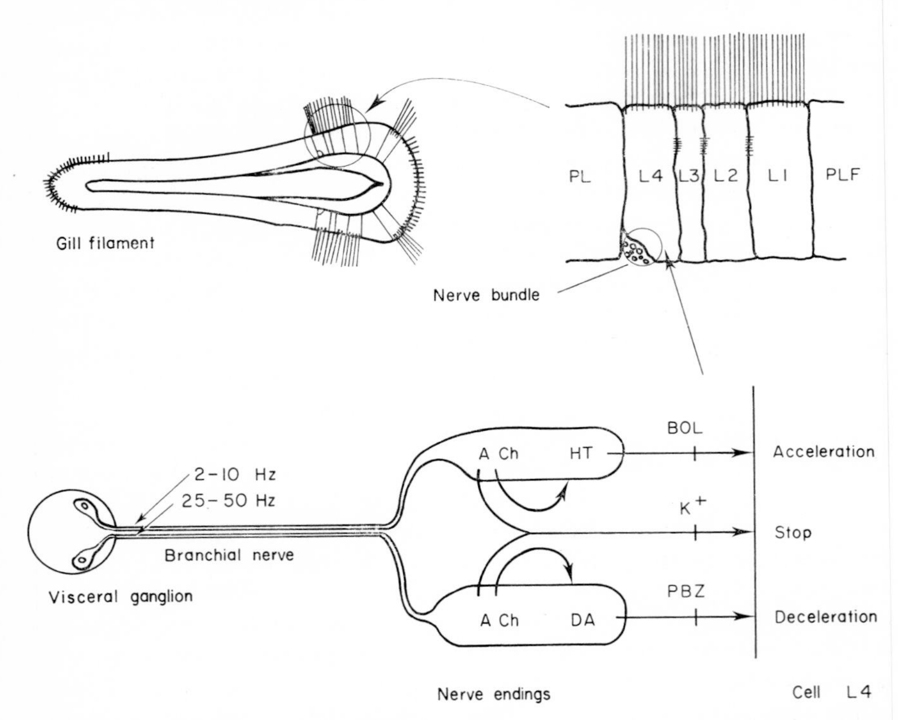

FIG. 5. Schematic view of a gill filament and its innervation in *M. edulis*, based on data discussed in the text. A nerve bundle lies between a lateral cell and the underlying supporting rod. The lateral cells are connected to each other by septate junctions. Cholinergic neurons arise in the visceral ganglion and travel in the branchial nerve to each filament. Electrical stimulation of this nerve causes lateral cilia to stop abruptly for a few seconds unless excess of potassium ions (K^+) is present. Stimulation at 2–10 Hz releases 5-hydroxytryptamine (HT) from some fibers which causes cilioacceleration unless bromolysergic acid (BOL) is present; stimulation at 25–50 Hz releases dopamine (DA) from different fibers causing ciliary slowing unless phenoxybenzamine (PBZ) is present. Cell types are: L = lateral; PL = postlateral, PLF = postlaterofrontal.

(1972), and others indicate that *Modiolus demissus* has similar but slightly different responses to 5-HT and dopamine. Compared to *M. edulis*, *M. demissus* is more apt to have spontaneous metachronal ciliary beating on excised gills, higher endogenous rates of anaerobic metabolism and greater anaerobic response to 5-HT, whereas *M. edulis* has a greater aerobic response to 5-HT and a more rapid conversion of 5-hydroxytryptophan to 5-HT, as described in the various papers cited above. Differences in dopamine content are revealed by the work of Paparo and Finch (1972) on *Elliptio complanatus* and Malanga *et al.* (1972) on *Modiolus modiolus*. Sweeney (1968) found yellow (5-HT) and green (dopamine) fluorescing fibers in the visceral ganglion but only yellow fibers in the gill of *Sphaerium*. Not enough species have been examined to make a general statement about mediators in the gill of bivalves in general. Lagerspetz *et al.* (1970) reported that ACh and eserine increase the activity of frontal cilia at low concentration and decrease it at high concentration in *Anodonta cygnea cellensis*. Endogenous ACh and ChE activity are only about 10% of the levels found in *Mytilus* and 5-HT has indefinite effects. These authors suggested that in this species ACh acts within the ciliated cells to control beating whereas 5-HT may be involved in nervous control. Despite these differences, it seems likely that all bivalves have some nervous mechanism for control of gill cilia. A suggested mechanism, aimed at utilizing all the data presently available on *M. edulis*, is presented in Fig. 5.

A different type of control has to do with the particle filtering efficiency of the latero-frontal cilia. Dral (1967) observed a change in the orientation of the latero-frontal cilia of *Mytilus edulis*, in that when the concentration of suspended matter was high, the cilia were directed more frontally, thereby allowing some particles to pass through the partially opened interfilamentar space. The mechanism of this controlled process was not elucidated. Other alterations in function that seem to indicate control of ciliary beating include the increase in pumping rate at lowered oxygen tensions observed by Bayne (1967).

The molecular control of ciliary beating in bivalve gill seems to be at least partly related to the availability of ATP. Schor (1965) showed that ATP in 10 mM concentration increased the cilioexcitatory effect of 10^{-7} M 5-HT on lateral cilia of *Elliptio complanatus*, even though ATP by itself had no effect. Child and Tamm (1963) found that 3 mM ATP activated ciliary beating on gills of *Modiolus demissus* that had been extracted with ethylene glycol and glycerol.

I. *Molluscan tissue other than bivalve gill*

In several other molluscan tissues the cilia may be under control. The palps of bivalves receive extensive innervation with nerve endings positioned between ciliated and mucous cells, but these are generally considered to be sensory (Gilloteaux, 1968). Extensive food sorting by the palps is supposedly accomplished by muscular control of ridges and grooves on the palps that contain ciliary food-sorting tracts. Merton (1923) noticed that ciliary activity on the lip of the snail *Physa* was activated by stimulation of the nerve innervating that tissue; beating stopped and started spontaneously in response to stimulation of the tentacle.

In a different kind of study Buznikov and Manukhin (1961) detected a 5-HT-like substance in the trochophore larvae of several nudibranch gastropods. Addition of 5 $\mu g\ ml^{-1}$ 5-HT increased the ciliary activity of the prototroch, as indicated by the speed of rotation of the trochophores. Velar cilia of veliger larvae were also stimulated by 5-HT. These cilia spontaneously stop and start again. Carter (1926) described the control of cilia in the veliger larvae of nudibranchs and diagrammed the innervation of the ciliated tracts. Careful scrutiny of unpublished electron micrographs (belonging to T. E. Thompson of Bristol) of the same general area in nudibranch veliger larvae revealed much detail of the ciliated cells but no nerve fibers to them.

III. Conclusion: The general nature of ciliary control mechanisms in Metazoa

As pointed out by preceding reviewers (compare Gray, 1928; Kinosita and Murakami, 1968; Sleigh, 1962) there is a general tendency for control to decrease in the higher forms. Ciliary beating is apparently autonomous in mammals and the only vertebrates with proven control are frogs, in which it takes predominantly the role of relatively sustained activation, perhaps through a cholinergic stimulation similar to that of glands and smooth muscle. The beating itself is probably autonomous, dependent on the availability of ATP and required ions, and if membrane potential is involved it has an overall influence on metabolism or rhythmicity, not a phasic change with each beat. One can speculate that intracellular levels of calcium and cyclic nucleotides might influence various enzymes involved in supplying ATP to the ciliary mechanism and in its hydrolysis there. The same kind of nervous control is apparent in several lower metazoans, perhaps amphioxus, and most probably in bivalve gill. In the latter, 5-HT is also involved but its relationship to ACh is not clear. In view of the ability of 5-HT to unlock "catch" in molluscan muscle it would be interesting to know if it has a direct influence on the ciliary mechanism. Bivalve gill also seems to have a dopaminergic mechanism which slows ciliary beating.

A more common type of control present in many metazoans including molluscs, ctenophores and lower chordates but not vertebrates is the sudden inhibition of active cilia. In most examples this seems to be brought about by a nervous reflex. Little is known about the mediator, and the possibility exists that the electrical activity associated with this phenomenon in some species occurs in the ciliated cell. In some groups, such as echinoderm larvae, hemichordates and ascidians, inhibition is accompanied or followed by some kind of reverse beating, making this somewhat analogous to ciliary reversal in Protozoa (Naitoh, 1966) especially if membrane potential reversal occurs. Ctenophores seem to be unique in that each beat is initiated, apparently by a depolarization, in response to an impulse conducted from the aboral sense organ, which itself generates an impulse from the bending of a cilium.

Although much discussion in the past was devoted to the possibility of nervous or neuroid control of metachronism, recent observations (Sleigh, 1971; Aiello and Sleigh, 1972) continue to offer largely mechanical explanations and it is likely that metachronal coordination is not normally under organismic control.

Acknowledgements

Figures 1 and 2 were supplied by A. Paparo of Cornell Medical School from a paper in *Biological Bulletin* (1972), with permission of the publisher. Figures 3 and 4 were supplied by G. Stefano from unpublished studies being conducted by him at Fordham University.

References

AIELLO, E. (1960). *Physiol. Zool.* **33**, 120–135.
AIELLO, E. (1962). *J. cell. comp. Physiol.* **60**, 17–21.
AIELLO, E. (1970). *Physiol. Zool.* **43**, 60–70.
AIELLO, E. and GUIDERI, G. (1964). *Science, Wash.* **146**, 1692–1693.
AIELLO, E. and GUIDERI, G. (1965). *Biol. Bull.* **129**, 431–438.
AIELLO, E. and GUIDERI, G. (1966). *J. Pharmac. exp. Ther.* **154**, 517–523.
AIELLO, E. and SLEIGH, M. A. (1972). *J. Cell Biol.* **54**, 493–506.
ALEXANDROV, V. Y. and ARRONET, N. I. (1956). *Dokl. Acad. Nauk, SSSR* **110,** 457–460. [In Russian.]
BABA, S. A. (1968). *J. Fac. Sci. Tokyo Univ.* IV, **2** (3), 385–393.
BAYNE, B. L. (1967). *Physiol. Zool.* **40**, 307–313.
BLANDAU, R. J. (1969). *In* "The Mammalian Oviduct" (E. Hafez and R. Blandau, eds), pp. 120–162. University of Chicago Press.
BONE, Q. (1960). *Phil. Trans. R. Soc. Ser. B* **243** (704), 241–269.
BOUFFARD, T. (1969). *Am. Zool.* **9**, 1108, abstr. #250.
BOUFFARD, T. and AIELLO, E. (1969). *Am. Zool.* **9**, 582, abstr. #214.

BRENNER, R. M. (1969). *In* "The Mammalian Oviduct" (E. Hafez and R. Blandau, eds), pp. 203–229. University of Chicago Press.
BRUNDIN, J. (1969). *In* "The Mammalian Oviduct" (E. Hafez and R. Blandau, eds), pp. 251–269. University of Chicago Press.
BULBRING, E., BURN, J. H. and SHELLEY, H. J. (1953). *Proc. R. Soc.* B **141**, 445–466.
BURDON-JONES, C. (1952). *Proc. R. Soc. Lond.* B **236**, 553–590.
BURN, J. H. and DAY, M. (1958). *J. Physiol.* **141**, 520–526.
BUZNIKOV, G. A. and MANUKHIN, B. N. (1961). *Zhurnal Obshchei Biologie* **22**, 226–232. [In Russian.]
CARTER, G. S. (1926). *J. exp. Biol.* **4**, 1–26.
CHILD, C. M. (1933). *J. comp. Neurol.* **57**, 199–252.
CHILD, F. M. and TAMM, S. (1963). *Biol. Bull.* **125**, 373–374 (abstr).
CORSSEN, G. and ALLEN, G. B. (1959). *J. appl. Physiol.* **14**, 901–904.
CRESS, H., SPOCK, A., and HEATHERINGTON, D. (1965). *J. Histochem. Cytochem.* **13**, 677–683.
DALHAMN, T. (1956). *Acta Physiol. scand.* **36**, Suppl. 123.
DRAL, A. D. G. (1967). *Neth. J. Sea Res.* **3**, 391–422.
ECKERT, R. and MURAKAMI, A. (1972). *In* "Contractility of Muscle Cells and Related Processes" (A. Podolsky, ed.), pp. 115–127. Prentice-Hall, New York.
FEDELE, M. (1926). *Riv. Biol.* **8**, 360–375.
FELTON, B. H. (1972). *Am. Zool.* **12**, XXXIV (abstr. #407).
FELTON, B. H. and AIELLO, E. (1971). *Am. Zool.* **11**, 665 (abstr. #236).
FRIEDRICSSON, B. (1969). *In* "The Mammalian Oviduct" (E. Hafez and R. Blandau, eds), pp. 311–332. University of Chicago Press.
GALT, C. P. and MACKIE, G. O. (1971). *J. exp. Biol.* **55**, 205–212.
GAUPP, E. (1904). "A. Ecker's und R. Wiedersheim's Anatomie des Frosches auf Grund eigener Untersuchung durchaus neu bearbeitet", Part 2, Vol. 3, p. 22.
GILLOTEAUX, J. (1968). *Annls Soc. r. zool. Belg.* **98**, 101–123.
GOSSELIN, R. E. (1961). *J. cell. comp. Physiol.* **58**, 17–26.
GOSSELIN, R. E. (1966). *Am. Rev. resp. Dis.* **93** (3, part 2), 41–59.
GOSSELIN, R. E., MOORE, K. E. and MILTON, A. S. (1962). *J. gen. Physiol.* **46**, 277–296.
GRAY, J. (1922). *Proc. R. Soc.* B **93**, 104–121.
GRAY, J. (1926). *Proc. R. Soc.* B **99**, 398–404.
GRAY, J. (1928). "Ciliary Movement". Macmillan, New York.
HILL, J. R. (1957). *J. Physiol.* **139**, 157–166.
HORRIDGE, G. A. (1965a). *Am. Zool.* **5**, 357–375.
HORRIDGE, G. A. (1965b). *Nature, Lond.* **205**, 602.
HORRIDGE, G. A. (1966). *In* "The Cnidaria and Their Evolution." *Symp. Zool. Soc. Lond.* **16**, 247–266.
HORRIDGE, G. A. and MACKAY, B. (1964). *Q. Jl microsc. Sci.* **105**, 163–174.
JØRGENSEN, C. B. (1966). "Biology of Suspension Feeding." Pergamon Press, New York.
KINOSITA, H. and MURAKAMI, A. (1967). *Physiol. Rev.* **47**, 53–82.
KNIGHT-JONES, E. W. (1952). *Proc. R. Soc.* B **236**, 315–354.

KRUEGER, A. P. (1962). *J. gen. Physiol.* **45** (4, part 2, suppl.), 233–241.
KRUEGER, A. P. and SMITH, R. F. (1960). *J. gen. Physiol.* **43**, 533–540.
LAGERSPETZ, K., LAENSIMIES, H., IMPIVAARA, H. and SENIUS, K. (1970). *Comp. gen. Pharmacol.* **1**, 152–154.
LASCHKOW, W. F. (1955). *Z. mikrosk.-anat. Forsch.* **61**, 229.
LOMMEL, F. (1908). *Dt. Arch. klin. Med.* **94**, 365–376.
LUCAS, A. M. (1935). *Am. J. Physiol.* **112**, 468–476.
LUCAS, A. M. and DOUGLAS, L. C. (1935). *Archs Otolaryng.* **21**, 285–296.
MACGINITIE, G. E. (1939). *Biol. Bull.* **77**, 443–447.
MACKIE, G. O. (1970). *Q. Rev. Biol.* **45**, 319–332.
MACKIE, G. O., SPENCER, A. N., and STRATHMANN, R. (1969). *Nature, Lond.* **223**, 1384–1385.
MALANGA, C. (1971). *Am. Zool.* **11**, 661 (abstr. #222).
MALANGA, C. and AIELLO, E. (1971). *Comp. gen. Pharmacol.* **8**, 856–868.
MALANGA, C. J. and AIELLO, E. (1972). *Comp. Biochem. Physiol.* **43B**, 795–806.
MALANGA, C., WENGER, G. and AIELLO, E. (1972). *Comp. Biochem. Physiol.* **43A**, 825–830.
MCDONALD, J., LEISURE, C., and LENNEMAN, E. (1927). *Proc. Soc. exp. Biol.* **24**, 968–970.
MERTON, H. (1923). *Pflüg. Arch. ges. Physiol.* **198**, 1–28.
MESSERKLINGER, W. (1958). *Arch. Ohr. Nas. Kehlkopfheilk.* **193**, 1.
MILTON, A. S. (1959). *Br. J. Pharmac. Chemother.* **14**, 323–326.
MOORE, K. E. and GOSSELIN, R. E. (1962). *J. Pharmac. exp. Ther.* **138**, 145–153.
MOORE, K. E., MILTON, A. S., and GOSSELIN, R. E. (1961). *Br. J. Pharmac. Chemother.* **17**, 278–285.
MURAKAMI, A. (1968). *J. Fac. Sci. Tokyo Univ.* Sect. IV, *Zool.* **11**, 373–384.
NAITOH, Y. (1966). *Science, Wash.* **154**, 660–662.
PAPARO, A. (1969). Ph.D. Dissertation, Fordham University, New York.
PAPARO, A. (1972). *Biol. Bull.* **143**, 592–605.
PAPARO, A. *Comp. gen. Pharmacol.* (in press).
PAPARO, A. and AIELLO, E. (1970). *Comp. gen. Pharmacol.* **1**, 241–250.
PAPARO, A. and FINCH, C. (1972). *Comp. gen. Pharmacol.* **3**, 303–309.
PARKER, G. H. (1928). *Proc. natn. Acad. Sci., U.S.A.* **14**, 713–714.
PARKER, G. H. (1931). *Phil. Trans. R. Soc.* **219**, 381.
PARKER, G. H. and MARKS, A. P. (1928). *J. exp. Zool.* **52**, 1–6.
RHODIN, J. (1966). *Am. Rev. resp. Dis.* **93** (3, part 2), 1–15.
SATIR, P. and CHILD, F. M. (1963). *Biol. Bull.* **125**, 390 (abstr.).
SATIR, P. and GILULA, N. (1970). *J. Cell Biol.* **47**, 468–487.
SATOH, K. (1959a). *Hirosaki med. J.* **10**, 9–19 (abstr. #2).
SATOH, K. (1959b). *Hirosaki med. J.* **10**, 196–201 (abstr. #29).
SCHOR, S. (1965). *Science, Wash.* **148**, 500–501.
SEO, A. (1931). *Jap. J. med. Sci. III. Biophysics* **2**, 47–75.
SEO, A. (1937). *Fukuoka Acta Medica* **30**, (4) 1–4.
SLEIGH, M. A. (1962). "The Biology of Cilia and Flagella." Pergamon Press, New York.
SLEIGH, M. A. (1968). *J. exp. Biol.* **48**, 111–126.
SLEIGH, M. A. (1971). *Endeavour* **30**, 11–17.

SLEIGH, M. A. MACKIE, G. O., SINGLA, C. L. and WILLIAMS, D. (personal communication, manuscript in preparation).

STEFANO, G. and AIELLO, E. (1972). *Am. Zool.* **12**, XXXVI (abstr. #413).

SWEENEY, D. (1968), *Comp. Biochem. Physiol.* **25**, 601–614.

TAKAHASHI, E. and MURAKAMI, A. (1968). *J. Fac. Sci. Tokyo Univ.* IV, **2**, 359–372.

VORHAUS, E. and DEYRUP, I. (1953). *Science, Wash.* **118**, 553–554.

WARDELL, J, CHARKIN, L. and PAYNE, B. (1970). *Am. Rev. resp. Dis.* **101**, 741–754.

Note added in proof

Several relevant observations have come to my attention since preparing this review. C. Malanga (in press, *Comp. Gen. Physiol.*) has found that dopamine stimulates frontal cilia while inhibiting lateral cilia on bivalve gill. This means that most of the published apparent correlations between ciliary activity and whole gill metabolism will have to be re-evaluated and the contribution of each cell type to total metabolism determined.

Regarding echinoderm larvae and the larvae of other invertebrates, R. Strathmann, T. Jahn and J. Fonseca (*Biol. Bull.* **142**, 505–519, 1972) have observed that the cilia which create water currents can reverse their effective stroke momentarily and in doing so remove particles from this stream of water and transfer them to the circumoral field from which they are passed to the mouth. The particles themselves seem to be the cause of the reversal but the mechanism requires further explanation.

Examples of the diversity of cilia and flagella

Chapter 14

Structural variants in invertebrate sperm flagella and their relationship to motility

DAVID M. PHILLIPS

The Population Council, The Rockefeller University, New York, N.Y. U.S.A

I. Introduction

Several kinds of approaches have been taken in studies of flagellar function. The chemistry, ultrastructure, and motility of the 9+2 flagellum have been studied. These are discussed in other chapters of this book. We would like to discuss another approach, that of comparing the structural and motile properties of exceptional flagella which do not conform to the 9+2 tubule pattern. The 9+2 arrangement of flagellar tubules has been remarkably stable throughout evolution. The ultrastructure of flagella (and cilia) is virtually the same in Protozoa as in man. A very few exceptional flagellar tubule patterns have been described, these mostly in insects. The significance of the general stability of flagellar form is obscure, as is the significance of those few variations that do occur. The idea behind our approach to understanding flagellar function has been to analyze flagella which differ in structure from the typical 9+2 pattern and to correlate existing differences in flagellar tubule pattern with differences in motile pattern.

II. Insect flagella

Most of the variants in tubule patterns of sperm flagella have been observed in insect species. Before discussing the relationship between structure and function in insect sperm which possess a variant type of flagellum, we will give some attention to norms of structure and function in the "typical" insect sperm flagellum.

The motile apparatus of most insect sperm is a flagellum with a 9+9+2 tubule pattern. That is, the motile apparatus consists of a more or less typical 9+2 flagellum surrounded by 9 peripherally situated tubules termed accessory tubules (Fig. 1). In some insect species, a small dense rod is situated between neighboring accessory tubules and juxtaposed to the doublet between subfibers A and B. Radial elements between the central pair and doublet tubules are generally prominent but appear to vary somewhat in morphology among species (Phillips, 1970a). This may be due in part to different fixation techniques employed by different investigators.

In addition to the motile apparatus, insect sperm contain one or two large mitochondrial derivatives which generally extend from the head nearly to the posterior end of the spermatozoon. Mitochondria in many species occupy most of the volume of the sperm. In some species which have sperm containing two mitochondrial derivatives, one of these may be more enlarged than the other in some regions or may extend further anteriorly than the other. In other insect species, both mitochondrial derivatives are of equal size and extension. Mitochondria of many species contain dense paracrystalline material.

The mitochondria, as well as the flagellar tubule pattern, must be considered if one is attempting to analyze sperm motility in terms of structure. Mitochondria may, and in some cases almost certainly do, impart stiffness to the flagellar beat, thereby affecting the form of the flagellar wave. Mitochondria also define a bilateral symmetry in the insect spermatozoon since they are located on one side of the flagellum.

We have recently been analyzing patterns of motility in sperm of a number of insect species and attempting to correlate sperm motility with ultrastructure. We study swimming spermatozoa on film taken with a high-speed motion picture camera (24–500 frames per second). For

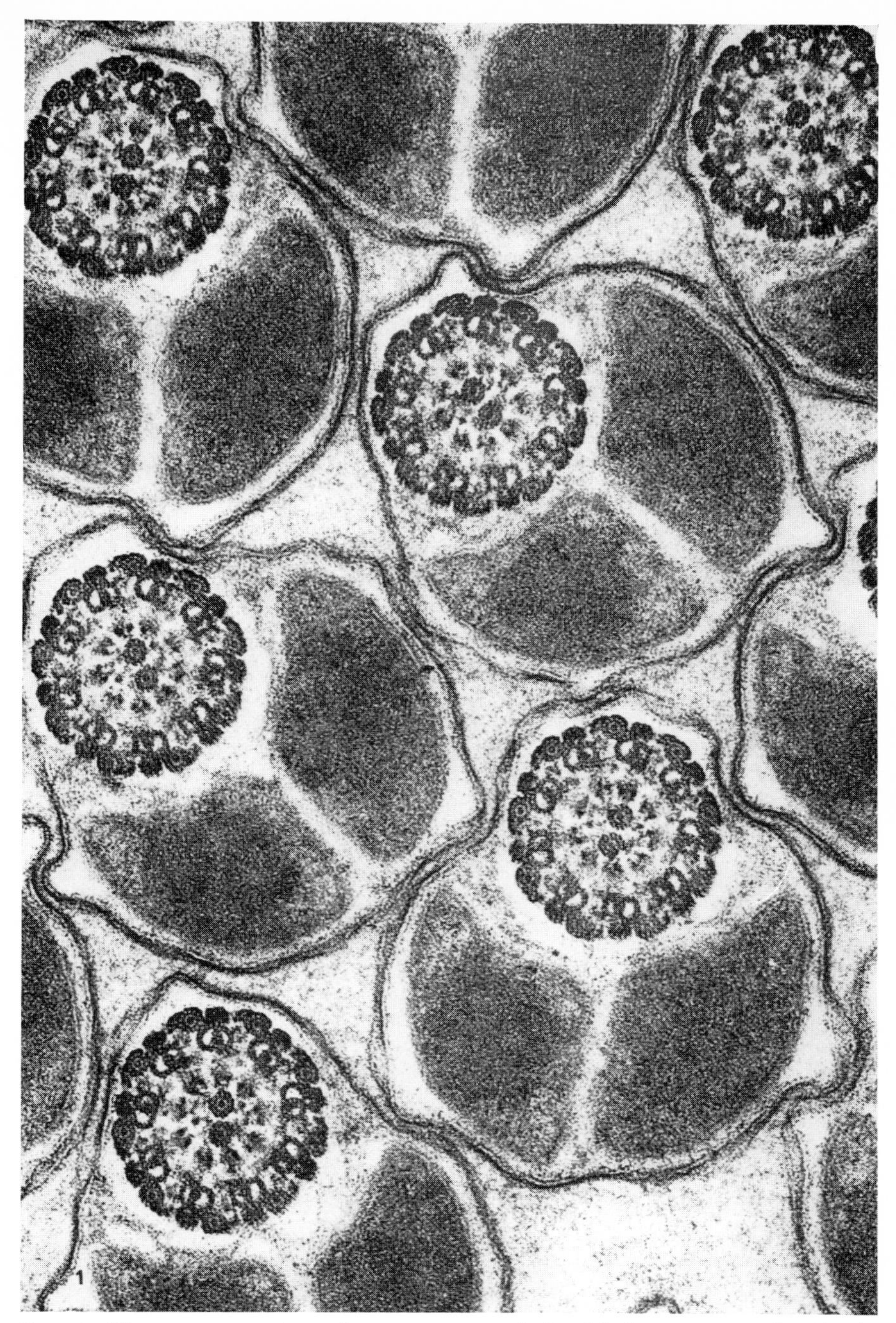

FIG. 1. Transverse section of spermatozoa of the black scavenger fly *Sepsis*, showing the typical 9+9+2 insect sperm flagellum. The central pair of tubules and peripheral singlet tubules are each composed of 13 protofibrils. The small tubule in the center of the tubules of the central pair and peripheral tubules is characteristic of spermatozoa of higher Diptera. ×115,000.

technical reasons, one cannot observe swimming spermatozoa in their natural environment in the female tract. We utilize the less ideal expedient of viewing motile spermatozoa in hanging drop preparations or on slides which have been elevated with petroleum jelly. We generally use Grace's

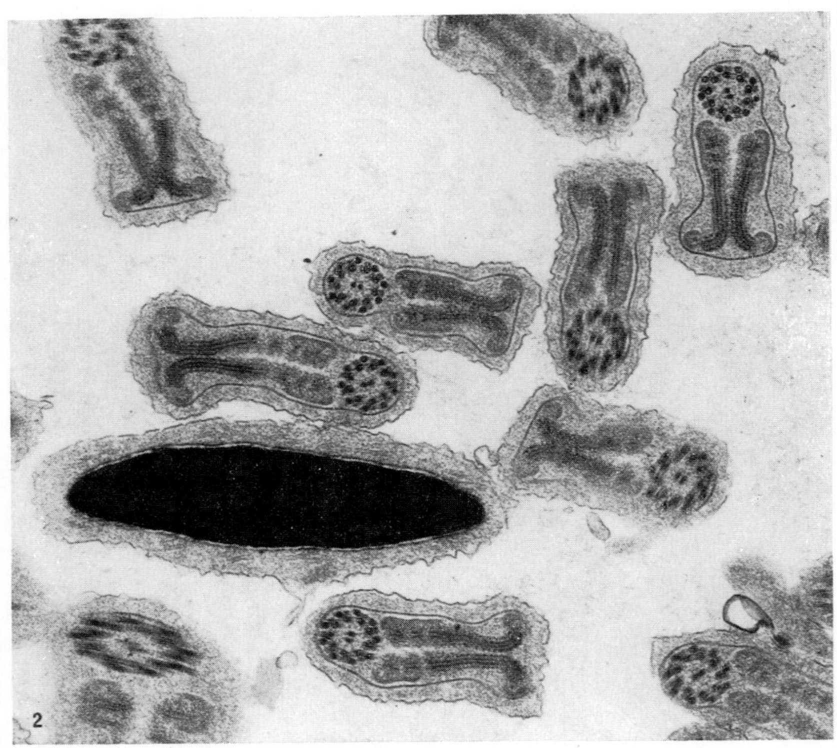

FIG. 2. Spermatozoa of the walking stick *Diapheromera femorata* are seen here in cross-section. The nucleus of sperm of this species, as illustrated in this cross-sectional view, is flattened enabling one to detect the rapid rotation of the spermatozoan during motility. Mitochondria of this species do not form a typical nebenkern and are sloughed from the spermatid cytoplasm during development. The structures seen adjacent to the sperm flagellum, although not mitochondrial derivatives, may serve a similar function. × 22,000.

insect medium (GIBCO) and mosquito culture medium (GIBCO) or Beadle-Ephrussi saline as fluids for studying motility. These produce environments which are, of course, very different from the natural environment in which spermatozoa swim; however, spermatozoa of many insect species swim readily and in a consistent manner under these conditions. Different media and pH do not noticeably affect the pattern of spermatozoan motility. We most often examine spermatozoa taken from the male

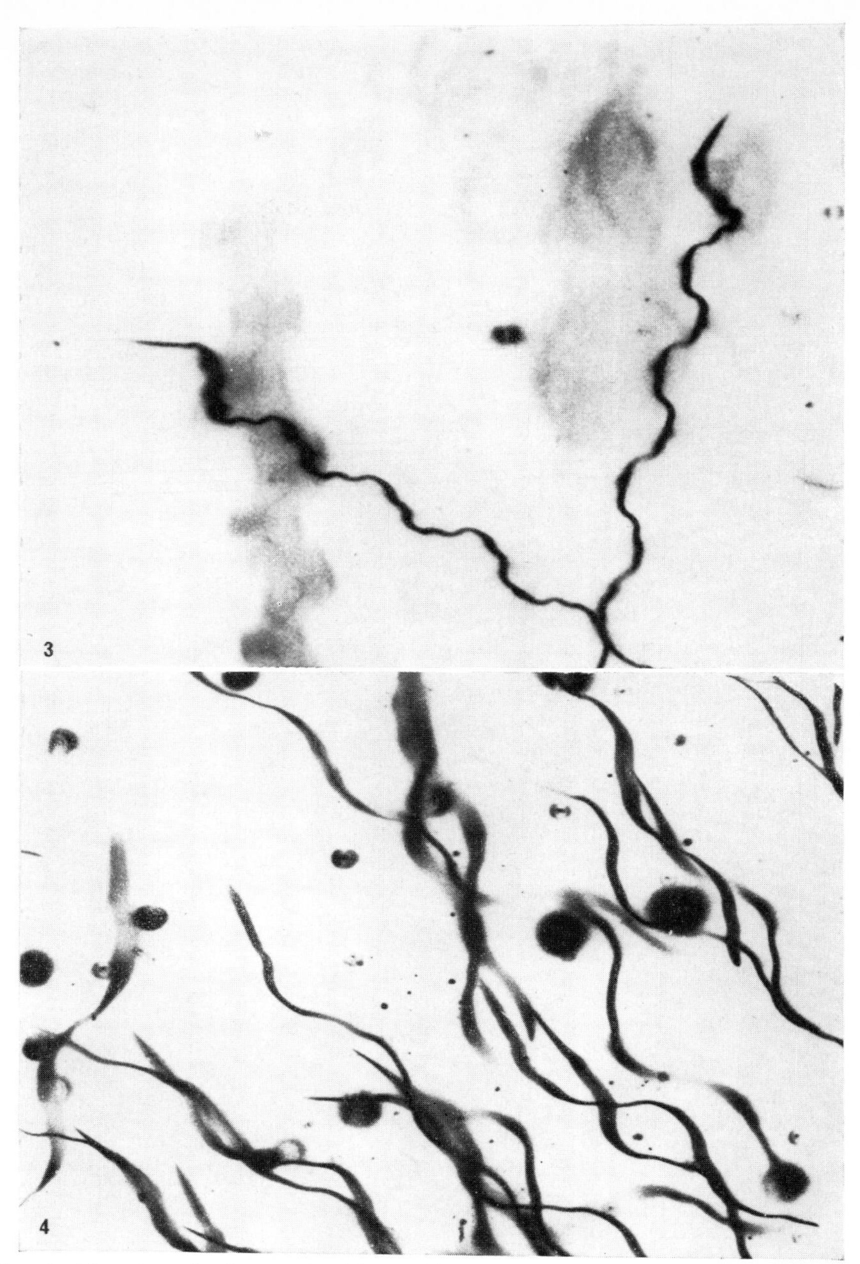

Fig. 3. Negative of a frame of a 16 mm film of mealworm (*Tenebrio*) spermatozoa taken at 50 frames per second in dark-field, showing the double sine wave beat pattern characteristic of spermatozoa of many insect species. ×1200.

Fig. 4. Negative of a frame of a milkweed bug (*Lygaeus* sp.) spermatozoa taken at 125 frames per second in dark-field optics. Spermatozoa of this species also show a double sine wave beat. The shorter wavelength wave has a frequency of approximately 110 beats per second—so fast that it can barely be detected at this frame rate. ×2000.

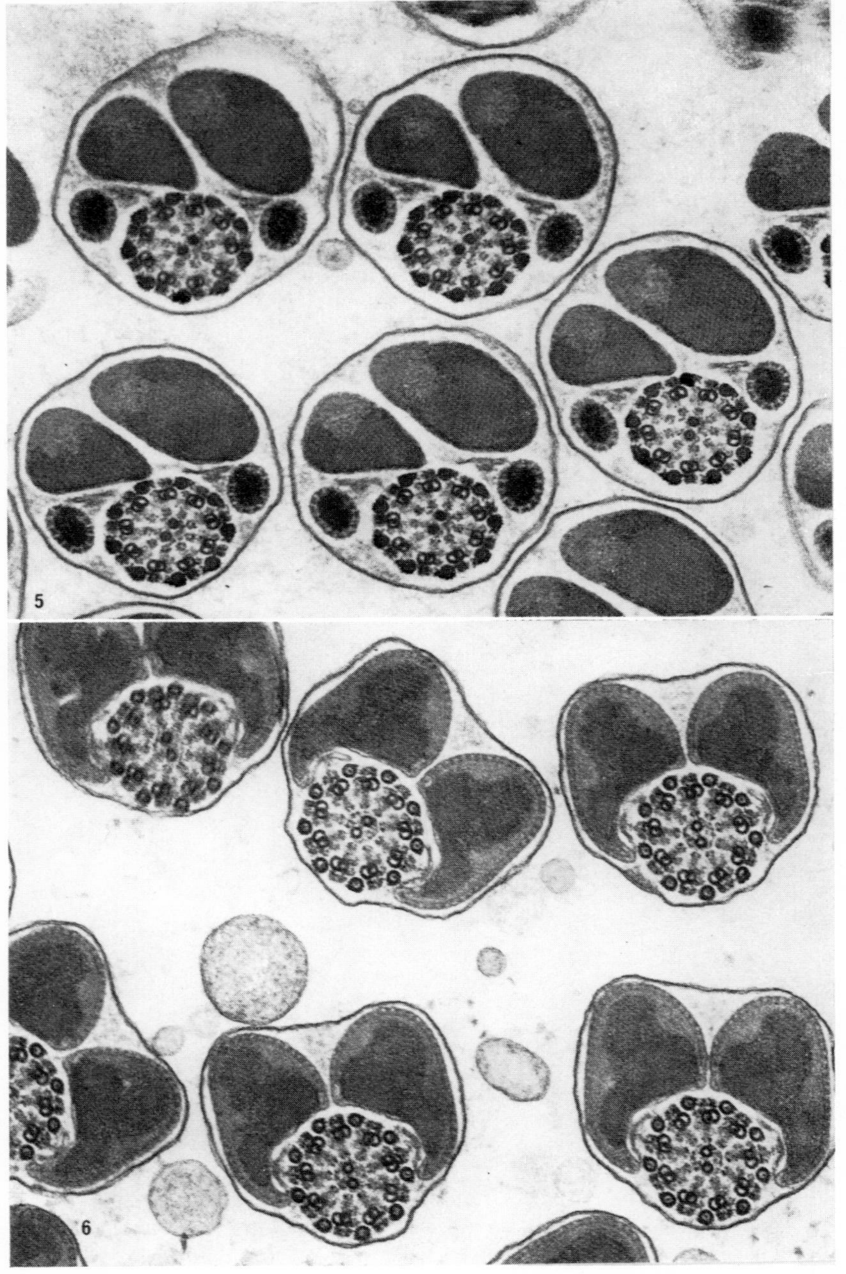

FIG. 5. Transversely sectioned spermatozoa of the mealworm, *Tenebrio*. ×61,000.

FIG. 6. Transverse sections of milkweed bug, *Lygaeus*, sperm. Although spermatozoa of *Tenebrio* and *Lygaeus* display a similar motile pattern, they are somewhat different in ultrastructure. ×61,000.

genital tract as they are far easier to obtain than spermatozoa from the genital tract of inseminated females. The patterns of motility of spermatozoa from male and female tracts cannot be presumed to be the same, because spermatozoa undergo structural changes in the female genital tract in at least some species of insects (Phillips, 1966b, 1971).

We have observed several types of patterns of motility among insect spermatozoa. One type of pattern exhibited by spermatozoa of many

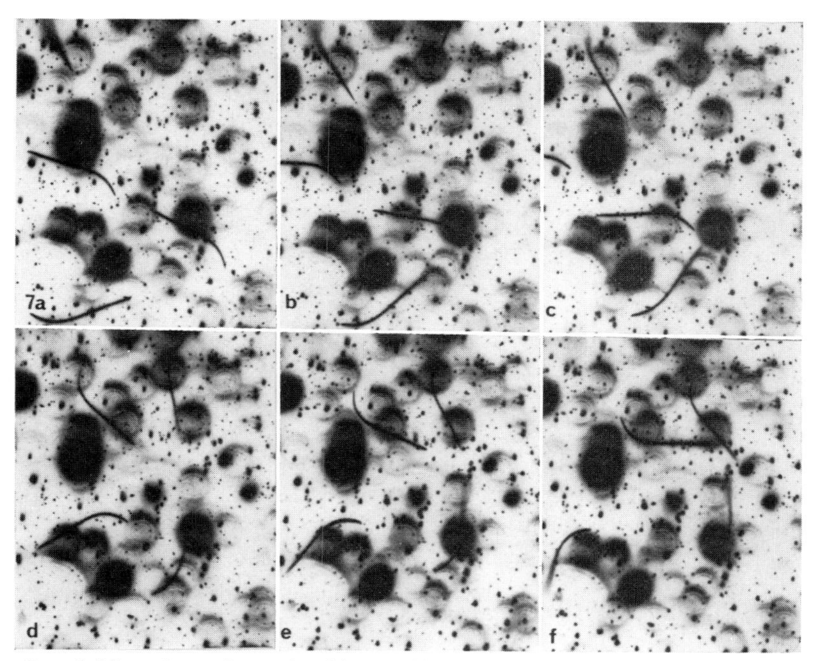

FIG. 7. a–f. Negative prints of a 16 mm film, showing motile spermatozoa of the long horned beetle *Prionus* taken with dark-field optics. The frames are at 3 s intervals and show "gliding" spermatozoa. A wave along the sperm axis can be detected in the films but does not show up well in these prints. ×200.

species is a non-planar, irregular beat in which the sperm rotates. We have observed this pattern of motility in two species of cockroach, several species of grasshoppers, and the walking stick, *Diapheromera femorata*. In some of these species the rotation of the head can be detected because the acrosome is slightly flattened in one plane. In sperm of the walking stick (Fig. 2), the head is flattened and one can easily discern that spermatozoa rotate rapidly as they progress. In some species the sperm head is cylindrical and rotation cannot be observed although it may occur.

In a second type of pattern, observed in spermatozoa of other insect

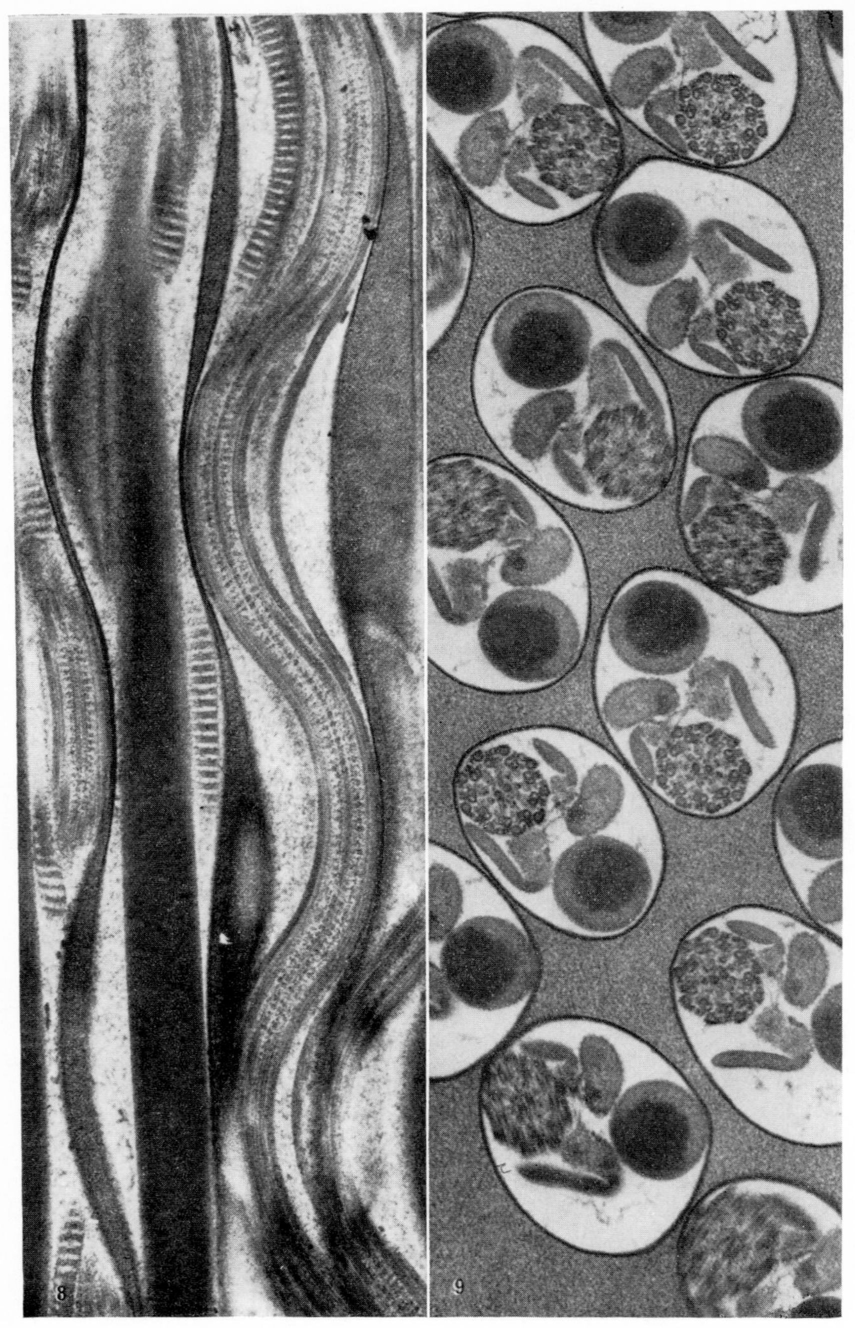

FIG. 8. Longitudinal section of *Prionus* spermatozoa. The sine wave shape of one flagellum can be followed in the section, demonstrating that it is reasonably planar, at least in "resting" spermatozoa in the testis. The paracrystalline material in the mitochondria shows a herringbone pattern. $\times 30,000$.

FIG. 9. The mitochondria of these transversely sectioned spermatozoa of *Prionus* contain dense paracrystalline material. $\times 43,000$.

species, spermatozoa display a two-component wave. The major wave appears essentially like a sine wave, though it is not planar. Superimposed on this major wave is a sine wave of much smaller amplitude and far greater frequency. We have observed this form of beat in sperm of the mealworm, *Tenebrio* (Fig. 3), and in several hemipteran species including the milkweed bug, *Lygaeus* (Fig. 4). In spermatozoa of the milkweed bug the minor wave has a frequency of about 130 waves per second. Ultrastructural analysis has yielded no clues as to the structural features involved in the two types of patterns. Cockroach and walking stick spermatozoa are morphologically very dissimilar although their patterns of motility are much alike. Spermatozoa of the mealworm and milkweed bug, both of which display a similar double wave pattern, and also not very similar in structure when viewed in the electron microscope (Figs 5, 6).

Spermatozoa of some insect species display what appears to be a gliding pattern of motility. This type of pattern is typical of spermatozoa of weevils, snout beetles, and long-horned beetles, as well as some other species of Coleoptera. Motile sperm appear straight and stiff; however, a wave of very low amplitude can be detected along the length of the cell (Fig. 7). In the electron microscope in longitudinal section, one can follow a planar sine wave flagellum along a straight, dense mitochondrial derivative (Fig. 8). In transverse section, it can be observed that the mitochondrion is filled with a straight dense rod (Fig. 9). It appears that the dense material in the mitochondrial derivative imparts the stiffness to the sperm. Spermatozoa of many amphibians have a similar motile pattern and an analogous structural feature, the axial rod, but in amphibians the dense element is derived in association with a flagellar doublet rather than a mitochondrion (Burgos and Fawcett, 1956; Barker and Biesele, 1967).

III. Variant sperm flagella

A. *Variations in the central elements*

1. *Absence of central elements*

The most common type of exception to the 9+2 tubule pattern of flagella is the occurrence of central elements other than the usual central tubule pair. Sperm flagella of mayflies lack central tubules completely and may be said to consist of a 9+9+0 tubule pattern (Phillips, 1969, 1970d), and when they are viewed in transverse section, a faint 800 Å central element can be discerned. In some mayfly species, connections of low electron density appear to radiate between the central element and the flagellar doublets (Fig. 11). We have found that mayfly spermatozoa are motile (Fig. 10). They exhibit a non-planar beat, and the sperm head rotates during motility. We have seen a similar type of beat in several mayfly species. In all species, nearly 100% of the spermatozoa taken from

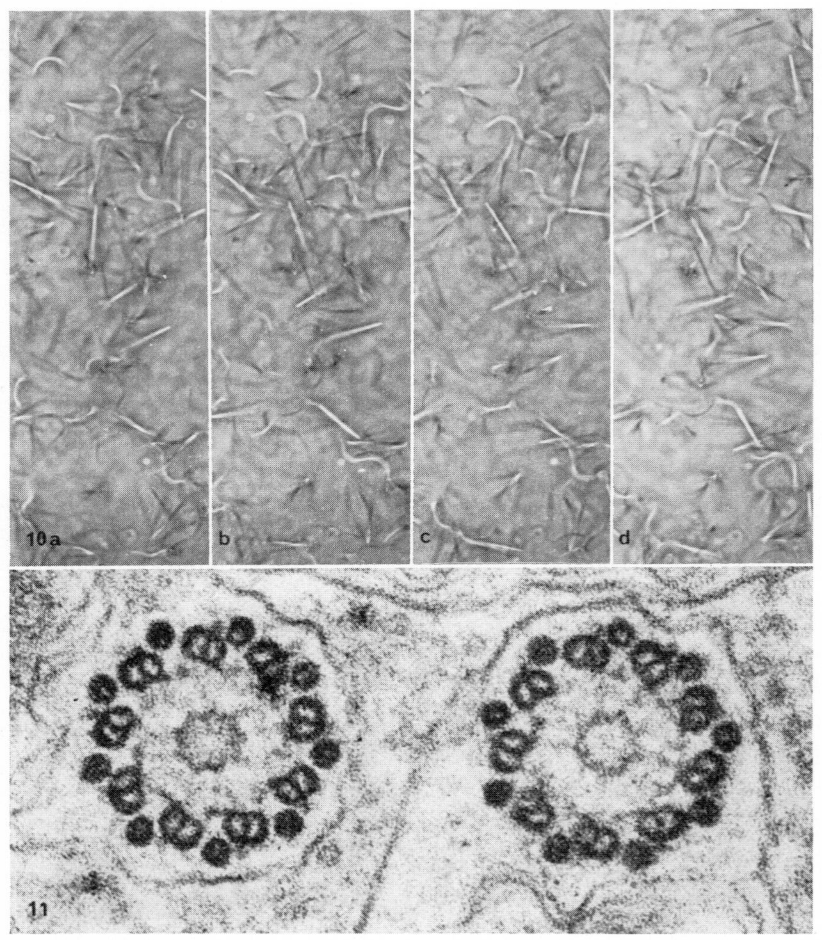

FIG. 10. a–d. Negative prints of swimming spermatozoa of the mayfly *Hexagenia* (phase contrast). Prints are at 0·05 s intervals. ×400.

FIG. 11. Spermatozoa of the mayfly *Hexagenia*. ×132,000.

male mayflies beat rapidly in any medium we put them in, even unbuffered 0·9% NaCl.

Recently Hood *et al.* (1972) have described a 9+0 flagellum in the sperm of the scorpion *Vejovis carolinianus*. According to these authors, a few spermatozoa exhibit motility in isotonic saline solution. We have examined the ultrastructure of spermatozoa of the hairy scorpion, *Hadurus hirsutus*. Spermatozoa of this species exhibit what may also be described as a 9+0 flagellar tubule pattern (Fig. 12). In the center of the sperm flagellum

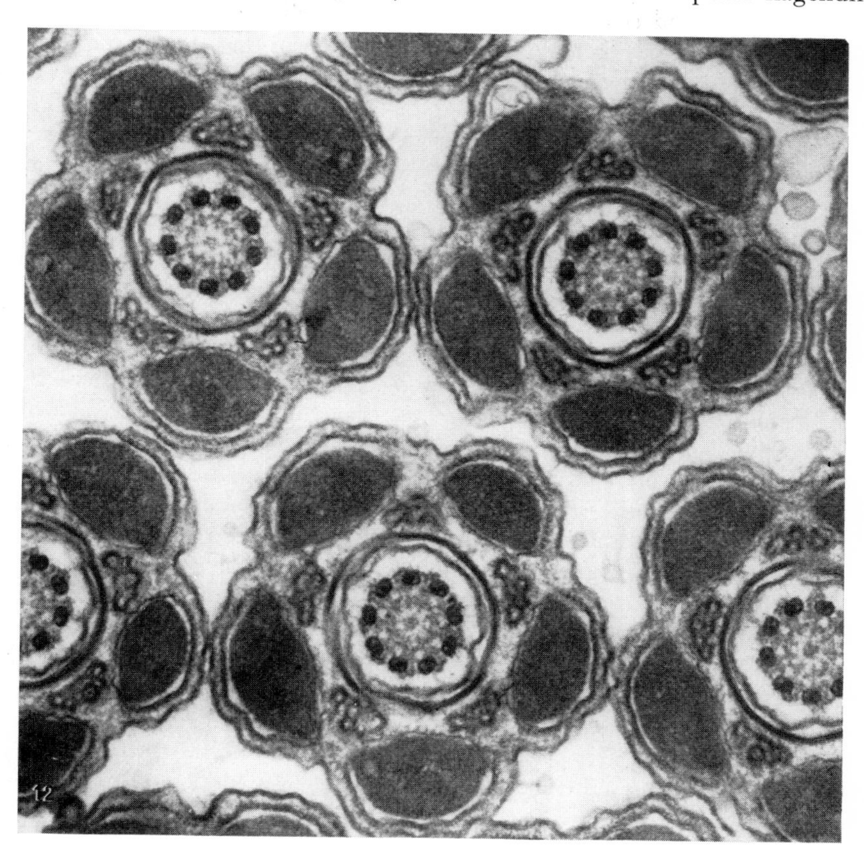

FIG. 12. Spermatozoa of the hairy scorpion *Hadurus hirsutus*. The spermatozoa of these species have no central tubules but possess some dense material in the center of the flagellum. The flagellum in this species is surrounded by 5 long mitochondria and 5 groups of structures that appear tubular. ×65,000.

there is a moderately electron dense area about 200 Å in diameter. Moderately electron dense structures of 9-fold symmetry are situated between the central structure and the doublets. As we have in some cases fixed the testis at room temperature and observed cytoplasmic microtubules in several cell types, it is unlikely that flagellar tubules more labile than the doublets are present in the flagellum of *Hadurus* but not preserved by the fixation. The flagellum of *Hadurus* has no accessory tubules or dense fibers. When *Hadurus* spermatozoa from a male animal are put into saline, they at first appear immotile but later a small percentage of the cells exhibit some slow (possibly abnormal) beating. They are, therefore, as reported by Hood *et al.* (1972), capable of motility. The fact that 9+0 scorpion spermatozoa beat so slowly does not mean that they do not swim rapidly *in vivo*. It is not unlikely that they require "activation" by the female or undergo changes in the female tract before they can swim. This is probably a prerequisite to motility in other species as well, as we have been unable to observe motility in spermatozoa of a number of insect species with typical 9+9+2 flagella.

In addition to the situation in the mayfly and in the scorpion, there have been two other reports of motile "9+0" flagella. These reports (Costello *et al.*, 1969; Desportes, 1966) describe motile spermatozoa of flatworm and gregarine protozoan species, respectively. Unfortunately in both of these reports there are very few micrographs of 9+0 sperm. The micrographs are not of sufficient detail to discern whether or not there is a moderately electron dense center in the flagellum such as we have observed in scorpion, and it is not clear that the micrographs are of mature spermatozoa and not spermatids. The latter point may be important since in sperm of the psocid which has a central element, the element forms late in spermiogenesis (Phillips, 1969). Central tubules also form after the flagellar doublets have formed in some species (Meyer, 1968).

Are there truly motile 9+0 flagella? Spermatozoa of mayflies do not have central tubules and most certainly are capable of motility, yet the 9 peripheral singlet tubules may serve the function of the central pair. Scorpion spermatozoa do not have a central pair of tubules, yet where resolution is sufficient, it is clear that they do have some central structure. They exhibit at best very poor motility and it is still unknown whether they are motile *in vivo*. Studies by Costello *et al.* (1969) and Desportes (1966) do not show entirely convincing 9+0 flagella in mature spermatozoa. The question arises as to what should be considered a 9+0 flagellum. Is a flagellum like that of the scorpion which has moderately electron dense material, possibly a rod, in the place of a central pair a 9+0 flagellum? If one considers such flagella which lack central tubules but have other material in the center of the axoneme to be 9+0 flagella, then it is fairly clear that these "9+0" flagella are motile. Convincing evidence for a motile flagellum with no central elements of any sort and without accessory tubules has not been demonstrated. It may be that the 9 doublet tubules may need to interact with some sort of central element for motility. Summers and Gibbons (1971) have recently demonstrated that flagellar doublets activated in glycerinated axonemes of sea urchin sperm are capable of sliding on one another without the involvement of central tubules. This observation would suggest that central elements may not be directly involved in sliding.

Related to this are the very interesting investigations of Sir John Randall and coworkers on paralyzed mutants of *Chlamydomonas* (Randall, 1969). These workers have isolated 21 non-motile mutants of *Chlamydomonas* in which the central pair of flagellar tubules is missing. These immotile flagella, termed 9+0 flagella by Randall, all contain dense amorphous material in the position of the central pair. Randall's results suggest that the central pair in its intact form is necessary for motility in *Chlamydomonas*.

2. *Single central elements*

Among the insects, all of the mosquito species that we have examined display a 9+9+1 flagellar tubule pattern (Fig. 14). Breland *et al.* (1968) have also observed 9+9+1 spermatozoa in mosquitos. Mosquito sperm rotate as they swim. The flagellar beat is planar over short distances but not along its entire length (Fig. 13). The type of motile pattern exhibited by mosquito spermatozoa is not obviously different from that exhibited by cockroach and walking stick spermatozoa.

Spermatozoa of many flatworm species have a 9+1 flagellum (Burton, 1967, 1968, 1972; Sato *et al.*, 1967; Silveira and Porter, 1964; Von Bonsdorff and Talkka, 1955). Active motility has been reported in some of these species. The central element of flatworm spermatozoa is not a simple tubule but rather an elaborate structure. Burton (1967) has carefully described the central element in spermatozoa of a frog lung fluke. In the center of the central element is a 300 Å structure surrounded by two rows of 125 Å subunits arranged in a double helix. This rather elaborate central element is 650–700 Å in diameter and is connected to subfiber A of each doublet by thin filaments. Nine+1 flagella have also been reported in the scorpion *Centroroides vittatus* (Hood *et al.*, 1972).

3. *Three central tubules*

A 9+3 flagellum has been described in the spiders, *Pisaurina* sp. (Reger, 1970), *Heptathela kimurai* (Osaki, 1969), and *Pholcus phalangioides* (Rosati *et al.*, 1970). Rosati *et al.* also report having seen 9+3 flagella in other spider species. The flagellum of spider spermatozoa is unusual. It does not project from the cells but becomes wound around inside the roughly spherical spermatids as it elongates. There is no membrane surrounding the flagellar axoneme and it has been suggested that the flagellum is incorporated into the cytoplasm of mature spermatozoa and is non-functional (Sharma, 1950; Sharma and Gupta, 1956).

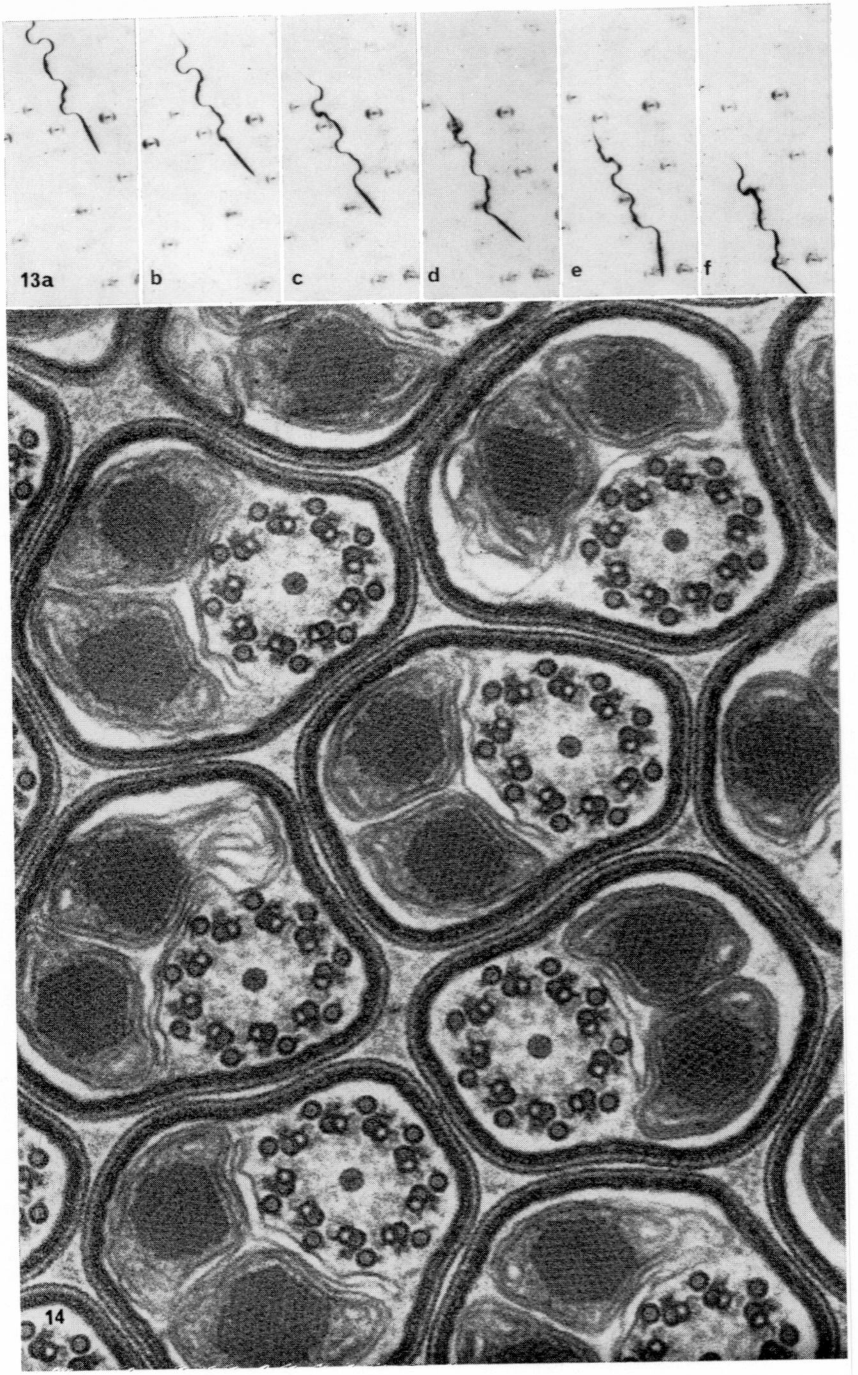

Fig. 13. a–f. Negative prints of 16 mm film of a swimming spermatozoan of the mosquito *Culex* sp. Prints are at 2 s intervals. ×350.

Fig. 14. Spermatozoa of *Culex* sp. ×96,000.

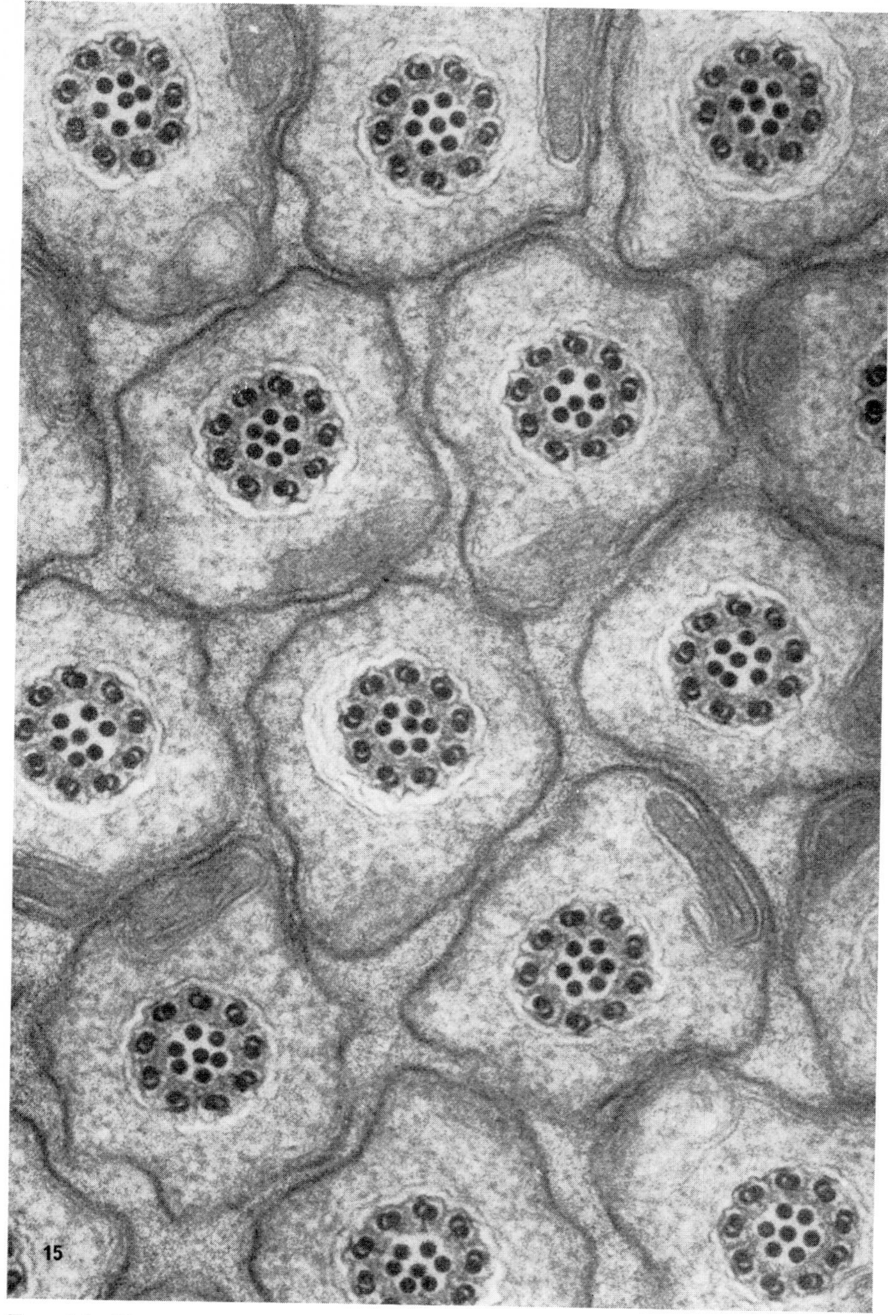

FIG. 15. Transverse sections of spermatozoa of a caddis fly *Hydropsyche* sp. ×65,000.

4. Seven central tubules

In two species of caddis flies which we have studied, we have observed seven central tubules in the position of the usual central pair (Phillips, 1969, 1970d). Six of the seven tubules are evenly spaced around a central tubule (Fig. 15). Spermatozoa of these two caddis fly species (*Hydropsyche* and *Polycentropus*) are peculiar in other respects. Unlike spermatozoa of other insect species, they do not contain nine peripheral singlet tubules. Mitochondria are situated at irregular intervals. The motility of these spermatozoa has not been examined.

B. Branching flagella

Spermatozoa of treehoppers have a most peculiar tubule pattern (Phillips, 1969, 1970d). The anterior portion of the flagellum is a typical 9+9+2 flagellum. Posteriorly the flagellum branches into four small flagellar tails. Three of the branches contain two doublet and two singlet tubules and the fourth contains three doublet and three singlet tubules. The central pair terminates at the level where the flagellum branches (Fig. 17). In some species the doublet tubule becomes partly dissociated in the flagellar tails. In transverse section the dissociated doublet appears as a circle and a "c" shaped profile. This phenomenon is common to other sperm flagella also (Phillips, 1970c), suggesting that subfiber A and B share a common wall. In sperm of some species, it can be ascertained that the "c" shaped element is derived from subfiber B and the complete tubule from subfiber A (Phillips, 1966c). We examined motility in seven species of treehoppers. Spermatozoa of all species are motile; however, in all seven species the four flagellar tails do not beat. The flagellar beat consists of a rapid sine wave of low amplitude and short wavelength. The wave ends abruptly at the region of the branching. The four flagellar tails are straight and exhibit no discernable wave (Fig. 16). This observation suggests that when the 9-fold symmetry of the flagellum is disarranged, the flagellum is incapable of motility.

C. Spiraling of flagellar tubules

The sperm flagellum of the psocid *Psocus* is peculiar in that the flagellar doublets and peripheral singlets are not parallel to the long axis of the cell but spiral around a single central 400 Å rod (Phillips, 1969, 1970d). Because of the spiral, in flagella that are cut so that the tubules on one side of the central element appear in transverse section as clearly defined circles, the tubules on the opposite side of the central element appear

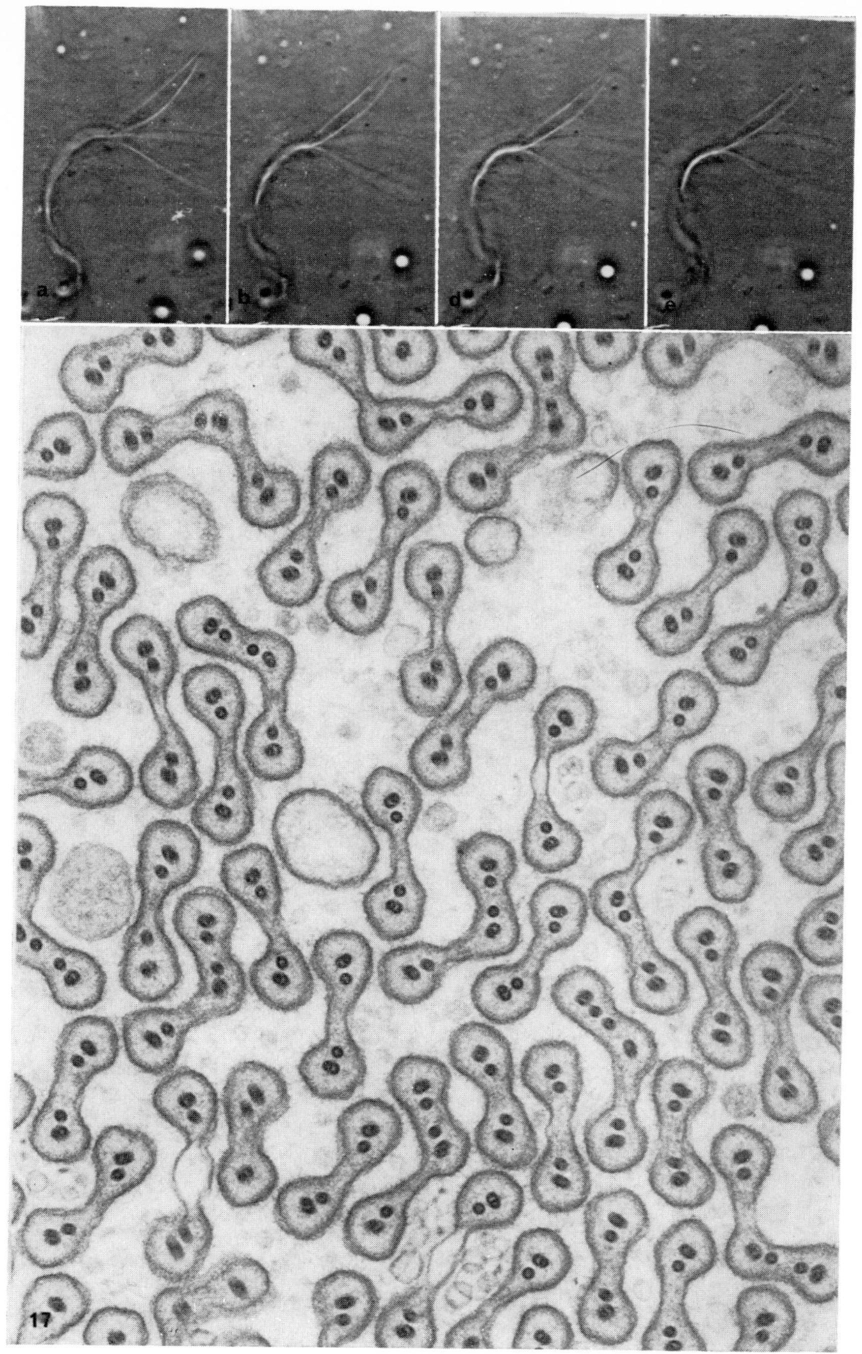

Fig. 16. Negative prints of 16 mm film of a swimming spermatozoan of the tree-hopper *Stictocephala* sp. taken with phase optics. Prints are at 0·08 s intervals. Waves move rapidly down the flagellum but stop abruptly at the point where the flagellar tails form. Flagellar tails remain straight and stiff. × 2000.

Fig. 17. Transverse sections of flagellar tails of *Stictocephala* spermatozoa. × 60,000.

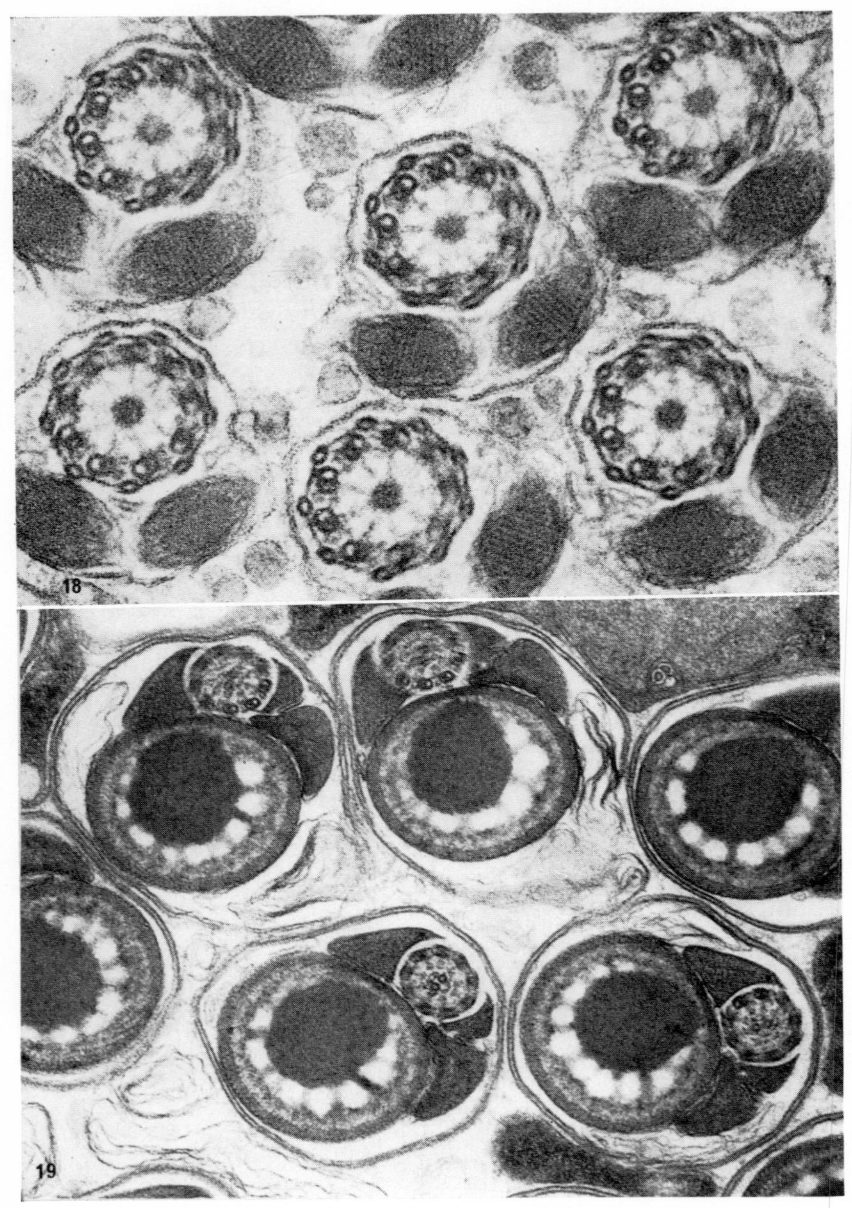

FIGS 18 and 19.—Captions on facing page.

blurred (Fig. 18), indicating that they are obliquely inclined with respect to the plane of section. In longitudinal sections of spermatozoa of *Psocus* sp., the flagellar tubules form an angle of about 8° with the long axis of the flagellum. In young spermatids of *Psocus*, all the flagellar tubules are parallel to the long axis of the cell and the central rod is missing. The spiraling of the flagellar tubules and the formation of the central rod is effected after the flagellar doublets and central pair have formed (Phillips, 1969).

Cross-sections of sperm flagella of the cat flea *Ctenocephalides felis* are reminiscent of those of *Psocus* in that some flagellar tubules always appear blurred when others in the same cell appear in true transverse section (Fig. 19). The flagellar tubules, therefore, spiral around the central elements. We have shown that the flagella of *C. felis* are morphologically more complicated than those in the spermatozoa of *Psocus* as the entire flagellum spirals around a large (0·5 μm) central structure (Phillips, 1969; Baccetti, 1968). Unlike spermatozoa of most other insect species, spermatozoa of *Ctenocephalides* lack peripheral singlet tubules. We have attempted to study motility of these peculiar spermatozoa. When put in saline or other medium, many sperm beat and move, but under these conditions there is so much variability among cells and so many non-motile cells that motility cannot be analyzed.

D. *Giant flagella*

Sciara and *Rhynchosciara* sperm flagella are composed of a very large number of doublet and peripheral singlet tubules. The large number of doublet and singlet tubules (some 60–90 in *Sciara* and about 360 in *Rhynchosciara*) appear to be a consequence of flagellar formation in association with giant centrioles (Phillips, 1966a, 1967). In *Sciara*, spermatid centrioles are composed of some 60–90 singlet tubules arranged in an oval configuration (Phillips, 1966b). Flagellar doublet tubules extend from

---

FIG. 18. In sections such as this one, which perpendicularly transect the tubules on one side of the sperm flagellum of *Psocus* so that the tubules appear as circular profiles, tubules on the opposite side of the flagellum appear to be oblique to the plane of section. This indicates that tubules describe a helical course. In this flagellum a dense rod occupies the position usually occupied by the central pair of tubules. ×85,000.

FIG. 19. Spermatozoa of the cat flea *Ctenocephalides felis*. Flagellar tubules on one side of the flagellum appear in cross-section while the flagellar tubules on the opposite side of the flagellum appear to be oblique to the plane of section. This indicates that the doublet tubules spiral around the central pair. ×44,000.

the giant centriole into the cytoplasm as an oval array. Later in spermiogenesis a singlet tubule develops peripherally to each doublet tubule and the oval becomes broken and takes on a spiral configuration in posterior regions of the spermatid. In the female genital tract, the spermatozoa

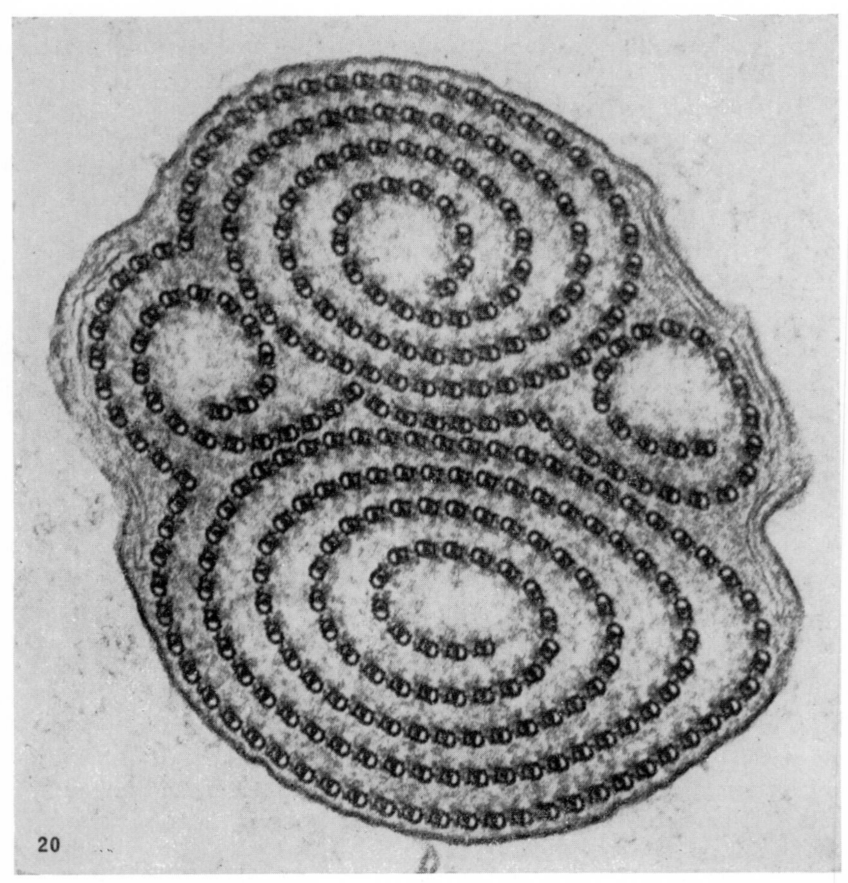

FIG. 20. Transverse section through the posterior region of a testicular spermatozoan of the fungus gnat *Rhynchosciara*. Dense material can be discerned between each subfiber A and subfiber B of the adjacent doublet. $\times 70{,}000$.

undergo gross morphological changes including sloughing off most of the mitochondrial material. The flagellar spiral is unwound and rewound into a different spiral configuration such that the outermost tubule of the spiral becomes the innermost tubule (Phillips, 1966b). Spermiogenesis in *Rhynchosciara* has recently been described by Shay (1972). It appears to be very similar to *Sciara* with the exception that the flagellum is composed of over 300 doublet and associated singlet tubules. Singlet tubules do not extend to the posterior regions of the flagellum (Fig. 20).

Motility of *Sciara* and *Rhynchosciara* spermatozoa has not been carefully analysed. We have observed that sperm of *Sciara* are motile if taken from the female genital tract at the time of oviposition. They then swim for a very short time with a very stiff beat. Since there are no central elements and *Sciara* spermatozoa are capable of movement, one might conclude that flagellar tubules can effect motility without interacting with a central element. The peripheral singlet tubules could, however, conceivably serve the function of the central pair in motility.

E. Disarranged flagellum

Baccetti *et al.* (1969) have recently described a strange type of flagellum in sperm of the thrip *Cryptothrips latus*. The flagellum consists of 18 doublet and four singlet tubules arranged in a completely disorganized manner. It is not known whether or not these spermatozoa are motile. It is possible that the disarrangement of tubules is a fixation artifact. We commonly observe random arrangements of flagellar tubules in the sperm flagella of *Sciara* with glutaraldehyde fixation whereas fixation with osmium alone produces flagella with regularly arranged tubules.

F. Coccid sperm—microtubule flagellar apparatus

Spermatozoa of coccids have a motile apparatus composed of some 20–250 singlet microtubules (depending on the species) arranged in a very regular array. The microtubules are very evenly spaced and arranged in circles, single or double spirals, or more complicated arrangements depending on the species (Robison, 1966, 1970; Ross and Robison, 1969) (Fig. 21). These spermatozoa have no mitochondria, acrosome, centrioles or 9+2 flagellum. The nucleus is situated in the center of the microtubular motile apparatus which extends the entire length of the cell. The formation of these unusual spermatozoa has recently been very carefully described by Moses and Wilson (1970). Motility of sperm bundles has been observed in several species (Robison, 1966; Ross and Robison, 1969). As microtubules are the only structure present which might be involved in motility, it is certain that the microtubules are the structures which effect motility. Whether or not the rows of microtubules are a true flagellum is somewhat a semantic argument. Most workers would probably not consider this motile apparatus a true flagellum as it does not arise from a centriole and does not have doublet tubules or 9-fold symmetry.

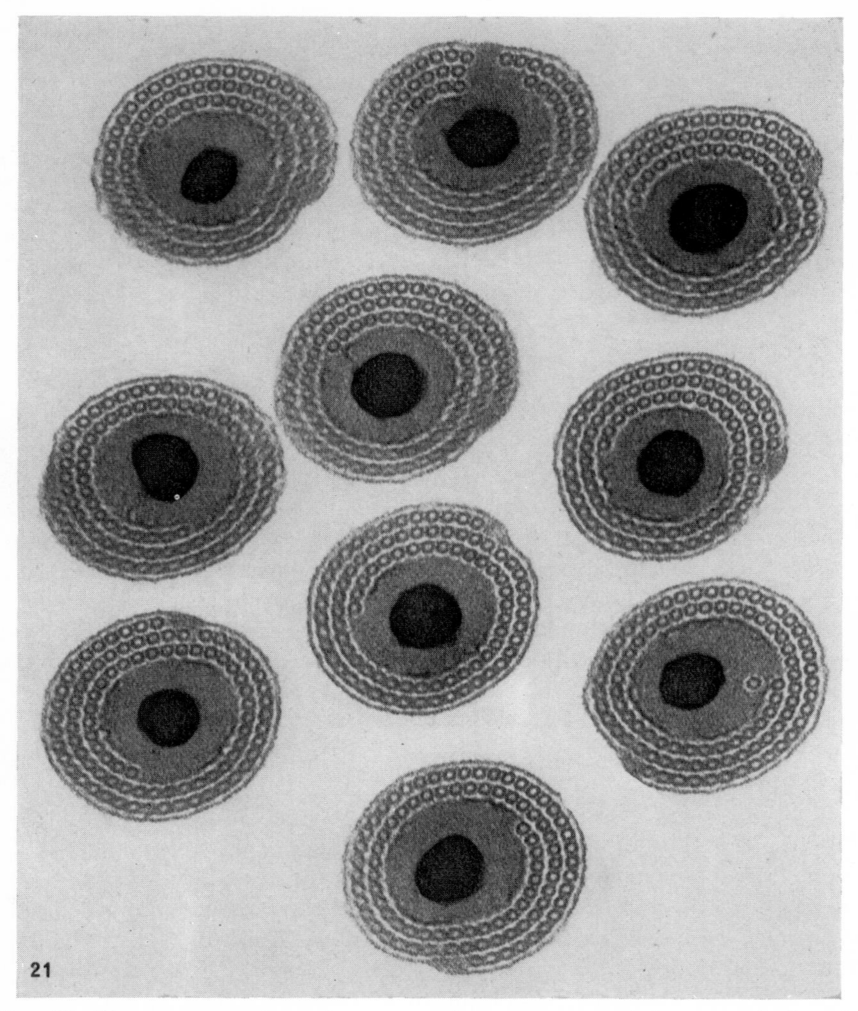

FIG. 21. Transverse section of a group of spermatozoa of the oystershell scale *Lepidosaphes ulmi.* Centrally located dense chromatin is surrounded by a motile apparatus composed of evenly spaced singlet microtubules. ×65,000.

IV. Conclusion

Comparative studies of sperm motility using sperm with exceptional flagella may lead to an understanding of how the form of the flagellar beat is established. In the preceding pages, exceptional flagella which have been studied to date have been described. Many more certainly exist.

Where possible, motile patterns have been considered in conjunction with analysis of flagellar structure; however, the lack of studies of motility of many exceptional flagella make it difficult as yet to pinpoint firm correlations between particular structural features and peculiarities in motile patterns. Hopefully, this chapter will point out the need for careful analysis of motility of exceptional flagella, as this will render information about sperm structure much more meaningful.

References

BACCETTI, B. (1968). *Redia* **51**, 153–158.

BACCETTI, B., DALLAI, R. and ROSATI, F. (1969). *J. Microscopie* **8**, 249–262.

BARKER, K. R. and BIESELE, J. J. (1967). *Cellule* **67**, 91–118.

BONSDORFF, C.-H. VON and TALKKA, A. (1965). *Z. Zellforsch.* **66**, 643–648.

BRELAND, O. P., BARKER, K. R., EDDLEMAN, C. D. and BIESELE, J. J. (1968). *Ann. ent. Soc. Am.* **61**, 1037–1039.

BURGOS, M. H. and FAWCETT, D. W. (1956). *J. biophys. biochem. Cytol.* **2**, 223–240.

BURTON, P. R. (1967). *J. ultrastruct. Res.* **19**, 166–172.

BURTON, P. R. (1968). *Z. Zellforsch. mikrosk. Anat.* **87**, 226–248.

BURTON, P. R. (1972). *J. Parasit.* **58**, 68–83.

COSTELLO, D. P., HENLEY, C. and AULT, C. R. (1969). *Science, Wash.* **163**, 678–679.

DESPORTES, I. (1966). *C. r. hebd. Séanc. Acad. Sci., Paris* ser. D **263**, 517–520.

HENLEY, C. and COSTELLO, D. P. (1969). *Biol. Bull.* **37**, 403 (abstr.).

HOOD, R. D., WATSON, O. F., DEASON, T. R. and BENTON, C. L. B., JR. (1972). *Cytobios* **5**, 167–181.

MEYER, G. F. (1968). *Z. Zellforsch. mikrosk. Anat.* **84**, 141–175.

MOSES, M. J. and WILSON, M. H. (1970). *Chromosoma* **30**, 373–429.

OSAKI, HARUKI. (1969). *Acta arachnologica* **22**, 1–13.

PHILLIPS, D. M. (1966a). *J. Cell Biol.* **30**, 477–497.

PHILLIPS, D. M. (1966b). *J. Cell Biol.* **30**, 499–517.

PHILLIPS, D. M. (1966c). *J. Cell Biol.* **31**, 635–638.

PHILLIPS, D. M. (1967). *J. Cell Biol.* **33**, 73–92.

PHILLIPS, D. M. (1969). *J. Cell Biol.* **40**, 28–43.

PHILLIPS, D. M. (1970a). *J. Cell Biol.* **44**, 243–277.

PHILLIPS, D. M. (1970b). *J. ultrastruct. Res.* **33**, 369–380.

PHILLIPS, D. M. (1970c). *J. ultrastruct. Res.* **33**, 381–397.

PHILLIPS, D. M. (1970d). *In* "Comparative Spermatology (B. Baccetti, ed.), pp. 263–273. Accademia Nationale dei Lincei, Rome, Italy. Academic Press, New York.

PHILLIPS, D. M. (1971). *J. ultrastruct. Res.* **34**, 567–585.

RANDALL, J. (1969). *Proc. R. Soc.* B **173**, 31–58.

REGER, J. F. (1970). *J. Morph.* **130**, 421–434.
ROBISON, W. G. (1966). *J. Cell Biol.* **29**, 251–265.
ROBISON, W. G., JR. (1970). *In* "Comparative Spermatology" (B. Baccetti, ed.), pp. 311–320. Accademia Nationale dei Lincei, Rome, Italy. Academic Press, New York.
ROSATI, F., BACCETTI, B. and DALLAI, R. (1970). *In* "Comparative Spermatology" (B. Baccetti, ed.), pp. 247–254. Accademia Nationale dei Lincei, Rome, Italy. Academic Press, New York.
ROSS, J. and ROBISON, W. G. (1969). *J. Cell Biol.* **40**, 426–445.
SATO, M., OH, M. and SAKODA, K. (1967). *Z. Zellforsch. mikrosk. Anat.* **77**, 232–243.
SHARMA, G. P. (1950). *Res. Bull. East Panjab Univ. Zool.* **5**, 67–80.
SHARMA, G. P. and GUPTA, B. L. (1956). *Res. Bull. Panjab Univ.* **84**, 1–5.
SHAY, JERRY W. (1972). *J. Cell Biol.* **54**, 598–621.
SILVEIRA, M. and PORTER, K. R. (1964). *Protoplasma* **59**, 240–265.
SUMMERS, K. E. and GIBBONS, I. R. (1971). *Proc. natn. Acad. Sci. U.S.A.* **68**, 3092–3096.

Chapter 15

Cilia in sense organs

V. C. BARBER

Department of Biology, Memorial University, St John's, Newfoundland, Canada

I. Introduction

Cilia, or modified cilia, occur in the majority of sense organs, the main exceptions being the eyes of many invertebrate species (see review by Eakin, 1972), and vertebrate taste receptors (see review by Murray, 1971). There is already a vast literature available on the structure and physiology of sense organs, and rather than make this chapter into a list of the occurrences of cilia, the emphasis in this account is on the possible function of the ciliary structures in the transduction process. Such an account naturally contains much hypothesizing, as information about the transduction process is still sparse. A real complication in this field is the uncertainty as to whether the ciliary structure is or is not important in the transduction process, and in some sense organs, cilia occur that are probably unconnected with the detection of the particular modality of stimulation in question.

II. Definitions

To help in the clarification of particular ciliary features mentioned in the text, the original description is used to define these structures.

The major components of the ciliary system that could be seen by light microscopy were the cilia, the "basal corpuscles" and the intracellular rootlets (a useful list of references to the older work is given by Fawcett and Porter, 1954). Electron microscopical studies enabled a clear definition of these various structures to be made and added ancillary ones to those already described.

The "basal corpuscle" was first termed the "basal body" by Gibbons and Grimstone (1960) and they defined it as "the whole intra-cytoplasmic part of the flagellar unit" (note that the term "basal body" was occasionally used by Fawcett and Porter, 1954, but was not clearly defined). Gibbons (1961) more clearly defined the basal body as an, "approximately cylindrical structure about 0·35 μm long formed by nine triplet outer fibres", and this is the definition used in the present account.

While earlier authors such as Rhodin and Dalhamn (1956) noticed the basal foot it was first well defined by Gibbons (1961) in the following terms: "Projecting from one side of the basal body is a dense, approximately conical structure that will be referred to as the basal foot. It is about 0·1 μm wide at the base and 0·15 μm long, and in favourable cross sections a delicate pattern of cross striations (overall periodicity 220 Å) may be observed running across it." In the present description this general definition is adhered to, the main features being the striated appearance and the club-shaped form.

The periodicity of the roots was first described by Fawcett and Porter (1954) and the figures they gave ranged from 550 to 700 Å. Other structures such as microtubules and fine fibres can also be found connected to the basal body region and these were mentioned by Gibbons and Grimstone (1960) and Gibbons (1961).

III. Eyes

A suggested scheme for classifying the invertebrate phyla was suggested many years ago and it divided the coelomate phyla into the Protostomia (the molluscs, annelids, arthropods, etc.) and the Deuterostomia (the vertebrates, echinoderms, tunicates, etc.), using such evidence as the differing embryological development of the animals. This division is valuable for many reasons, and covers various features including the structure of the eyes. Eakin (1963) showed that the eyes could be divided into what he termed the annelid line (protostomes), where the photoreceptor membrane is formed from extensions of the outer cell membrane to produce an array of microvilli, the rhabdom, and the echinoderm line (deuterostomes), where the photoreceptor membrane is formed from a modified cilium or cilia. At that time only one exception to this pattern was known, in the eyes of the bivalve mollusc, *Pecten* (Miller, 1958). Later reviews (Eakin, 1965, 1968) extended the information further and enumerated many exceptions to this evolutionary idea. All eyes in the annelid line are certainly not rhabdomal but eyes can still be usefully divided into those where cilia form the membrane array and those where the array is formed from microvilli (Eakin, 1972).

The main occurrence of ciliary-based eyes is in the vertebrates, and in certain molluscs and annelids. Each of these groups will be considered in turn.

A. *Ciliary-based eyes in the molluscs and annelids*

As has already been indicated, the first ciliary-based eye to be found in this group was described in the scallop, *Pecten* (Miller, 1958, 1960). A later electron microscopical account confirmed Miller's findings and showed that there were two distinct types of photoreceptive cells, the distal ones with photoreceptive membranes composed of modified cilia, and the proximal ones with an array of microvilli (note: one cilium does occur on each cell of the proximal retina) (Barber *et al.*, 1967). In those species so far examined the eye of *Pecten* is unique in that it has a multi-layered concave reflector at the back of the eye. This forms a real image by reflection, approximately in the position of the array of cilia of the distal cells (Land, 1965). Electrophysiological recordings from the separate nerves of the distal and proximal retinae showed that the physiological responses of the two layers were different (Hartline, 1938; Land, 1966). The proximal retina was a monitor of light intensity, a graded response occurring on illumination, while the distal retina gave an "off" response when a darkening of the visual field occurred, which is probably used to detect the approach of a predator. Using intracellular electrodes it was confirmed that the two types of responses occurred in the respective cell layers. In the proximal cells there was a graded depolarizing response to illumination, and in the distal cells hyperpolarization occurred. Action potentials occurred during illumination in proximal cells and following illumination in distal cells (the "off" response) (Gorman and McReynolds, 1969; McReynolds and Gorman, 1970a). Anatomical (Barber *et al.*, 1967) and physiological studies (Land, 1966; McReynolds and Gorman, 1970a, 1970b) confirmed that these responses were not produced by synaptic interaction between the two layers. Thus it was concluded that the ciliary-based cells, comprising the distal retina, gave rise to the "off" response.

So far little has been said about the type of cilium that occurs in the distal retina cells in *Pecten*. Each receptor bears numerous cilia. Each cilium arises from a basal body that bears a basal foot but no root system.

The cilium has a 9+0 internal tubule structure. The nine peripheral tubules are double but do not have arms (Barber *et al.*, 1967). The position of the flattened ciliary array below the base of the lens suggests that they are not motile, and the typical sensory 9+0 tubule system, with the absence of the ATPase containing arms on the peripheral tubules gives support for this idea (see Gibbons, 1965).

Receptor cells with a ciliary structure were also found in the tentacular eyes of another lamellibranch mollusc, *Cardium* (Barber and Wright, 1969a) (Figs 1, 2). Electrophysiological recordings from the tentacular nerve showed that these eyes also exhibited "off" responses on a darkening of the visual field (Barber and Land, 1967). The dorsal eyes of the nudibranch mollusc *Onchidium* are also ciliary in nature (Yanase and Sakamoto, 1965). Whilst there are no electrophysiological recordings in this case, electroretinograms have been made and these again support the likely presence of "off" responses (Fujimoto *et al.*, 1966).

The finding that the ciliary-based receptors in these molluscs gave rise to "off" responses led to the suggestion that such a response may be in some way correlated with the derivation of the photoreceptor membrane from the modified membranes of cilia (Land, 1968; Thurm, 1969). While this may be so, recent electrophysiological studies on the photoreceptors of primitive chordates have shown that it is possible for both ciliary-based and microvilli-based photoreceptors to produce hyperpolarizing responses, so the relationship is not exclusive (Gorman *et al.*, 1971).

The examples already given are the only ones where there is information available on both the physiology and fine structure, although several other species of molluscs bear ciliary-based eyes. In some of these examples the main photoreceptive area is composed of microvilli so the function of the cilia or ciliated cells is uncertain. Eyes where the main photoreceptive membrane surface is composed of modified cilia have been found in *Arca* (Levi and Levi, 1971), and *Pterotrachea* (Dilly, 1969b), and eyes where cilia are also present but are of uncertain photoreceptive function have been found in *Onithochiton* (Boyle, 1969a, 1969b), *Littorina* (Owen and Charles, 1966) and *Aplysia punctata* (Hughes, 1970) (*Aplysia californica* is apparently essentially without cilia, Jacklet *et al.*, 1972). Photoreceptive ciliated epithelial cells have been found in *Nassarius* (Crisp, 1971, 1972).

So far no mention has been made of ciliated eyes in annelids. Such eyes do occur but no electrophysiological studies on them have been made. Ultrastructural studies have shown that eyes where the photoreceptor surface is purely ciliary in nature occur in the annelids, *Branchiomma* (Krasne and Lawrence, 1966; Lawrence and Krasne, 1965), *Dasychone* (Kerneis, 1966, 1968) and in *Potamilla* (Kerneis, 1971). Cilia are present, although not thought to be photoreceptive, in the eyes of *Lumbricus*

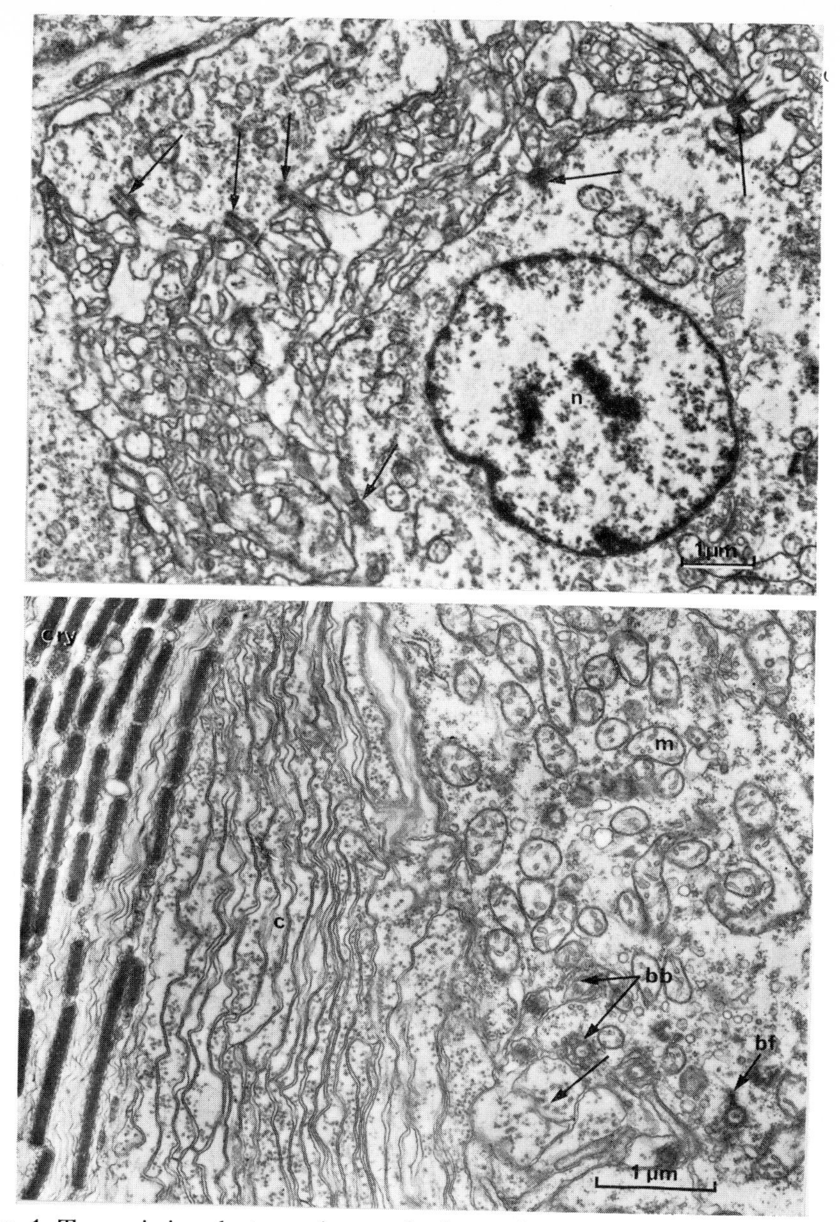

FIG. 1. Transmission electron micrograph of a portion of the eye-cup in *Cardium*. Many cilia (arrows) are present and these give rise to an array of photoreceptive membranes. n, nucleus.

FIG. 2. Transmission electron micrograph of part of a receptor cell in the eye of *Cardium*. The eye cup is enclosed by a crystal reflector (cry). Cilia give rise to a membrane array (c) adjacent to the crystals. Note the 9+0 cilium (arrow). bb, basal body; bf, basal foot; m, mitochondrion.

(Rohlich *et al.*, 1970) and *Phascolosoma* (Hermans and Eakin, 1969). Eye-like structures with cilia are also present in the trochophore larvae of *Harmothoë* (Holborow, 1971; Holborow and Laverack, 1972).

B. Ciliary-based eyes in invertebrate phyla other than molluscs and annelids

Before going on to describe vertebrate retinal cells it is necessary to list a few other examples of ciliated retinal cells in other invertebrate species; a more comprehensive discussion of ciliated and non-ciliated eyes and further references will be found in the review by Eakin (1972). In the larva of the bryozoan *Bugula*, a protostome coelomate, the probable photoreceptor organs have 9+2 cilia as the potential light receptive elements (Woollacott and Zimmer, 1972). Another presumed ciliated photoreceptor has been described in the ctenophore *Pleurobrachia* (Horridge, 1964). The cilia comprising these cells have a 9+0 tubular content and form lamellated bodies reminiscent of those in the eyes of some annelids and molluscs.

So far we have considered receptor cells where there were numerous comparatively unmodified cilia arising from each cell. There are a few examples, leading eventually to the vertebrates, where a single cilium-like process gives rise to a greatly elaborated membrane surface. An example of this is seen in the ocellus of the coelenterate, *Polyorchis*, where the receptor cells end distally in a cilium-like process, whose outer membrane possesses numerous tubular branches that make a complex membrane array (Eakin and Westfall, 1962). It should be noted that the ctenophores and coelenterates are non-coelomate and therefore neither deuterostomes nor protostomes.

Numerous cilia and microvilli form the photoreceptor array in the ocelli of some echinoderms such as *Asterias*; the cilia have either a 9+0 or 9+2 tubule structure (see Eakin, 1968). In such chaetognaths as *Sagitta*, each receptor cell bears at its distal end a 9+0 cilium which swells out to form a large bulbous structure containing numerous tubules (see Eakin, 1968).

Because of the phylogenetic importance of the protochordates such as *Ciona* and *Amphioxus* the eyes of several examples of such species have been examined. In the ascidian tadpole larva, which is thought to be near the main line of evolution of the vertebrates, the presumed photoreceptive membranes arise from a modified cilium of 9+0 internal tubule content (Dilly, 1961, 1964, 1969a; Eakin and Kuda, 1971). The basal body bears a striated root. Some similarity to the visual receptors of vertebrates can clearly be seen in these comparatively primitive ascidians.

C. *Vertebrate eyes*

The retinal receptor cells, the rods and cones, are similar in structure in all vertebrates (see review by Crescitelli, 1972) (Fig. 3). Studies of the development of the cells has shown that the membrane stack arises by the invagination of the outer membrane of a single cilium (Sjöstrand,

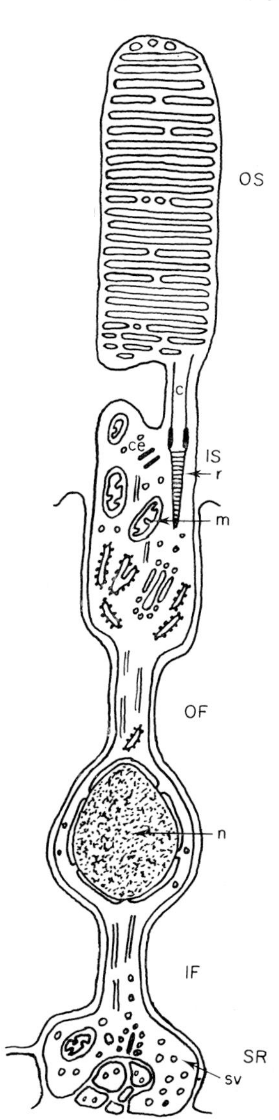

FIG. 3. Diagram of a rod cell from a vertebrate retina. The outer segment (OS) of the cell is composed of a membrane stack at the base of which is a 9+0 ciliary segment (c). The inner segment (IS) from which the outer segment arises contains the usual organelles and is connected to the nuclear portion (n) by the outer fibre zone (OF). The synaptic region (SR) of the cell is connected to the nuclear region by the inner fibre zone (IF). The cell would receive light from the bottom of the page. ce, centriole; r, root; m, mitochondrion; sv, synaptic vesicles.

1959). A similar structure and mode of development occurs in the photoreceptors of the amphibian, lizard and cyclostome "third eye" (parietal eye) (see Eakin, 1968; Eakin and Westfall, 1961).

The importance of the ciliary structure in the vertebrate retinal cells is not understood. Clearly the formation of the lamellated structure, which is orientated so that maximal numbers of photons of light will be intercepted by it, is critical for the light collecting function of the cells. The use of a cilium as the source of the membrane stack could have only evolutionary significance. However, the point of contact between the membrane stack, where the transduction of the stimulus to a generator potential occurs, and the perikaryon, where the generator potential is amplified to produce an action potential in the component nerve, is by means of a short portion of cilium, which has a 9+0 tubule content. So as well as producing a large membrane surface area, the ciliary connection must have some conductive function.

Recent studies on the retinae of frogs have provided some evidence as to how the light photons being received by the photosensitive pigment produce the change in ion permeability that occurs on the production of a generator potential. It is suggested that rhodopsin molecules are located in the membranes of the receptor stack (these molecules in the frog have a molecular weight of 40,000 and a spherical shape of about 40–46 Å. They would thus take up a considerable portion of the thickness of the receptor membrane). The effect of light on the rhodopsin molecule is thought to be to cause it to rotate and produce an effective channel for ion movement, thus producing the generator potential (Brown, 1972; Cone, 1972). This idea at least provides some information as to the possible transduction mechanism but still gives no information as to the function of the ciliary components of the retinal cells.

IV. Olfactory organs

Odoriferous molecules can be detected by a wide variety of animals. As with the eyes, the olfactory mucosae have a similar structure in all vertebrates, being composed of receptor cells, supporting cells and basal cells (see Graziadei, 1971; Moulton and Beidler, 1967, for reviews). They will be considered first.

A. *Vertebrate olfactory organs*

The receptor cells of the vomero-nasal (Jacobson's) organ do not bear cilia, although certain centrioles and precursor bodies of cilia can be present (Bannister, 1968; Kolnberger, 1971; and other references in Graziadei, 1971). The receptor cells in the main olfactory regions of vertebrates all bear cilia (additional sense cells without cilia have been

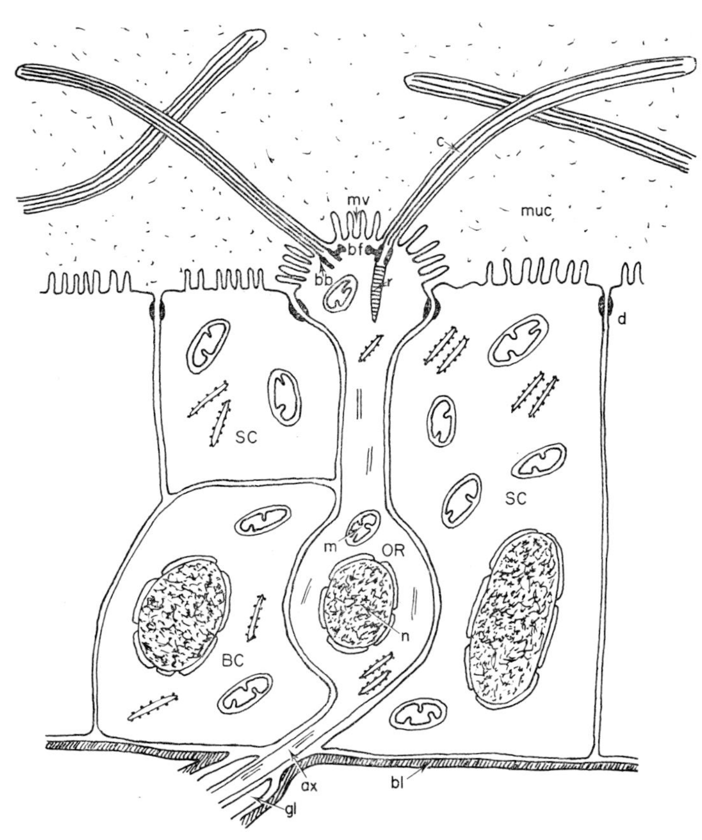

FIG. 4. Diagram of cells in a generalized vertebrate olfactory epithelium. The olfactory receptor cell (OR) bears 9+2 cilia (c) and microvilli (mv) at its distal end. The cilia arise from basal bodies that generally bear basal feet and often roots (r). An axon (ax) arises from the proximal end of the cell. Basal cells (BC) and supporting cells (SC) are also present. bl, basal lamina; d, desmosome; gl, glial cell; muc, mucus.

reported in the olfactory epithelium of the teleost fishes *Gasterosteus* and *Phoxinus* by Bannister, 1965; and in the eel *Anguilla* by Schulte, 1972). The cilia are variable in number, varying from five or less in the mole, *Talpa* (Graziadei, 1966), to 100 or 150 cilia per cell in the dog (Okano *et al.*, 1967). The cilia have a 9+2 internal tubule content at their origin, and the basal bodies generally bear basal feet and sometimes roots (see

Andres, 1969; Okano *et al.*, 1967). Their structure suggests that they may be motile and in fact the cilia of at least some olfactory epithelia can beat asynchronously (Graziadei, 1971; Reese, 1965).

Cilia vary in length according to the species examined, being as short as 3 μm in the rabbit (Barber and Boyde, 1968), up to 80 μm long in the cat (Andres, 1969) and up to 200 μm long in the frog (Reese, 1965). The number of sub-tubules at the distal ends of these long cilia is less than $9+2$; in the cat, for example, there are only two microtubules (Andres, 1969), and in the frog, nine microtubules (Reese, 1965). Compound cilia, in which many $9+2$ tubular arrays are enclosed in one outer ciliary membrane, have been described in the goldfish *Carassius* (Wilson and Westerman, 1967) and in the eel *Anguilla* (Schulte, 1972). In receptors with a few cilia arranged in a ring around the distal end of the cells, the basal feet generally point towards the middle of the cell (Bannister, 1965) (Fig. 4).

B. *Transduction mechanism in the vertebrate olfactory receptor*

Vertebrate olfactory organs are sensitive to amounts of certain odoriferous molecules as small as 10^{-14} g, which for example in man detecting the odorant β-ionone, amounts to a response to as little as 10^8 molecules. With such a small input, amplification of the input signal by a factor of several thousand must occur (see the review by Davies, 1971). One might expect some correlation between the number of odorant molecules necessary for threshold stimulation and the number of cells in the olfactory epithelium, and in man there are about 2×10^7 olfactory cells, so for β-ionone only a few molecules of odorant per cell (perhaps only one) are sufficient to cause the production of a generator potential in the receptor cells (Davies, 1971).

The discussion now concerns the electro-olfactogram (EOG), whether its site of origin is in the ciliary region, and whether or not it is the generator potential. The EOG certainly has the characteristics of a generator potential; it is a graded non-conducted response, the time course and amplitude of which are closely related to the parameters of olfactory stimulation, so it is reasonable to assume that it may be the generator potential (see Davies, 1971, and Moulton and Beidler, 1967, for detailed discussions on this point). Assuming that the EOG is the generator potential, where does it originate? It is only obtained from the olfactory and not respiratory region (these regions are easily differentiated in the frog because of the yellow pigment in the sensory regions) (Ottoson, 1971). The potential probably arises from the receptor cells as was demonstrated by sectioning the olfactory nerves of bull-frogs and recording the EOG at various stages of retrograde degeneration of the olfactory epithelium. The EOG decreased in parallel with the reduction in the number of sensory cells, and disappeared 8–16 days after nerve section (Takagi and Yajima, 1965) (other explanations involving the supporting cells are possible, see Moulton and Beidler, 1967). Ottoson (1956) showed that the EOG became progressively smaller as the recording electrode was advanced into the epithelium from the outside, which suggested that the potential originated in structures close to the surface of the epithelium. Although the evidence is not conclusive it seems probable that the EOG is the generator potential and that it arises at the distal end of the receptor cells possibly on the cilia.

There are various theories of olfactory stimulation and the one described here is the "penetration and puncturing" theory as described by Davies (1971). The theory is that an odorant molecule is adsorbed and penetrates the lipid region of the cell membrane. The odorant molecule diffuses rapidly through the bimolecular lipid membrane and leaves a hole behind it. Ions pass through the channel and initiate the generator current. Thus, the cilia and tips of the olfactory cells, if this theory is correct, would become permeable to ions when the odorant molecules penetrate and this would set up a generator current in this region, which eventually gives rise to the nerve impulse at the point of origin of the olfactory nerve of the olfactory receptor cell. The frequency of the nerve impulse would depend on the intensity of the generator current, which depends on the number of channels formed in any given membrane (Davies, 1971). A slightly modified theory suggests that conformational changes occur in the receptor membrane on odorant adsorption. Some evidence for this is that an increase in surface tension occurs when odorants are added to monolayers of lipids spread on the surface of Ringer solution (Koyama and Kurihari, 1972). The lipids forming the monolayer in these experiments were extracted from bovine olfactory epithelia.

In these theories it is clearly the membranes that are of importance, and presumably the ciliary membrane would be the site of the membrane changes, as its surface area represents by far the majority of the membranes at the distal end of most olfactory receptor cells. The de-ciliation studies of Tucker (1967) showed that cilia are not essential for olfactory functioning, and presumably their main purpose is to increase greatly the membrane area available for the detection of odoriferous molecules. The movement of the cilia through the mucus would also help in this detection. As mentioned earlier in this account, some olfactory receptors do not bear cilia (Bannister, 1965; Schulte, 1972), and so in these cases presumably the surface membrane of the cells serves as the site of transduction.

C. Insect olfactory organs

Insects generally detect scent with a pair of segmented head appendages, the antennae. The appendages also bear sense organs for other modalities such as taste, touch, sound, etc. The antennae can be very branched to produce a large extended area of as much as 1 cm^2 in some saturniid moths, and such a structure can filter out most of the odoriferous molecules from an air stream (Kaissling, 1971).

The olfactory receptors of insects are not arranged as epithelia but in units termed sensilla. However, the sensilla can be densely packed to form a structure equivalent to an olfactory epithelium. Each sensillum is covered with a particular differentiation of the cuticular exoskeleton, usually hair-, peg- or plate-shaped. Each unit is composed of sensory, trichogen and tormogen cells. The trichogen and tormogen cells produce the cuticular part of the sensillum (see Kaissling, 1971; Lewis, 1970; Schneider and Steinbrecht, 1968; Slifer, 1970; Steinbrecht, 1969a, for recent reviews). All the cells originate in embryological development from one epidermal mother cell by a particular series of cell divisions (Henke and Rönsch, 1951). There are generally from two to six receptor cells in each sensillum although more or less are possible. The receptors are bipolar primary sense cells, which send their axons to the central nervous system. Neither fusion of the very small axons nor synaptic contact at the periphery occurs (Steinbrecht, 1969b). At the distal end of the sense cell a short cilium with a $9+0$ internal tubule content arises from a basal body. Another basal body lies proximal to the one that gives rise to the ciliary structure. A process, either unbranched, but more usually branched, arises from the cilium and passes up into the liquor-filled space inside the cuticular portion of the sensillum (see Ernst, 1972, for details of the development of these processes) (Fig. 5). The odoriferous molecules reach these ciliary processes via pores in the cuticular covering (Schneider and Steinbrecht, 1968). It seems certain that the odoriferous molecules produce their effects on the membranes of these processes. Whilst the cilium probably has a conductive function, the purpose of its processes would appear to be to increase the surface area to enable the detection of the minimal amount of odorant.

Olfactory receptor cells in insects can be divided into two general cell types: odour generalists and odour specialists (Schneider and Steinbrecht, 1968). Odour generalists respond to a particular spectrum of odours whereas odour specialists are highly sensitive to a few biologically important odours such as pheromones, warning odours and specific food odours.

It is not intended to review the electrophysiological data concerning insect olfactory organs as this has been adequately done in a recent review

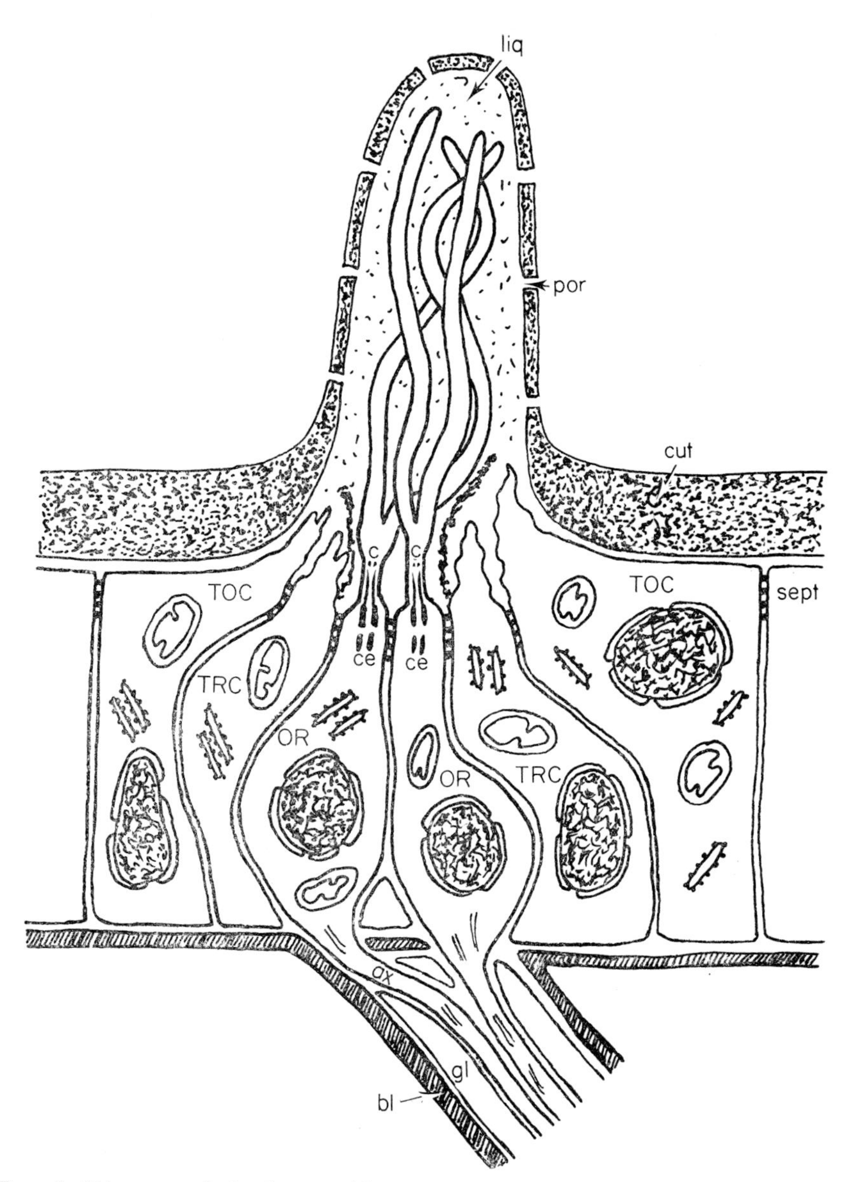

FIG. 5. Diagram of the insect olfactory organ. In this example there are two olfactory receptor cells (OR). 9+0 ciliary segments give rise to the processes of the outer segment. Odoriferous molecules reach the outer segment via pores in the cuticle (por). Surrounding the olfactory cells are the trichogen (TRC) and tormogen (TOC) cells. cut, cuticle; liq, liquor surrounding the outer segment; sept, septate desmosome.

(Kaissling, 1971). The response of the cells is phasic-tonic, the frequency of the response rapidly reaches a peak, and then rapidly declines to a steady level which persists until the stimulus is removed. The "penetration and puncturing" theory of sensory transduction enumerated earlier in this account is applicable at least to the odour generalists, although some specific receptor molecules or speciality in the membranes would need to be present to explain the functioning of the odour specialist receptor cells.

D. *Olfactory organs in other phyla*

There is little electrophysiological data on possible olfactory organs in animals other than vertebrates and insects. Consequently the sense organs described in this section are classed as olfactory mainly on structural, positional, and behavioural criteria. Therefore some of those classed in this section may subsequently have to be re-classified in the light of later experimental work.

An example of such a probable olfactory organ is that of the rhinophore of the cephalopod mollusc *Nautilus*. These receptor organs are situated below each eye and each is a blind-ending sac of cells that opens to the exterior by a pore. Three types of cells are present, flask-shaped ciliated cells (presumed to be the receptors), other ciliated cells (presumed to bear motile cilia that move the mucus), and mucous cells (Barber and Wright, 1969b). Only circumstantial evidence suggests that the flask-shaped ciliated cells are chemoreceptors. Such evidence is the flask-shaped form of the cells, and the sac-like form of the rhinophore with its narrow pore opening to the exterior, which makes it unlikely that the cells could respond to other than olfactory stimuli. The structure of the rhinophore in *Nautilus* is very similar to that of the so-called "olfactory organ" in *Octopus* and other cephalopods (Graziadei, P., unpublished; Watkinson, 1909). However, it has not been determined whether or not the "olfactory organ" is a chemoreceptor although it is possible that it samples water in the respiratory stream (Watkinson, 1909). Nautili certainly locate their food by olfaction rather than by sight but it is possible that some of the many tentacles bear olfactory receptor cells (Bidder, 1962).

Presumed chemoreceptor cells are also present in the suckers of *Octopus*, and the lip of *Sepia* (Graziadei, 1964, 1965). These receptor cells have some similarities to the flask-shaped cells in the rhinophore of *Nautilus* in that they are ciliated, and that the cilia have access to odoriferous molecules via pores to the exterior.

Osphradia, which are present in most molluscs, also have structural similarities to the rhinophores of *Nautilus*. Whilst there is considerable variation in the size and complexity of the osphradia there is the basic similarity in that there are sensory cells, generally bearing cilia, cells that bear motile cilia (and presumably move mucus and fluid over the sensory cells), and mucous cells (Bailey and Benjamin, 1968; Benjamin, 1971; Benjamin and Peat, 1971; Storch and Welsch, 1969; Welsch and Storch, 1969). The structure of the ciliated sensory cells in the species so far studied is variable, there being just a few cilia and microvilli at the distal end of a sensory nerve ending arising from cells below the epithelium in *Planorbis* (Bailey and Benjamin, 1968; Benjamin and Peat, 1971), and sensory cells bearing numerous cilia in *Buccinum* (Welsch and Storch, 1969). Four other types of non-ciliated, presumed receptor cells are also present in the osphradium of *Buccinum* and two of these are thought by the investigators to be additional olfactory receptors (Welsch and Storch, 1969).

Electrophysiological data on osphradia are still comparatively sparse. Studies on the carnivorous *Buccinum* show that the receptor cells in the osphradium of this animal respond to a restricted range of chemical stimuli, the most pronounced response being produced on exposure to extracts of *Mytilus* (normally preyed on by *Buccinum*) (Bailey and Laverack, 1963, 1966). They were insensitive to mechanical stimulation produced by particulate matter suspended in the water. However, an osmoreceptor function has been shown for the small osphradium of the herbivorous *Aplysia*, so it cannot be assumed that all osphradia have a simple chemoreceptor function (Stinnakre and Tauc, 1969). Also, it has been suggested that mechanoreceptor cells are present in some osphradia, and certainly the variety of sense cells in *Buccinum*, including ones deep inside the epithelium, suggests that sensory modalities other than chemical ones might well be received (Welsch and Storch, 1969).

V. Acoustico-vestibular systems

With a few exceptions, acoustico-vestibular systems in animals have ciliary units in them. In vertebrate acoustico-vestibular systems the connection between structural orientation of the ciliary structures and the functional polarity of the cells has been demonstrated for over a decade, although the biophysical significance of this correlation is still not determined. So, more than perhaps with the other sense organs so far considered, study of the acoustico-vestibular systems concerns the study of ciliary structures and their significance.

A. Vertebrate acoustico-vestibular systems

As with the vertebrate olfactory organs, the receptor cells of the vertebrate acoustico-vestibular systems have a similar structure, varying comparatively little in such diverse systems as the organ of Corti of man and the lateral line organs in fishes.

Vertebrate hair cells are not primary receptor cells, and consequently, unlike invertebrate hair cells, they do not bear axons but are innervated by neurons. A variety of both afferent and efferent endings innervate the hair cells (see reviews by Flock, 1965, 1971; Spoendlin, 1968; Wersäll *et al.*, 1965).

In the vestibular and lateral line organs the sensory hair bundles of each receptor cell are composed of a large number of stereocilia and one kinocilium (sometimes two kinocilia, see Flock, 1964; Lowenstein and Thornhill, 1970). (Kinocilia are the usual 9+2 cilia; stereocilia are similar to microvilli but are generally longer, have an electron-dense core at their bases, and are often of a larger diameter at their distal ends.) The internal tubules of the 9+2 kinocilium are orientated so that a line joining the two central tubules of the cilium is at right angles to the axis of the basal foot, and so that four pairs of peripheral tubules lie on the basal foot side of the central tubules (nos 4, 5, 6 and 7, numbered according to Afzelius, 1959) and five pairs lie on the other side (nos 1, 2, 3, 8 and 9). Another feature of polarization, well shown in scanning electron microscopical preparations of the epithelia, is that the stereocilia adjacent to the kinocilium are longer than those on the other side of the group, thus forming a group of stereocilia with graded heights (Hillman, 1969; Hillman and Lewis, 1971; Lewis, 1972; Lewis and Nemanic, 1972; Lim, 1969; Lim and Lane, 1969) (Fig. 6). Note that the kinocilium can be very much longer than the longest stereocilium or it can be the same length. Although a kinocilium is present in the embryonic organ of Corti it is generally shed before birth (Duvall *et al.*, 1966), so the hair cell group usually lacks a kinocilium, although a basal body is generally present (the basal body can be absent in adults, as for example in the cat; Spoendlin, 1968). A basal foot is not present in the organ of Corti. The stereocilia in the organ of Corti are arranged in three rows in a W-shaped array with the basal body being positioned at the base of the "W" (Flock *et al.*, 1962).

Lowenstein and Wersäll (1959) first suggested (in fish) that the polarization already described had a functional correlate. Their theory was later extended to include the basal foot (Flock and Duvall, 1965; Lowenstein *et al.*, 1964). The theory suggested that the sense cells were depolarized when the sensory hairs were displaced towards the kinocilium of the cell and its basal foot (or basal body in the organ of Corti). The cells were thought to be hyperpolarized when displacement of the hairs occurred in the opposite direction. Nerve fibres innervating the cells increased their firing rate during the depolarization phase and decreased it on hyperpolarization (see Flock, 1971, for a review). In fact, the directional sensitivity of the hair cells approximates to a cosine function of the direction

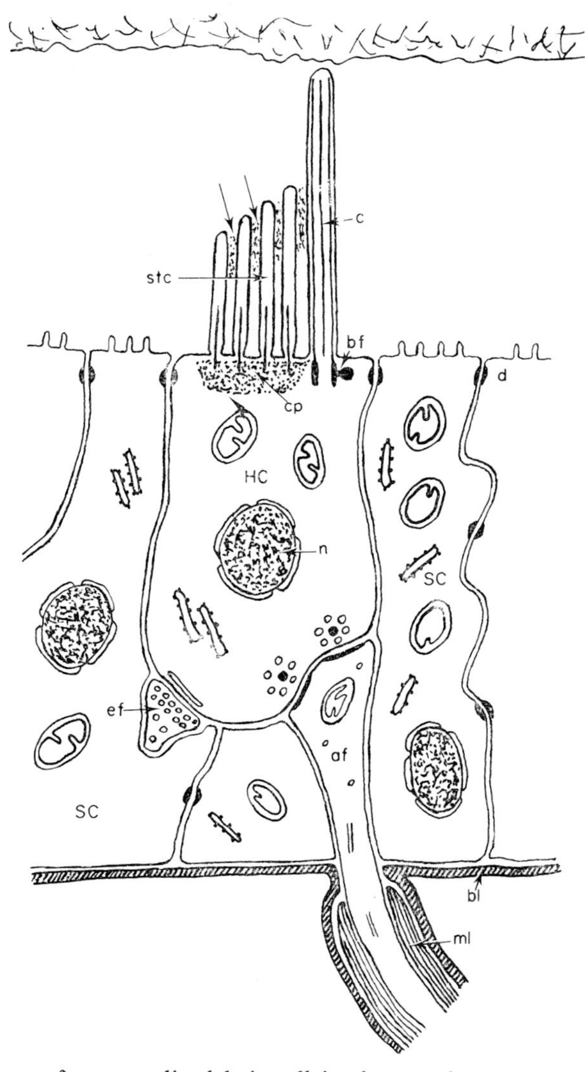

FIG. 6. Diagram of a generalized hair cell in the vertebrate acoustico-vestibular system. The hair cell (HC) bears a kinocilium (c) and numerous stereocilia (stc) at its distal end. The hairs are probably connected together by a mucopolysaccharide coat (arrows). Overlying the hair group is an otolith or similar structure and the method of attachment of the hairs to this structure will vary. The hair cell is innervated by efferent (ef) and afferent axons (af). Surrounding the hair cell are supporting cells. cp, cuticular plate; ml, myelin sheath.

of stimulation, the output varying as a cosine of the angle between the direction of maximum sensitivity and the applied displacement (see Flock, 1971). The possible biophysical mechanisms of the transduction process will be considered in a later section.

B. Molluscan statocysts

In the invertebrates the greatest variability in statocyst structure is found in the molluscs. The statocysts range from the simple ones of the majority of molluscs to the highly differentiated ones of the coleoid cephalopods, which have systems for the reception of linear and angular acceleration (see Barber, 1968, for a review).

The fine structure of the hair cells in a variety of molluscs other than cephalopods, is fairly well known (see Barber and Dilly, 1969, *Pecten* and *Pterotrachea*; Coggeshall, 1969, *Aplysia*; Geuze, 1968, *Lymnaea*; Laverack, 1968, *Helix*; Quattrini, 1967, *Helix*; Vinnikov *et al.*, 1971, various; Wolff, 1969, *Arion* and *Limax*). The static sac of these statocysts is composed of hair cells and supporting cells. The numbers of hair cells range from some ten in *Lymnaea* (Geuze, 1968) to very many more in *Pecten* (Barber and Dilly, 1969). In contrast to the vertebrate systems the hair cells in mollusc statocysts are primary receptor cells, which send their axons to the cerebral ganglion. Synaptic profiles are present at the bases of the hair cells, at least in *Arion* and *Limax* (Wolff, 1969) and *Pecten* (Barber and Dilly, 1969), and electrophysiological recordings give evidence that these endings are probably the endings of efferent fibres (Wolff, 1970b). Further electrophysiological studies show that the static receptor cells are of the phasic-tonic type responding as gravity receptors only (Wolff, 1968, 1970a).

Whilst the numbers and arrangements vary, the hair cells bear kinocilia with a $9+2$ internal tubular content. Various accounts describe the presence of basal feet attached to the basal bodies, but in some of these examples the structures illustrated are probably not basal feet as defined earlier in this account but are lateral roots (Geuze, 1968; Wolff, 1969). Main roots attached to the basal bodies are also present.

Except for *Nautilus*, which has simple statocysts of the non-cephalopod type, the statocysts of cephalopods are highly differentiated into a macula with statolith, and a crista, which is arranged approximately in three planes. In the decapod cephalopods a further complication occurs in that in many species the macula is divided into three sections (see Barber, 1968; Budelmann *et al.*, 1973; Hamlyn-Harris, 1903; Klein, 1931; Young, 1960, 1971; for further details and references).

As in other molluscan statocysts the sensory epithelium in both the macula and crista is composed of hair cells and supporting cells. The hair

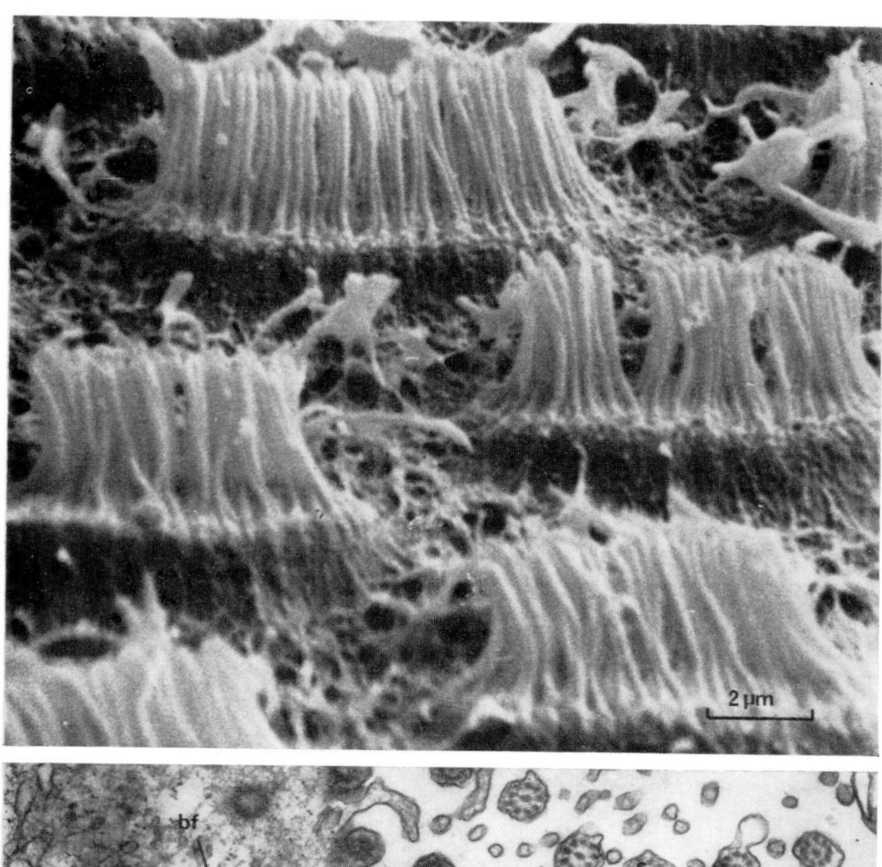

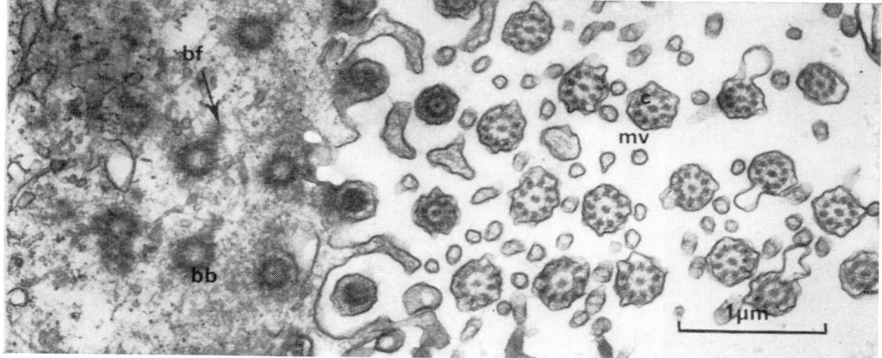

FIG. 7. Scanning electron micrograph of several groups of cilia from the macula of *Octopus*. Each ciliary group arises from a single hair cell and can contain over a hundred cilia. To remove the statolith the epithelium underwent mild hydrochloric acid treatment.

FIG. 8. Transmission electron micrograph of part of a hair cell ciliary group from the crista of *Octopus*. Notice the orientation of the ciliary tubules and the basal feet. The direction of excitation is thought to be in the direction of the basal feet.

cells are primary receptor cells. Electron microscopical studies have shown that each hair cell bears numerous kinocilia and microvilli at its distal end, which are arranged to form an elongated group (Barber, 1965, 1966a, b, 1968; Vinnikov *et al.*, 1967, 1971; Fig. 7) (for further scanning electron microscopical pictures of cephalopod statocysts see Barber and Boyde, 1968; Boyde and Barber, 1969; Budelmann *et al.*, 1973). The number of cilia borne by each cell varies but is generally about 100–130 and they have a maximum length of about 10 μm. Each kinocilium has the usual 9+2 internal tubule content and is morphologically polarized by the line connecting its two central tubules being approximately in line with the long axis of the ciliary group. Further polarization is obtained as each cilium bears a basal foot, which projects at right angles to this long axis (see Barber, 1968; Fig. 8). Because there are nine pairs of peripheral tubules there is asymmetry, as four tubules are on the basal foot side and five on the opposite side of the cilium. At the base of each cilium is a ring of about eight short (2 μm long) microvilli. In the plexus below the hair cells are synaptic profiles, which are probably inhibitory endings of efferent axons (see Budelmann *et al.*, 1973). The orientation of the hair cell groups in the macula and crista will not be detailed here, but has been determined in *Octopus*, *Sepia* and *Loligo* (see Barber, 1968; Budelmann *et al.*, 1973).

C. *Structural features to be considered in theories of transduction mechanisms in hair cells of vertebrates and molluscs*

At this point in our knowledge the transduction mechanisms can only be discussed in outline; however, certain points can be determined and some tentative conclusions reached. The first thing to decide is whether the cilia are the essential organelles involved in transduction. In vertebrate acoustico-vestibular systems there is generally only one cilium present, although as indicated earlier it can be very much longer than the longest stereocilium (Lewis, 1972) (no cilium is present in the organ of Corti). In these situations it is reasonable to suggest that the function of the basal body or cilium is to act as an organizer for the growth and arrangement of the stereocilia, and recent developmental studies of the vestibular system of the lamprey (*Lampetra*) give some support for this idea in that the development of the cilium precedes that of the stereocilia (Thornhill, 1972). The basal foot in the latter case did not develop until after the development of the stereocilia. Interestingly enough a single cilium usually occurs in other arrays of microvilli such as those comprising the rhabdoms in certain invertebrate eyes (see Eakin, 1972). If this organizing function is correct, the graded heights of the stereocilia, the shortest being furthest away from the basal body, could be explained by differential diffusion of some chemical substance that stimulated the elaboration of the hair cell outer membrane to form stereocilia.

Cilia are of major importance in molluscan statocysts because of their large numbers and the absence of any other processes approaching their length (many microvilli are present and cannot be ignored, but they are much shorter than the cilia). The orientation of cilia in cephalopod hair cells is identical with the orientation of beating cilia. Gibbons (1961) showed that in the gills of the lamellibranch mollusc *Anodonta*, the beating stroke of these motile cilia was towards peripheral fibres 5 and 6, and hence towards the basal foot of the cilium. However, in the ctenophore comb plate, Afzelius (1961) showed that the cilia beat in the opposite direction (basal feet were not described in this case). More recent studies on *Opalina* give further evidence that the plane of beating is at right angles to the line joining the two central tubules (Tamm and Horridge, 1970). It seems likely that the non-motile cilia in the statocyst of cephalopods have a polarity of passive bending established by their internal tubule and basal foot orientation, and it has been postulated that passive movement of these cilia in the direction of their basal feet is excitatory (see Barber, 1968).

It can now be seen that in the systems considered, a polarity of bending possibly occurs, caused either by ciliary orientation or the graded height of the stereocilia.

D. *Membrane distortion theories of transduction*

In the vestibular system of the frog the end of the cilium is bulbous and the stereocilia are attached to it by thin filaments to form a compact hair group (Hillman, 1969; Hillman and Lewis, 1971). It was suggested that displacement of the hairs deformed the membrane at the base of the cilium by a plunger-like action. Note that the stereocilia, but not the kinocilium, are attached to an intracellular cuticular plate, a differentiation that might serve to increase this membrane distortion. Electron microscopical studies on the organ of Corti using ruthenium red staining, which is thought to stain mucopolysaccharides, showed that the spaces between the stereocilia were filled with stained material, which also indicates attachment of the stereocilia to each other in this system (Spoendlin, 1968).

It was postulated that distortion of the membranes produces changes in the molecular organization of the excitable area which cause changes in ion permeability leading to a depolarization of the hair cell (Hillman and Lewis, 1971). Similar distortion mechanisms in transduction have also been suggested in, for example, the Pacinian corpuscle (Gray, 1959). An alternative possibility is that this distortion of the membrane may initiate an enzyme system whose initial velocity is dependent on the intensity of the stimulus and which produces the change in membrane permeability (Duncan, 1963).

E. *Piezo-electric theory of transduction*

The probable presence of mucopolysaccharides around the cilium and stereocilia already described in the previous section (see Hillman, 1969; Spoendlin, 1968) can also be used to support a physical theory of transduction. Deformations of mucopolysaccharides adsorbed in narrow diameter tubes cause electrostatic changes and give rise to potential differences, which, when produced by oscillatory displacement, resemble microphonic potentials recorded from the ear (Christiansen, 1964; Vilstrup and Jensen, 1961). This suggests that a film of mucopolysaccharide molecules covering the cilia and/or stereocilia form the basis for a condenser-like function of an electrostatically charged membrane, where the charge is modulated by mechanical displacement of the sensory hairs; these charges would act as a generator potential. One attractive feature of this theory is that all that is required is the appropriate distortable membrane-bound processes and the mucopolysaccharide coat, so that either stereocilia or kinocilia could perform this function. This theory therefore encompasses the variations in sensory hair composition between the various acoustico-vestibular and static systems.

F. *Ciliary mechanosensitivity theory of transduction*

The idea that passive movement of the cilium or cilia causes electrical changes in the hair cells is the reverse of what happens in motile cilia. When motile cilia beat, electrical changes occur in the cells. Horridge (1965a) considered that it was reasonable to conclude from his study of the ctenophore comb plate, that depolarization of the ciliated cells set off the beat of the cilia of those cells. Studies by Thurm (1968) showed that displacement, of as little as 0·5 μm, of an anaesthetized motile cilium near to its basal body caused it to beat once. These experiments were done on gill cilia in *Mytilus* and not on hair cells, but nevertheless suggested that there was an inherent mechanosensitivity in cilia and that passive displacement of the cilium or cilia might initiate electrical changes in hair cells. Circumstantial evidence supporting this theory is that in hair cells the direction of excitatory passive displacement of the cilia is in the morphological plane of beating as if the cilia were motile.

From these considerations it can be seen that at the moment no definitive theory of the mechanisms of transduction can be made. One problem that does arise in the mechanosensitivity theory is of course the absence of cilia in the organ of Corti, although it is possible that a different mechanism could be operating in this system. The deformation and piezo-electric theory have the advantage that they can be applied to all systems. The direction of passive displacement could be determined largely by the graded heights of the stereocilia in the vertebrate systems and by the ciliary orientation in the cephalopod statocysts.

G. *Other static and auditory receptors*

The structure of a variety of auditory and static receptors other than those already described have been determined. In many cases, as one might expect, little information is available about their physiology.

The original description of the fine structure of an insect auditory receptor was given by Gray (1960). It was shown that each sensillum contained a sensory neuron that gave rise to a single process that had a 9+0 ciliary structure along part of its length. Sound causes vibration in a chitinous tympanum and these vibrations are then transmitted to the sensory process, which must be the stimulated portion of the cell.

While these structures have an auditory function, the statocysts of Crustacea are gravity receptors. The ingenious classic experiment that demonstrated this was done on the shrimp statocyst (Kreidl, 1893). It was realized that after moulting the shrimps have to refill their statocysts with granules to make the statolith; shrimps were therefore kept in a tank with iron filings on the bottom. The statocysts were re-filled with these filings, and experiments with magnets showed that the statocysts were organs of equilibrium and did not have an acoustic function. Each statocyst consists of an uncalcified chitinous sac, which fills the basal antennular segment and opens to the exterior by means of a small aperture. In the lobster there are about 470 hairs and overlying them is a statolith (Cohen, 1960). These "hairs" are hollow chitinous shafts and at their bases are dilations called ampullae. Processes of the sensory cells connect to a chorda, which in turn is connected to the ampulla. In the crayfish, *Astacus*, there are three sense cells per hair and three processes connected to the chorda. Along part of their length they have a 9+0 ciliary structure (Schöne, 1971; Schöne and Steinbrecht, 1968). The chitinous hair shaft presumably serves as a lever to exert mechanical forces on the sensory nerve terminal. The 9+0 ciliary structure, and the likely stretching of the sensory endings suggest that the mechanosensitivity theory is unlikely to be relevant in this system and that simple distortion or piezo-electric current generation would be the transduction processes. It is a feature of note that the receptor cells of arthropod chordotonal proprioceptor organs have a similar organization to that described in the statocysts of *Astacus*, having a similar 9+0 ciliary structure at the base of the sensory process (see the review by Howse, 1968).

Apart from noting a brief report on the statocyst of the annelid *Protodrilus* (Merker and von Harnack, 1967), the remaining results to be considered in this section concern the statocysts of medusoid coelenterates, and ctenophores. The statocyst in the coelenterate *Corymorpha* lacks cilia (Campbell, 1972) but the receptor cells in the other statocysts so far examined bear at least one kinocilium per cell (Horridge, 1969). In the statocysts of medusae the number of cilia borne by the sensory cells varies from several per cell to only one per cell, and in the latter case, as in *Rhopalonema* and *Geryonia*, there are also many stereocilia present around the cilium (Horridge, 1969). The statocyst of the ctenophore is situated at the centre of the apical organ. Each sensory cell of this statocyst bears numerous kinocilia (termed "balancer" cilia as the statolith rests on them) (Horridge, 1965b). There is little physiological information concerning the functioning of these organs but the passive displacement of the cilia must be the effective stimulus.

VI. Receptors whose modality of excitation is uncertain

In this section a selection has been made from the large variety of studies of other ciliated sense organs whose modality of excitation is not known.

Many of the research papers already referred to contained information on other receptors, such as for example possible chemoreceptors in the eye-bearing tentacles of *Cardium* (Barber and Wright, 1969a), the tentacles of *Nautilus* (Barber and Wright, 1969b), and the tentacles of *Nassarius* (Crisp, 1971), and possible mechanoreceptors in the sucker of *Octopus* (Graziadei, 1964), and the osphradia of *Buccinum* (Welsch and Storch, 1969).

Various potential mechanoreceptors have been described. For example, possible mechanoreceptor cells, generally bearing one 9+2 cilium, with a specialized intracellular basal region, were described in the epithelium of finger-like processes in the ctenophore, *Leucothea* (Horridge, 1965c). Receptors with a more definite mechanoreceptor function were described by the same author in the chaetognath, *Spadella* (Horridge and Boulton, 1967). Each sense cell again bears a single 9+2 cilium. Further results confirm the ability of chaetognaths to detect vibrations, giving additional circumstantial evidence that these are indeed mechanoreceptors (Newbury, 1972). An unusual probable mechanoreceptor was described in the coelenterate polyp of *Coryne*. Only the basal part of the structure was ciliary, having a 9+0 internal tubule structure, but there were no tubules in the extended distal portion of the process (Tardent and Schmid, 1972).

A variety of other sense organs have been described that have either a mechanoreceptor or chemoreceptor function, or both types of function. For example, two types of ciliated receptor cells have been described in the epidermis of the earthworm *Lumbricus* (Knapp and Mill, 1971). It was suggested that the uniciliated receptors were probably chemoreceptors and that the multiciliated cells were mechanoreceptors, but the present author, by comparison with other mechanoreceptors, suggests that the reverse is probably the case. A probable mechanoreceptor has been described in the epidermis of *Priapulus* and of the oligochaete *Rhynchelmis* (Moritz and Storch, 1971). These receptors bear one 9+2 cilium per cell and this cilium is surrounded by a ring of microvilli. This arrangement is reminiscent of the receptor cells in some medusan statocysts as mentioned earlier (see Horridge, 1969). Ciliated and non-ciliated free nerve endings have recently been described in the epidermis of the freshwater snails *Lymnaea* and *Biomphalaria* (Zylstra, 1972); it was suggested that these cells probably have a chemoreceptor and/or a mechanoreceptor function. Receptor cells bearing either one or two cilia have been described in the receptor processes of the planarian *Dugesia*, but whether they were mechano- or chemoreceptors was uncertain (MacRae, 1967). Two types of ciliated receptor cells, which were probably mechano- and chemoreceptors were described in the nemertine, *Lineus* (Storch and Moritz, 1971). One type bore a single cilium (mechanoreceptor?) and the other type bore between 20 and 40 cilia. A ciliated receptor cell bearing 9+2 cilia protrudes into the lumen of the oviduct of *Aplysia* but whether these latter cells are mechano- or chemoreceptors is not known (Coggeshall, 1971). Probable chemoreceptors bearing numerous 9+2 cilia and microvilli have also been described in the tentacle of the mollusc, *Vaginulus* (Renzoni, 1968).

As was indicated earlier, most arthropod sense organs contain ciliated receptor cells which are remarkably similar in overall structure in the diverse receptor systems so far studied. While cilia have not been found in the retinal cells of arthropods, even here the presence of centrioles and roots has been reported in the rhabdoms of certain beetles (Home, 1972). For further details concerning arthropod receptors the reader is referred to the review references given earlier. Three additional recent references concern the CO_2 receptors of the mosquito (McIver, 1972), the palpal receptors of the tick (Foelix and Chu-Wang, 1972), and the chemoreceptors in copepods (Elofsson, 1971).

Unlike the ordinary receptor cells of the lateral line, the ampullary electro-receptor cells in the gymnotid fishes do not bear cilia or contain ciliary derivatives (see Lissmann and Mullinger, 1968, for details). However, a recently published account of the ampullae of *Polyodon* describes probable electro-receptor cells each of which bears a single 9+0 cilium at its tip (Jørgensen *et al.*, 1972). An extra basal body is found below the basal body of the cilium. The importance of this arrangement is unknown.

VII. Conclusion

Ciliated receptor systems have now been described and as one has seen there is considerable structural variation in different organs. While a great deal of this structural detail is known, adequate physiological studies of the systems are generally not available. Consequently, it is often not possible to decide to which sensory modality the sense organ is responding. There are many theories of transduction mechanisms but in no case have these theories been fully substantiated.

It is often not even possible to determine whether or not the cilium or ciliary structure is important in the transduction process. In many cases, such as in the ciliated retinal cells and olfactory organs, the ciliary membrane is expanded to form a large surface area for trapping photons of light or odoriferous molecules, and in these cases it may be essentially irrelevant whether or not the membrane extension is derived from cilia. In other systems, such as in the vertebrate acoustico-vestibular system, there is a correlation between ciliary orientation and the physiological responses of the cells. In this case the cilia must be involved directly in the transduction process, although even here, if the piezo-electric theory is correct, their only function would be to provide membranes to carry the distortable mucopolysaccharide molecules, a function that would be analogous to that found in retinal and olfactory receptors.

The problems involved are clearly ones that are difficult to solve and more work will be needed before any real answers to the problems of transduction can be obtained.

Acknowledgements

The author would like to thank Ms J. M. Barber for secretarial and Ms S. West for photographic assistance. This chapter was written whilst the author was in receipt of a National Research Council of Canada operating research grant.

Studies in Biology from Memorial University of Newfoundland No. 320.

References

AFZELIUS, B. (1959). *J. biophys. biochem. Cytol.* **5**, 269–278.
AFZELIUS, B. (1961). *J. biophys. biochem. Cytol.* **9**, 383–394.
ANDRES, K. H. (1969). *Z. Zellforsch. mikrosk. Anat.* **96**, 250–274.
BAILEY, D. F. and BENJAMIN, P. R. (1968). *Symp. zool. Soc. Lond.* **23**, 263–268.
BAILEY, D. F. and LAVERACK, M. S. (1963). *Nature, Lond.* **200**, 1122–1123.
BAILEY, D. F. and LAVERACK, M. S. (1966). *J. exp. Biol.* **44**, 131–148.
BANNISTER, L. H. (1965). *Q. Jl microsc. Sci.* **106**, 333–342.
BANNISTER, L. H. (1968). *Nature, Lond.* **217**, 275–276.
BARBER, V. C. (1965). *J. Microscopie.* **4**, 547–550.
BARBER, V. C. (1966a). *Z. Zellforsch. mikrosk. Anat.* **70**, 91–107.
BARBER, V. C. (1966b). *J. Anat.* **100**, 685–686.
BARBER, V. C. (1968). *Symp. zool. Soc. Lond.* **23**, 37–62.
BARBER, V. C. and BOYDE, A. (1968). *Z. Zellforsch. mikrosk. Anat.* **84**, 269–284.
BARBER, V. C. and DILLY, P. N. (1969). *Z. Zellforsch. mikrosk. Anat.* **94**, 462–478.
BARBER, V. C., EVANS, E. M. and LAND, M. F. (1967). *Z. Zellforsch. mikrosk. Anat.* **76**, 295–312.
BARBER, V. C. and LAND, M. F. (1967). *Experientia* **23**, 677–678.
BARBER, V. C. and WRIGHT, D. E. (1969a). *J. Ultrastruct. Res.* **26**, 515–528.
BARBER, V. C. and WRIGHT, D. E. (1969b). *Z. Zellforsch. mikrosk. Anat.* **102**, 293–312.
BENJAMIN, P. R. (1971). *Z. Zellforsch. mikrosk. Anat.* **117**, 485–501.
BENJAMIN, P. R. and PEAT, A. (1971). *Z. Zellforsch. mikrosk. Anat.* **118**, 168–189.
BIDDER, A. M. (1962). *Nature, Lond.* **196**, 451–454.
BOYDE, A. and BARBER, V. C. (1969). *J. Cell Sci.* **4**, 223–239.
BOYLE, P. R. (1969a). *Nature, Lond.* **222**, 895–896.
BOYLE, P. R. (1969b). *Z. Zellforsch. mikrosk. Anat.* **102**, 313–332.
BROWN, P. K. (1972). *Nature New Biology* **236**, 35–38.
BUDELMANN, B. U., BARBER, V. C., and WEST, S. (1973). *Brain Res.* **56**, 25–41.
CAMPBELL, R. D. (1972). *Nature, Lond.* **238**, 49–51.
CHRISTIANSEN, J. A. (1964). *Acta. oto-lar.* **57**, 33–49.
COGGESHALL, R. E. (1969). *J. Morph.* **127**, 113–131.
COGGESHALL, R. E. (1971). *Tissue Cell* **3**, 637–648.
COHEN, M. J. (1960). *Proc. R. Soc.* B **152**, 30–49.
CONE, R. A. (1972). *Nature New Biology* **236**, 39–43.
CRESCITELLI, F. (1972). *In* "Handbook of Sensory Physiology. VII/1. Photochemistry of Vision" (H. J. A. Dartnall, ed.), pp. 245–363. Springer-Verlag, Berlin.
CRISP. M. (1971). *J. mar. biol. Ass. U.K.* **51**, 865–890.
CRISP, M. (1972). *J. mar. biol. Ass. U.K.* **52**, 437–442.
DAVIES, J. T. (1971). *In* "Handbook of Sensory Physiology. IV/1. Olfaction" (L. M. Beidler, ed.), pp. 322–350. Springer-Verlag, Berlin.
DILLY, P. N. (1961). *Nature, Lond.* **191**, 786–787.
DILLY, P. N. (1964). *Q. Jl microsc. Sci.* **105**, 13–20.
DILLY, P. N. (1969a). *Z. Zellforsch. mikrosk. Anat.* **96**, 63–65.

DILLY, P. N. (1969b). *Z. Zellforsch. mikrosk. Anat.* **99**, 420–429.
DUNCAN, C. J. (1963). *J. theor. Biol.* **5**, 114–126.
DUVALL, A. J., FLOCK, Å. and WERSÄLL, J. (1966). *J. Cell Biol.* **29**, 497–505.
EAKIN, R. M. (1963). *In* "General Physiology of Cell Specializations" (D. Mazia and A. Tyler, eds), pp. 393–425. McGraw-Hill, New York.
EAKIN, R. M. (1965). *Cold Spring Harb. Symp. quant. Biol.* **30**, 363–370.
EAKIN, R. M. (1968). *In* "Evolutionary Biology. II." (T. Dobzhansky, M. K. Hecht and W. C. Steers, eds), pp. 194–242. Appleton-Century-Crofts, New York.
EAKIN, R. M. (1972). *In* "Handbook of Sensory Physiology. VII/1. Photochemistry of Vision" (H. J. A. Dartnall, ed.), pp. 625–684. Springer-Verlag, Berlin.
EAKIN, R. M. and KUDA, A. (1971). *Z. Zellforsch. mikrosk. Anat.* **112**, 287–312.
EAKIN, R. M. and WESTFALL, J. A. (1961). *Embryologia* **6**, 84–98.
EAKIN, R. M. and WESTFALL, J. A. (1962). *Proc. natn. Acad. Sci., U.S.A.* **48**, 826–833.
ELOFSSON, R. (1971). *Acta zool. Stockh.* **52**, 299–315.
ERNST, K. D. (1972). *Z. Zellforsch. mikrosk. Anat.* **129**, 217–236.
FAWCETT, D. W. and PORTER, K. R. (1954). *J. Morph.* **94**, 221–282.
FLOCK, Å. (1964). *J. Cell Biol.* **22**, 413–431.
FLOCK, Å. (1965). *Cold. Spring Harb. Symp. quant. Biol.* **30**, 133–145.
FLOCK, Å. (1971). *In* "Handbook of Sensory Physiology. I. Principles of Receptor Physiology" (W. R. Loewenstein, ed.), pp. 396–441. Springer-Verlag, Berlin.
FLOCK, Å. and DUVALL, A. J. (1965). *J. Cell Biol.* **25**, 1–8.
FLOCK, Å., KIMURA, R., LUNDQUIST, P.-G. and WERSÄLL, J. (1962). *J. acoust. Soc. Am.* **34**, 1351–1355.
FOELIX, R. F. and CHU-WANG, I.-W. (1972). *Z. Zellforsch. mikrosk. Anat.* **129**, 548–560.
FUJIMOTO, K., YANASE, T., OKUNO, Y. and IWATA, K. (1966). *Mem. Osaka Gakugei Univ.* **15**, 98–108.
GEUZE, J. J. (1968). *Neth. J. Zool.* **18**, 155–204.
GIBBONS, I. R. (1961). *J. biophys. biochem. Cytol.* **11**, 179–205.
GIBBONS, I. R. (1965). *Archs Biol., Liège* **76**, 317–352.
GIBBONS, I. R. and GRIMSTONE, A. V. (1960). *J. biophys. biochem. Cytol.* **7**, 697–716.
GORMAN, A. L. F. and MCREYNOLDS, J. S. (1969). *Science, Wash.* **165**, 309–310.
GORMAN, A. L. F., MCREYNOLDS, J. S. and BARNES, S. N. (1971). *Science, Wash.* **172**, 1052–1054.
GRAY, E. G. (1960). *Phil. Trans. R. Soc.* B **243**, 75–94.
GRAY, J. A. B. (1959). *In* "Handbook of Physiology. I. Neurophysiology. I." (J. Field, ed.), pp. 123–145. Waverley Press, Baltimore.
GRAZIADEI, P. (1964). *Z. Zellforsch. mikrosk. Anat.* **64**, 510–522.
GRAZIADEI, P. (1965). *Cold Spring Harb. Symp. quant. Biol.* **30**, 45–57.
GRAZIADEI, P. (1966). *J. Zool., Lond.* **149**, 89–94.
GRAZIADEI, P. P. C. (1971). *In* "Handbook of Sensory Physiology. IV/1. Olfaction" (L. M. Beidler, ed.), pp. 27–58. Springer-Verlag, Berlin.

Hamlyn-Harris, R. (1903). *Zool. Jb.* (*Anat.*) **18**, 327–358.
Hartline, H. K. (1938). *J. cell. comp. Physiol.* **11**, 465–477.
Henke, K. and Rönsch, G. (1951). *Naturwissenschaften* **38**, 335–336.
Hermans, C. O. and Eakin, R. M. (1969). *Z. Zellforsch. mikrosk. Anat.* **100**, 325–339.
Hillman, D. E. (1969). *Brain Res.* **13**, 407–412.
Hillman, D. E. and Lewis, E. R. (1971). *Science, Wash.* **174**, 416–419.
Holborow, P. L. (1971). *In* "Fourth European Marine Biology Symposium" (D. J. Crisp, ed.), pp. 237–246. Cambridge University Press.
Holborow, P. L. and Laverack, M. S. (1972). *Mar. Behav. Physiol.* **1**, 139–156.
Home, E. M. (1972). *Tissue Cell* **4**, 227–234.
Horridge, G. A. (1964). *Q. Jl microsc. Sci.* **105**, 311–317.
Horridge, G. A. (1965a). *Nature, Lond.* **205**, 602.
Horridge, G. A. (1965b). *Am. Zool.* **5**, 357–375.
Horridge, G. A. (1965c). *Proc. R. Soc.* B **162**, 333–350.
Horridge, G. A. (1969). *Tissue Cell* **1**, 341–353.
Horridge, G. A. and Boulton, P. S. (1967). *Proc. R. Soc.* B **168**, 413–419.
Howse, P. E. (1968). *Symp. zool. Soc. Lond.* **23**, 167–198.
Hughes, H. P. I. (1970). *Z. Zellforsch. mikrosk. Anat.* **106**, 79–98.
Jacklet, J. W., Alvarez, R. and Bernstein, B. (1972). *J. Ultrastruct. Res.* **38**, 241–261.
Jørgensen, J. M., Flock, Å. and Wersäll, J. (1972). *Z. Zellforsch. mikrosk. Anat.* **130**, 362–377.
Kaissling, K. E. (1971). *In* "Handbook of Sensory Physiology. IV/1. Olfaction" (L. M. Beidler, ed.), pp. 351–431. Springer-Verlag, Berlin.
Kerneis, A. (1966). *C. r. hebd. Séanc. Acad. Sci., Paris* **263**, 653–656.
Kerneis, A. (1968). *Z. Zellforsch. mikrosk. Anat.* **86**, 280–292.
Kerneis, A. (1971). *C. r. hebd. Séanc. Acad. Sci., Paris* **273**, 372–375.
Klein, K. (1931). *Z. Zellforsch. mikrosk. Anat.* **14**, 481–516.
Knapp, M. F. and Mill, P. J. (1971). *Tissue Cell* **3**, 623–636.
Kolnberger, I. (1971). *Z. Zellforsch. mikrosk. Anat.* **122**, 53–67.
Koyama, N. and Kurihari, K. (1972). *Nature, Lond.* **236**, 402–404.
Krasne, F. B. and Lawrence, P. A. (1966). *J. Cell Sci.* **1**, 239–248.
Kreidl, A. (1893). *Sber. Akad. Wiss. Wien* **102**, 149–174.
Land, M. F. (1965). *J. Physiol., Lond.* **179**, 138–153.
Land, M. F. (1966). *J. exp. Biol.* **45**, 83–99.
Land, M. F. (1968). *Symp. zool. Soc. Lond.* **23**, 75–96.
Laverack, M. (1968). *Symp. zool. Soc. Lond.* **23**, 299–326.
Lawrence, P. A. and Krasne, F. B. (1965). *Science, Wash.* **148**, 965–966.
Levi, P. and Levi, C. (1971). *J. Microscopie* **11**, 425–432.
Lewis, C. T. (1970). *In* "Insect Ultrastructure" (A. C. Neville, ed.), pp. 59–76. Blackwell Scientific Publications, Oxford.
Lewis, E. R. (1972). *In* "Proceedings 13th Annual EMSA Meeting" (C. J. Arceneaux, ed.), pp. 64–65. Claitor's Publishing Division, Baton Rouge.
Lewis, E. R. and Nemanic, P. (1972). *Z. Zellforsch. mikrosk. Anat.* **123**, 441–457.
Lim, D. J. (1969). *Acta. oto-lar.* Suppl. **255**, pp. 1–38.
Lim, D. J. and Lane, W. C. (1969). *Archs Otolar.* **90**, 283–292.

LISSMANN, H. W. and MULLINGER, A. M. (1968). *Proc. R. Soc.* B **169**, 345–378.
LOWENSTEIN, O., OSBORNE, M. P. and WERSÄLL. J. (1964). *Proc. R. Soc.* B **160**, 1–12.
LOWENSTEIN, O. and THORNHILL, R. A. (1970). *Proc. R. Soc.* B **176**, 21–42.
LOWENSTEIN, O. and WERSÄLL, J. (1959). *Nature, Lond.* **184**, 1807–1808.
MACRAE, E. K. (1967). *Z. Zellforsch. mikrosk. Anat.* **82**, 479–494.
MCIVER, S. B. (1972). *Can. J. Zool.* **50**, 571–576.
MCREYNOLDS, J. S. and GORMAN, A. L. F. (1970a). *J. gen. Physiol.* **56**, 376–391,
MCREYNOLDS, J. S. and GORMAN, A. L. F. (1970b). *J. gen. Physiol.* **56**, 392–406.
MERKER, G. and VON HARNACK, M. V. (1967). *Z. Zellforsch. mikrosk. Anat.* **81**. 221–239.
MILLER, W. H. (1958). *J. biophys. biochem. Cytol.* **4**, 227–228.
MILLER, W. H. (1960). *In* "The Cell" (J. Brachet, and A. E. Mirsky, eds), Vol. IV, pp. 325–364. Academic Press, New York.
MORITZ, K. and STORCH, V. (1971). *Z. Zellforsch. mikrosk. Anat.* **117**, 226–234.
MOULTON, D. G. and BEIDLER, L. M. (1967). *Physiol. Rev.* **47**, 1–52.
MURRAY, R. G. (1971). *In* "Handbook of Sensory Physiology. IV/2. Taste" (L. M. Beidler, ed.), pp. 31–50. Springer-Verlag, Berlin.
NEWBURY, T. K. (1972). *Nature, Lond.* **236**, 459–460.
OKANO, M., WEBER, A. F. and FROMMES, S. P. (1967). *J. Ultrastruct. Res.* **17**, 487–502.
OTTOSON, D. (1956). *Acta. physiol., scand.* **35**, Suppl. **122**, 1–83.
OTTOSON, D. (1971). *In* "Handbook of Sensory Physiology. IV/1. Olfaction" (L. M. Beidler, ed.), pp. 95–131. Springer-Verlag, Berlin.
OWEN, G. and CHARLES, G. H. (1966). Quoted in "Physiology of Mollusca. II." (K. M. Wilbur and C. M. Yonge, eds), pp. 455–521. Academic Press, New York.
QUATTRINI, D. (1967). *Boll. Soc. ital. Biol. sper.* **43**, 785–786.
RENZONI, A. (1968). *Z. Zellforsch. mikrosk. Anat.* **87**, 350–376.
REESE, T. S. (1965). *J. Cell Biol.* **25**, 209–230.
RHODIN, J. and DALHAMN, T. (1956). *Z. Zellforsch. mikrosk. Anat.* **44**, 345–412.
RÖHLICH, P., AROS, B. and VIRAGH, SZ. (1970). *Z. Zellforsch. mikrosk. Anat.* **104**, 345–357.
SCHNEIDER, D. and STEINBRECHT, R. A. (1968). *Symp. zool. Soc. Lond.* **23**, 279–297.
SCHÖNE, H. (1971). *In* "Gravity and the Organism" (S. A. Gordon and M. J. Cohen, eds), pp. 223–235. University of Chicago Press.
SCHÖNE, H. and STEINBRECHT, R. A. (1968). *Nature, Lond.* **220**, 184–186.
SCHULTE, E. (1972). *Z. Zellforsch. mikrosk. Anat.* **125**, 210–228.
SJÖSTRAND, F. S. (1959). *Ergebn. Biol.* **21**, 128–160.
SLIFER, E. H. (1970). *A. Rev. Ent.* **15**, 121–142.
SPOENDLIN, H. (1968). *In* "Hearing Mechanisms in Vertebrates" (A. V. S. de Reuch and J. Knight, eds), pp. 89–119. Churchill, London.
STEINBRECHT, R. A. (1969a). *In* "Olfaction and Taste" (C. Pfaffmann, ed.), pp. 1–21. Rockefeller University Press, New York.
STEINBRECHT, R. A. (1969b). *J. Cell Sci.* **4**, 39–53.
STINNAKRE, J. and TAUC, L. (1969). *J. exp. Biol.* **51**, 347–361.

STORCH, V. and MORITZ, K. (1971). *Z. Zellforsch. mikrosk. Anat.* **117**, 212–225.
STORCH, V. and WELSCH, U. (1969). *Z. Zellforsch. mikrosk. Anat.* **97**, 528–536.
TAKAGI, S. F. and YAJIMA, T. (1965). *J. gen. Physiol.* **48**, 559–569.
TAMM, S. L. and HORRIDGE, G. A. (1970). *Proc. R. Soc.* B **175**, 219–233.
TARDENT, P. and SCHMID, V. (1972). *Expl Cell Res.* **72**, 265–275.
THORNHILL, R. A. (1972). *Proc. R. Soc.* B **181**, 175–198.
THURM, U. (1968). *Symp. zool. Soc. Lond.* **23**, 199–216.
THURM, U. (1969). *In* "Processing of Optical Data by Organisms and by Machines" (W. Reichardt, ed.), pp. 236–256. Academic Press, New York.
TUCKER, D. (1967). *Fedn Proc. Fedn Am. Socs exp. Biol.* **26**, 544, abstract 1609.
VILSTRUP, TH. and JENSEN, C. E. (1961). *Acta oto-lar.* Suppl. **163**, 42–46.
VINNIKOV, YA. A., GASENKO, O. G., BRONSTEIN, A. A., TSIRULIS, T. P., IVANOV, V. P. and PYATKINA, G. A. (1967). *In* "Symposium on the Neurobiology of Invertebrates", pp. 29–48. Publishing House of the Hungarian Academy of Sciences, Budapest.
VINNIKOV, YA. A., GASENKO, O. G., TITOVA, L. K., BRONSTEIN, A. A., TSIRULIS, T. P., REVZNIER, R. A., GOVARDOVSKII, V. I., GRIBAKIN, F. G., IVANOV, V. P., ARONOVA, M. Z. and TCHEKHONADSKII, N. A. (1971). *In* "Problems in Space Biology. 12" (V. N. Tchernigovskii, ed.). pp. 1–523. "Nauka", Leningrad. [In Russian.].
WATKINSON, G. B. (1909). *Jena. Z. Naturw.* **44**, 353–414.
WELSCH, U. and STORCH, V. (1969), *Z. Zellforsch. mikrosk. Anat.* **95**, 317–330.
WERSÄLL. J., FLOCK, Å. and LUNDQUIST, P.-G. (1965). *Cold Spring Harb. Symp. quant. Biol.* **30**, 115–132.
WILSON, J. A. F. and WESTERMAN, R. A. (1967). *Z. Zellforsch. mikrosk. Anat.* **83**, 196–206.
WOLFF, H. G. (1968). *Experientia* **24**, 848–849.
WOLFF, H. G. (1969). *Z. Zellforsch. mikrosk. Anat.* **100**, 251–270.
WOLFF, H. G. (1970a). *Z. vergl. Physiol.* **69**, 326–366.
WOLFF, H. G. (1970b). *Z. vergl. Physiol.* **70**, 401–409.
WOOLLACOTT, R. M. and ZIMMER, R. L. (1972). *Z. Zellforsch. mikrosk. Anat.* **123**, 458–469.
YANASE, T. and SAKAMOTO, S. (1965). *Zool. Mag., Tokyo* **74**, 238–242.
YOUNG, J. Z. (1960). *Proc. R. Soc.* B **152**, 3–29.
YOUNG, J, Z. (1971). "The Anatomy of the Nervous System of *Octopus vulgaris*." Clarendon Press, Oxford.
ZYLSTRA, U. (1972). *Neth. J. Zool.* **22**, 283–298.

Basal structures attached to cilia

Chapter 16

Basal bodies and root structures

DOROTHY R. PITELKA

Department of Zoology and its Cancer Research Laboratory, University of California, Berkeley, California, U.S.A.

I. Introduction

An intracellular basal body, or kinetosome, lies at the origin of every cilium or flagellum. It serves at least two obvious functions: the peripheral microtubules of the cilium or flagellum originate from a pre-existing basal body by elongation of some of its fibrillar components, and the active motor organelle must remain firmly rooted in the cytoplasm in order that the movement of its free length may exert force on its surroundings. In addition to these direct developmental and mechanical roles, a number of more complex and subtle functions have been attributed to basal bodies over the past century by careful observers of their distribution and behavior. These putative functions range from control of the initiation and pattern of ciliary or flagellar beating to obligate self-replication and the genesis of numerous other organelles.

A continuing flow of new information from morphological and biochemical studies of microtubules has stimulated a series of recent reviews of predominantly microtubular organelles, including cilia and flagella, centrioles, and the mitotic apparatus (e.g. Porter, 1966; Stubblefield and Brinkley, 1967; de Harven, 1968; Pickett-Heaps, 1969; Pitelka, 1969c;

Fauré-Fremiet, 1970; Stephens, 1971; Fulton, 1971; Tilney, 1971; Warner, 1972). Most of these consider basal bodies in passing, but only one review (Wolfe, 1972) has focused specifically on them. In fact, basal bodies are structurally more various than either the cilia and flagella to which they give rise or the centrioles with which they may be interconvertible. The present review attempts to relate current information on the diversity of their structure and morphogenesis with current and older evidence of their function. Because ciliated or flagellated cells are distributed in almost every group of eukaryotic organisms and have been studied from many points of view, the survey will include a broad sampling but by no means a complete coverage of the voluminous and widely scattered literature. To limit its scope, the review will not include the basal bodies of animal sperm cells, which are so highly specialized for their particular function that they need separate consideration (see Phillips, Chapter 14, this volume). It will deal only incidentally with basal bodies in the large array of receptor cells that bear modified cilia.

Kinetosome as a synonym of *basal body* has the advantages of a term applying exclusively to this organelle and easily transliterated into any language, but it has not achieved wide usage in English among investigators of metazoan basal bodies. In this review the two terms will be used interchangeably.

II. Morphology of basal bodies

The common denominator of mature kinetosome structure is a cylinder whose wall is formed by nine triplet microtubules (Figs 1–4); within and around the wall of the cylinder, filamentous or amorphous electron-dense material is arranged in variable amounts and patterns. The nine peripheral, doublet microtubules of the shaft (axoneme) of the cilium or flagellum are direct continuations of doublet components of the kinetosomal triplets (Fig. 9). Intimately associated with the walls of most basal bodies are one or more appendages composed of filaments, microtubules, amorphous dense material, or combinations of these.

Asymmetry and polarity are inherent in the structure and arrangement of kinetosomal microtubules. Each of the nine triplets is a band of three microtubules that share their common walls; one, the A-tubule, appears as a complete cylinder 20–26 nm in diameter, whereas the B- and C-tubules have indented or crescentic profiles in cross-section (Fig. 1). The transverse axis of the triplet row does not lie on the circumference of the cylinder but forms an angle with a tangent to the circumference (the "triplet angle" of Anderson, 1972) that diminishes from 30–45° at the base of the kinetosome to 10–20° at the level where the C-tubule of the triplet terminates (compare Figs 1 and 5). If one views a cross-section of the basal body from its basal, or proximal, side, the inward slant of the triplet axes is always clockwise, with the A-tubule the innermost in each triplet. It is always from the opposite or distal end of the basal body, identified as such by the direction of slant of the triplet axes, that the A- and B-tubules extend as the peripheral doublets of the axoneme.

The two ends of a basal body, even when it is not bearing a cilium, are usually differentiated by added structures. Amorphous dense material commonly is more abundant around the proximal than the distal end. During the formation of a basal body, but not always persisting in the mature organelle, a delicate cartwheel structure (Figs 1, 2, 9) consisting of a cylindrical hub and one or more tiers of nine radiating spokes is present at the proximal end. The spokes attach to hooked foot processes extending inward from the A-tubules. At the distal end of the kinetosome, its lumen usually is partly or completely closed off by an annulus or transverse plate of variable density (Figs 5, 6, 9, 10).

The structure thus far described is nearly universal, not only for basal bodies that at some time bear motile cilia or flagella, but also for those that bear immotile cilia modified for sensory reception and for the centrioles that appear to serve as mitotic centers in most animal and some plant cells. Most of the known exceptions to the common pattern are found in animal sperm cells, especially of insects, not discussed here (see Phillips, Chapter 14).

Superimposed on the overwhelmingly constant kinetosomal pattern is a wide range of variations in proportion or embellishment of various parts and associated root structures. A comprehensive survey of the structure and distribution of known variations is impossible in a brief review; some examples, grouped according to the part of the kinetosome affected, will be considered.

A. Orientation of kinetosomal triplets

In what is perhaps the most painstaking analysis of three-dimensional kinetosome structure yet reported, Anderson (1972) compared electron micrographs of thin-sectioned basal bodies in the rhesus monkey oviduct with plastic scale models constructed according to several alternative microtubule orientations; some of the models were embedded in plastic and sliced in various planes to mimic two-dimensional sections. He concluded that each microtubule triplet: (1) leans slightly inward from base to apex, so that the inside diameter of the basal body decreases somewhat from 150 to 130 nm; (2) follows a slightly helical path, pitched 10–15° counterclockwise from base to apex; and (3) rotates centripetally on the

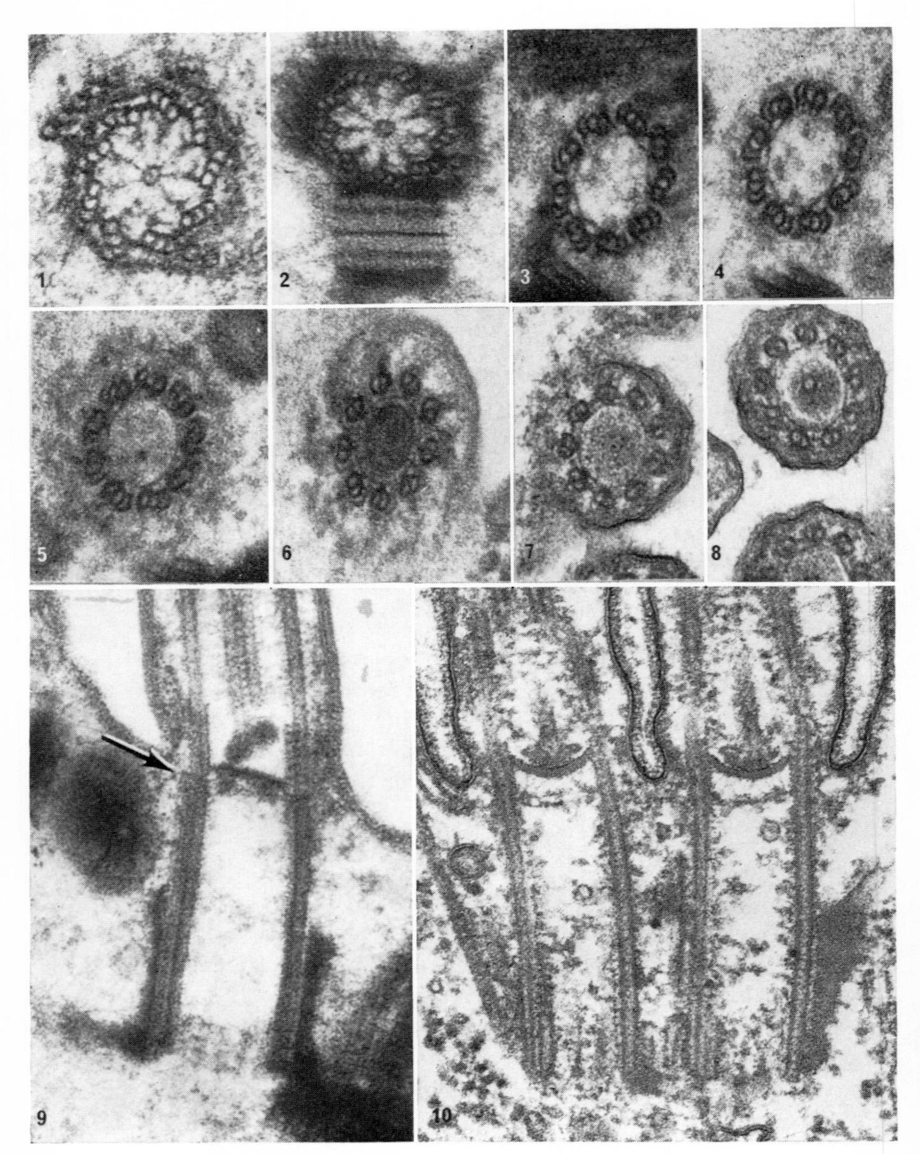

FIGS. 1 to 10.—Captions on facing page.

FIGURES 1–8 are cross-sections of basal bodies cut at successively higher levels; all are type I basal bodies of ciliate Protozoa and are viewed from the base toward the apex. The triplet angle decreases from about 40° in Figs 1 and 2 to about 20° in Figs 5 and 6. Some C-tubules are incomplete in Figs 3–5. A–C links are conspicuously present in Figs 1 and 2 and become tenuous and inconsistent in Figs 3–5. Filaments that connect all the A-tubules in a nine-sided figure are visible in Figs 3 and 4; in Figs 5–8, showing the transition zone, a circular filament (or cylinder?) is linked by short bridges to all of the A-tubules. The central cartwheel and peripheral dense material are evident in Figs 1 and 2. Associated microtubule ribbons appear at lower right and upper left in Fig. 1, and a striated band connecting basal bodies in a row is seen in Fig. 2. Figure 5, cut barely below the cell membrane, shows the moderately dense terminal plate filling the kinetosome lumen and extending out around the microtubules as an expanded disc (which attaches to the cell membrane). In Figs 6–8, wedges of dense material connect doublets to the constricted ciliary membrane. Figures 6 and 7 include the dense plate just beneath the axosome; Fig. 8 shows the axosome with its single central axonemal microtubule. Figure 1, *Didinium nasutum*, × 90,000; Figs 2–6, *Lacrymaria marina*, × 80,000; Figs 7 and 8, *Condylostoma magnum*, × 80,000.

FIG. 9. Longitudinal section of a basal body in *Didinium nasutum*. The proximal half of the organelle is encircled by a sleeve of dense material. In the lumen at the basal end is the delicate cartwheel. Distally, the abrupt ending of a C-tubule in the left wall is marked by an arrow, at the same level as the lower and less dense of two transverse plates. Above the upper plate is the asymmetric axosome, in which the left axonemal microtubule inserts. Just internal to the microtubules in this zone are vertical lines that represent the circular filament or cylinder seen in Figs 5–8. × 80,000.

FIG. 10. Longitudinal section through a pair of kinetosomes of *Condylostoma magnum*. The section plane includes the entire kinetosomal length of a single microtubule in the right wall of each; since the slight curvature of this tubule in both cases is parallel to the curvature of the left wall, it probably is not indicative of a helical twist. In the transition zone, a thin transverse plate lies proximal to a thicker plate cupped beneath the hemispherical axosome. Dense material between microtubules and the constricted ciliary membrane in the transition zone is visible here and in Fig. 9. × 80,000.

longitudinal axis of its A-tubule so that, as the triplet angle decreases from base to apex, the outside diameter of the basal body diminishes more than the inside diameter, from 250 to 165 nm. As effects of these several inclinations, the basal body seen in longitudinal section either tapers distally or looks barrel-shaped, and cross-sections that are not perfectly perpendicular to the kinetosomal axis show triplets on one side in sharp profile and those 180° opposite less distinctly.

It is likely that basal bodies in most vertebrate epithelia, and perhaps in other organisms and cell types as well, share features of microtubule orientation described by Anderson. Similar profiles in section have been noted repeatedly before, and Anderson concluded from examination of published micrographs that many are explainable by the same three-dimensional configuration. But in many protozoan basal bodies, as Anderson also noted, where matrix material in the wall is often scant and microtubule contours relatively clean, departure of the triplets from a true axial orientation appears less pronounced. Published micrographs of fields of basal bodies in ciliates and flagellates (e.g. Gibbons and Grimstone, 1960; Tucker, 1971; many others) demonstrate that kinetosome cross-sections showing all nine microtubule triplets in equally sharp profile are not at all as rare as they should be if the triplets follow helical paths of a 10–15° pitch. It is also not uncommon to find longitudinal sections including the whole length of a single kinetosomal microtubule (Fig. 10). If differences do exist in orientation of mature kinetosomal microtubules in different cell types, these need to be measured carefully; such variants might be related to functional or developmental features of the basal bodies and their cilia.

For a short distance below their termination, the C-tubules in some basal bodies are incomplete; their walls attach to the B-tubules medially but not peripherally (Figs 3, 4). C-tubules—whether for this reason or not—may be more vulnerable to isolation procedures than A- and B-tubules; Wolfe (1970) found consistently short C-tubules in isolated kinetosomes of *Tetrahymena*, whereas micrographs of thin sections (Allen, 1969; Munn, 1970) indicate that they are present for the length of the kinetosome.

B. *Other components of the wall*

At various levels of the basal body, adjacent microtubule triplets are linked by structures that have been interpreted either as sheets (Anderson, 1972; Tucker, 1971) or as filaments (Stubblefield and Brinkley, 1967; Dippell, 1968). At the proximal end of the organelle, and sometimes continuing distally, bridges are regularly present between the A-tubule of each triplet and the C-tubule of the next adjacent one (Figs 1, 2). Around or above the middle of the basal body, the A-tubules of all nine triplets are linked by a thin band (or filament) circumscribing the kinetosomal lumen (Figs 3, 4). Other slender bridges in different configurations have been shown in some kinetosomes (e.g. Gibbons and Grimstone, 1960).

The amount and density of amorphous material embedding the triplets in the kinetosome wall vary greatly—where it is very dense, as is often the case, the existence and distribution of links among microtubules become difficult to discern. A–C links and filaments joining the A-tubules have been described for centrioles (see Fulton, 1971), and it is possible that similar structures are present in all basal bodies. Certainly the majority of centrioles and basal bodies are strictly cylindrical (compression during thin sectioning may make them appear elliptical), although the cytoplasm in which they lie may be very mobile, as in many ciliates and flagellates that undergo frequent change in body shape. The wall matrix and intertriplet connections of various kinds may help to maintain the overall shape of the organelle and the spacing and triplet angle of the microtubules. In the mouth region of *Condylostoma* is a long row of nonciliated basal bodies paralleled by rows of ciliated ones. The barren ones are elliptical in cross-section (not necessarily true of other barren kinetosomes); they have the same A–C links and the same relative triplet angles as their cylindrical neighbors, but appear to lack the distal A–A interconnections. Slight curvature of the entire kinetosome axis is frequently observed in protozoa (Fig. 10).

C. Content of the lumen

In several vertebrate eipthelia, mature basal bodies contain no proximal cartwheel (Kalnins and Porter, 1969; Anderson, 1972); in adult *Condylostoma* the cartwheel is so shallow as to be almost undetectable. Mature basal bodies in many protists and metazoans, however, contain cartwheels extending 10–20% of their length, with several tiers of spokes (Figs 4, 13). In the ciliate *Nassula* (Tucker, 1971) and the flagellates *Trichonympha* and *Pseudotrichonympha* (Gibbons and Grimstone, 1960), cartwheels fill much of the length of the kinetosome. The kinetosome is up to 5 μm long in the two flagellates, but is of average length, about 0·5 μm, in *Nassula*.

Apart from the cartwheel, the content of the kinetosome lumen varies greatly. In Metazoa generally and in many protists it looks like the cytoplasmic matrix, containing small amounts of flocculent material (Fig. 9). Or the enclosed material may be in the form of delicate cylinders (*Pseudotrichonympha*, Gibbons and Grimstone, 1960), a dense axial rod (a trypanosome flagellate: Anderson and Ellis, 1965; *Tetrahymena*: Allen, 1969;

Munn, 1970), dense granules (*Paramecium*: Fig. 12; Pitelka, 1965; Dippell, 1968; *Condylostoma*, Fig. 26), or glycogen (Hollande and Carruette-Valentin, 1971). The variations observed are so diverse that no pattern of distribution is discernible.

D. *The transition zone*

The transition from basal body to flagellum or cilium may be described as the interval between termination of the C-tubules and origin of the central tubules of the axoneme. Although structural variations in this zone are too numerous to detail, they generally fall into two classes (Fig. 11). In type I basal bodies (Fig. 9), a transverse plate near the site of termination of the C-tubules is approximately level with the cell membrane where the latter is deflected upward to become the ciliary membrane, and the distance from the plate to the origin of central microtubules is short, usually less than 0·1 μm. In type II basal bodies (Figs 13, 14), C-tubules also terminate at the level of deflection of the cell membrane, but the doublet microtubules continue upward for 0·2 μm or more before a transverse plate appears and the central axonemal tubules begin. The long transition zone here emerges above the immediately adjacent cell surface, but pictures of this zone (Satir, 1965) do not suggest that it participates in movements of the cilium or flagellum.

In both types, the flagellar or ciliary base sometimes is withdrawn more or less deeply below the general cell surface in a flagellar pocket or circumciliary depression, but the deflection of the cell membrane lining the depression outward to enclose the flagellum remains a reliable landmark. In both types, one (rarely both) of the central axonemal tubules may originate in a dense granule, called the axosome, lying above or merging with the distal transverse plate (Figs 8–10). In both types, two kinds of close attachment of microtubules to the adjacent plasma membrane are present: nine structures identified as triangular sheets (alar sheets of Anderson, 1972) or fibrils (transitional fibrils of Gibbons and Grimstone, 1960, and most later authors) usually attach the kinetosomal triplets at the ends of the C-tubules to the adjacent cell membrane, and doublets in the transition zone are connected to a more or less tightly constricted membrane collar by nine bands or wedges of dense material. And in both types, doublets in the transition zone are interconnected in a circle by sheets or filaments that in some fashion circumscribe the lumen (Figs 5–9). These doublet linkages are particularly complex in the type II transition zones of many algal flagellates (Lang, 1963; Manton, 1963; Ringo, 1967).

Published pictures indicate that type I basal bodies are characteristic of ciliated cells in birds and mammals (e.g. Doolin and Birge, 1966;

Kalnins and Porter, 1969; Anderson, 1972), of some plant and animal flagellates (e.g. some fungi: Fuller, 1966; ameboflagellates: Schuster, 1963; Dingle and Fulton, 1966; Outka and Kluss, 1967; trichomonad flagellates: Honigberg *et al.*, 1971; Joyon *et al.*, 1969; hypermastigid flagellates: Gibbons and Grimstone, 1960; Hollande and Carreutte-Valentin, 1971), and of ciliates (Figs 9, 10, 12). Type II basal bodies appear more widespread;

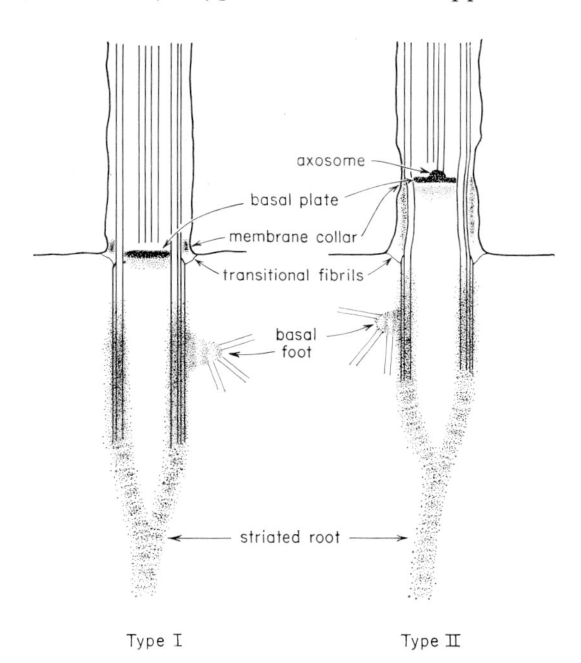

FIG. 11. Diagram showing the differences in the transition zones of type I and type II basal bodies as they typically appear in Metazoa. The intracytoplasmic cylinder of triplet microtubules is similar in the two types, except that it usually is longer in type I (about 0·5 μm) than in type II (about 0·35 μm). See text for further explanation.

they are shown in micrographs of amphibians (Reese, 1965; Steinman, 1968) and fish (Flock and Duvall, 1965), in most if not all invertebrate Metazoa (e.g. amphioxus: Olsson, 1962; Welsch and Storch, 1969; ascidians: Olsson, 1962; molluscs: Fig. 14; Gibbons, 1961; Satir, 1965; an echinoderm: Tilney and Gibbins, 1968; an annelid: Reger, 1967; rotifers: Lansing and Lamy, 1961; Warner, 1969; flatworms: Dorey, 1965; Wilson, 1969; Bresciani and Køie, 1970; coelenterates: Szollosi, 1964; orthonectids: Kozloff, 1971), in kinetoplastid flagellates (Anderson and Ellis, 1965; Vickerman, 1969; Brooker, 1971), in some fungi (Fuller, 1966) and in most plant flagellates (Joyon and Mignot, 1969).

The distribution of the two types offers little clue to their significance. Neither can be identified with a particular phylogenetic line, a particular developmental mode (see below), or a particular type of ciliary activity. Type I occurs in avian and mammalian epithelia, where ciliary movement

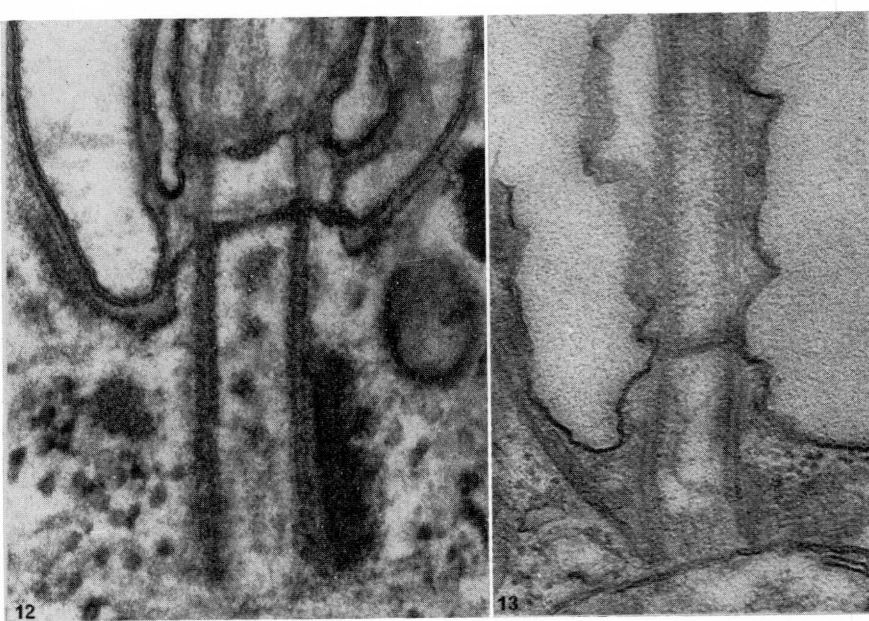

FIG. 12. The modified type I basal body characteristic of ciliates with multiple pellicular membranes is shown here in a longitudinal section from *Paramecium multimicronucleatum*. The dense terminal plate that closes off the lumen (and is present in temporarily or permanently non-ciliated, mature kinetosomes) is continuous laterally with a layer of dense material that parallels the innermost membrane of the pellicle. Immediately above the terminal plate is a very thin plate, and about 100 nm above this is a cupped plate beneath the axosome; the ciliary membrane forms a constricted collar at this level. Dense granules are present in the kinetosome lumen. × 70,000.

FIG. 13. The simple type II basal body of *Bodo saltans*, a kinetoplastid flagellate. Central axonemal tubules, not included in this section, originate just above the upper of the two transverse plates; the constricted membrane collar is at the level of these plates. A cartwheel occupies approximately a quarter of the length of the lumen. × 55,000.

is reduced to a fixed direction in a fixed plane, and in ciliates, which have almost infinitely versatile ciliary beating patterns. Type II is found in lower vertebrate epithelia, with equally simplified and stereotyped ciliary movement, and in euglenoid and other plant flagellates whose flagella are capable of almost any contortion possible for a thread attached at one end. One major phylogenetic line that shows internal consistency is that including the green algae and higher plants, where all flagellated cells for which adequate information is available have a distinctive double-star configuration of doublet linkages in the transition zone (see Manton, 1963; Ringo, 1967; Joyon and Mignot, 1969; Carothers and Kreitner, 1968).

E. Transverse plate (basal plate, terminal plate)

The partition that crosses the lumen in the transition zone shortly below the site of origin of the two central axonemal microtubules has been called by most authors the *basal plate*, the implication being that it marks the beginning of the cilium. Occasionally the plate is tenuous or diffuse. In ciliates and in most type II kinetosomes there typically is more than one transverse plate. A proximal one in ciliates, at the level of C-tubule termination, extends laterally beyond the kinetosomal microtubules to merge with an amorphous layer of the pellicle (Figs 5, 12); this connection apparently substitutes in ciliates for the transitional fibrils of other kinetosomes (a similar plate appears in the complex transition zone of the fungus *Saprolegnia*, Heath and Greenwood, 1971). This plate has been called the *terminal plate* (Pitelka and Child, 1964) because in many ciliates it forms well before the basal body sprouts a cilium (Dippell, 1968; Tucker, 1971) and persists if the cilium is shed (Pitelka and Child, 1964); hence it appears to be part of the mature kinetosome. More distal plates or septa may be analogous to the basal plate in simpler kinetosomes. In the type II basal bodies of *Chlamydomonas*, the entire region containing the double-star pattern and including the transverse partition goes with the flagella when the latter are detached prior to cell division (Johnson and Porter, 1968). The significance of a partition across the lumen is quite unknown. Since it remains stationary in the transition zone while the axonemal microtubules are growing out, it cannot be a site of microtubule assembly for this growth, which probably is distal (Tamm, 1967; Rosenbaum and Child, 1967). It may be mechanical, helping, along with doublet links, to stabilize the cylindrical shape of the transition zone or barring the passage of luminal or cytoplasmic particles into the axoneme.

III. Composition of basal bodies

Kinetosomes and centrioles have never been successfully separated from other cell constituents without loss of such components as wall matrix and luminal contents. The controlled stepwise dismemberment that has proved so fruitful for chemical identification of components of cilia and flagella (see Chapter 3 by Stephens, this volume) has therefore not been applicable to them. Wolfe (1972) concluded from acrylamide gel electrophoresis of isolated oral apparatuses of *Tetrahymena* (containing kinetosome microtubules and a reticulum of fine filaments) that a unique microtubule protein, in addition to those found in cilia, may be present in the basal bodies. Further analysis of the microtubular proteins of basal bodies will undoubtedly proceed rapidly, but identification of such structural components as cartwheels, intertriplet links, or wall matrix remains elusive.

A major effort has been made during the last decade to learn whether DNA is present in kinetosomes or centrioles. Randall and Disbrey (1965) and Smith-Sonneborn and Plaut (1967, 1968) used light microscope fluorescence and autoradiographic techniques to demonstrate DNA at cortical sites corresponding to the distribution of kinetosomes in *Tetrahymena* and *Paramecium*. Attempts to confirm this work in the same ciliates by electron microscope autoradiography or by extraction of DNA from cortical fractions have led several subsequent workers (Pyne, 1968; Hufnagel, 1969; Sonneborn, Dippell and Grimes, cited in Sonneborn, 1970; Flavell and Jones, 1971; Younger *et al.*, 1972) to conclude that any true DNA present represents mitochondrial or nuclear contamination. The generally unsatisfactory state of the art of kinetosome analysis has been discussed repeatedly in recent reviews (see Fulton, 1971; Wolfe, 1972) and does not warrant reiteration here.

IV. Root structures

A. *Morphology*

Appendages of basal bodies are made up of amorphous or finely filamentous dense material, bundles of packed filaments that may or may not show periodic cross-banding, or microtubules. The range of variations in pattern and organization in these roots is relatively restricted in metazoan ciliated cells. In plant and animal protists, where the locomotor apparatus is often the dominant morphological and functional feature of the organism, the number and arrangement of root structures become more varied and usually more complex, culminating in the intricate interweaving of kinetosome-associated fibril systems in higher zooflagellates and ciliates.

The commonest root structures (aside from transitional fibrils considered above) of metazoan basal bodies are basal feet and striated roots. Both of these typically merge with the dense material of the kinetosome wall matrix. The basal foot is a short, dense, banded cone projecting laterally from about the mid-region of the basal body (Fig. 11). Gibbons (1961) determined that the basal feet on kinetosomes of three cell types in the mussel gill extended in the direction of the effective stroke of their cilia, and in most epithelia all basal feet are uniformly oriented. The striated root originates in dense material around or below the kinetosome base and in almost all metazoa extends down into the cell interior toward

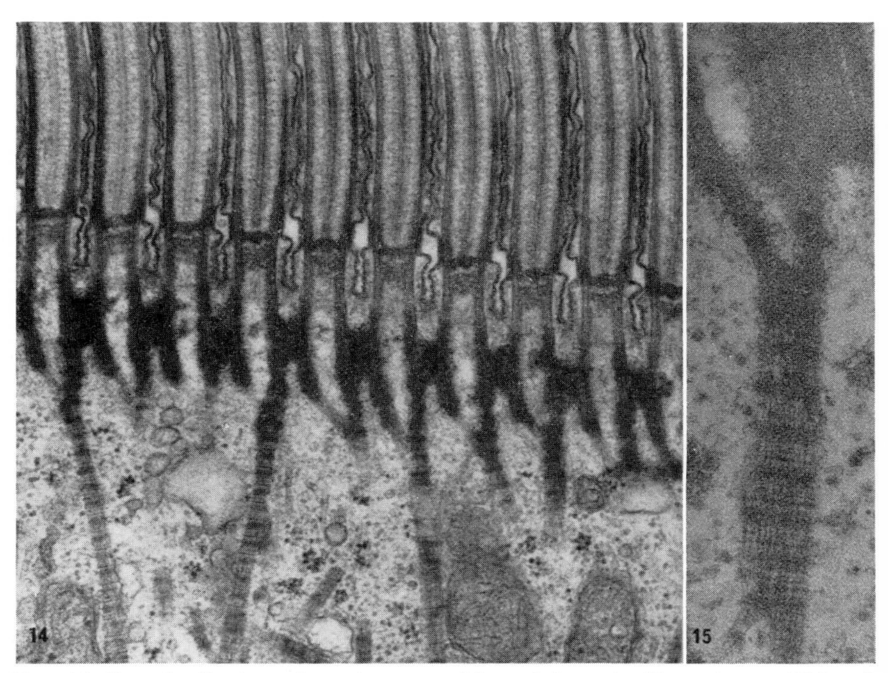

FIG. 14. Longitudinal section of a row of lateral-frontal cilia and type II basal bodies of the freshwater mussel *Elliptio complanatus*. The dense transverse plate has a central thickening from which one or both central axonemal microtubules seem to arise; the ciliary membrane forms a distinct collar in this region. Below the plate is a less dense plate or mass of material. Basally, the dense wall matrix grades imperceptibly into the striated roots, which extend downwards at various angles. The close-packed cilia beat in unison as compound organelles, and their basal bodies are interconnected proximally by masses of dense material continuous with the wall matrix. ×47,000.

FIG. 15. Proximal end of a basal body with its dense wall matrix and striated root. The filamentous substructure and complex banding pattern of the root are evident. ×90,000. Figures 14 and 15 from F. D. Warner, unpublished.

or around the nucleus (Figs 14, 15). The tapering root(s) from each basal body may be delicate or massive, single or multiple, straight or bifurcated, short or long. The striation is complex, with several sub-bands and a repeating period of up to 900 nm. The root probably always has a substructure of longitudinally arranged filaments.

In some instances among metazoa, a relationship of the striated root with other organelles is evident. Olsson (1962) and Welsch and Storch (1969) found in the endostyle of amphioxus rows of mitochondria appressed against the ciliary roots, their cristae stacked with a periodicity corresponding to that of the fibers. Striated roots originating laterally on kinetosomes in some cells of flatworms (Wilson, 1969; Bresciani and Køie, 1970), orthonectids (Kozloff, 1971) and dicyemid mesozoa (Bresciani and Fenchel, 1967) extend in a uniform direction just below and parallel to the ciliated surface.

Microtubules frequently radiate from the tip of the basal foot, much as they may radiate from satellite bodies associated with interphase centrioles. These generally are not highly conspicuous or evidently patterned, and their course, in cytoplasm that may contain other microtubules not kinetosome-based, is usually not identifiable.

In epithelia, filaments commonly are present in the apical cytoplasm. Tracts of similar filaments may weave around the basal bodies (Frisch and Reith, 1966), but they do not usually appear to be attached to them as rootlets.

In all ciliated or flagellated cells of algae, fungi, and protozoa, striated, filamentous and/or microtubular appendages are associated with each other, with single or grouped kinetosomes, and often with other organelles in very precise and species-specific patterns. These are far too varied and complex for adequate summary here; only some general trends and a few examples will be mentioned. Reviews of kinetosome-associated structures in protists have been published by Grain (1969), Joyon and Mignot (1969), Pitelka (1969b) and Didier (1970a).

Where kinetosomes are arranged in pairs, clusters, or close-packed rows, they usually are interconnected by striated bands or by strands of filamentous or amorphous material (Fig. 2). Striated roots comparable to those of metazoa are common in all protistan groups. In the flagellates they typically depart from the base of the kinetosome and descend into the cytoplasm, where they may adhere laterally to the nucleus, Golgi bodies, mitochondria, or chloroplasts; sometimes the adhesion is transitory, related to a limited stage in the cell cycle. In most ciliates, striated fibrils attach to the periphery of the kinetosome base and extend to the right and anteriorly, overlapping with similar fibrils from other kinetosomes to form a composite fiber accompanying each longitudinal row of basal bodies.

Rows of microtubules arising at specific locations at or near the bases of the kinetosomes are present in all ciliates, probably all algal and fungal flagellates, and many animal flagellates. For example in *Chlamydomonas* (Ringo, 1967), four ribbons, of four microtubules each, arise close to a striated band linking the two apical kinetosomes and radiate out and posteriorly for most of the length of the cell, their constituent microtubules gradually diverging to become fairly evenly dispersed under the plasma membrane. Similar peripheral rootlets in other phytoflagellates may be composite, consisting of parallel striated and microtubular bands (e.g. *Oedogonium*, Hoffman, 1970). More elaborate systems of microtubules oriented immediately beneath the cell membrane are characteristic of euglenoid flagellates, ameboflagellates, trypanosomes and their free-living relatives, and flagellate stages of some slime molds. The formidably complex patterns of basal bodies and fibrils in the higher zooflagellates have

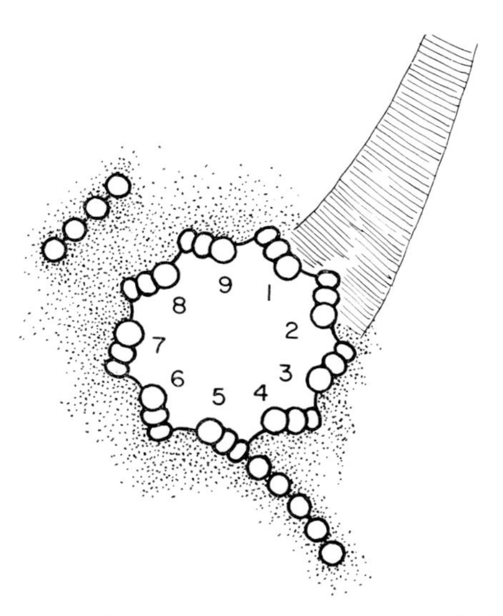

FIG. 16. Diagram showing the common root structures of ciliate basal bodies: a striated fibril originating at the right, anterior surface of the kinetosome base; a ribbon of microtubules extending the transverse axis of triplet 5; and a ribbon of microtubules rising near the left, anterior surface of the kinetosome base.

been described by Honigberg *et al.* (1971), Joyon *et al.* (1969), Grimstone and Gibbons (1966), and Hollande and Carruette-Valentin (1971).

The specificity of attachment of root fibrils is nowhere better illustrated than in the ciliates. The most common basic pattern, some or all elements of which are identifiable in all ciliates, is shown diagrammatically in Fig. 16. The striated root fibril is broadly attached in the region of triplets 1 to 4—that is, to the anterior right surface of the basal body (viewed from the interior of the cell)—and extends to the right and anteriorly. A ribbon of microtubules arises on a curving extension of the axis of triplet 5, extends up to the inner surface of the pellicle, and twists to the right and posteriorly; it may consist of only two or three short microtubules or of a broad ribbon of 30 or more that continues for a long distance posteriorly, overlapping with similar ribbons from other kinetosomes (Figs 17, 18).

Another ribbon of microtubules originates roughly concentric to triplets 7–9 and passes up to the pellicle and leftward, anteriorly, or posteriorly. The orientation at their origin of these appendages is so constant that they reliably identify right, left, anterior, and posterior directions in simple ciliature; even in highly modified kinetosome groupings the origin of the

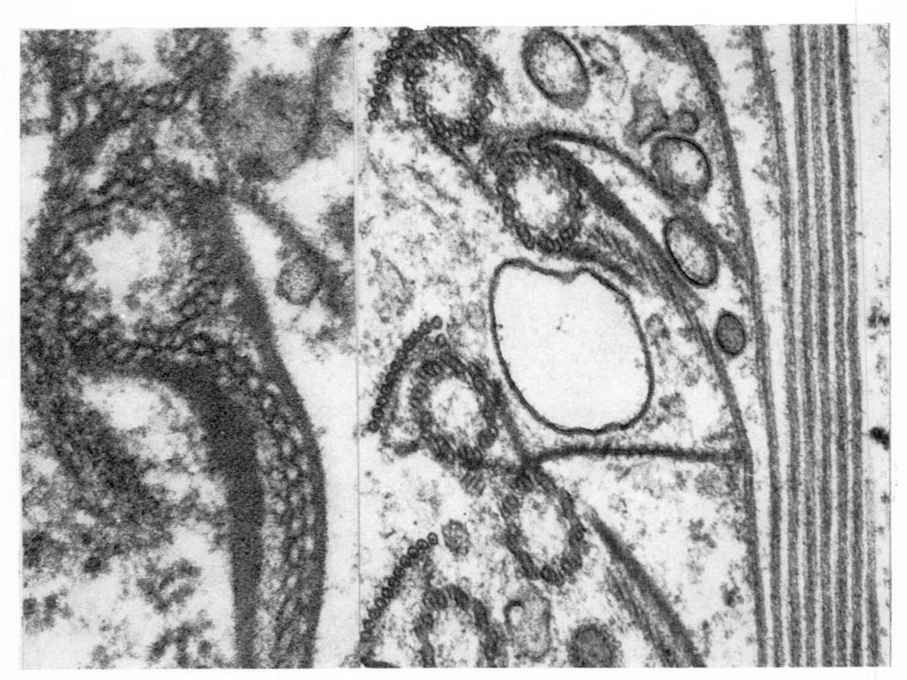

FIG. 17. Cross-section of a basal body of *Condylostoma magnum* showing the wide ribbon of microtubules arising on an extension of the transverse axis of triplet 5; the ribbon is accompanied proximally by heavy strands of dense material similar to the wall matrix of the mussel kinetosome in Fig. 15. $\times 100{,}000$.

FIG. 18. Section through two pairs of kinetosomes and part of a third from a ciliary row in *Condylostoma*. The antero-posterior axes of the paired basal bodies are tilted about 40° from the axis of the row. The overlapping posterior microtubule ribbons are stacked on the right side of the row. Several other fibrillar and microtubular structures are associated with the kinetosomes. $\times 64{,}000$.

ribbon on triplet 5 identifies the developmentally or phylogenetically primitive orientation of the kinetosome. A fourth kinetosomal appendage very common in oral ciliature is an orderly bundle of a few to several dozen microtubules originating in a dense plaque below the base of the kinetosome and extending generally downward. Microtubules in ribbons or in bundles are linked to one another by slender dense bridges.

B. Composition

Despite their common occurrence in metazoans as well as protists, striated roots have received little serious attention. Their striated, fibrillar character, suggestively similar to collagen, has led to the assumption that they are protein; recent analysis by Goodwin and Samuels (1972) of the heavy, striated costa of trichomonad flagellates confirms the assumption in this case.

Chemical analysis of isolated microtubular root structures also has only begun. Mooseker and Tilney (1973) reported evidence that the proteins of isolated axostyles of the oxymonad flagellate *Saccinobaculus* were similar, in several properties measured, to axonemal proteins, including both tubulin and a dynein component presumably responsible for demonstrated ATPase activity (see Stephens, Chapter 3, this volume, for discussion of axonemal proteins).

It is apparent that microtubules fall into two groups, distinguished by their susceptibility to antimitotic agents such as colchicine, cold, and pressure. Kinetosomes themselves, axonemal microtubules, and the highly organized, kinetosome-based microtubule aggregates of ciliates survive exposures to these agents that do cause disassembly of the mitotic apparatus and of many cytoplasmic microtubules (Tilney and Gibbins, 1968; Rosenbaum and Carlson, 1969; Tartar and Pitelka, 1969). Probably the stable microtubules remain so because they do not undergo rapid, spontaneous exchange with a subunit pool in the cytoplasm. However, development and growth of all of these resistant organelles is blocked by treatment with colchicine and similar agents (Rosenbaum and Carlson, 1969; Tartar and Pitelka, 1969; Steinman, 1970; Turner, 1970). Moore (1972) has shown that 10-min treatment of *Tetrahymena* with a pressure of 7000 psi may cause extensive resorption of complete cilia, kinetosomes and roots in oral complexes that were in advanced stages of differentiation during treatment, but not in non-differentiating body ciliature. As Moore emphasizes, resorption of oral organelles is an active process that can be initiated by various metabolic inhibitors (and that occurs normally in many ciliates); hence the pressure effect here may have been to trigger a physiologically controlled disassembly process rather than to act directly on the microtubules.

C. Functions

The obvious fact that mechanical stresses are created at the site of attachment of a lashing flagellum or cilium suggests that any root structures will help to dissipate these stresses through the cytoplasm, whatever else they may do. Such an anchoring function has generally been assumed for striated roots; however their association with basal bodies of non-motile cilia (e.g. Eakin and Kuda, 1971) and their frequent appearance on non-ciliated centrioles (Sakaguchi, 1965; Lauweryns and Boussauw, 1972) indicate that dissipation of force is not always their role.

Tethering of other organelles to kinetosomes at the apex or kinetic center of the cell is a common function of striated fibers in flagellates. The polarization of organelles thus enforced may merely serve to protect them in a cell body subject to much plastic deformation, it may facilitate interaction of organelles or their products with each other, or it may help to ensure their appropriate disposition when the cell divides. The alignment of striated fibers along kinetosome rows in some lower metazoans and in ciliates may play a mechanical role in the establishment and maintenance of the rows themselves (see discussion of ciliary spacing below).

The period in any given kind of striated root is not necessarily constant. Variation (27–55 nm) within a single parabasal fibril of the flagellate *Trichonympha* has been observed (Pitelka and Schooley, 1958), and Simpson and Dingle (1971) found periodicities ranging from about 14 to 25 nm in roots in the ameboflagellate *Naegleria* when fixed *in situ*, but little variation in isolated roots. They suggested that the fiber is either contractile or elastic. Elasticity would seem compatible with the functions of absorbing stresses, holding organelles in place, or skeletal support. Active contraction has yet to be proven for any of these striated protein fibers in cells, but evidence has recently been presented (Mattern and Honigberg, 1971) that the costa of trichomonad flagellates may undulate independently.

The roles of skeletal support and participation in cytoplasmic movement, commonly attributed to microtubules (Porter, 1966; Tilney, 1971), are quite evidently filled by many of the kinetosome-based microtubular structures of protists. Cortical microtubules that are relatively abundant or accompanied by striated roots probably help to maintain the shape of the cell, at least locally. Transition from ameboid to flagellate phases in *Naegleria* (Schuster, 1963; Dingle and Fulton, 1966) and *Tetramitus* (Outka and Kluss, 1967) involves the deployment of various root fibrils and the simultaneous acquisition of characteristic shape and polarity by the cell. Experimental treatments causing reversible disappearance of cortical microtubules in the chrysomonad flagellate *Ochromonas* also cause reversible loss of cell shape (Brown and Bouck, 1973). In many ciliates, stout microtubular rods originating (developmentally, but often detached in adult cells; Tucker, 1970; Didier, 1970b) from kinetosome bases visibly provide support in the part of the cell where ingestion takes place. In this region food organisms and membrane-wrapped food vacuoles are moved into and through the cytoplasm; it is possible that the microtubules here may also play a role in the movement of cytoplasm, membranes, or particles past their surfaces (Tucker, 1968, 1972).

Active movement on the part of kinetosome-associated microtubular organelles other than cilia and flagella has been demonstrated in at least two cases. The axostyle of oxymonad flagellates, a long, curving band composed of stacked ribbons of microtubules, undulates actively (Grimstone and Cleveland, 1965; Mooseker and Tilney, 1973; McIntosh, 1973; McIntosh *et al.*, 1973). The long, overlapping microtubule ribbons accompanying kinetosome rows of the ciliate *Stentor* (and such relatives as *Condylostoma*, Fig. 18) cause extension of the contracted body by rapid sliding over one another (Huang and Pitelka, 1973). The active contraction of the ciliate cell, however, is accomplished by shortening of bundles of filaments that are not directly kinetosome-based.

An important combination of the roles of skeletal support and participation in morphogenetic movements has been attributed to the various rootlets diverging from kinetosomes in ciliated cells by several authors (e.g. Lwoff, 1950; Dorey, 1965; Pitelka, 1969a, 1969b). The number, length, and spacing of cilia and the contour of the ciliated surface together determine the kinds of motor effects that beating of the cilia can achieve. Hence mechanisms to ensure accurate duplication at each fission and maintenance in the free-swimming organism of the specific pattern of kinetosome distribution and orientation are of prime importance to the species. Roots may hold kinetosomes together or keep them apart; they may maintain alignment in rows during active or passive distortion of the ciliated surface; they may guide the movements of newly formed kinetosomes to their definitive sites. A transected *Condylostoma* in which microtubule growth is inhibited by antimitotic agents is unable to remodel its shape until the agent is washed out (Tartar and Pitelka, 1969).

V. Origin and morphogenesis of basal bodies

At least three modes of kinetosome production have been identified. The first involves a pre-existing centriole or kinetosome and usually occurs in cells producing a small number of kinetosomes or a number not greatly exceeding that of the population already present. It will be called the kinetosomal mode (centriolar pathway of Anderson and Brenner, 1971). The second, described only in the differentiation of ciliated epithelia in vertebrates, occurs in cells that initially contain one pair of centrioles and proceed to produce a large number of kinetosomes; this will be called the deuterosomal mode. The third occurs in cells in which no pre-existing kinetosome or centriole has been detected and is the *de novo* mode.

A. Kinetosomal mode

In conventional centriole replication (Gall, 1961; Murray *et al.*, 1965; Robbins *et al.*, 1968), a short procentriole, consisting of a central cartwheel and nine triplets in a dense wall matrix, forms at right angles to the proximal end of a mature interphase centriole. The pair, called a

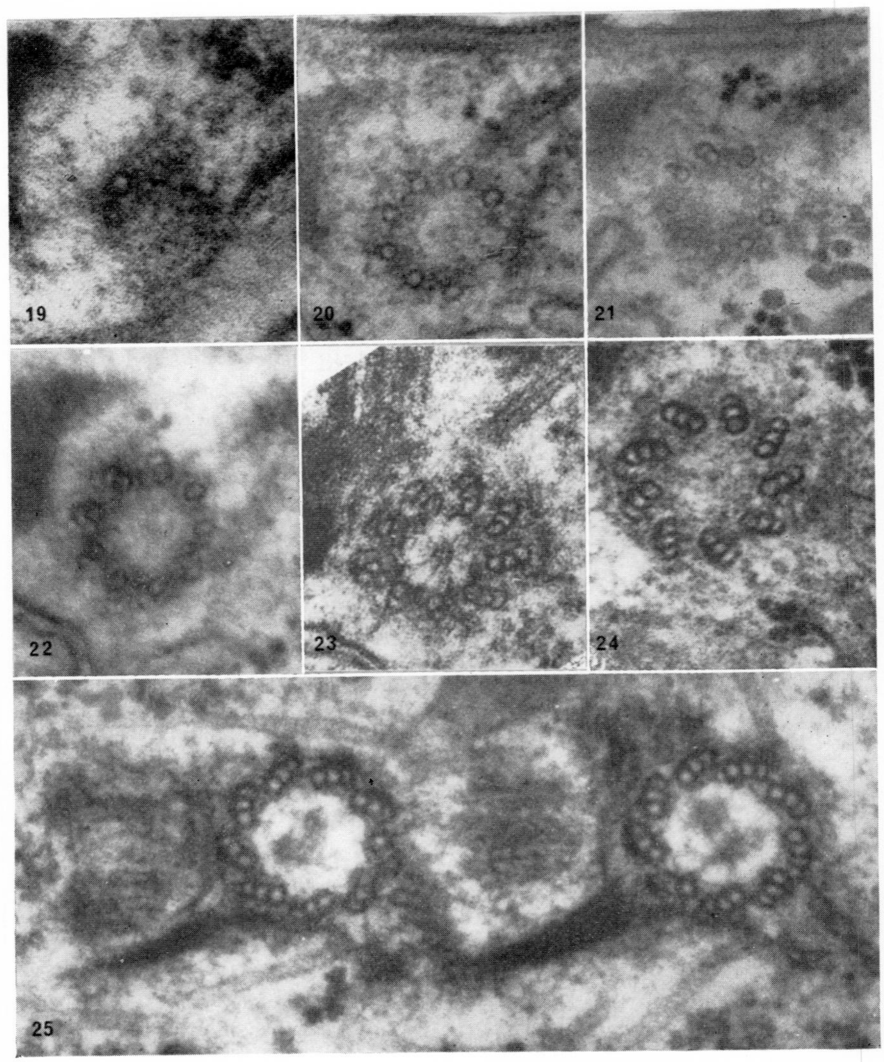

FIGS 19 to 25.—Captions on facing page.

diplosome, maintains this perpendicular relationship and close association while the young organelle elongates, through the next mitosis and into interphase, when they separate and another generation of procentrioles may form. Under appropriate circumstances, one or both of the diplosomal centrioles may become basal bodies.

The same appearance and orientation of single prokinetosomes with respect to older ones has been seen in many flagellates and ciliates (e.g. Johnson and Porter, 1968; Paulin, 1969; Allen, 1969; Millecchia and Rudzinska, 1970). Stages in the morphogenesis of a prokinetosome (Figs 19–25) were first identified by Dippell (1968) in *Paramecium*. The earliest prokinetosome is detectable anterior to an older (often cilium-bearing) kinetosome as a flat disc of filamentous material, resembling and possibly connected with the wall matrix of the more mature organelle. Single A-microtubules develop sequentially, equally spaced around a circle. B-tubules appear on the peripheral walls of the A-tubules as outgrowths that curve around and re-attach to the walls more medially, acquiring their characteristic fat crescent cross-section. C-tubules form similarly to complete a short (around 70 nm) cylinder of triplets, whose angles are greater (60–70°) than in mature kinetosomes. At this time the central cartwheel becomes distinct, A–C links form, and the prokinetosome elongates, moves anteriorly, and tilts into position perpendicular to the cell surface. Although only one prokinetosome per basal body has been seen at any one time, several generations of new kinetosomes may form in rapid succession (Fig. 26), permitting the abundant, localized proliferation that occurs during differentiation of oral ciliary complexes.

Development of prokinetosomes in association with cilium-bearing basal bodies apparently is rare in vertebrates. Sorokin (1968) and Anderson and Brenner (1971) reported it as a minor source of prokinetosomes in

---

FIGURES 19–24 are stages in the morphogenesis of basal bodies in *Paramecium aurelia*; cross-sectioned prokinetosomes are viewed from the base, with the cell surface toward the top of the page. Figures 19 and 20 show the sequential formation of singlet A-tubules in a disc or annulus of moderately dense material. In Figs 20–22, the formation of B-tubules as outgrowths from the peripheral surface of the A-tubules is seen. Figure 23 shows the proximal end of a complete prokinetosome, with central cartwheel now distinct and a triplet angle of about 70°. Figure 24 is the distal end of a complete prokinetosome. ×150,000. Figures 19–21, 23 and 24 from micrographs by R. V. Dippell first published in Dippell, 1968. Figure 22 from R. V. Dippell, unpublished.

FIG. 25. A section cut parallel to the surface of *Paramecium aurelia*, showing two mature kinetosomes in cross-section, viewed from the distal end, with a prokinetosome, longitudinally sectioned, perpendicular to the base of each. The anterior direction is toward the left; prokinetosomes always appear anterior to older ones in the body ciliature of many ciliates. ×150,000. From R. V. Dippell, unpublished.

cells producing large numbers by the deuterosomal mode; in both cases, a cluster of prokinetosomes appeared around a diplosomal kinetosome bearing a rudimentary cilium. Such clusters forming around mature

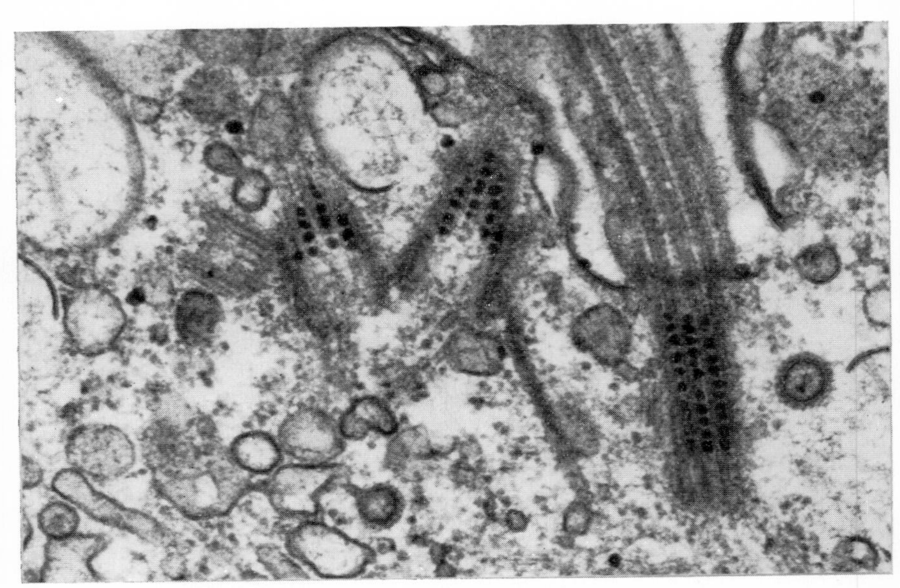

FIG. 26. Kinetosomes in the regenerating oral area of a transected *Condylostoma*. The mature, ciliated basal body on the right was originally part of a row of normal body cilia. The three kinetosomes to its left are out of the row and not perpendicular to the cell surface, indicating that they are recently formed; the two oldest of them have elongated and contain dense luminal granules; the short prokinetosome at the far left of the group does not. × 50,000.

centrioles had originally been described by Gall (1961) in atypical spermatogenesis in a snail. Detailed accounts of kinetosome production in invertebrate cells other than spermatocytes apparently have not been published.

B. Deuterosomal mode

Kinetosome production not directly associated with pre-existing centrioles has been described for respiratory epithelia of mammals (Stockinger and Cireli, 1965; Dirksen and Crocker, 1966; Frasca *et al.*, 1967; Frisch and Farbman, 1968; Sorokin, 1968), the chick (Kalnins and Porter, 1969; Kalnins *et al.*, 1972), and the toad *Xenopus* (Steinman, 1968); oviduct of mammals (Dirksen, 1971; Anderson and Brenner, 1971); and epidermis of *Xenopus* (Steinman, 1968). In all of these the differentiating cell initially contains one pair of centrioles and within about 24 h produces hundreds of cilia. Prokinetosomes develop perpendicular to the surface of a spherical or elongate dense structure, usually 100 nm or more in diameter, that is not a centriole. The dense structure has been given many names, perhaps the most useful being Sorokin's *deuterosome*. It may be associated with 1–14 prokinetosomes and in most cases appears to diminish in size or density as they grow; ultimately they become free in the cytoplasm, migrate to the cell apex, and sprout cilia. Morphogenesis of the prokinetosome is essentially identical to that described by Dippell, except that the cartwheel appears before any singlet microtubules are visible.

Events leading up to the appearance of deuterosomes are less distinct. In chick trachea (Kalnins and Porter, 1969), some of the cylindrical deuterosomes appear to originate in contact with the diplosomal centrioles. In all of the tissues examined, clusters of dense granules 40–60 nm in diameter, usually in a filamentous matrix, are present in the region where deuterosomes and prokinetosomes later appear. Most authors suggest that these granules contribute precursor materials to deuterosomes or directly to procentrioles and perhaps later to growing axonemes. All the prokinetosomes visible in any one cell usually appear to be in the same developmental state, but Dirksen and Crocker (1966), Dirksen (1971) and Kalnins *et al.* (1972) present evidence that more than one wave of prokinetosome formation may occur. Repeated kinetosome formation in what appears to be a deuterosomal mode is indicated in rat choroid cells (Martinez-Martinez and Daems, 1968).

C. De novo *mode*

The appearance of centrioles or kinetosomes in cells that previously have lacked them is now too well documented to be seriously questioned (sea-urchin eggs: Dirksen, 1961; ameboflagellates: Schuster, 1963; Outka and Kluss, 1967; Fulton and Dingle, 1971; the rhizopod *Labyrinthula*: Perkins, 1970; the ciliate *Oxytricha*: Grimes, 1973; plants: Manton *et al.*, 1970, Mizukami and Gall, 1966; see reviews by Pickett-Heaps, 1969, 1971). The earliest stages of morphogenesis have been identified as accumulations of dense material, which sometimes themselves act as mitotic centers (Perkins, 1970; Manton *et al.*, 1970); cartwheels and microtubules ultimately appear in the dense material and development proceeds as in the kinetosomal and deuterosomal modes.

D. *Centriole–kinetosome transformation*

It appears that the products of conventional centriole replication frequently become basal bodies, fortuitously or as part of normal development, upon making contact distally with an appropriate membrane. This is assumed to have occurred when electron micrographs show a cilium-bearing kinetosome and adjacent centriole in diplosomal orientation in a cell in which no other centrioles are seen. Such single cilia, lacking central microtubules, become specialized parts of sensory receptors in most animals; they also have been observed as adventitious structures of no known function in a great many cell types in developing and adult vertebrates and in cell culture (Barnes, 1961; Scherft and Daems, 1967; Sorokin, 1968; Wheatley, 1969). In other cases, single cilia produced by diplosomes have the 9+2 pattern and are normally motile, as in echinoderm embryos examined by Tilney and Gibbins (1968). Occasionally both members of a diplosome may produce rudimentary cilia (Wheatley, 1969; Lin and Chen, 1969). Experimental stimulation of the centriole–kinetosome transformation has been reported by Stubblefield and Brinkley (1966) in colcemid-treated, cultured fibroblasts, and by Milhaud and Pappas (1968) in pargyline-treated cat brain. In the latter case multiple cilia with 9+2 axonemes appeared; apparently centriole replication as well as axoneme formation had been stimulated.

Structural alterations associated with centriole–kinetosome transformation have not been clearly defined. Most centrioles and developing basal bodies have a layer of moderately dense material at the distal end of the lumen; whether this is related to a more discrete transverse plate in the transition zone of the mature kinetosome is unknown. Several instances have been reported of the presence of transitional fibrils on typical centrioles (e.g. O'Hara, 1970). These fibrils probably function in the distal attachment of a young kinetosome to the membrane where its cilium will be produced; their presence might indicate a centriole that is preconditioned to become a basal body.

The reverse transformation, from kinetosome to active centriole, has been much less frequently shown or discussed. Light microscopy suggests that zygote centrioles come from the flagellated sperm in species where the egg cell lacks them, and that kinetosomes in some flagellates may serve as mitotic centers (see Went, 1966, for examples and discussion). These observations apparently have not been confirmed by electron microscopy; reconstructing the path of a sperm centriole through egg cytoplasm would be a difficult undertaking at best. Several studies of mitosis in flagellates however, have demonstrated that, although basal bodies may be foci of radiating microtubules before and during mitosis, they do not themselves occupy the spindle poles (Manton, 1964; Johnson and Porter, 1968; Bouck and Brown, 1973). Fritz-Niggli and Suda (1972) show that spermatogonial centrioles in *Drosophila* move to the cell surface and sprout flagella, which are retained while their basal bodies behave as meiotic centrioles, and which eventually become sperm flagella.

Perhaps the best circumstantial evidence for reverse transformation lies in the very frequency with which diplosomal centrioles apparently produce adventitious cilia in developing tissues and cell cultures. It has been pointed out (Dingemans, 1969; Rash *et al.*, 1969) that these are not seen in mitotic cells, but Fonte *et al.* (1971) find them frequently in cell populations considered to be in the division cycle. Wheatley (1969) found cilia associated with at least a third of the centrioles observed in cultured rat embryo fibroblasts. The possibility exists that ciliated centrioles in vertebrate tissues do not revert; if so, these cells need not be blocked from further mitosis if one member of the diplosome has remained nonciliated and upon appropriate stimulation can initiate centriole replication. In any event, it may be true (Rash *et al.*, 1969) that cilium formation by diplosomes is a means of diverting cells at least temporarily from the division cycle.

VI. Functions of basal bodies

Apart from self-evident roles in the genesis and cytoplasmic attachment of cilia and flagella, most of the other functions proposed for basal bodies have been, to say the least, thrown into doubt by observations made in recent years. Motive force for flagellar beating is provided within the flagellum itself (see Goldstein, Chapter 6, this volume), not in the basal body. Coordination of beat is a combination of external hydrodynamic and surface electrochemical phenomena (see Chapters 10 by Machemer, 11 by Sleigh, and 12 by Naitoh and Eckert), not channeled through the basal body. The *de novo* genesis of kinetosomes and centrioles, their probable lack of DNA, the frequent origin of cytoplasmic microtubules at sites remote from kinetosomes or centrioles, and the fact that the mitotic apparatus in most protists and higher plants is independent of centrioles or kinetosomes—all have combined to evoke serious questioning (Pickett-Heaps, 1969, 1971; Friedlander and Wahrman, 1970; Fulton, 1971; Hollande, 1972; Wolfe, 1972) of the role of the centriole as a mitotic center and, indeed, of the capacity of either organelle to generate anything whatever except flagella.

A. What are centrioles for?—a recurrent view

The picture emerging from this reconsideration of centriole dogma is approximately as follows. A eukaryotic cell contains microtubule-organizing centers (MTOCs, Pickett-Heaps, 1969, 1971), which may be too diffuse to be seen or may appear as variously shaped masses of moderately dense material. These initiate the assembly of all microtubular structures, including centrioles, kinetosomes, and the mitotic spindle. Centrioles are inert passengers on the spindle, which distributes them, as well as chromosomes, to daughter cells. The only real function of centrioles is to become basal bodies when required. In animals, where the requirement is frequent, they permanently accompany the spindle; in ciliates and most flagellates, methods of distribution of kinetosomes and of chromosomes are independent; in multicellular plants, centrioles may appear *de novo* a few cell generations before they are needed to produce flagella.

This argument is basically a return to the view of Chatton and Lwoff, proposed in the 1920s and '30s (see Lwoff, 1950), that the phylogenetic and primary function of centrioles was to produce flagella. They believed that the multiple potential of the basal body as an organizer of fibrous structure secondarily equipped it to manage the mitotic apparatus. They argued strongly for the genetic continuity of kinetosomes and for the importance of microenvironment in determining which of its potentialities a kinetosome expresses at any given time.

In the process of abandoning the now untenable theory of genetic continuity, some of the fundamental observations on which the Chatton–Lwoff kinetosome concept was based have been neglected. Light microscopists showed, with many different materials, that the kinetosome physically occupies a center of extensive morphogenetic activity, relatively easily observed in generation after generation of protistan cells, less easily in Metazoa. Electron microscopy confirms the outgrowth of a profusion of fibrous structures from or near the surfaces of basal bodies.

The current consensus would attribute this morphogenetic activity to dense material—MTOCs, and perhaps other centers for other fibrils—previously regarded as part of the basal body but now as separate entities. The considerable heuristic value of this view is in its emphasis on MTOCs and their capability to generate kinetosomes and other structures in the absence of any preformed model.

Two interrelated questions about kinetosome function seem particularly intriguing in the light of the contemporary concept. First, MTOCs and kinetosomes tend to travel together, and sequences of morphogenetic events are associated with this combination; do the kinetosomes themselves have any morphogenetic significance apart from cilium production? Second, since *de novo* formation is possible, why do centrioles and basal bodies in so many cells and organisms originate near pre-existing ones if any are present, often at the price of elaborate subsequent migration? Putting it another way, why have millions of generations of animal-cell spindles been burdened with the distribution of potential kinetosomes when relatively few cells need them?

In the search for answers, a salient fact may be that those organisms that produce centrioles or kinetosomes *de novo* generally use them for a single purpose: motility of a short-lived reproductive or dispersive cell. By contrast, basal bodies are needed in a single animal to produce sperm flagella, cilia for a variety of sensory receptors, and epithelial cilia of several kinds, in addition to "adventitious" cilia of unknown significance; each of these has its characteristic structure and appendages. Ciliate kinetosomes develop different appendages and cilia of different lengths on different parts of the single cell.

These facts suggest the possibility that, where more than one option for the developmental fate of basal bodies is required, morphogenetic economy calls for a formed organelle with multiple potentialities—what Lwoff (1950) calls a polyvalent kinetosome—to be copied in all parts of the body.

We need to consider how a kinetosome might possess multiple potentialities, and, if it can, why the kinetosomal mode seems to be the method of evolutionary choice for copying it.

B. *The morphogenetic significance of basal body geometry—a hypothesis*

The initiation of microtubule assembly, the imposition of a pattern on the developing microtubule aggregate, and the subsequent orderly integration of other microtubular or fibrous structures represent three differing levels of organization. The evidence given in section V indicates that an MTOC can generate both microtubules and the cartwheel that may impose the 9-part kinetosomal pattern on them; the MTOC in this case controls the first two of these organizational levels. The specific structure of the basal body itself may be instrumental in the achievement of the third level.

The inherent asymmetry and polarity of the kinetosome's microtubular skeleton make sense as a framework for the addition of numerous, highly individual, asymmetric structures at specific sites on its surface. Only one initial "landmark" site is necessary to specify all the others as unique. The microtubule ribbon extending from one triplet of the ciliate basal body (Fig. 16) distinguishes this triplet from all the others, and the slant of the triplets specifies directions both around the kinetosome and along its axis.

Geometrically specific sites for outgrowth, attachment, or interaction with any asymmetry in the environment may be exploited in positioning the kinetosomes with respect to each other and to cell topography and in determining such properties of cilia as the fixed or preferred direction of their effective beat. If, under the control of nuclear genes, MTOCs or receptors for MTOC attachment are incorporated into the kinetosome at distinctive sites, activation of specific ones in a given microenvironment would result in formation of appropriate rootlets and ciliary structures. Polyvalent basal bodies capable of differential reaction to a variety of environments could be constructed.

Examples are available among ciliates to illustrate both the determination of a new kinetosome's organization by pre-existing structures and the modifiability of this organization when local conditions change.

When Beisson and Sonneborn (1965) experimentally rotated by 180° a strip of the cortex of *Paramecium*, new kinetosomes appearing in the ciliary rows in that strip developed root fibrils (Fig. 16) in the same orientation as the rotated old ones. The upside-down rows—recognizable in life because the reversed ciliary beat in the strip caused aberrant swimming—elongated, divided at fission, and were inherited by subsequent cell generations just like normal rows (Sonneborn, 1970). Here the orientation of rootlet sites and the preferred beating direction of the new cilia were determined by pre-existing local polarity. The landmark site might have been imprinted (by an unknown mechanism!) on each prokinetosome while its proximal end lay next to a mature kinetosome.

Development of a complete new oral complex in a ciliate such as *Tetrahymena* (Williams and Frankel, 1973), *Stentor* (Paulin and Bussey, 1971), or *Condylostoma* (Pitelka and Tartar, unpublished) may involve the conversion of pre-existing ordinary basal bodies into oral ones, as well as the production of a large population of new ones. Specific rootlets of the old kinetosomes are resorbed and new kinds formed, while new kinetosomes develop only the new kinds. Both the orientation and the spacing of young kinetosomes bearing short cilia and microtubular rootlets are anarchic at first; apparently the site on the kinetosome and not its position in a field gradient has determined the location of the growing rootlet. Only later are all basal bodies in the field aligned in groups with their microtubule rootlets uniformly oriented. In this case the induction of oral morphogenesis causes old and new kinetosomes to develop specialized appendages and then become aligned in specialized groupings. Young basal bodies are not constrained to stay in a row with their "parents", but the opportunity for determining the position of a landmark site is present during prokinetosome formation.

A similar random dispersal of young kinetosomes bearing basal feet, with subsequent orientation, is described in embryonic mouse nasal epithelium (Frisch and Farbman, 1968), where kinetosomes are produced by the deuterosomal mode. Although true centrioles precede and may influence the construction of the deuterosomes (Dirksen and Crocker, 1966; Dirksen, 1971; Kalnins and Porter, 1969) this mode resembles *de novo* genesis.

Orderly alignment of kinetosomes and associated fibrils is conspicuously achieved after *de novo* production in plant cells; both here and in ciliated epithelia, the kinetosomes finally differentiated in the motile cell are a dead end. The adaptive advantages that must be ascribed to the kinetosomal mode may have to do with efficient production of polyvalent kinetosomes or with the capacity for infinite sequential copying. It is not at all clear why animals supply all their cells with identical polyvalent basal bodies unless it is important, for unsuspected reasons, that most or all cells be capable of producing "adventitious" cilia. The obvious possibility that coupling of spindle MTOCs to the centriole offers enough selective advantage to have justified its perpetuation seems difficult to defend, unless it adds a critical margin to an accumulation of other benefits in animals. Even those multicellular plants that produce functional centrioles during spermatogenesis have not exploited the opportunity to maintain them in vegetative cells.

It is unlikely that many of the outstanding questions of kinetosome function can be answered without much more detailed chemical and experimental analysis than is feasible now. One would like, for example, to be able to dissect basal bodies chemically and then reimplant variously denuded versions in kinetosome-free cytoplasm, or to transplant viable basal bodies between species. The existence of an enormous diversity of flagellated and ciliated cells whose kinetosomes show a wide gamut of structures and behaviors, only sampled in this review, provides an opportunity for analytical comparative anatomy that is open right now.

Acknowledgements

I am indebted to R. V. Dippell and F. D. Warner for providing electron micrographs for inclusion here. My work was supported by United States Public Health Service grants CA05388 and CA05045.

References

Allen, R. D. (1969). *J. Cell Biol.* **40**, 716–733.
Anderson, R. G. W. (1972). *J. Cell Biol.* **54**, 246–265.
Anderson, R. G. W. and Brenner, R. M. (1971). *J. Cell Biol.* **50**, 10–35.
Anderson, W. A. and Ellis, R. A. (1965). *J. Protozool.* **12**, 483–499.
Barnes, B. G. (1961). *J. Ultrastruct. Res.* **5**, 453–467.

BEISSON, J. and SONNEBORN, T. M. (1965). *Proc. natn. Acad. Sci. U.S.A.* **53**, 275–282.

BOUCK, G. B. and BROWN, D. L. (1973). *J. Cell Biol.* **56**, 340–359.

BRESCIANI, J. and FENCHEL, T. (1967). *Ophelia* **4**, 1–18.

BRESCIANI, J. and KØIE, M. (1970). *Ophelia* **8**, 209–230.

BROOKER, B. E. (1971). *Bull. Br. Mus. nat. Hist. Zool.* **22**, 89–102.

BROWN, D. L. and BOUCK, G. B. (1973). *J. Cell Biol.* **56**, 360–378.

CAROTHERS, Z. B. and KREITNER, G. L. (1968). *J. Cell Biol.* **36**, 603–616.

DIDIER, P. (1970a). *Ann. Stat. Biol. Besse* **5**, 1–274.

DIDIER, P. (1970b). *C. r. Séanc. Soc. Biol.* **164**, 313–317.

DINGEMANS, K. P. (1969). *J. Cell Biol.* **43**, 361–367.

DINGLE, A. D. and FULTON, C. (1966). *J. Cell Biol.* **31**, 43–54.

DIPPELL, R. V. (1968). *Proc. natn. Acad. Sci. U.S.A.* **61**, 461–468.

DIRKSEN, E. R. (1961). *J. biophys. biochem. Cytol.* **11**, 244–247.

DIRKSEN, E. R. (1971). *J. Cell Biol.* **51**, 286–302.

DIRKSEN, E. R. and CROCKER, T. T. (1966). *J. Microscopie* **5**, 629–644.

DOOLIN, P. F. and BIRGE, W. J. (1966). *J. Cell Biol.* **29**, 333–345.

DOREY, A. E. (1965). *Q. Jl microsc. Sci.* **106**, 47–172.

EAKIN, R. M. and KUDA, A. (1971). *Z. Zellforsch.* **112**, 287–312.

FAURÉ-FREMIET, E. (1970). *Ann. Biol.* **9**, 1–61.

FLAVELL, R. A. and JONES, I. G. (1971). *J. Cell Sci.* **9**, 719–726.

FLOCK, Å. and DUVALL, A. J. (1965). *J. Cell Biol.* **25**, 1–8.

FONTE, V. G., SEARLS, R. L. and HILFER, S. R. (1971). *J. Cell Biol.* **49**, 226–229.

FRASCA, J. M., AUERBACH, O., PARKS, V. R. and STOECKENIUS, W. (1967). *Expl molec. Pathol.* **7**, 92–104.

FRIEDLANDER, M. and WAHRMAN, J. (1970). *J. Cell Sci.* **7**, 65–89.

FRISCH, D. and FARBMAN, A. I. (1968). *Anat. Rec.* **162**, 221–231.

FRISCH, D. and REITH, E. J. (1966). *J. Ultrastruct. Res.* **15**, 490–495.

FRITZ-NIGGLI, H. and SUDA, T. (1972). *Cytobiologie* **5**, 12–41.

FULLER, M. S. (1966). *Colston Papers* **18**, 67–84.

FULTON, C. (1971). *In* "Results and Problems in Cell Differentiation" (J. Reinert and H. Ursprung, eds), Vol. 2 "Origin and Continuity of Cell Organelles", pp. 170–221. Springer-Verlag, New York.

FULTON, C. and DINGLE, A. D. (1971). *J. Cell Biol.* **51**, 826–836.

GALL, J. G. (1961). *J. biophys. biochem. Cytol.* **10**, 163–194.

GIBBONS, I. R. (1961). *J. biophys. biochem. Cytol.* **11**, 179–205.

GIBBONS, I. R. and GRIMSTONE, A. V. (1960). *J. biophys. biochem. Cytol.* **7**, 697–716.

GOODWIN, F. L. and SAMUELS, R. (1972). *J. Protozool* **19** (Suppl.), 31.

GRAIN, J. (1969). *Ann. Biol.* **8**, 53–97.

GRIMES, G. W. (1973). *J. Protozool.* **20**, 92–104.

GRIMSTONE, A. V. and CLEVELAND, L. R. (1965). *J. Cell Biol.* **24**, 387–400.

GRIMSTONE, A. V. and GIBBONS, I. R. (1966). *Phil. Trans. R. Soc. Ser. B* **250**, 215–242.

DE HARVEN, E. (1968). *In* "The Nucleus" (A. J. Dalton and F. Hagenau, eds), pp. 197–227. Academic Press, New York.

HEATH, I. B. and GREENWOOD, A. D. (1971). *Z. Zellforsch.* **112**, 371–389.

Hoffman, L. R. (1970). *Can. J. Bot.* **48**, 189–196.
Hollande, A. (1972). *Ann. Biol.* **11**, 427–466.
Hollande, A. and Carruette-Valentin, J. (1971). *Protistologica* **7**, 5–100.
Honigberg, B. M., Mattern, C. F. T. and Daniel, W. A. (1971). *J. Protozool.* **18**, 183–198.
Huang, B. and Pitelka, D. R. (1973). *J. Cell Biol.* **57**, 704–728.
Hufnagel, L. (1969). *J. Cell Sci.* **5**, 561–573.
Johnson, U. G. and Porter, K. R. (1968). *J. Cell Biol.* **38**, 403–425.
Joyon, L. and Mignot, J. P. (1969). *Ann. Biol.* **8**, 1–52.
Joyon, L., Mignot, J. P., Kattar, M. R. and Brugerolle, G. (1969). *Protistologica* **5**, 309–326.
Kalnins, V. I., Chung, C. K. and Turnbull, C. (1972). *Z. Zellforsch.* **135**, 461–472.
Kalnins, V. I. and Porter, K. R. (1969). *Z. Zellforsch.* **100**, 1–30.
Kozloff, E. N. (1971). *J. Parasitol.* **57**, 585–597.
Lang, N. J. (1963). *J. Cell Biol.* **19**, 631–634.
Lansing, A. I. and Lamy, F. (1961). *J. biophys. biochem. Cytol.* **9**, 799–812.
Lauweryns, J. M. and Boussauw, L. (1972). *Z. Zellforsch.* **131**, 417–427.
Lin, H. S. and Chen, I. L. (1969). *Z. Zellforsch.* **96**, 186–205.
Lwoff, A. (1950). "Problems of Morphogenesis in Ciliates." John Wiley, New York.
Manton, I. (1963). *Jl R. microsc. Soc.* **82**, 279–285.
Manton, I. (1964). *Jl R. microsc. Soc.* **83**, 317–325.
Manton, I., Kowallik, K. and von Stosch, H. A. (1970). *J. Cell Sci.* **7**, 407–443.
Martinez-Martinez, P. and Daems, W. T. (1968). *Z. Zellforsch.* **87**, 46–68.
Mattern, C. F. T. and Honigberg, B. M. (1971). *Trans. Am. microsc. Soc.* **90**, 309–313.
McIntosh, J. R. (1973). *J. Cell Biol.* **56**, 324–339.
McIntosh, J. R., Ogata, E. W. and Landis, S. C. (1973). *J. Cell Biol.* **56**, 304–323.
Milhaud, M. and Pappas, G. D. (1968). *J. Cell Biol.* **37**, 599–609.
Millecchia, L. L. and Rudzinska, M. A. (1970). *J. Cell Biol.* **46**, 553–563.
Mizukami, I. and Gall, J. (1966). *J. Cell Biol.* **29**, 97–112.
Moore, K. C. (1972). *J. Ultrastruct. Res.* **41**, 499–518.
Mooseker, M. S. and Tilney, L. G. (1973). *J. Cell Biol.* **56**, 13–26.
Munn, E. A. (1970). *Tissue Cell* **2**, 499–512.
Murray, R. G., Murray, A. S. and Pizzo, A. (1965). *J. Cell Biol.* **26**, 601–619.
O'Hara, P. T. (1970). *J. Ultrastruct. Res.* **31**, 195–198.
Olsson, R. (1962). *J. Cell Biol.* **15**, 596–599.
Outka, D. E. and Kluss, B. C. (1967). *J. Cell Biol.* **35**, 323–346.
Paulin, J. J. (1969). *Trans. Am. microsc. Soc.* **88**, 400–410.
Paulin, J. J. and Bussey, J. (1971). *J. Protozool.* **18**, 201–213.
Perkins, F. O. (1970). *J. Cell Sci.* **6**, 629–653.
Pickett-Heaps, J. D. (1969). *Cytobios* **1**, 257–280.
Pickett-Heaps, J. D. (1971). *Cytobios* **3**, 205–214.
Pitelka, D. R. (1965). *J. Microscopie* **4**, 373–394.

PITELKA, D. R. (1969a). *In* "Progress in Protozoology", Proc. 3rd Internat. Congr. Protozool., Leningrad, pp. 44–46. Nauka, Leningrad.
PITELKA, D. R. (1969b). *In* "Research in Protozoology" (T. T. Chen, ed.), Vol. 3, pp. 280–388. Pergamon Press, Oxford.
PITELKA, D. R. (1969c). *In* "Handbook of Molecular Cytology" (A. Lima-de-Faria, ed.), pp. 1199–1218. North-Holland, Amsterdam.
PITELKA, D. R. and CHILD, F. M. (1964). *In* "Biochemistry and Physiology of Protozoa" (S. H. Hutner, ed.), Vol. 3, pp. 131–198. Academic Press, New York.
PITELKA, D. R. and SCHOOLEY, C. N. (1958). *J. Morph.* **102**, 199–246.
PORTER, K. R. (1966). *In* "Principles of Biomolecular Organization" (G. E. W. Wolstenholme and M. O'Conner, eds), pp. 308–345. Churchill, London.
PYNE, C. K. (1968). *C. r. hebd. Séanc. Acad. Sci., Paris.* **267**, 755–757.
RANDALL, J. and DISBREY, C. (1965). *Proc. R. Soc.* B **162**, 473–491.
RASH, J. E., SHAY, J. W. and BIESELE, J. J. (1969). *J. Ultrastruct. Res.* **29**, 470–484.
REESE, T. S. (1965). *J. Cell. Biol.* **25**, 209–230.
REGER, J. F. (1967). *J. Ultrastruct. Res.* **20**, 451–461.
RINGO, D. L. (1967). *J. Cell. Biol.* **33**, 543–571.
ROBBINS, E., JENTZCH, G. and MICALI, A. (1968). *J. Cell Biol.* **36**, 329–339.
ROSENBAUM, J. L. and CARLSON, K. (1969). *J. Cell Biol.* **40**, 415–425.
ROSENBAUM, J. L. and CHILD, F. M. (1967). *J. Cell Biol.* **34**, 345–364.
SAKAGUCHI, H. (1965). *J. Ultrastruct. Res.* **12**, 13–21.
SATIR, P. (1965). *J. Cell Biol.* **26**, 805–834.
SCHERFT, J. P. and DAEMS, W. T. (1967). *J. Ultrastruct. Res.* **19**, 546–555.
SCHUSTER, F. L. (1963). *J. Protozool.* **10**, 297–313.
SIMPSON, P. A. and DINGLE, A. D. (1971). *J. Cell Biol.* **51**, 323–328.
SMITH-SONNEBORN, J. and PLAUT, W. (1967). *J. Cell Sci.* **2**, 225–234.
SMITH-SONNEBORN, J. and PLAUT, W. (1968). *J. Cell Sci.* **5**, 365–372.
SONNEBORN, T. M. (1970). *Proc. R. Soc.* B **176**, 347–366.
SOROKIN, S. P. (1968). *J. Cell Sci.* **3**, 207–230.
STEINMAN, R. M. (1968). *Am. J. Anat.* **122**, 19–55.
STEINMAN, R. M. (1970). *J. Ultrastruct. Res.* **30**, 423–440.
STEPHENS, R. E. (1971). *In* "Biological Macromolecules" (S. N. Timasheff and G. D. Fasman, eds), Vol. 4, pp. 355–387. Marcel Dekker, New York.
STOCKINGER, L. and CIRELI, E. (1965). *Z. Zellforsch.* **68**, 733–740.
STUBBLEFIELD, E. and BRINKLEY, B. R. (1966). *J. Cell Biol.* **30**, 645–652.
STUBBLEFIELD, E. and BRINKLEY, B. R. (1967). *In* "Formation and Fate of Cell Organelles" (K. B. Warren, ed.), pp. 175–218. Academic Press, New York.
SZOLLOSI, D. (1964). *J. Cell Biol.* **21**, 465–480.
TAMM, S. L. (1967). *J. exp. Zool.* **164**, 163–186.
TARTAR, V. and PITELKA, D. R. (1969). *J. exp. Zool.* **172**, 201–218.
TILNEY, L. G. (1971). *In* "Results and Problems in Cell Differentiation" (J. Reinert and H. Ursprung, eds), Vol. 2, "Origin and Continuity of Cell Organelles", pp. 222–260. Springer-Verlag, New York.
TILNEY, L. G. and GIBBINS, J. R. (1968). *Protoplasma* **65**, 167–179.
TUCKER, J. B. (1968). *J. Cell Sci.* **3**, 493–514.

TUCKER, J. B. (1970). *J. Cell Sci.* **6**, 385–420.
TUCKER, J. B. (1971). *J. Cell Sci.* **9**, 539–568.
TUCKER, J. B. (1972). *J. Cell Sci.* **10**, 883–903.
TURNER, F. R. (1970). *J. Cell Biol.* **46**, 220–234.
VICKERMAN, K. (1969). *J. Protozool.* **16**, 54–69.
WARNER, F. D. (1969). *J. Ultrastruct. Res.* **29**, 499–524.
WARNER, F. D. (1972). *Adv. cell. molec. Biol.* **2**, 193–236.
WELSCH, U. and STORCH, V. (1969). *Z. Zellforsch.* **102**, 432–446.
WENT, H. A. (1966). *Protoplasmatologia* **6**, G1, 1–109.
WHEATLEY, D. N. (1969). *J. Anat.* **105**, 351–362.
WILLIAMS, N. E. and FRANKEL, J. (1973). *J. Cell Biol.* **56**, 441–457.
WILSON, R. A. (1969). *J. Parasit.* **55**, 124–133.
WOLFE, J. (1970). *J. Cell Sci.* **6**, 679–700.
WOLFE, J. (1972). *Adv. cell. molec. Biol.* **2**, 151–192.
YOUNGER, K. B., BANERJEE, S., KELLEHER, J. K., WINSTON, M. and MARGULIS, L. (1972). *J. Cell Sci.* **11**, 621–637.

Author Index

Numbers in *italics* indicate those pages where references are given in full

C

T

Subject Index

B

C

D

N

O

T

U

V

W

X

Y